Hydraulic Systems Volume 4

Hydraulic Fluids Conditioning

Dr. Medhat Kamel Bahr Khalil, Ph.D, CFPHS, CFPAI.
Director of Professional Education and Research Development,
Applied Technology Center, Milwaukee School of Engineering,
Milwaukee, WI, USA.

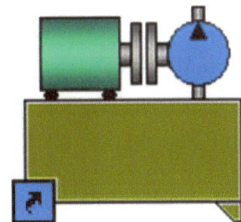

CompuDraulic LLC
www.CompuDraulic.com

CompuDraulic LLC

Hydraulic System Volume 4

Hydraulic Fluids Conditioning

ISBN: 978-0-9977634-8-5

Printed in the United States of America
First Published by Sept. 2022
Revised by Feb. 2024

Disclaimer

It is always advisable to review the relevant standards and the recommendations from the system manufacturer. However, the content of this book provides guidelines based on the author's experience.

Any portion of information presented in this book might not be suitable for some applications due to various reasons. Since errors can occur in circuits, tables, and text, the author/publisher assumes no liability for the safe and/or satisfactory operation of any system designed based on the information in this book.

The author/publisher does not endorse or recommend any brand name products by including such brand name products in this book. Conversely the author/publisher does not disapprove any brand name product not included in this book. The publisher obtained data from catalogs, literatures, and material from hydraulic components and systems manufacturers based on their permissions. The author/publisher welcomes additional data from other sources for future editions. This disclaimer is applicable for the workbook (if available) for this textbook.

Hydraulic Systems Volume 4
Hydraulic Fluids Conditioning

PREFACE

In addition to hydraulic fluids contamination control, topics for hydraulic fluids conditioning are essential for hydraulic fluids to perform effectively within hydraulic systems. Hydraulic fluids and contamination control are discussed in Volume 3 of this series of textbooks. This textbook focuses only on the topics of hydraulic fluids hosting, transmission, sealing, temperature control, and filtration.

With 30+ years of experience in teaching fluid power for industry professionals, the author had effectively applied his solid understanding to the subject and his post-doctoral level of academic education in developing this book.

The author has a goal of supporting fluid power professional education by developing the following series of volumes:

- Hydraulic Systems Volume 1: Introduction to Hydraulics for Industry Professionals.
- Hydraulic Systems Volume 2: Electro-Hydraulic Components and Systems.
- Hydraulic Systems Volume 3: Hydraulic Fluids and Contamination Control.
- Hydraulic Systems Volume 4: Hydraulic Fluids Conditioning.
- Hydraulic Systems Volume 5: Safety and Maintenance.
- Hydraulic Systems Volume 6: Troubleshooting and Failure Analysis.
- Hydraulic Systems Volume 7: Hydraulic Systems Modeling and Simulation for Application Engineers.
- Hydraulic Systems Volume 8: Design Strategies of Hydraulic Systems.
- Hydraulic Systems Volume 9: Design Strategies of Electro-Hydraulic Systems.
- Hydraulic Systems Volume 10: Hydraulic Components Modeling and Simulation.

ACKNOWLEDGEMENT

All praise is to Allah who granted me the knowledge, resources and health to finish this work.

To the soul of my parents who taught me the values of ISLAM

To my family: wife, sons, daughters in law, and grandchildren

To my best teachers and supervisors

The author wishes to thank the following gentlemen for their effective support in developing this book:

- Kamara Sheku, Dean of Applied Researches at Milwaukee School of Engineering.
- Tom Wanke, CFPE, Director of Fluid Power Industrial Consortium and Industry Relations at Milwaukee School of Engineering.
- Paul Michael, Research Chemist, Fluid Power Institute at MSOE.

The author thanks the following companies (listed alphabetically) for permitting him to use portions of their copyrighted literatures in this book.

- American High-Performance Seals
- American Technical Publishers
- Argo-Hyots
- Assofluid
- Bosch Rexroth
- Brennan Industries
- C.C. Jensen Inc
- CD Industrial Group Inc.)
- CEJN
- Donaldson
- Gates
- Hallite Seals
- Hydac
- Hydraulic and Pneumatic Magazine
- Hydraulic Specialist Study Manual, IFPS
- MP Filtri
- MFP Seals
- Noria Corporation
- Pall Corporation
- Parker Hannifin
- Schroeder
- Smart Reservoirs
- Trelleborg
- Womack

Lastly, the author extends his thanks to the following sources of public information used to enrich the contents of the book.

- www.powermotiontech.com
- https://helgesen.com
- www.americanmobilepower.com
- www.efficientplantmag.com
- www.luneta.com
- www.compudraulic.com
- www.stauffusa.com
- www.hyvair.com
- www.sapphirehydraulics.com
- www.epha.com
- www.sealsaver.com)
- www.hydrotechnik.com
- mac-hyd.com
- https://flanges-pipe.com
- www.grainger.com
- www.hydrotechnik.com
- www.daman.com
- www.applerubber.com
- www.marcorubber.com
- www.hydrapakseals.com
- news.ewmfg.com
- www.wyomingtestfixtures.com
- www.schoolcraftpublishing.com
- www.skf.com
- www.SouthwestThermal.com
- https://ej-bowman.com/knowledge-centre/zinc-anodes
- www.marinedieselbasics.com
- www.universalhydraulik.com
- www.savree.com

ABOUT THE BOOK

Book Description:

The book is targeting students and professionals who are looking to advance their fluid power careers. The book is colored and has the size of standard A4. The book is associated with a separate-colored workbook. The workbook contains printed power point slides, chapter reviews and assignments. This book is the fourth in a series that the author plans to publish to offer a complete and comprehensive educational curriculum for fluid power industry. This book is an attempt to fill the gap between the very academic style and the very commercial style of fluids power textbooks that are produced by fluid power manufacturers basically to promote their products.

The book presents different types of hydraulic reservoirs, transmission lines, sealing elements, and heat exchangers. For each of these elements, the book introduces the methods of operations, construction, sizing, and selection.

The book contains a total of eight chapters distributed over approximately 400 pages with very demonstrative figures and tables. The contents of the book are brand non-biased and intends to introduce the latest technologies related to the subject of the book.

Book Objectives:

Chapter 1: Introduction
In addition to hydraulic fluids contamination control, conditioning of hydraulic fluids is an essential process for a hydraulic system to perform properly and reliably. This chapter introduces the scope of hydraulic fluids conditioning.

Chapter 2: Hydraulic Reservoirs
Unless a hydraulic reservoir is designed, constructed, installed and maintained properly, the reliability of the entire system will be adversely affected. This chapter presents an overview of the different types of hydraulic reservoirs for industrial and mobile machines. This chapter also provides guidelines for design and sizing of hydraulic reservoirs.

Chapter 3: Hydraulic Transmission Lines
This chapter focuses on browsing the construction and the features of the three main hydraulic transmission lines, Pipes, Tubes, and Hoses. For each transmission line, the following topics are presented: sizing, material, construction and pressure rating. This chapter also presents information about fittings and manifolds.

Chapter 4: Hydraulic Sealing Elements

This chapter provides a knowledge base for fluid power users to become familiar with the commonly used seals in hydraulic components. This chapter presents an overview of hydraulic sealing elements including seal functions, classifications, and materials. This chapter also presents 15 various properties of hydraulic seals and the relevant standard test methods. This chapter also presents sealing solutions for cylinders and rotational shafts.

Chapter 5: Hydraulic Heat Exchangers

This chapter overviews various types of heat exchangers including air-type, water-type, and plate-type. Construction, operation, features, applications, and sizing calculations are discussed.

Chapter 6-Introduction to Hydraulic Filters

This chapter presents an overview of hydraulic filters including the contribution of filters in hydraulic systems, ISO1219 symbols, construction and operating principles. The chapter also presents various types of filters based on application in which the filter is used, type of connection to the circuit, body style of the filter, placement in the hydraulic circuit. The chapter also discusses the added accessories to the filter such as bypass valve and clogging indicators. Examples from industry are presented.

Chapter 7-Filter Media and Filtration Mechanisms

This chapter presents an overview of filter elements including the construction and material of the filter media. This chapter discusses surface filters versus depth filters. The chapter discusses also the principles of various filtration mechanisms that are applicable in hydraulic filters such as direct interception, absorption, adsorption, and magnetic separation.

Chapter 8-Filter Selection Criteria

This chapter presents a selection checklist as a guide for selecting proper filters. The chapter also discusses briefly the concepts for cost-effective filtration and selecting a filter cleanliness level based on system requirements. This chapter presents several examples of filtration solution for hydraulic systems.

Note: you may notice that there are some duplications in the figures and body text between chapter 2 and chapter 6. The reason is that the author wants to make each subject is a standalone chapter that can be taught independent from the other chapters.

Book Statistics:

The table shown below contains interesting statistical date about the textbook:

Chapter #	Pages	Figures	Tables	Words	Editing Time (Hours)
Chapter 1	4	1	0	626	100
Chapter2	63	63	2	11450	205
Chapter 3	104	122	15	13159	224
Chapter 4	105	120	11	14279	374
Chapter 5	42	44	2	6764	147
Chapter 6	56	69	0	6897	189
Chapter 7	20	26	0	3160	181
Chapter 8	12	8	2	1247	165
	406	453	32	57582	1,585 Hour = 66 Days

ABOUT THE AUTHOR

Medhat Khalil, Ph.D. is Director of Professional Education & Research Development at the Applied Technology Center, Milwaukee School of Engineering, Milwaukee, WI, USA. Medhat has consistently been working on his academic development through the years, starting from bachelor's and master's Degrees in Mechanical Engineering in Cairo Egypt and proceeding with his Ph.D. in Mechanical Engineering and Post-Doctoral Industrial Research Fellowship at Concordia University in Montreal, Quebec, Canada. He has been certified and is a member of many institutions such as: Certified Fluid Power Hydraulic Specialist (CFPHS) by the International Fluid Power Society (IFPS); Certified Fluid Power Accredited Instructor (CFPAI) by the International Fluid Power Society (IFPS); Member of Center for Compact and Efficient Fluid Power Engineering Research Center (CCEFP); Listed Fluid Power Consultant by the National Fluid Power Association (NFPA); and Listed Professional Instructor by the American Society of Mechanical Engineers (ASME). Medhat has balanced academic and industrial experience. Medhat has vast working experience in Fluid Power teaching courses for industry professionals. Being quite aware of the technological developments in the field of fluid power,

Medhat had worked for several world-wide recognized industrial organizations such as Rexroth in Egypt and CAE in Canada. Medhat had designed several hydraulic systems and developed several analytical and educational software. Medhat also has considerable experience in modeling and simulation of dynamic systems using Matlab-Simulink. Medhat has been selected among the inductees for Pioneers in fluid Power by NFPA (2012) and Hall of Fame in fluid Power by IFPS (2021).

Chapter 1
Introduction

Objectives

In addition to hydraulic fluids contamination control, conditioning of hydraulic fluids is an essential process for a hydraulic system to perform properly and reliably. This chapter introduces the scope of hydraulic fluids conditioning.

Brief Contents

1.1- Scope of Hydraulic Fluids Conditioning

1.2- Scope of the Textbook

Chapter 1: Introduction

1.1- Scope of Hydraulic Fluids Conditioning

Hydraulic fluid *Conditioning* is a term that inherently contains all the actions required for hydraulic fluid to perform optimally in a hydraulic system.

The main conditioning action for hydraulic fluids is to remove various types of contamination such as:

Fluid Analysis
Particulate removal is usually done with mechanical filters. A well-designed reservoir that allows settling of contaminants will also help in keeping particulates out of the mainstream fluid. For ferrous particulates and rust, reservoir magnets or strainer band magnets can also be used. Other methods such as centrifuging or electrostatic filtration units can also be used.

Removal of Air
Getting air out of the system is best done by adding a 100-micron screen in the reservoir, approximately 30° from horizontal to capture entrained air bubbles. Large bubbles rise easily to the surface and dissipate there. Additional actions help getting rid of air such as reservoir design, layout of suction and return lines, and use of foam suppressing additives.

Removal of Water
Several techniques exist to prevent water or moisture ingression or to remove water once it is present in a hydraulic or lube oil system. The best choice for a specific technique of water removal depends on the form of water content (dissolved or free), and also the quantity of water. For example, moisture can be reduced or prevented from entering a fluid reservoir using absorptive breathers. However once free water is present in small quantities, water absorbing filters or active venting systems usually provide adequate removal means. For large quantities of water, vacuum dehydration, and centrifuges are appropriate techniques for its removal.

Removal of Chemicals
Removal of acids, sludge, gums, varnishes, soaps, oxidation products and other chemicals generally requires special absorbent (active) filters.

All the above-mentioned actions of contamination control are discussed in the Volume 3 of this textbook's series under the title of "*Hydraulic Fluids and Contamination Control*".

1.2- Scope of Textbook

This textbook focuses on the other conditioning actions for hydraulic fluids such as fluids hosting, transmission, sealing, temperature control, and filters. Figure 1.1 shows that hydraulic fluids are handled within the system as follows:

- **Hosting (Hydraulic Reservoirs):** Hydraulic fluids are hosted in hydraulic reservoirs that must be sized and designed properly to help fluids get rid of heat, water, foam, and particulate contaminants.

- **Sealing (Hydraulic Sealing Elements):** Hydraulic fluids work under high pressure. Therefore, without proper sealing, hydraulic fluids can leak internally or externally.

- **Transmission (Hydraulic Transmission Lines):** Hydraulic fluids travel within the system through transmission lines that must be sized and designed to secure proper and efficient flow patterns.

- **Temperature Control (Heat Exchangers):** Hydraulic systems generate heat and receives heat from various sources. Therefore, without a proper temperature control system, hydraulic fluids lose properties and hydraulic systems became inefficient.

- **Removal of Particulate Contaminants and Silt (Filters):** Particulate removal is usually done with mechanical filters. A well-designed reservoir that allows settling of contaminants will also help in keeping particulates out of the mainstream fluid. For ferrous particulates and rust, reservoir magnets or strainer band magnets can also be used. Other methods such as centrifuging or electrostatic filtration units can also be used. Silt, defined as very fine particles under 5 μm in size, requires very fine filtration that is referred to "oil polishing".

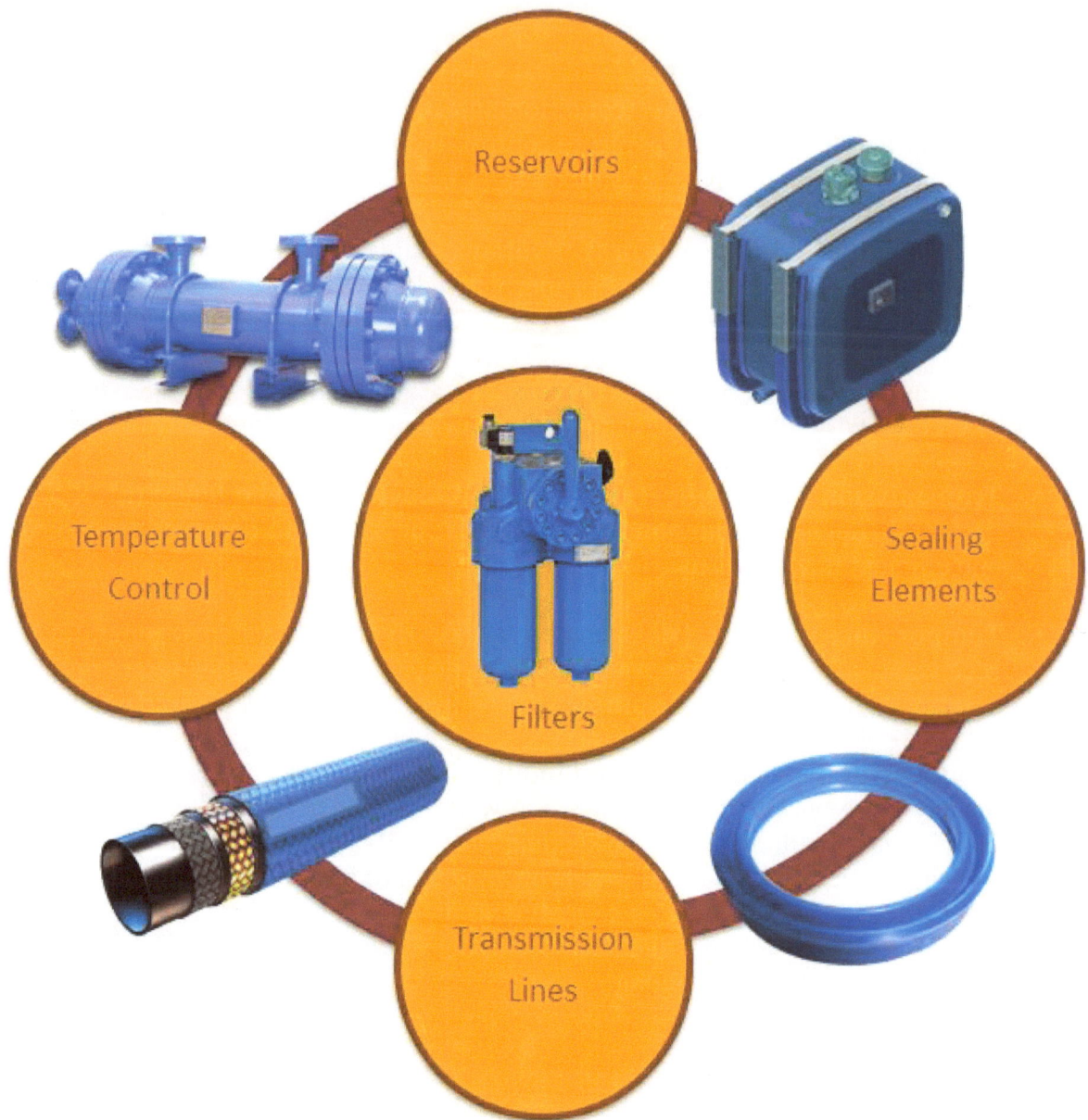

Fig. 1.1 – Hydraulic Fluids Conditioning

Chapter 2

Hydraulic Reservoirs

Objectives

Unless a hydraulic reservoir is designed, constructed, installed, and maintained properly, the reliability of the entire system will be adversely affected. This chapter presents an overview of the different types of hydraulic reservoirs for industrial and mobile machines. This chapter also provides guidelines for design and sizing of hydraulic reservoirs.

The following topics are discussed in Chapter 9 in Volume 5 "Safety and Maintenance" of this series of textbooks:
- Reservoirs Selection and Replacement
- Reservoirs Maintenance Scheduling
- BP-Reservoirs Installation and Maintenance

The following topics are discussed in Chapter 9 in Volume 6 "Troubleshooting and Failure Analysis" of this series of textbooks:
- Hydraulic Reservoirs Inspection
- Hydraulic Reservoirs Troubleshooting
- Hydraulic Reservoirs Failure Analysis

Brief Contents

2.1- Contribution of Hydraulic Reservoirs
2.2- Configurations of Hydraulic Reservoirs
2.3- Construction of Hydraulic Reservoirs
2.4- Design of Hydraulic Reservoirs
2.5- Hydraulic Reservoir Design Case Study

Chapter 2: Hydraulic Reservoirs

2.1- Contribution of Hydraulic Reservoirs

Hydraulic oil Reservoir is also referred to as "Oil Tank". Referring to Fig. 2.1 A hydraulic reservoir contributes to the performance of hydraulic system as follows:

Main Contribution (1):
A hydraulic reservoir in a hydraulic circuit act like a grounding point in an electrical circuit. The main contribution of a hydraulic reservoir is to host the oil volume required to operate the system, and to receive all the return oil from the various hydraulic components.

Auxiliary Duties:
- Remove Contaminants (2): Helps in settling solid and liquid contaminants at the bottom.

- Remove Air (3): Helps in removing air bubbles that come back with the return oil and suppressing foam that is generated in the reservoir.

- Remove Heat (4): Helps in dissipating some of the generated heat through the walls. Properly sized reservoir works with the heat exchanger to maintain working temperature approximately 49 $^{\circ}$C (120 $^{\circ}$F).

- Equipment Base (5): Used as a base to mount hydraulic components.

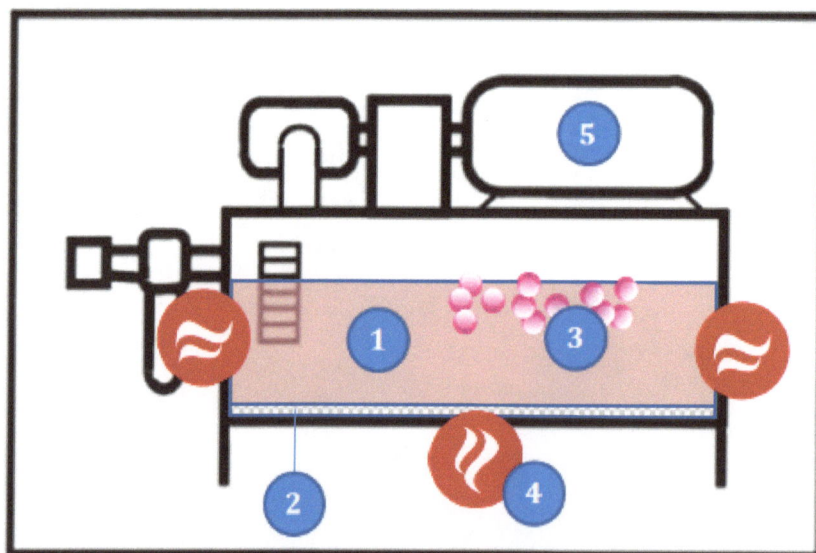

Fig. 2.1 – Contribution of Hydraulic Reservoirs

2.2- Configurations of Hydraulic Reservoirs

Hydraulic reservoirs are configured in various ways as shown in Table 2.2:

Pressure	Shape	Volume	Application	Function	Material	Size
Open	Parallelogram	Fixed Volume	Standard (Industrial)	Main Reservoir	Traditional (Steel)	Compact & Small
Closed	Cylindrical (Custom)	Variable Volume	Custom (Mobile)	Add-On Reservoir	Hybrid	Medium & Large

Table 2.2 – Configurations of Hydraulic Reservoirs

2.2.1- Open versus Closed Hydraulic Reservoirs

As show in Fig. 2.2, based on the pressure on the surface of the fluid inside the reservoir, hydraulic reservoirs are either open or closed.

Open Reservoirs: The symbol is as shown in the figure. In an *Open Tank*, the oil surface is subjected to atmospheric pressure. through a breather filter. Obviously Open tanks are easy to build. Open tanks are widely used for conventional applications where the environmental pressure isn't changing from atmospheric or where the contamination is within the acceptable limits.

Closed Reservoirs: The symbol is as shown in the figure. In a *Closed Tank*, the tank is isolated from the atmosphere and the oil contained is subjected to a pressure slightly higher than the atmospheric pressure. The oil inside the tank is pressurized by several ways such as placing a Nitrogen Bladder inside the tank. The advantages of using a closed tank over an open one is that the closed one boosts the intake pressure of the pump and consequently avoids cavitation. Additionally, it helps reducing the contamination that might get into the tank through the breather filter in open tanks. On the other hand, using a closed tank will add back pressure against the return oil. The use of Closed Tanks is mandatory in some applications such as aerospace industry, highly contaminated work environment, and some mobile applications.

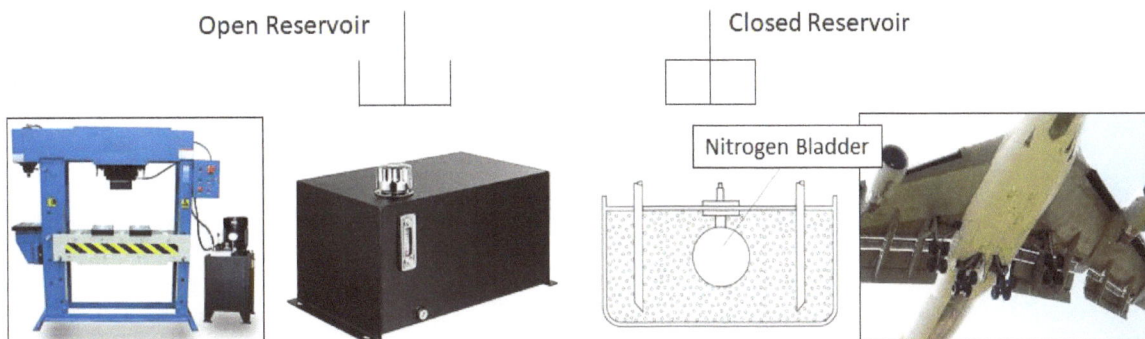

Fig. 2.2 – Open versus Closed Hydraulic Reservoirs

2.2.2- Parallelogram-Shaped versus Rounded Reservoirs

Instead of the traditional parallelogram-shaped reservoirs, *rounded reservoirs* offer the following features:

Promote Cleanliness: As shown in Fig. 2.3, the reservoir is structured as vertical cylindrical shape with conical bottom. The suction connection is located in the lowermost part of the reservoir, at the apex of the inverted cone. The return flow inlet is tangential to the cylindrical wall. Dynamics of the return flow forces the fluid to rotate around the center. As a result, contaminants concentrate at the center of rotation. From there, contaminants will be trapped by the filter instead of settled at the bottom of a flat surface reservoir.

Cost Effective Design: Vertically oriented cylindrical reservoirs are better utilization of the materials than with conventional designs, especially when larger reservoirs are required.

In conventional reservoirs, hydrostatic pressure due to fluid column on sharp corners has to be considered to prevent the reservoir's side plates from deflecting outward. This is usually done by welding ribs inside or outside the reservoir. Welded ribs, however, are expensive to install. Furthermore, locating them inside the reservoir creates corners that collect contamination. However, locating them outside the reservoir is a losing proposition too, because they consume additional floor space.

In a cylindrical reservoir, the hydrostatic pressure itself presents no practical impact on the thickness of the side plate. This is because the geometry of a cylinder distributes stress from the hydrostatic pressure evenly around the circumference of the reservoir.

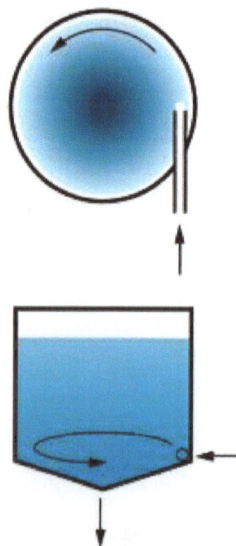

Fig. 2.3 – Cylindrical Hydraulic Reservoirs (www.powermotiontech.com)

2.2.3- Fixed versus Variable Volume Reservoirs

Concept: As shown in Fig. 2.4, hydraulic systems very often rely on conventional *Fixed Volume Reservoirs* (FVR) that are too large and too heavy.

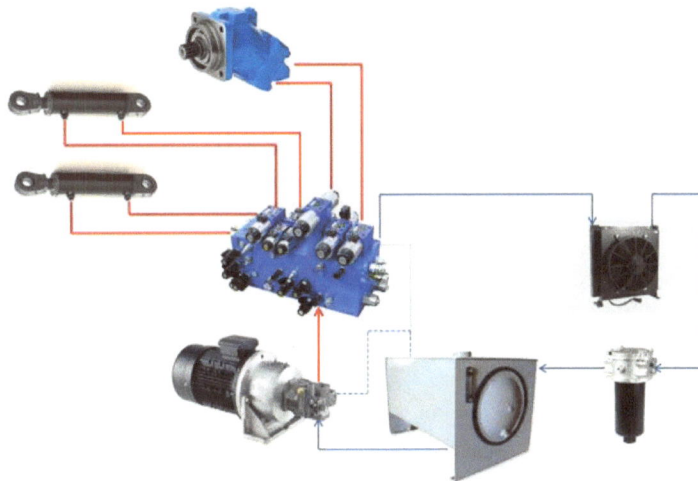

**Fig. 2.4 – Typical Simplified Hydraulic Circuit with Conventional Fixed Volume Reservoir
(Courtesy of Smart Reservoirs)**

The *Variable Volume Reservoir* (VVR) is a concept derived from the aerospace "bootstrap" reservoir design. As shown in Fig. 2.5, a VVR is installed near the pump and only see differential oil volume due to thermal expansion or rod volume in a differential cylinder.

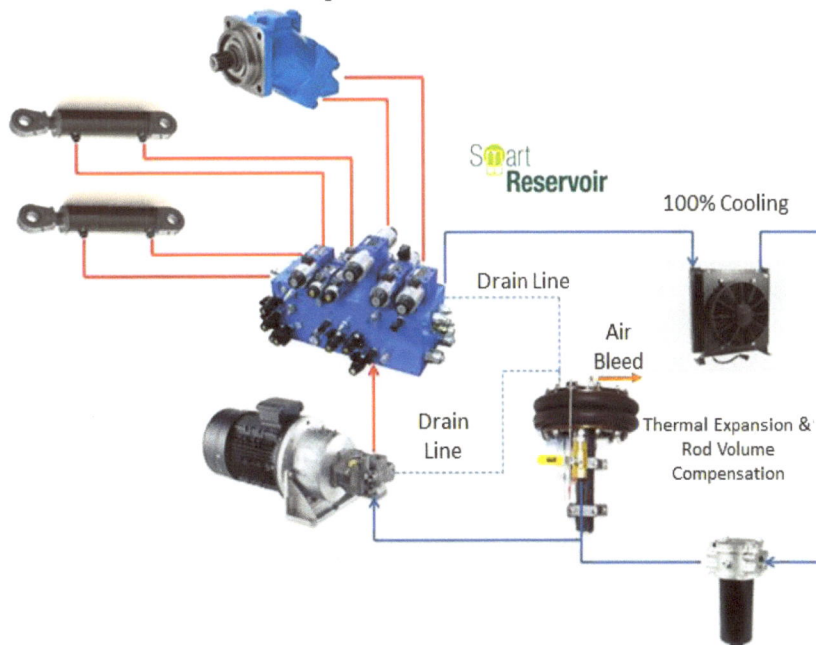

**Fig. 2.5 – Typical Simplified Hydraulic Circuit with Variable Volume Reservoir
(Courtesy of Smart Reservoirs)**

Applications: The VVR purpose is to replace the "classic" atmospheric reservoir (tank) on selected applications. The VVR targets industrial, mobile, military, marine and offshore fluid power systems. It can serve open circuit systems to hydrostatic drives. Weight reduction can be up to 100:1 compared to classic solutions.

Operating Principle: As shown in Fig. 2.6, in a closed circuit that drives a hydraulic motor, it allows expansion and contraction to compensate for fluid thermal temperature changes and boosts the intake side of the pump. The pressurized fluid boosts the low-pressure side of the closed circuit without a need for a separate boost pump. As shown in Fig. 2.7, in an open circuit, VVR reacts to the differential volume caused by rod motion of a differential cylinder.

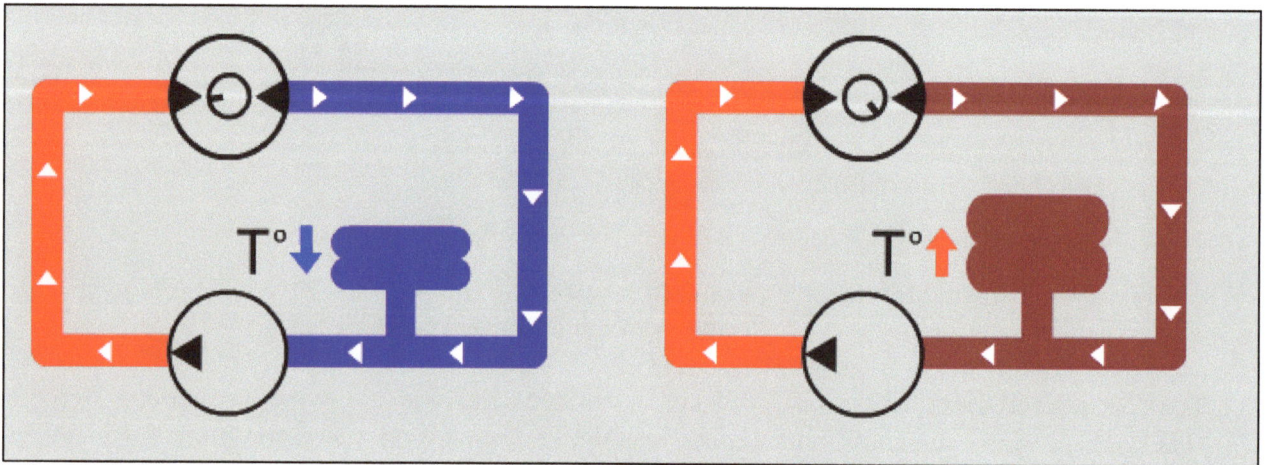

**Fig. 2.6 - Operating Principle of Variable Volume Reservoirs in a Closed Hydraulic Circuit
(Courtesy of Smart Reservoirs)**

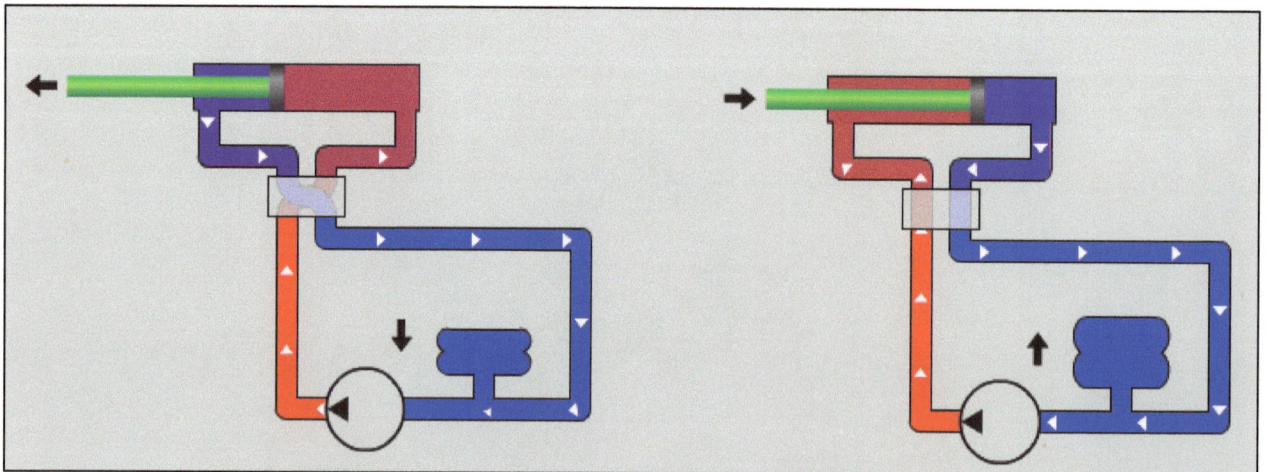

**Fig. 2.7 - Operating Principle of Variable Volume Reservoirs in an Open Hydraulic Circuit
(Courtesy of Smart Reservoirs)**

Construction: as shown in Fig. 2.8, the VVR standard model is constructed as follows:

1- The main part is an airless bellows slightly pressurized to 0.1 to 0.6 bar (1 to 9 psi).

2- Anodized aluminum cover (A) with built-in low-level switch (B), an air bleed valve (C), and a filtration/bleed return with isolation ball valve [D].

3- Anodized aluminum base manifold (A) with a connection port (B), a VVR coupling fitting (C), an air bleed valve (D), a filling coupler (E), an adjustable temperature switch (F), a visual temperature indicator (G), a pressure gauge (H), and a relief valve (I).

4- Full assembly of VVR with the instrumentation manifold.

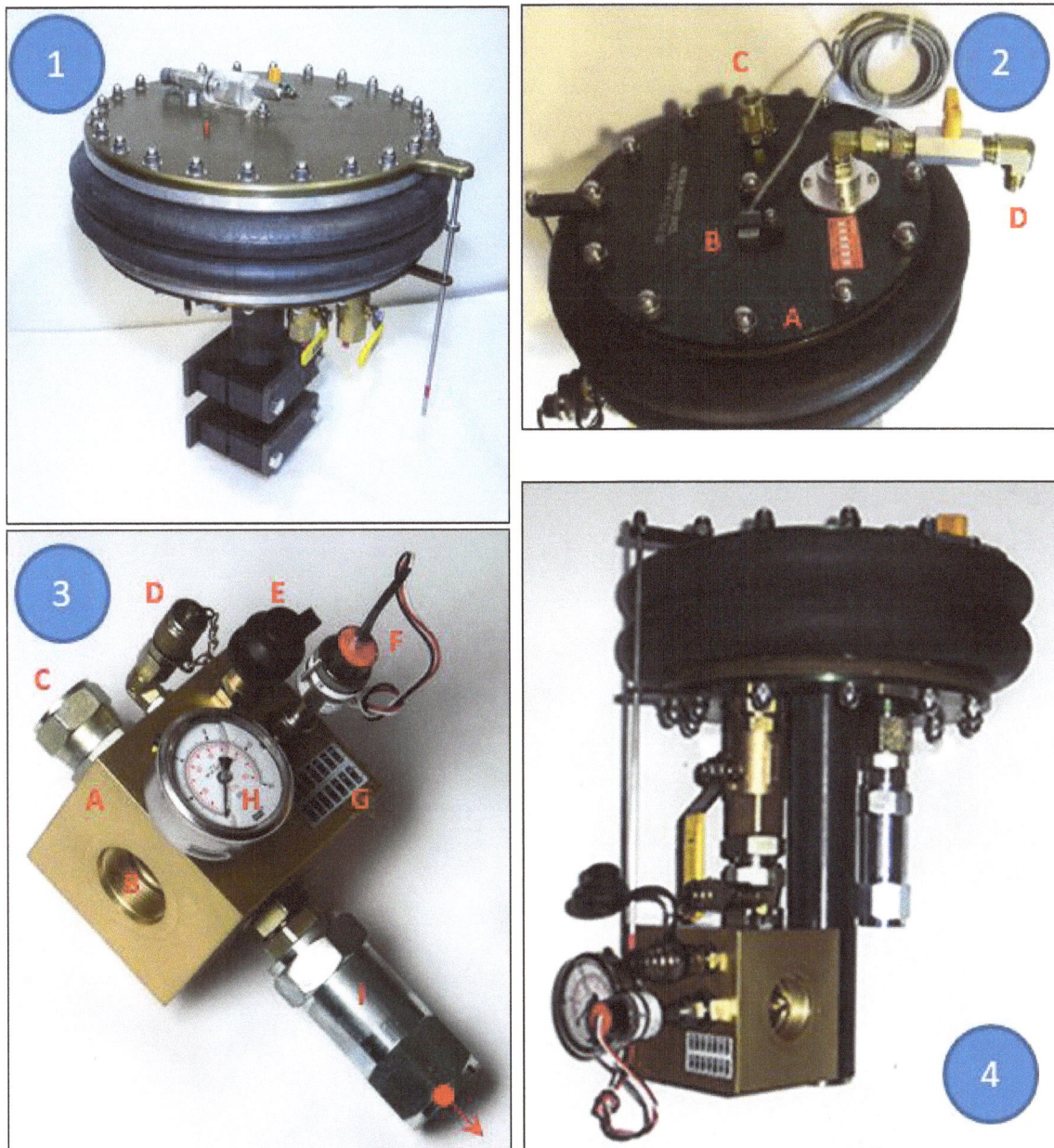

Fig. 2.8 - Construction of Variable Volume Reservoir (Courtesy of Smart Reservoirs)

Installation of VVR: Figure 2.9 shows typical circuit diagrams for using VVR with multiple variable displacement pumps. As shown in the figure, a hand pump is used to fill in the system with the required volume of oil. When the system works, all the pump and motors case drain will be directed to the VVR after passing through an in-line filter. Return line is directly routed to the pump inlet. With this configuration returning air is forced towards the VVR. Accumulated air will gather to be purged out through air bleed valve.

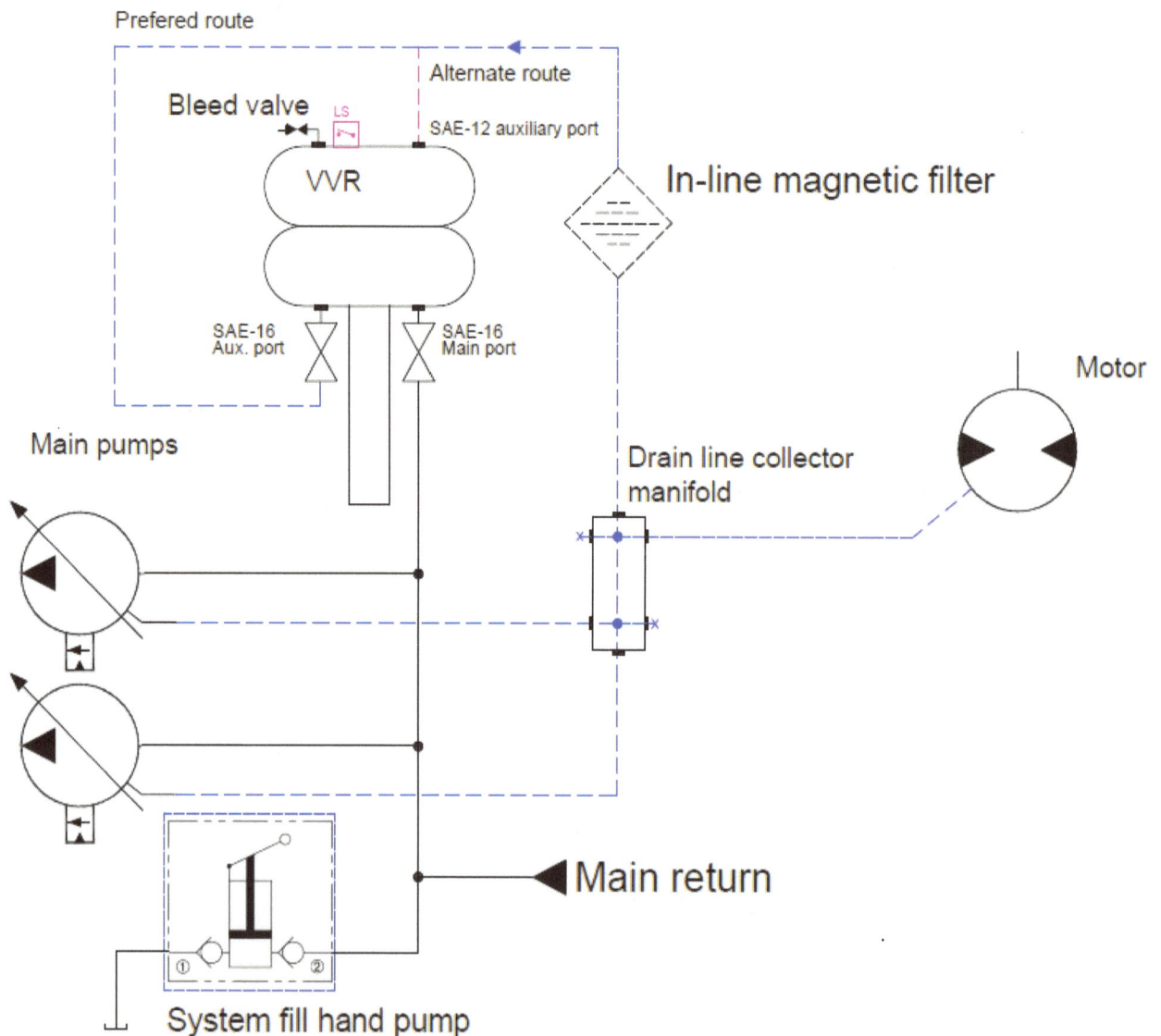

Fig. 2.9 – Typical Simplified Circuit Diagram using VVR with Variable Displacement Pumps (Courtesy of Smart Reservoirs)

Sizing of VVR: In conventional systems, the reservoir oil capacity is sized in relation to the pump flow rate (typically 2-3 times). The VVR is not affected or dependent of pump flow since the main return line is directly routed to the pump inlet. Synchronous cylinders and hydraulic motors are not regenerating flow difference if their motions are reversed. Differential cylinders and accumulators provide supplemental flow. Therefore, a VVR is sized based on the volume of the differential cylinder rod volume and the network fluid volume thermal expansion. As shown in Fig. 2.10, one VVR can feed many pumps or one per pump to isolate circuits. Multiple VVR's can be installed in series or parallel to increase volume.

Fig. 2.10 - Sizes of Variable Volume Reservoir (Courtesy of Smart Reservoirs)

VVR versus FVR:

- **Reduced Weight and Space:** Since the VVR use a small volume, it can cut weight up to a ratio of 100:1 and space up to a ratio of 20:1. This is a significant advantage for mobile applications.
- **Pump Supercharge:** VVR's are slightly pressurized thus improve pump inlet conditions regardless of the operating altitude and orientation. They can even work under water thus making an excellent pressure compensator for subsea applications.
- **Pump Suction Line Size:** Since the VVR is slightly pressurized, avoid cavitation and the pump suction line can be reduced by half.
- **Reduced Contaminant Ingestion:** Being sealed and airless, the VVR isolates the system from harsh environments like solid particles, moisture and water ingestion.
- **Lower Maintenance Cost:** Since the fluid and system are isolated from the environment, it will consequently reduce overall maintenance cost including fluid and filter replacement.
- **Environmental-Friendly:** The small VVR volume promotes the use of expensive biodegradable fluids at a significant reduced cost. Further, fluid leaks are detected earlier without much spillage and costly consequences.
- **Fluid Compatibility:** The VVR can be used with most mineral and biodegradable fluids and water glycol.
- **VVR is Airless:** Being sealed and airless, this feature requires to use a specific procedure when filling and bleeding the system before starting up.

VVR versus Accumulators: The first impression about the VVR may reveal that it performs like an accumulator, which is the case however the VVR offers benefits that accumulators can't offer such as:

- **Pressure Sensitivity:** The VVR reacts to fractional pressure variations where piston accumulators' seals friction just can't provide this sensitivity.
- **Temperature Variation:** VVR performance is not affected by temperature variations while gas charged accumulator's performance is affected by temperature changes.
- **Weight and Volume:** For same fluid displacement, the VVR weight is much less than an accumulator (fluid + gas volume in heavy shell).
- **Simplicity:** The VVR is leak free, gas free and requires no maintenance.
- **Performance Characteristics:** The VVR provides linear output. Accumulators performance follow the nonlinear gas laws and whether it's isothermal or adiabatic charging/discharging process.

2.2.4- Standard (Industrial) versus Custom (Mobile) Reservoirs

Design: To ensure a mobile hydraulic machine performs optimally, regardless of cost, intelligent design of the hydraulic reservoir must be considered. As shown in Fig. 2.11, unlike industrial applications, mobile applications have their own design rules on size and shape of the reservoir. These rules should consider space limitation, kinematic and the dynamic motion of the equipment so that the pump intake must be covered by oil in all phases of machine operation to avoid cavitation.

Shape: Hydraulic reservoirs for mobile applications may have special dimensions or irregular shape to meet the requirements. Figure 2.12 shows examples for hydraulic reservoirs for typical mobile applications.

Fig. 2.11 – Reservoirs for Mobile Applications

Fig. 2.12 – Reservoirs for Mobile Applications (https://helgesen.com)

Material: Figure 2.13 shows examples of different materials from which hydraulic reservoirs for mobile applications are manufactured.

Steel Tanks   Polyethylene Tanks   Aluminum Tanks

Transfer Tanks   Reefer Tanks

Upright Tanks   Sidemount Tanks   Saddlemount Tanks

Fig. 2.13 – Reservoirs for Mobile Applications (www.americanmobilepower.com)

2.2.5- Main versus Add-On Reservoirs

A main reservoir in a hydraulic system is the one that the pump is connected to. However, As shown in Fig. 2.14, an *Add-On* reservoir is an extension to the main reservoir. This solution might be needed due to space limitations when a large volume of oil is required. Add-On reservoirs have the following advantages:

- Increases the duration the oil stays in the reservoir.
- Increases the heat dissipation area.
- Increases the possibility of dirt settlement.
- Increases the possibility of air removal.

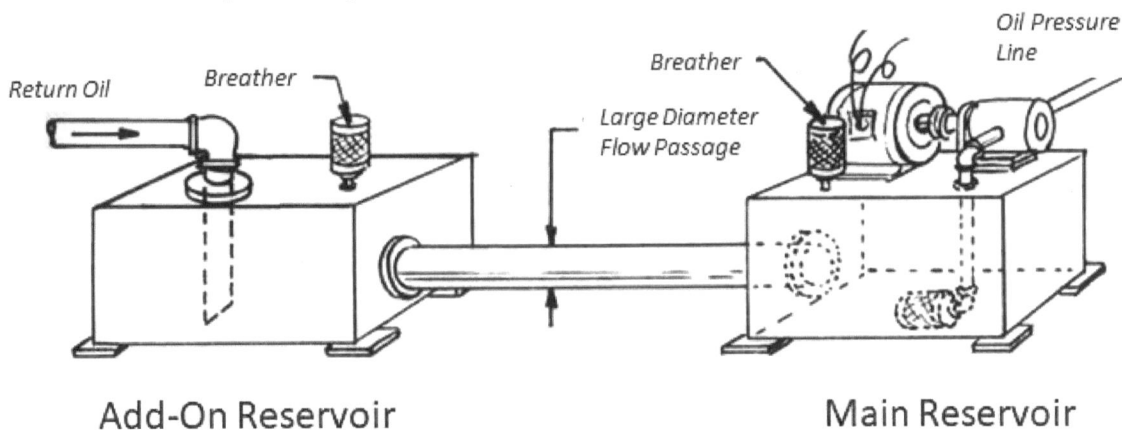

Fig. 2.14 – Main versus Add-On Reservoirs (Courtesy of Womack)

The following are best practices for setting up Add-On Tanks

- **Capacity:** Could be different from the main tank and should consider the volume of return oil.

- **Connecting Line:** Must connect both tanks at the lowest points.

- **Size of Connection Line:**
  - Rule-of-thumb 1: Sized to keep flow velocity to 1 foot/s.
  - Rule-of-thumb 2: Sized as one squared inch area for each 3 GPM of pump flow. Example, if the pump flows is 30 GPM, the connection line area is 10 squared inches.

- **Suction:** Pump sucking oil from the main tank.

- **Return:** Main return flow into the add-on tank.

- **Venting:** Each tank must be independently vented to atmosphere through a breather.

2.2.6- Traditional (Steel) versus Modern (Hybrid) Reservoirs

Traditional hydraulic reservoirs are made of steel with other hydraulic components are fixed on the top and sides of the reservoir by welding, bolting, etc. Modern technology is used to manufacture what is called "Hybrid Reservoirs or Tanks"

Product Description: As shown in Fig. 2.15, the *Hybrid Tank* is a ready-to-install complete module. All required tank functions are already integrated.

Production Technology: The hybrid tank, using Polyamide material, intelligently combines the two manufacturing technologies of rotational molding and injection molding. The unique combination of innovative production processes with a modular kit makes the hybrid tank economically feasible even for small quantities.

Connections with other Components: Quick-Connect fittings allow fault-free and tool-free hose mounting on the tank and can also be dismantled at any time. Numerous components (such as filter cover, Quick-Connect fitting, etc.) come from an existing modular system and do not required any special tool costs to the customer. Since the filter housing is part of the tank, there are no sealing points and therefore no risk of leakage. Filter element replacement is easily ensured.

Technical Data:
Tank Volume: up to 150 Liters.
Temperature: from -30 °C to +100 °C. The high thermal strength of the Polyamide material used allows the tank to be used even at higher operating temperatures.
Fluids: Mineral oil and Environmentally Friendly hydraulic fluids HEES and HETG.

Fig. 2.15 – Hybrid Integrated Tank (Courtesy of Argo-Hyots)

Construction: Figure 2.16 shows the basic construction of the hybrid tank

Design Features: The hybrid integrated tank has the following design features:

- Installation of a complete module in the machine.
- Weight reduction.
- High mechanical strength and thermal stability by using Polyamide.
- There is no risk of leakage between the filter head and the tank.
- Excellent corrosion resistance.
- Multiple ports save for flexible plumbing.
- Tool-free assembly of Quick-Connect fittings.
- Cost savings compared to traditional tank solutions.
- Complex geometries can be realized, see Fig. 2.17

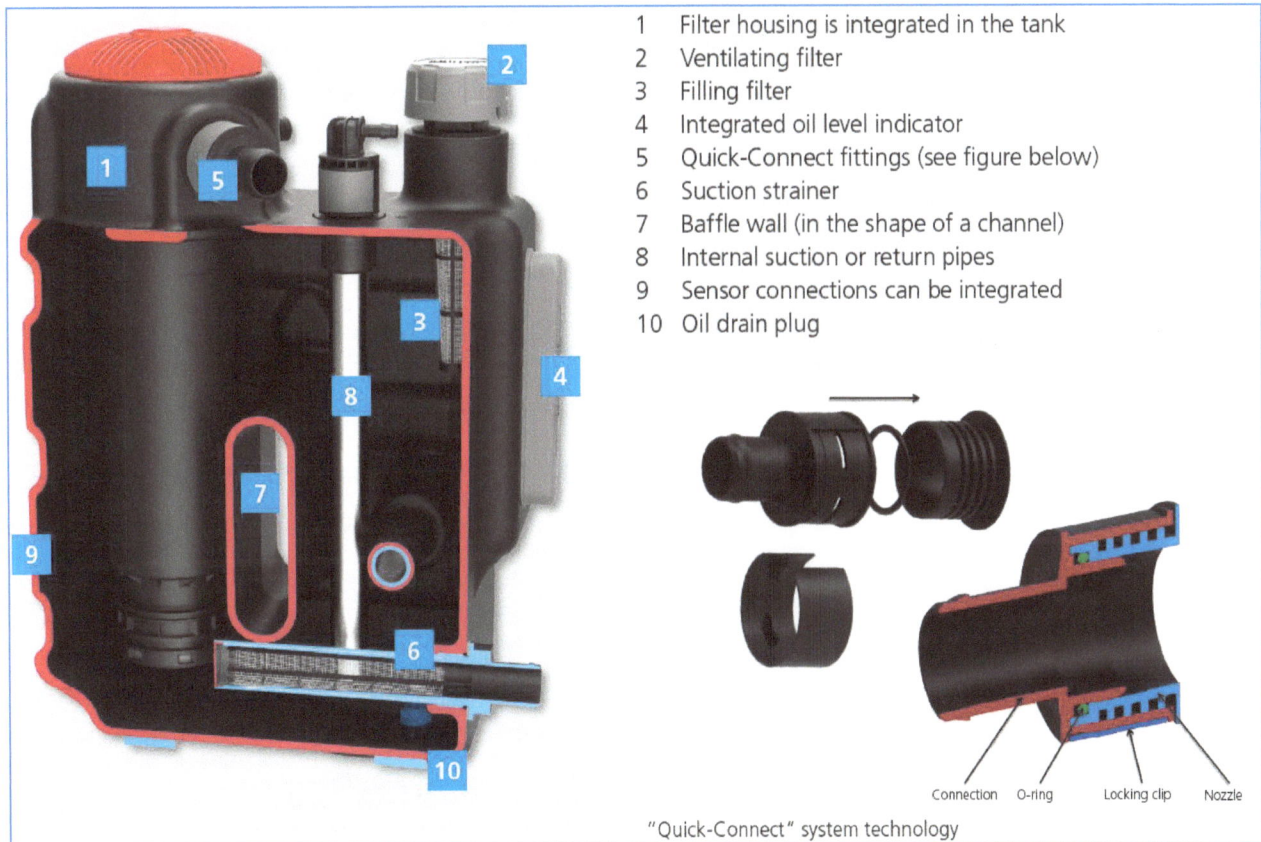

1   Filter housing is integrated in the tank
2   Ventilating filter
3   Filling filter
4   Integrated oil level indicator
5   Quick-Connect fittings (see figure below)
6   Suction strainer
7   Baffle wall (in the shape of a channel)
8   Internal suction or return pipes
9   Sensor connections can be integrated
10   Oil drain plug

Connection   O-ring   Locking clip   Nozzle

"Quick-Connect" system technology

Fig. 2.16 – Construction of Hybrid Integrated Tank (Courtesy of Argo-Hyots)

As shown in Fig. 2.17, the tank geometry can be adapted as required by the installation situation which s it make much more desirable for mobile applications.

Fig. 2.17 – Complex Shapes of Hybrid Integrated Tank (Courtesy of Argo-Hyots)

2.2.7- Reservoirs for Compact Power Units

Concept: As shown in Fig. 2.18, unlike hybrid tanks where all components are molded, reservoirs for compact power units are integrated with the rest of the components on a mounting block. The pump is immersed inside the reservoir. These compact power units are widely used in mobile applications and standalone hydraulic axes.

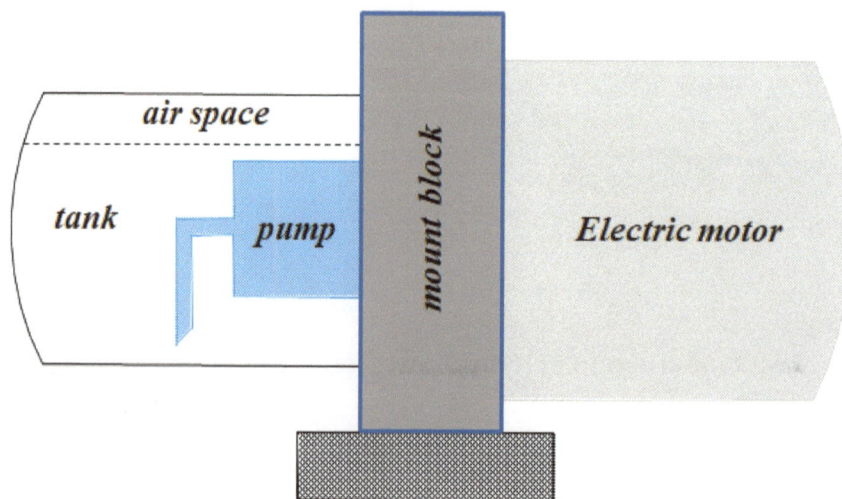

Fig. 2.18 – Concept of Compact Power Unit

Figure 2.19 shows examples of compact power units where the tank could be metallic or nonmetallic. As shown in the figure, some have just basic equipment, and some are fully equipped to drive multiple circuits. The following are common specifications of the compact power units:

The prime mover: It could be pneumatic, DC electric (12-24 Volt DC) or AC electric (one- or three-phase) type depending on the requirements.

The Pump: Depending on the displacement and the number of revolutions of the prime mover, the pump of standard extremal gear versions provides flow rates from a tenth of a liter to 10 l/min. Some special versions can sustain pressures even higher than 1000 bar. Some compact power units have a double pump to select pressure/flow and the motor or the directional valve can be operated via a remote control.

Reservoir: Apart from standard welded steel versions, tanks are made of plastic or a light alloy with external cooling fins.

Applications: Compact power units are used in a number of applications such as civil automations for gates, doors, lifts, dentist's reclining chairs, pleasure boats, firefighting equipment, mobile machines for small forklifts, tailboards and dump bodies of self-propelled machines, machine tools and light presses.

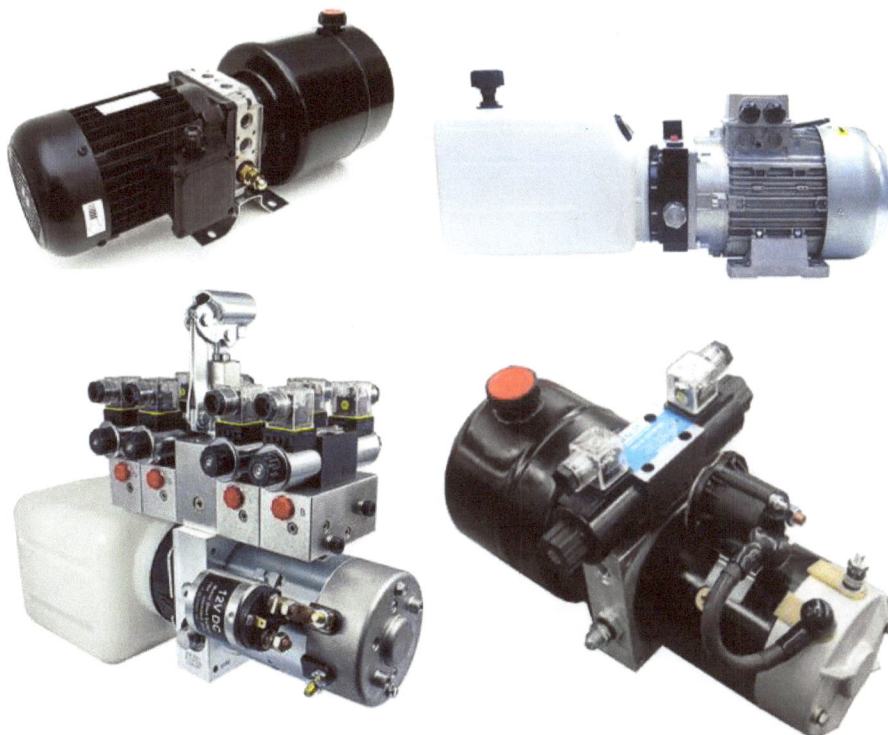

Fig. 2.19 – Examples of Compact Power Unit

Example: Figure 2.20 shows a particular example of a compact power unit provided by Rexroth and named CytroPac Hydraulic Power Unit. As shown in the figure, this special compact power unit contains the following elements:

1. Oil tank with motor-pump group (optional cooling packages).
2. Central plate (integrated heat exchanger).
3. Return flow filter.
4. Filter contamination sensors.
5. Filling level and temperature sensor.
6. Filling and breathing filters.
7. Cover with frequency converter below.
8. Electrical connections (see Fig. 2.21).
9. Visual oil level check and hydraulic fluid draining.
10. Clip (for removal of the hydraulic fluid hose for hydraulic fluid draining).

Product Description: The CytroPac is a frequency-controlled (variable speed) plug-and-play hydraulic power unit consists of a frequency converter, an electric motor, a hydraulic pump, a ring-shaped oil tank, a cooling system and sensor technology.

Fields of Applications: The CytroPac is a compact drive system for hydraulic machines and particularly machine tools and assembly lines under limited space conditions. The unit serves for regulated generation of hydraulic power on continuous or intermittent demand.

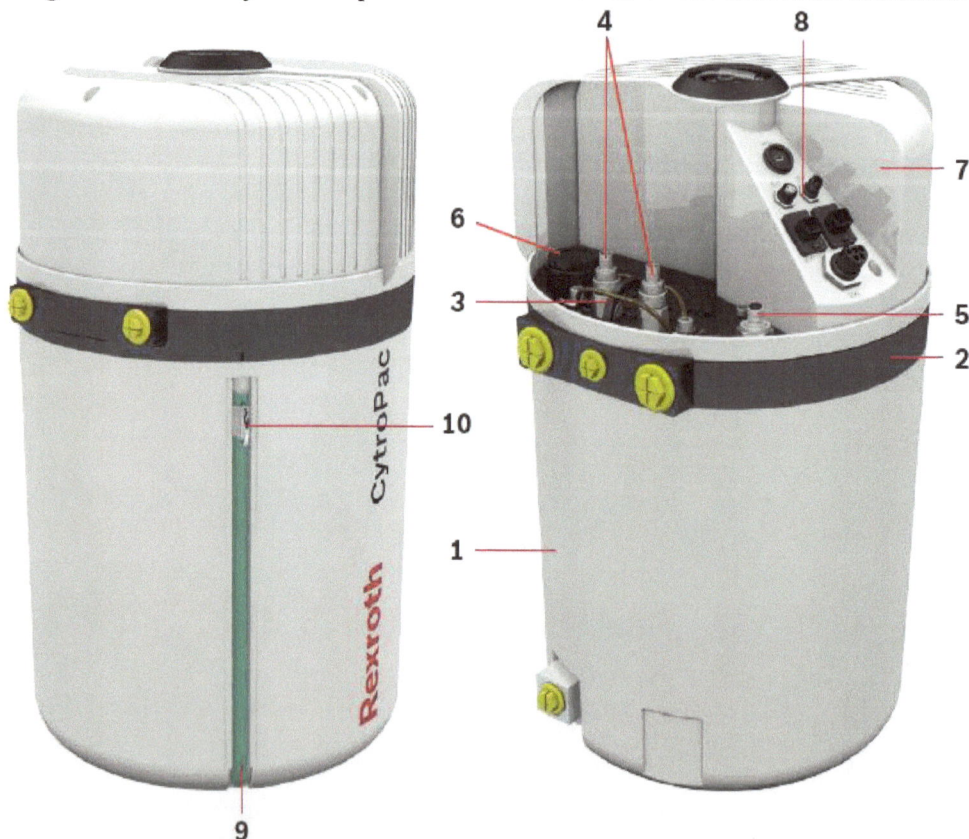

Fig. 2.20 – CytroPac Compact Power Unit (Courtesy of Bosch Rexroth)

Circuit Diagram: Figure 2.21 shows the circuit diagram for CytroPac.

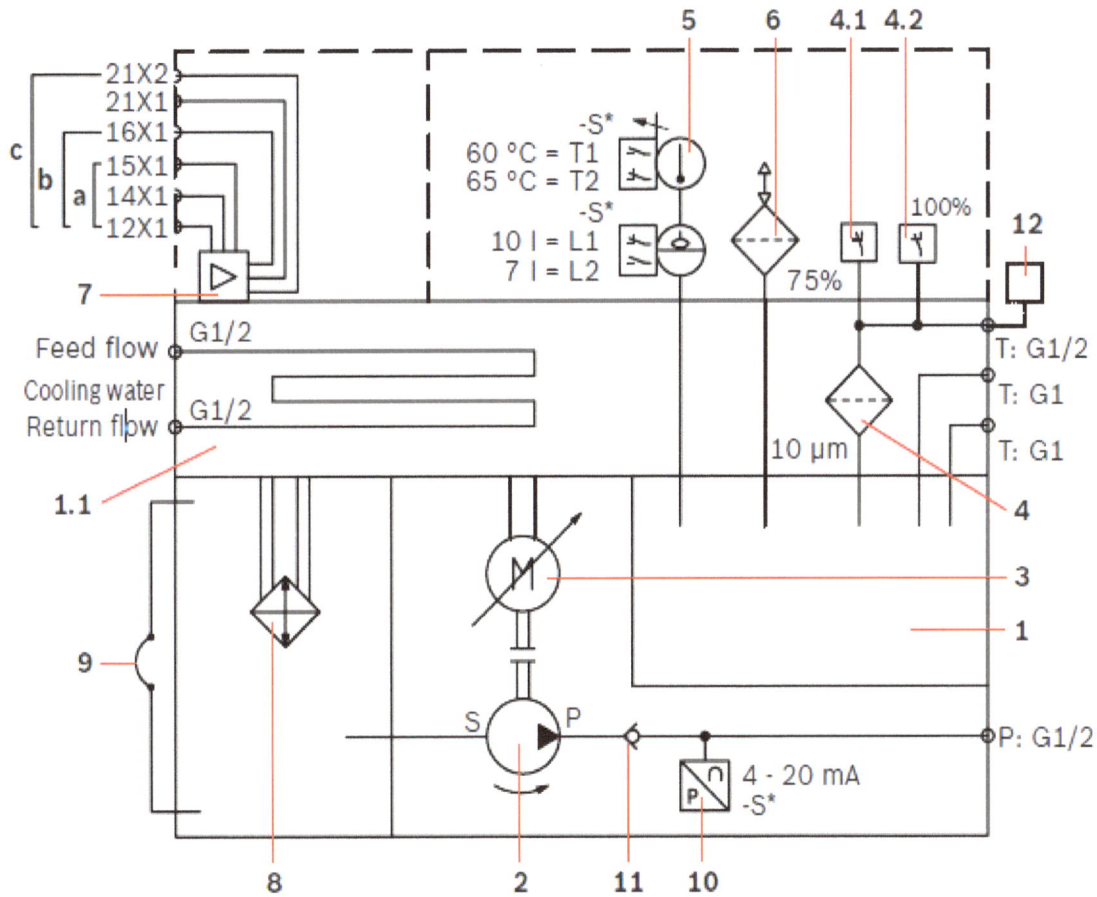

Fig. 2.21 – Circuit Diagram for CytroPac Compact Power Unit (Courtesy of Bosch Rexroth)

1 Oil tank	
1.1 Central plate (integrated heat exchanger)	
2 Pump	7 Frequency converter
3 Motor	8 Cooling package (optional)
4 Return flow filter	9 Visual oil level check and hydraulic fluid draining
4.1 Filter contamination sensor 75%	
4.2 Filter contamination sensor 100%	10 Pressure load cell
5 Filling level and temperature sensor	11 Check valve
6 Filling and breathing filters	12 Filling coupling (optional)

Design Features: As reported by the unit's manufacturer, it has the following features

- **Power Level:** Units are available with 1.5 - 4 kW (2 – 5.36 HP) with identical frame size and interface.

- **Low Noise Level:** The CytroPac is particularly quite due to plastic enclosure of all noise sources.

- **Cost-Effective Operation:** The frequency converter of the variable speed pump drive ensures dynamic adjustment of the power/speed and the flow to current requirements to reduce the operating costs.

- **Various Configurations:**
  o **Basic Configuration:** The filling level, temperature and filter contamination sensors must be wired by the customer for evaluation via the customer-side machine control system.

  o **Advanced Configuration:** All sensors and the motor in the hydraulic power unit are wired to the frequency converter The frequency converter serves as sensor node and bundles all measured values of filling level, temperature and filter contamination sensors and forwards these values via a mating connector to the customer-side machine control system. The status of the hydraulic power unit is also indicated via the integrated LED strip.

  o **Premium Configuration:** Unit performance is forwarded via a multi-standard Ethernet interface to the machine control system. This enables easy integration in Industry.

- **Prestart Control:** By means of a control signal, the drive unit is already accelerated before hydraulic actuators are connected. This reduces the collapse of pressure without a hydraulic accumulator.

- **Sleep Function:** By means of the integrated pressure monitoring, the hydraulic power unit is automatically switched off if the command pressure is reached maximum value, **e.g.** during accumulator charging operation. The unit is automatically switched on if the pressure is dropping. This increases the energy efficiency.

- **Warning Signals:** Early warning signals in case of faults regarding oil level, temperature, return flow filter and frequency converter.

Characteristics: Figure 2.22 shows the characteristic curve for 4 kW unit with various size pumps (4 – 14 cc). For example, for 4 cc pump size, it works under maximum pressure of 240 bar as long as the flow below 10 l/min. Once the flow demand increases above 10 l/min, pressure must decrease accordingly to meet the maximum power requirements. By calculating the power at 240 bar and 10 l/min, it is 3.6 kW. This means the unit keeps 0.4 kW to cope with the pressure spikes during start or changing the direction of motion of the actuators.

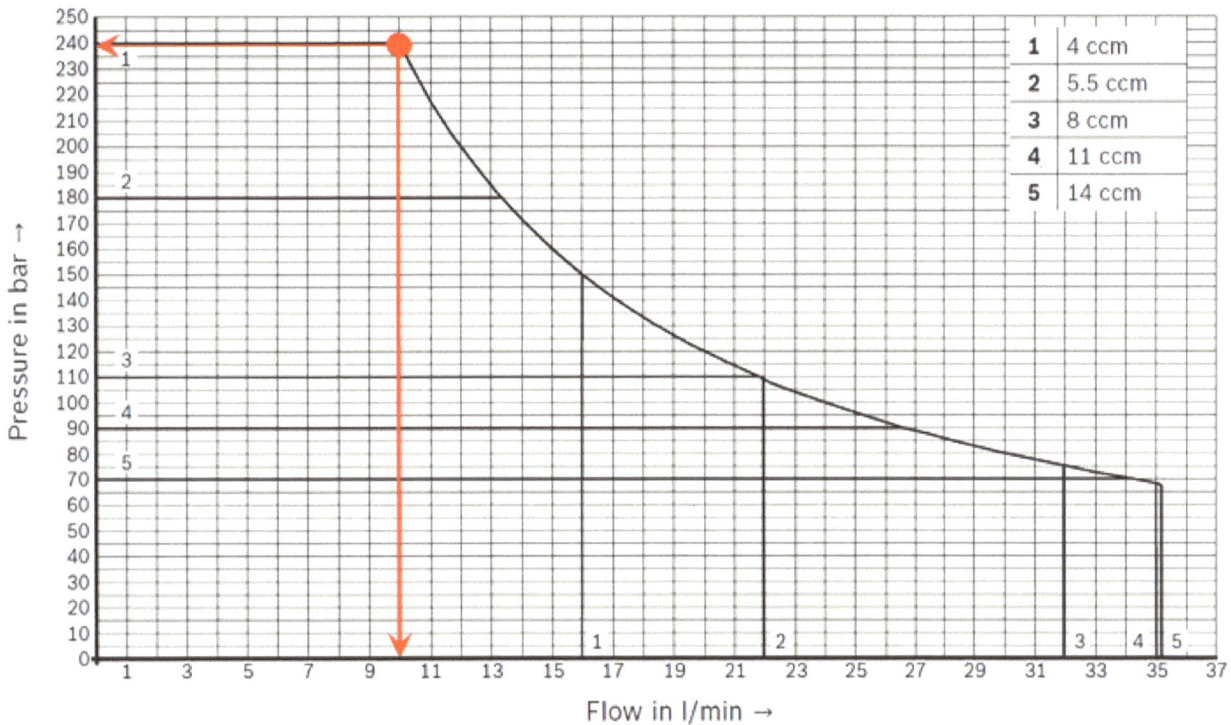

Fig. 2.22 - CytroPac Compact Power Unit 4 kW Characteristic Curves (Courtesy of Bosch Rexroth)

2.2.8- Foot-Mounted Reservoirs for Small Size Power Units

As shown in Fig. 2.23, the reservoir is placed on the ground and stands on two feet. That is why it is referred to as *Foot-Mounted* reservoirs. Such reservoirs are most common for small size industrial hydraulic power units.

The upper cover of the reservoir is commonly used as a base for the control valves and other accessories such as accumulators, filters, and heat exchangers. The pump can be immersed in the tank (1) for better cooling, but it will be difficult for pump replacement and troubleshooting. Alternatively, pump-motor unit can be assembled on top of the reservoir (2).

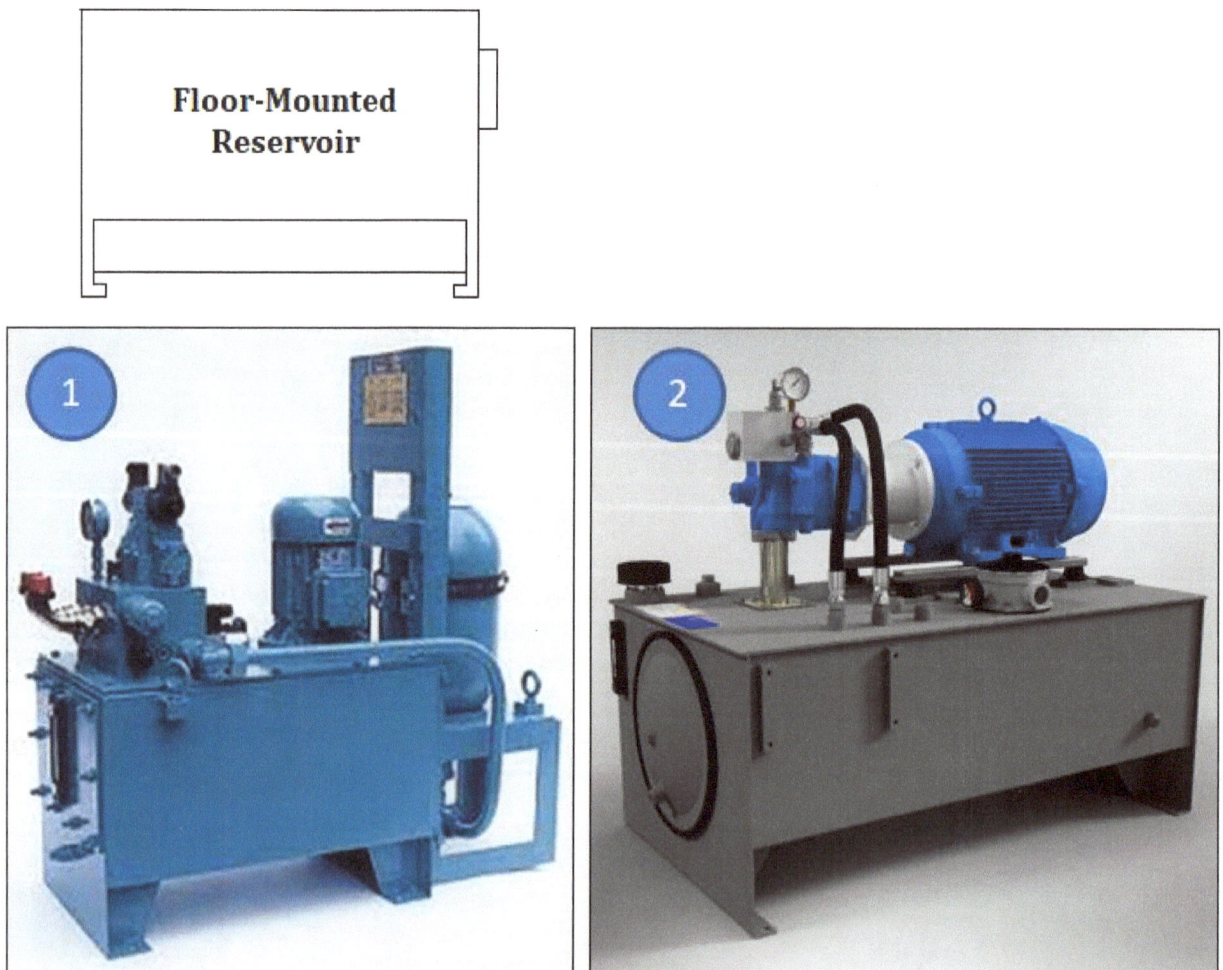

Fig. 2.23 Foot-Mounted Hydraulic Reservoirs

2.2.9- L-Shaped Reservoirs for Medium Size Power Units

As shown in Fig. 2.24, *L-Shaped* reservoirs are given such a name because the tank and the base form letter "L". Such type of reservoirs is recommended for medium size industrial hydraulic power units, where it is too heavy to place the pump and electric motor on top of the reservoir.

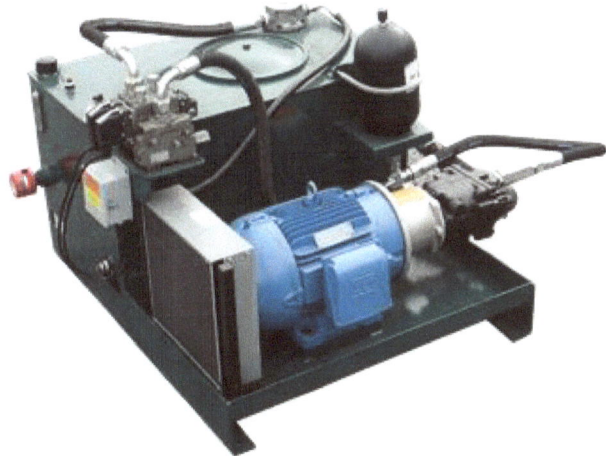

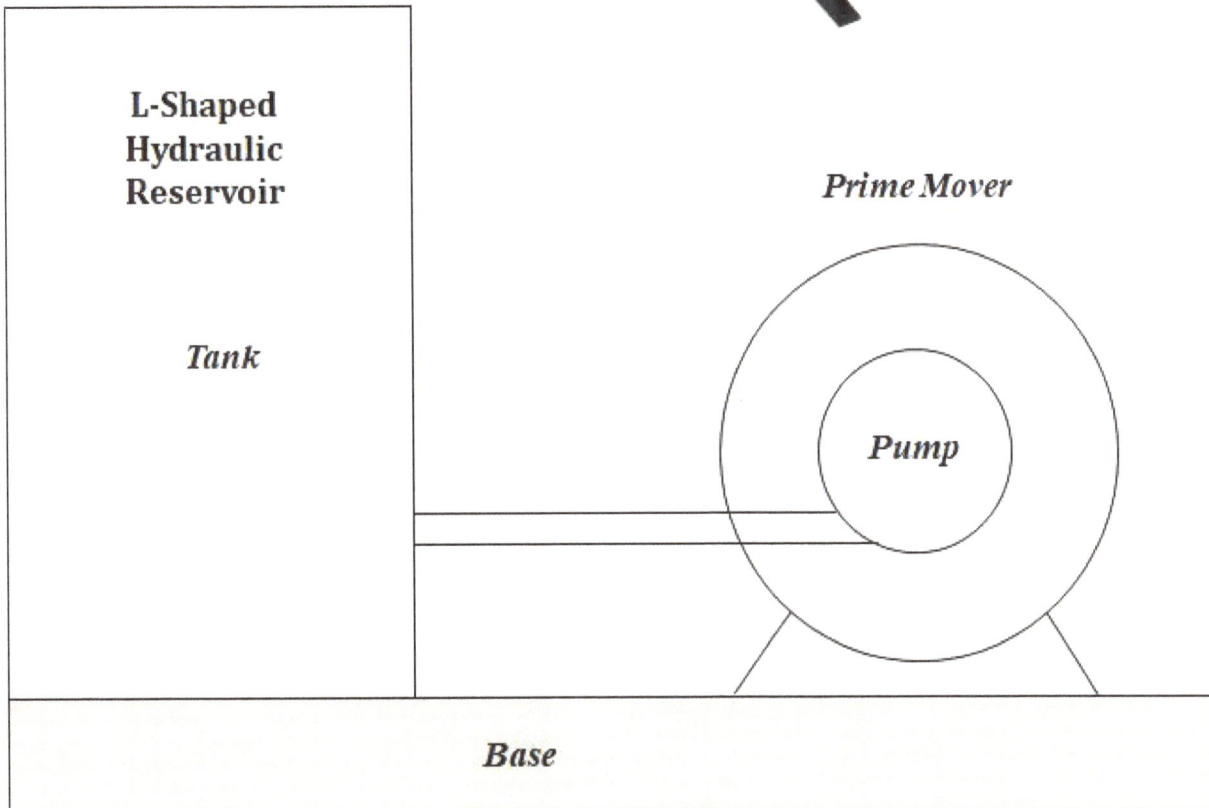

Fig. 2.24 - L-Shaped Mounted Hydraulic Reservoirs

2.2.10- Overhead Reservoirs for Large Size Power Units

As shown in Fig. 2.25, an *Overhead* Reservoir is placed on a level higher than the level of the pump so that it helps build positive pressure at the pump inlet (1 psi @2.5 ft for a typical petroleum oil). Such type of reservoirs is recommended for large size hydraulic power units where the pump flow is large and/or the pump is driven at high speed. Likely such pump subjected to cavitation if the intake line pressure isn't boosted. A typical application of using such reservoirs, as shown in the figure, are power units of hydraulic presses. For large presses, the tank is often mounted above the press (on a second-floor balcony or platform). This enables gravity flow from the tank to fill the volume above a single-acting ram.

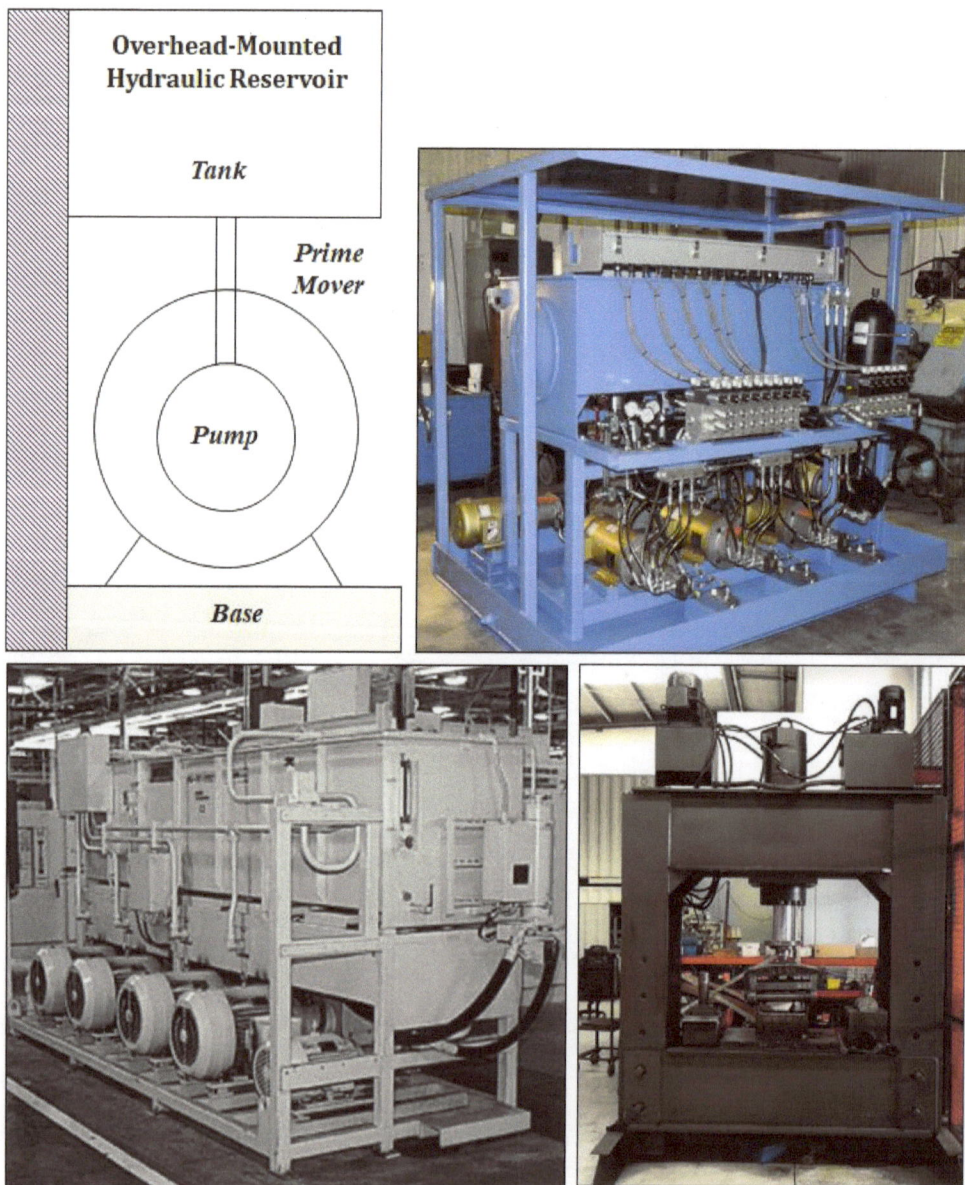

Fig. 2.25 - Overhead-Mounted Hydraulic Reservoirs

2.3- Construction of Hydraulic Reservoirs

In most stationary hydraulic circuits, the tank is part of a group of components (known as *'Hydraulic Power Unit'*). As shown in Fig. 2.26, the main part of the power unit is the hydraulic reservoir that consists of the main structure and accessories.

The main structure of the reservoir includes the steel box that hosts the fluid volume, baffle plate, deaeration screen, cleaning holes. A reservoir is equipped with auxiliary components named *Reservoir Accessories* such as a cap, a breather. etc.

A typical power unit consists of the reservoir, reservoir accessories and the following basic components:
- A pump and electric motor.
- An electric control panel that includes an on/off buttons and emergency stop button.
- A suction filter and a return filter.
- Shut-off valves, auxiliary connections and pressure gauges to measure pressure at different points.

In addition, depends on the system design, one or more of the following components are included:
- An accumulator.
- Heat exchanger.
- Off-line filtration unit

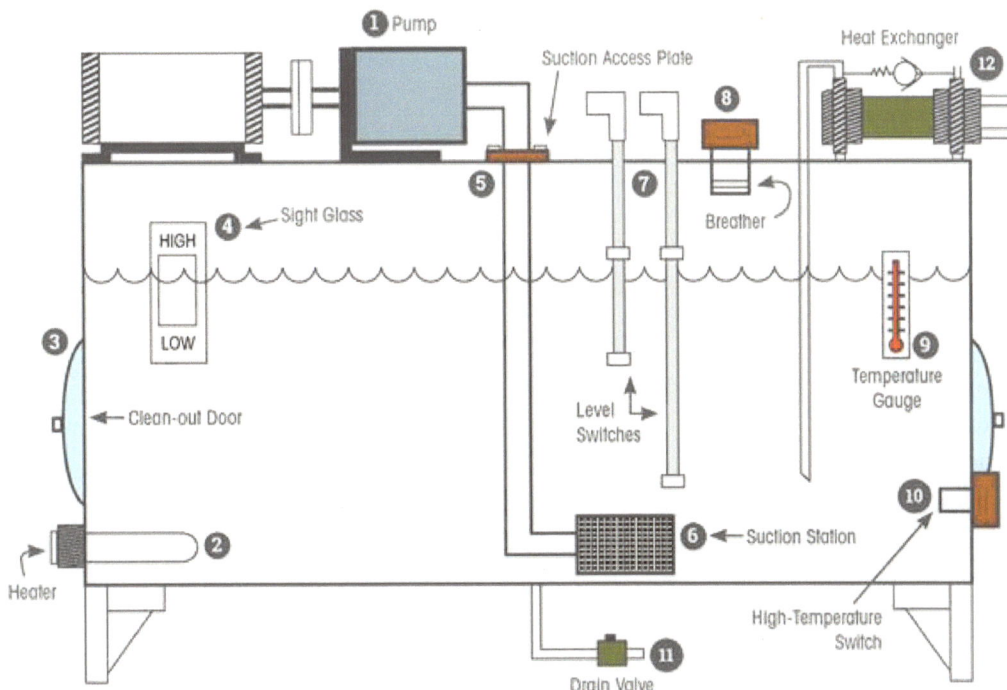

Fig. 2.26 – Construction of Hydraulic Reservoirs (www.efficientplantmag.com)

2.4- Design of Hydraulic Reservoirs for Industrial Applications

Improper design of a hydraulic reservoirs could lead to one or all of the following malfunctions in the System:
- Oil aeration and foaming.
- Pump cavitation.
- Heating up the System.
- Building fluid contamination.
- Leakage and environmental consequences.

Design Standards: the following *standards* contain guidelines for designing hydraulic reservoirs. It is highly recommended that they be reviewed before the design process.
- SAE Aerospace Standard (AS5586) for General Requirements for Hydraulic Reservoir.
- NFPA -T3.16.2-1969.
- ANSI -B93.18-1973 (R1987).
- ISO 4413:2010(E) 16 © ISO 2010.

Design Considerations: Designing of reservoirs for *mobile applications* may be more complex and requires specific simulation software that considers vehicle dynamics during all phases of operation. Reservoir design for mobile applications is out of scope of this textbook.

Designing a hydraulic reservoir for *industrial applications* must consider the requirements of the reservoir's contribution in hydraulic systems as follows:

- ❖ Body Construction
    1- Sizing.
    2- Heat Dissipation.
    3- Baffle Plates.
    4- Air Removal.
    5- Cleaning Holes.
    6- Drainage and Dirt Collection.
    7- Corrosion Protection.
    8- Structural Integrity.
- ❖ Hydraulic Lines
    9- Suction Lines.
    10- Return Lines.
    11- Drain Lines.
- ❖ Accessories and Attachments
    12- Filling Caps.
    13- Air Breathers.
    14- Oil Level and Temperature Indicators.

2.4.1- Sizing of Hydraulic Reservoirs

Sizing hydraulic reservoirs is vital for reliable performance of hydraulic systems. Referring to Fig. 2.27, The following set of oil volumes are to be considered when sizing a hydraulic reservoir:

- **Vsmax** = Maximum oil volume required to operate the system.
- **Vfill** = Oi filling volume.
- **Vth** = Oil volume due to thermal expansion.
- **Vmax** = Maximum oil volume in the reservoir.
- **Vas** = Aire space require on top of the oil volume.
- **Vo** = Overall reservoir volume.
- **Vmin** = Minimum absolute oil volume.

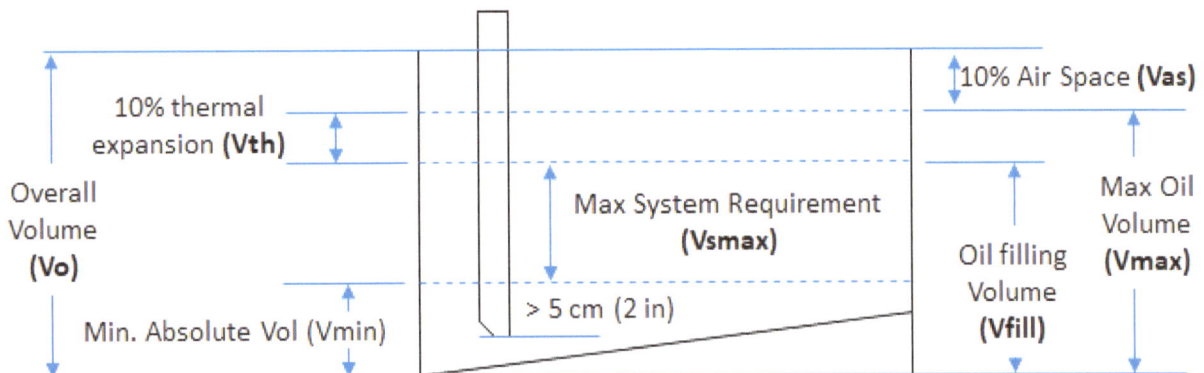

Fig. 2.27 - Rules for Sizing a Hydraulic Reservoir

❖ **Maximum System Requirement (Vsmax):** As shown in Fig. 2.28, during a machine operation, oil is withdrawn and return to tank continuously due to cylinders movement and accumulators charging/discharging, A system demands various fluid volumes up to maximum system requirements **Vsmax**.

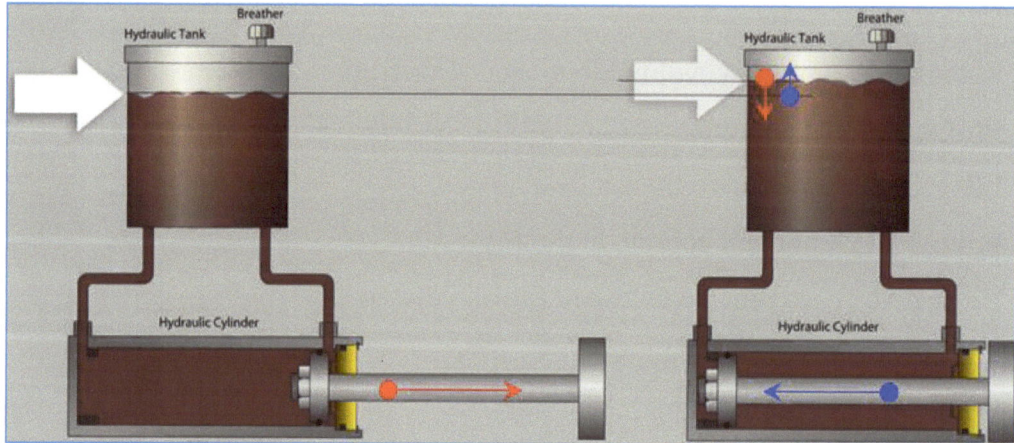

**Fig. 2.28 - Oil Level Change in the Reservoir during Machine Operation
(Courtesy of CD Industrial Group Inc.)**

❖ **Reservoir Filling Oil Volume (Vfill):**
  ▪ For Open Hydraulic Circuits:
  ▪ In *open circuits* for *industrial* applications, since there are no restrictions on the space, it is then preferable to use a large oil volume increasing the settling time in the reservoir to improve cooling, deaeration and contamination settlement. However, oil is expensive. Therefore, the applicable rule of thumb is that the oil filling volume **Vfill** = 2-3 times the pump flow **Qp**. This amount should provide a slow recirculating velocity that allows for the release of entrained air, dissipating heat with the reservoir walls, and the settlement of heavy contaminants.

  ▪ **Vfill = (2-3) Qp**   2.1A

  ▪ **Note:** Fluid volume calculated by Eq. 2.1 includes Vsmax.

  ▪ or Closed Hydraulic Circuits (Hydrostatic Transmissions): One of the important advantages of a *closed circuit* over an *open circuit* is that a much smaller oil volume is being circulated between the system and the reservoir; then a smaller reservoir can be used without sacrificing reliability or performance. Because the system is prefilled, reservoir shall contain the circulating oil volume of the charge pump, not the main pump. Charge pump in a closed circuit circulates the leakage from the pump and the motor between the low-pressure side of the circuit and the reservoir. Therefore, reservoir is sized based on oil volume of (10-20) % of the main pump.

- ▪ **Vfill = 0.2 x Qp**   **2.1B**

❖ **Thermal Expansion (Vth):** Oil volume increases as the working temperature increases. Therefore, 10% of oil volume should be left as a space for volume change due to thermal expansion. Equation 2.2 is used to determine **Vth**.

- ▪ **Vth = 0.1 x Vfill**   **2.2**

❖ **Maximum Oil Volume (Vmax):** Oil volume when Vsmax return to tank and Vfill oil volume subjects to thermal expansion.

- ▪ **Vmax = Vfill + Vth**   **2.3**

❖ **Air Space (Vas):** 10% addition on top of the maximum oil volume should be left as a space for the reservoir to breath and to help entrained air to escape to atmosphere.

- ▪ **Vas = 0.1 x Vmax**   **2.4**

❖ **Reservoir Overall Volume (Vo):** Is the structural volume of the reservoir, based on which the walls will be cut.

- ○ **Vmax + Vas = Vfill + Vth + Vas**   **2.5**

❖ **Absolute Minimum Oil Volume (Vmin):** No matter what the reservoir volume is, a hydraulic reservoir shall maintain an absolute minimum oil volume to secure the following:
  - ▪ Safe working height to allow sufficient fluid access to supply suction lines during all operating cycles.
  - ▪ Cover the suction line by at least 5-10 cm (2-4 inches) when all cylinders are in their extended positions and accumulators are fully charged by fluid.
  - ▪ Avoid pump cavitation.
  - ▪ Consider the reservoir inclination in mobile applications.

- ▪ **Vmin = Vfill - Vsmax**   **2.6**

2.4.2- Heat Dissipation from Hydraulic Reservoirs

One of the auxiliary contributions of a hydraulic reservoir is partially cooling of the system. A reservoir should be able to provide passive system cooling. When *passive cooling* is not sufficient, *active cooling* shall be provided using a heat exchanger. The following best practices are considered to maximize the heat dissipation from the tank to the surrounding area.
A. Shape and Dimension of the Reservoir.
B. Reservoir Ground Clearance.
C. Material of the Reservoir.
D. Cooling Fins.
E. Calculation of the Cooling Capacity.

A. Shape and Dimension of the Reservoir: As shown in Fig. 2.29, a parallelogram-shaped tank has a larger wall area, hence better heat dissipation will occur. As a rule of thumb, the height **H** of the tank is 85% of the width **W** that is also 85% of the length **L**. Knowing the overall volume of the reservoir, Eq. 2.7 is used to determine the reservoir length.

Volume (Vo) = H x W x L= 0.85W x 0.85L x L = 0.85^2L x 0.85L x L = $(0.85xL)^3$　　2.7

Mostly this calculation will result in nonstandard reservoir length. Nearest standard length can then be optionally considered. Knowing the reservoir length, then Eq. 2.8 is used to determine the reservoir wedth. Footprint area of the reservoir is calculated using Eq. 2.9.

Reservoir Width (W) = 0.85 L　　2.8

Reservoir Footprint Area (AR) = L x W　　2.9

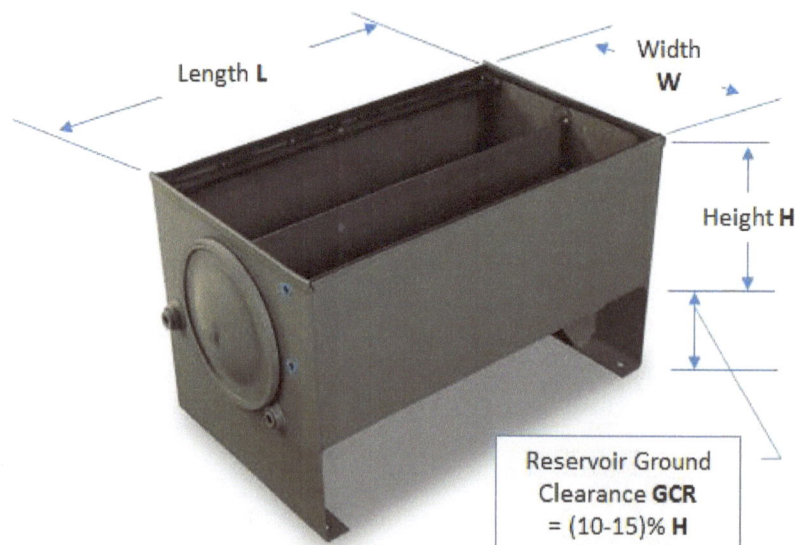

Fig. 2.29 - Recommended Shape and Dimensions of a Reservoir for better Heat Dissipation

Referring to Fig. 2.30, knowing the maximum, minimum, and overall oil volume of the reservoir, Eq. 2.10, 2.11, and 2.12 are used to determine minimum oil hight, maximum oil height, and the height of the reservoir.

Minimum Oil Height (Hmin) = Vmin/AR   **2.10**

Maximum Oil Height (Hmax) = Vmax/AR   **2.11**

Reservoir Height (H) = Vo/AR   **2.12**

B. Ground Clearance of the Reservoir (GCR): Tanks must be constructed to allow air flow underneath the tank. So, the bottom of the tank must be elevated from the ground by a minimum of 15 cm (6 inches). As shown in the figure, a rule of thumb, foot-mounted tanks should be elevated from the ground by (10-15) % of the reservoir height. Equation 2.13 is used to determine the reservoirs *ground clearance*.

Reservoir Ground Clearance (GCR) = 0.15 x H   **2.13**

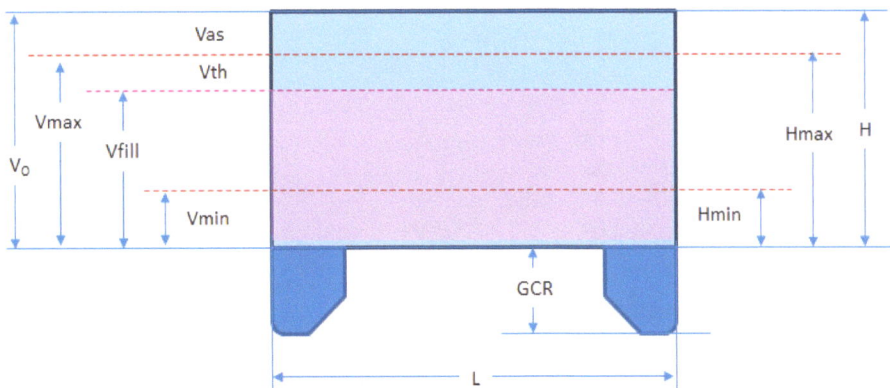

Fig. 2.30 - Fluid Hight in the Reservoir during Machine Operation

C. Material of the Reservoir: Hydraulic reservoirs should be produced using *materials* of higher *heat transfer* coefficient, such as *Steel* or *Aluminum*.

D. Cooling Fins: Figure 2.31 shows cooling fins added to improve the heat transfer from the tank to surroundings by welding fins (ribs) to the side walls of the reservoir. As shown in the figure, fins should be vertically welded with open ends at top and bottom to allow vertical convection air currents between them.

Fig. 2.31 - Cooling Fins Welded to the Side Walls to Improve Cooling Capacity (Courtesy of Womack)

E. Calculation of the Cooling Capacity (CC): Heat is dissipated from reservoir walls to the surrounding air if the reservoir walls temperature is higher than the temperature of the surrounding air; and vise-versa.

Cooling capacity is function of the reservoir contact area with the surroundings **AC**. It is recommended to include only the area in contact with the oil from the side walls only. Surface area of the tank that faces the ground may be excluded as a safety factor. Even when considering the area from the side walls, consider the lower level of the fluid in the reservoir during the machine operation. So, Eq. 2.14 is used to determine the contact area.

Contact (Wet) Area (AC) = 2 x Hmin x (L + W)   **2.14**

Equations 2.15A and 2.15B are used to calculate the reservoir cooling capacity in metric and English system of units; respectively. As shown in the equations, the rate at which the heat passes is a function of the wall material, the amount of air circulating, the fluid type and the difference in temperature between the air and the fluid.

Cooling Capacity (kW) = AC (m^2) x ΔT (^{o}C) x K   **2.15A**

Cooling Capacity (HP) = AC (ft^2) x ΔT (^{o}F) x K   **2.15B**

Where:
- **1 HP = 2545 BTU/HR**
- **1 kW = 3413 BTU/HR**
- **1 kW = 1.341 HP (i.e. 1 HP = 0.746 kW)**
- **AC = Area in contact with oil.**
- **ΔT = Differential temperature (Fluid Temp. – Air Temp.).** Worst ambient conditions (minimum temperature differential) should be considered.
- **K = Heat Transfer Factor.** It depends on the reservoir material as follows:
  - **For Eq. 2.15A:** Stainless Steel (K = 0.0058), for Carbon Steel (K = 0.0144), and for Aluminum (K = 0.0693)
  - **For Eq. 2.15B:** Stainless Steel (K = 0.0004), for Carbon Steel (K = 0.001), and for Aluminum (K = 0.0048).

2.4.3- Baffle Plates

As shown in Fig. 2.32, a *Baffle Plate* is placed to divide to large size reservoirs into two connected approximately equal chambers. The purpose of the baffle plate is to elongate the time the return oil stays in the reservoir before it is sucked again by the pump. Hence, baffle plate maximizes the time for de-aeration, heat dissipation, and settlement for particulate contaminants. As shown in the figure, the height of the baffle plate **BPH** is slightly higher than the maximum oil level **Hmax**. At one end of the baffle plate, a connecting area **BPCA** should be left to connect the two compartments. As a rule of thumb, this area is sized to have 1 cm² for every 2 lit/min (1 in² for every 3 GPM) of pump flow. In other words, the connecting area in cm² = half the pump flow in lit/min. of the pump flow. The width of the baffle plate connecting area **BPCAW** is calculated, in worst conditions, based on minimum oil level **Hmin**. The following equations are used to determine the previous parameters.

Baffle Plate Height (BPH) (cm) = Hmax (cm) + (1-5) cm   2.16

Baffle Plate Connecting Area (BPCA) (cm²) = Qp (lit/min)/2   2.17

Width of Baffle Plate Connecting Area (BPCAW) (cm) = BPCA (cm)/Hmin (cm)   2.18

Other tips of designing the baffle plate are as follows:
- A baffle plate should separate the return fluid from pump inlet lines.
- A baffle plate shall not prevent thorough cleaning of the reservoir.

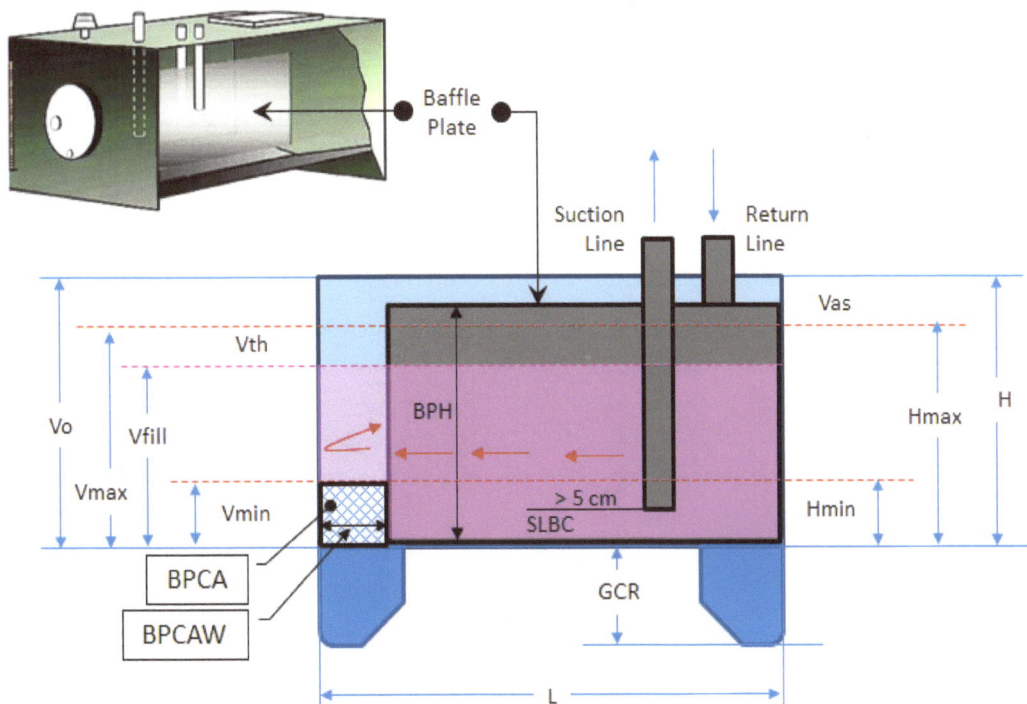

Fig. 2.32 - Baffle Plate in Hydraulic Reservoir

2.4.4- Air Removal

As shown in Fig. 2.33, *air removal* can be achieved by using 100-micron mesh size screen. Such a screen is placed inside the reservoir with 30-45 degrees inclination angle. As shown in the figure, return oil arrives at the upper side of the screen and the suction is taken from the lower side of the screen. The screen works like a bubble filter. The screen encourages the bubbles to merge and form bigger bubbles so that it can float easily up and dissipate at the surface.

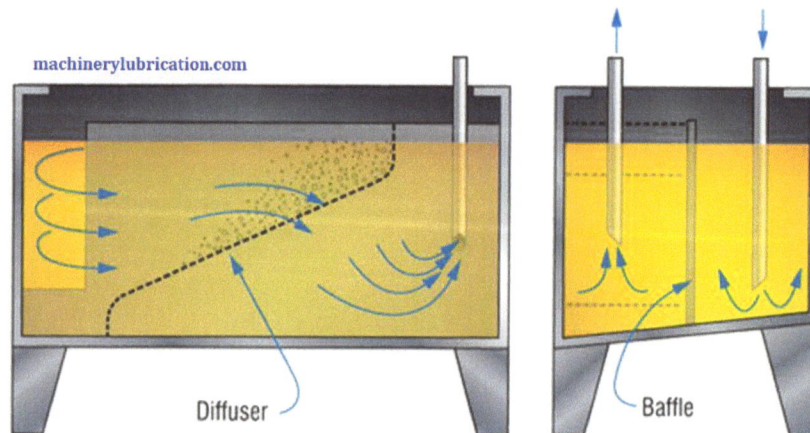

Fig. 2.33 - Air Removal Screen in Hydraulic Reservoir

2.4.5- Cleaning Holes

As shown in Fig. 2.34, for cleaning large reservoirs, there should be a side *Cleaning Hole*, also referred to as *Manhole*. It should be big enough to have a maintenance person get inside the reservoir through it. As shown in the figure, a gasket must be used to prevent external leakage.

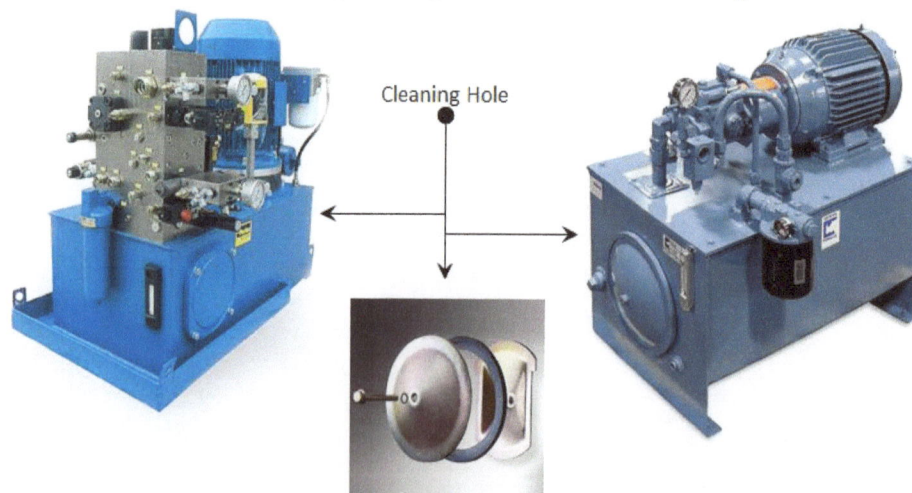

Fig. 2.34 - Cleaning Hole in Hydraulic Reservoirs

2.4.6- Drainage and Dirt Collection

Reservoirs should be designed to allow:
- Complete oil draining in place.
- Easy removal of settled water and contamination.
- Easily accessible for removal or cleaning of suction strainers, return diffusers and other replaceable internal reservoir components.

Figure 2.35 shows how to design a reservoir o maintain such a requirement.

Bottom of the reservoir: As shown in the front view, the bottom of the tank is slightly concaved. As shown in the side view, the bottom of the tank is slightly inclined towards the drain plug. As shown in the figure, every hydraulic reservoir should have one or two drain ports.

Drain Plugs: A magnetic drain plug is used to attract metallic such as iron and steel particles.

Drain Valves: A drain valve should be installed on each of these drain ports. Drain valves are used for reservoir draining, remove settled water, connection to an external filtering system, etc. It is recommended to use lockable *drain valves* to avoid accidental drainage.

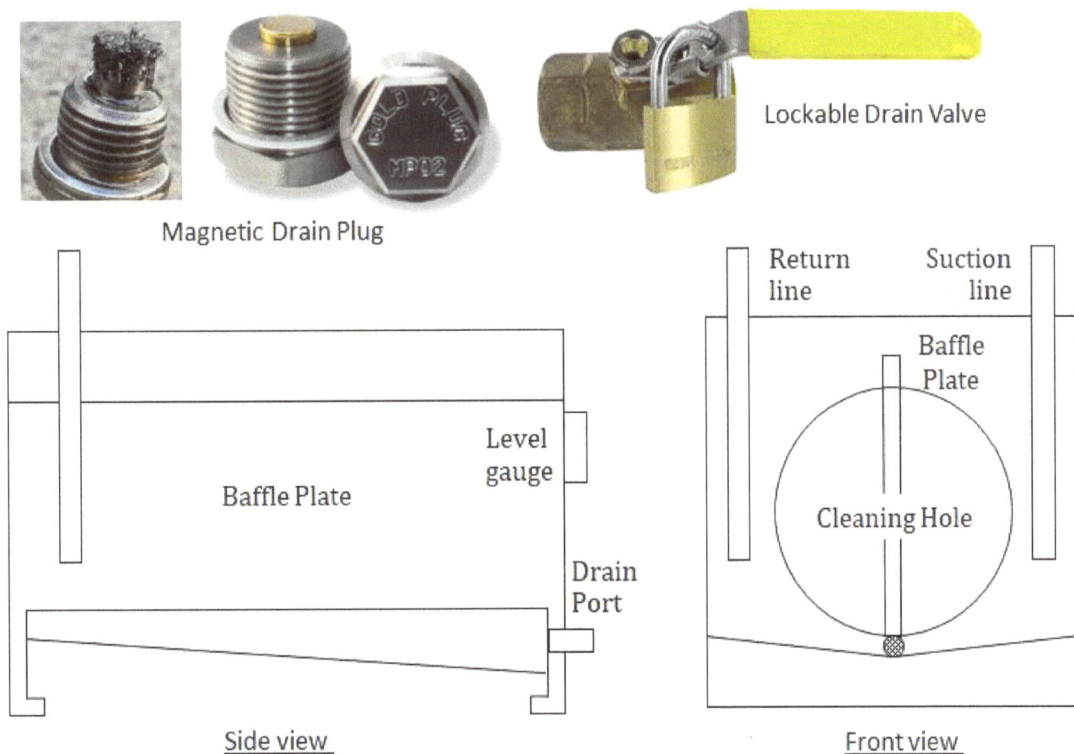

Magnetic Drain Plug

Lockable Drain Valve

Return line   Suction line

Baffle Plate

Level gauge

Baffle Plate

Cleaning Hole

Drain Port

Side view

Front view

Fig. 2.35 - Reservoir Design for Better Drainage and Dirt Collection

2.4.7- Corrosion Protection

Hydraulic reservoirs that are made of steel are vulnerable to *rust* from outside due to environmental conditions and from inside due to water content in the fluid, particularly if water-based fluids are used. Therefore, reservoir inside and outside painting is required. Otherwise, rust particles will flake off in the oil and will cause damage to the system if they are small enough to pass through the pump suction strainer. It is to be noted that the inside of the reservoir must be coated with a paint which is compatible with the fluid used. For example, phosphate ester is not compatible with ordinary paint. So, when changing fluid to phosphate ester, previous paints must be stripped and replaced with one that is compatible. However, Electrostatic Powder Spraying should be seriously considered because it creates a hard-finished skin that is tougher than conventional paint.

2.4.8- Structural Integrity of Hydraulic Reservoirs

Hydraulic reservoirs shall be designed to provide adequate *structural integrity* that considers the following conditions:

- **Pressure Change:** Reservoirs are exposed to positive and negative pressures caused by withdrawal or return of fluid at rates required by the system. So, the strength considerations should include the ability to withstand any internal pressure that may be developed during operation.
- **Weight of the Fluid:** The weight of the fluid determined in the sizing calculation must be considered in determining the strength of the reservoir body.
- **Weight of the Components:** Strength of the reservoir should consider components that are assembled on top of the reservoir such as the pump, prime, mover, accessories, accumulators, pipes, manifolds, etc.
- **Lifting:** Reservoirs should be equipped with sufficiently robust lifting points for transporting the hydraulic system. Lifting points should consider distribution of components on the reservoir and the overall center of gravity. As shown in Fig. 2.36, alternative to lifting points, forklift usage carrying option may be considered.
- **Structure Analysis:** After considering all the weight to be supported by the reservoir, a structural analysis should be made to determine the minimum material sizes for the walls, bottom, top and supporting members or base structure.

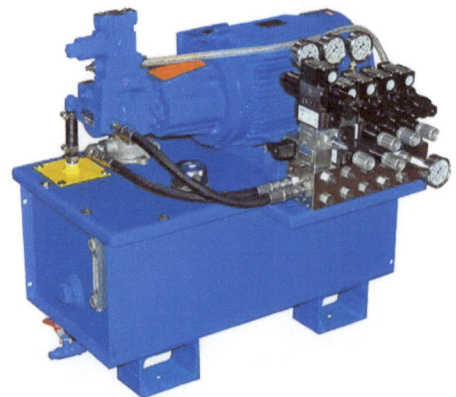

Fig. 2.36 – Forklift Carrying Option (hyvair.com)

2.4.9- Suction Lines

Suction Line Diameter DS: The *suction line* of a hydraulic pump is the most critical one among the other transmission lines in a hydraulic system. Sizing a suction line is vital for safe pump operation. Undersizing the suction line may cause immediate pump cavitation. Therefore, diameter of the suction line should be checked through the following steps:

- Pump Suction Port: Suction line diameter shouldn't be less than the pump suction port. The pump manufacturer sizes the pump intake port based on the pump flow, that is function of the pump size and driving speed, to avoid cavitation.

- Flow Speed in the Line: The rule of thumb is to keep the flow speed within the range of (0.6-1.2) m/s = (2-4) ft/s. That is to minimize the pressure losses and avoid pump cavitation. Knowing the pump flow **(Q)**, Eq. 2.19A in metric systems (or Eq. 2.19B in English system) are used to solve for the transmission line crossection area **(A)** assuming the flow speed **(v)** is within the recommended range:

$$Q\left[\frac{\text{lit}}{\text{min}}\right] = \frac{v\,[\text{cm/s}] \times A[\text{cm}^2] \times 60}{1000} \qquad 2.19A$$

$$Q[\text{gpm}] = \frac{v\,[\text{fps}] \times A[\text{in}^2]}{0.321} \qquad 2.19B$$

- Reynold's Number: To avoid turbulent flow, *Reynold's Number* should be less than 2000. Knowing the pump flow **(Q)**, and fluid viscosity **(v)**, use Eq. 2.20A in metric systems (or Eq. 2.20B in English system) to solve for the transmission line minimum inner diameter **(DS)** assuming the Reynold's number **(v)** equals 2000:

$$D[\text{mm}] = \frac{21231\,Q\left[\frac{l}{\text{min}}\right]}{v\,[\text{Cst}] \times R_e} \qquad 2.20A$$

$$D[\text{mm}] = \frac{3164\,Q[\text{gpm}]}{v\,[\text{Cst}] \times R_e} \qquad 2.20B$$

- Validation using HCSC Software: use the Hydraulic Component Sizing Calculator (HCSC) or any other simulation software to validate the calculations. HCSC can be ordered from (https://www.compudraulic.com/software).

- <u>Example:</u> For a pump that discharges 100 lit/min (25 gpm), Fig. 27.x shows an example of undersized intake line that causes turbulent flow. When the intake line is properly sized, flow speed is reduced, flow becomes laminar, and cavitation is avoided.

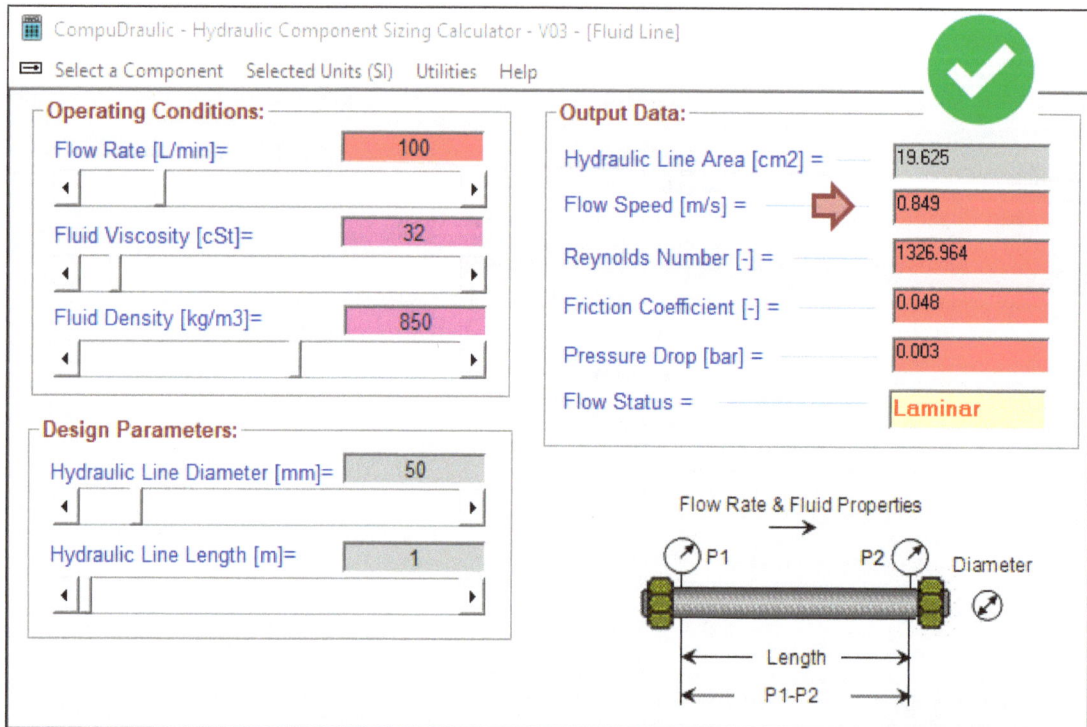

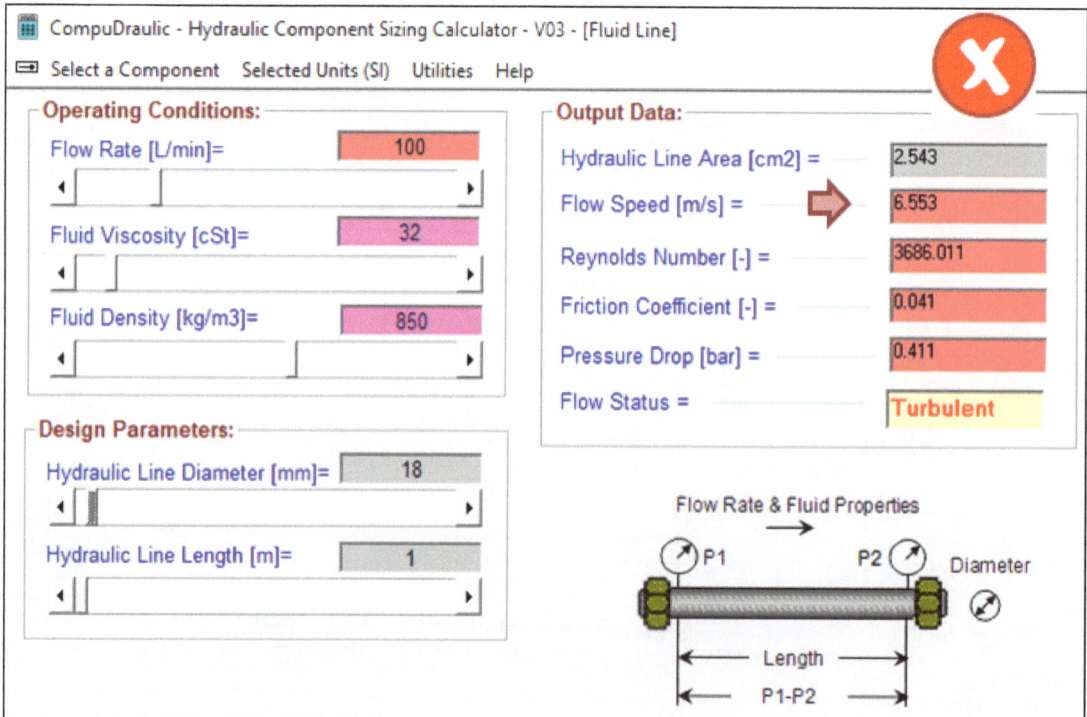

Fig. 2.37 - Proper Sizing of Intake Line (www.compudraulic.com)

Suction Line Placement: As shown in Fig. 2.38, Placement of the suction and return lines should consider maximizing the oil travel distance between them. However, suction line should not in any case be near the point of dirt collection.

Suction Line Bottom Clearance (SLBC): The figure shows some tips to minimize the losses in suction line and provide better suction conditions:

- Entrance of the oil at the bottom part of the suction line shall be edged on 45 degrees in order to enlarge the opening to reduce the pressure losses.
- Entrance of the oil shall be directed to the wall to make it difficult to suck dirt.
- Equation 2.21 shows a rule of thumb to determine the suction line *bottom clearance.*

Suction Line Bottom Clearance (SLBC) = (2-3) x Suction Line Diameter (DS)   **2.21**

Suction Line Covered Bottom (SLCB): Equation 2.22 shows a rule of thumb to check on the suction line *covered bottom.* It supposed to be greater than 5-10 cm (2-4 in). Otherwise, minimum oil volume should be adjusted accordingly.

Suction Line Covered Bottom (SLCB) = Hmin - SLBC   **2.22**

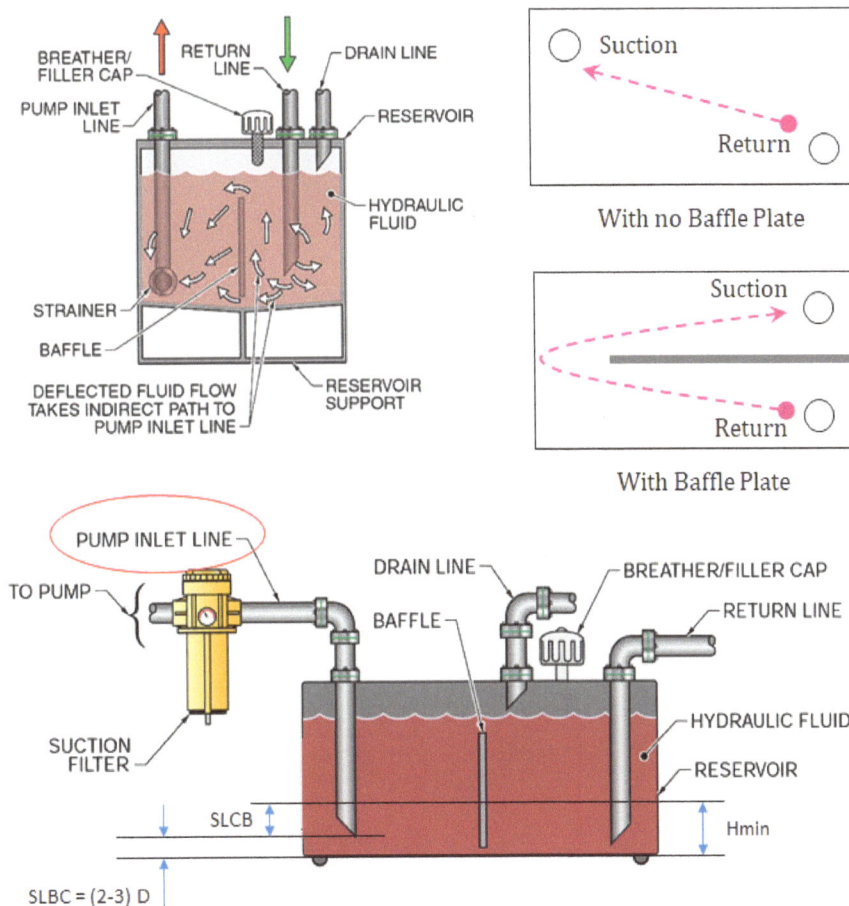

Fig. 2.38 - Suction Line (Courtesy of American Technical Publisher)

Suction Line Length Ls: Suction line length should neither be very long in order to avoid cavitation, nor very short to avoid bringing turbulence into the pump intake port. As shown in Fig. 2.39, regardless the positioning of the pump with respect to the reservoir, length (**Ls**) of a suction line is defined based on the result of applying the following three rules of thumbs. It is to be noted that first rule of thumb has lower priority than the second, and the second rule has lower priority than the third.

- **First:** Suction line minimum length is at least 10 times the line diameter.
- Suction Line Length **Ls = 10 DS**   **2.23**

- **Second:** Suction line minimum length is at least 25 cm (10 inches) from last turbulent point (such as a strainer or a bend in the line) before the pump intake port. This is to ensure that the flow is laminar before the pump port.

- **Third:** Suction line maximum length that develops maximum permissible pressure losses in the line.

- **Example 1:**
  o Assuming applying 1st rule → Ls = 10 cm and
  o Assuming applying 2nd rule → Ls >= 12 cm,
  o Then so far, Ls >= 12 because second rule overrides the first one.
  o Assuming applying 3rd rule → Ls <= 15 cm.
  o Then (12 cm >= Ls <15 cm) because if it becomes longer than 15 cm, pressure drop in the line increases above maximum permissible.

- **Example 2:**
  o Assuming applying 1st rule → Ls = 15 cm and
  o Assuming applying 2nd rule → Ls >= 13 cm,
  o Then so far Ls >= 13 because second rule overrides the first one.
  o Assuming applying third rule → LS <= 10 cm.
  o Then then LS <=10 cm because if it becomes longer, pressure drop in the line increases above maximum permissible. Notice that third rule overrides 1st and 2nd rules.

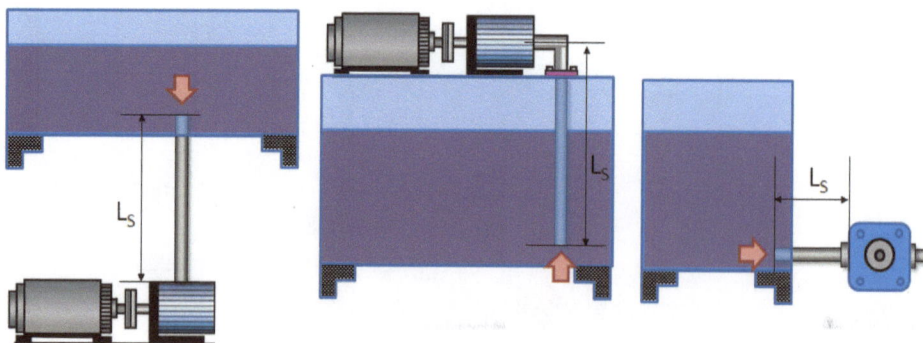

Fig. 2.39 - Suction Line Length

Suction Head and Suction Pressure: Placing the pump above the oil surface creates negative suction pressure. In designing a power unit, it is greatly important to review the pump manufacturer regarding the maximum allowable negative suction gauge pressure. If no information is found, suction pressure at the suction port must not fall below 0.8 bar (12 psia) absolute or -0.2 bar gauge (-3 psig) during operation and during cold start. Alternatively, for cavitation avoidance, placing the pump below the oil surface creates positive suction pressure due to the weight of the fluid in the suction line. Equations 24.A and 24.B are used to calculate the suction head in metric and English system of units; respectively.

$$Hs\,(m) = \frac{SP\,(Pa)}{SG \times \gamma_w(N/m^3)} = \frac{10^5 \times SP\,(bar)}{SG \times 9.81 \times \gamma_w(kg/m^3)}$$

$$\gamma_w\left(\frac{kg}{m^3}\right) = 1000 \rightarrow Hs\,(m) = \frac{100 \times SP\,(bar)}{SG \times 9.81} \qquad 2.24A$$

$$HS\,(ft) = \frac{SP\,(psi)}{SG \times \gamma_w(lb/ft)} = \frac{SP\,(psi)}{SG \times 62.4 \times \gamma_w(lb/lb^3)} = \qquad 2.24B$$

Where (Refer to Fig. 2.40):
- **Hs** = Suction Head
- **SP = PSP** = Positive Suction Gauge Pressure (corresponds to positive suction head when the pump is placed below the surface of the oil).
- **SP = NSP** = Negative Suction Gauge Pressure (corresponds to negative suction head when the pump is placed above the surface of the oil).
- γ_w = Specific Weight of water & **SG** = Specific Gravity of the hydraulic fluid.

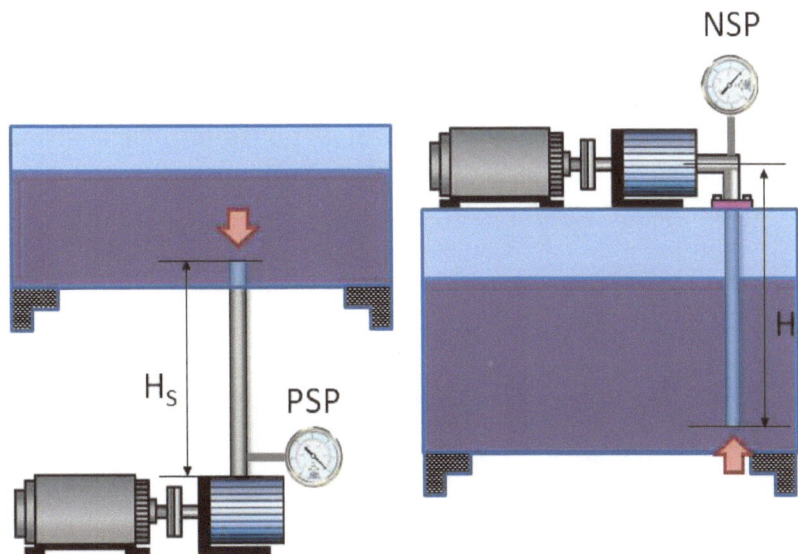

Fig. 2.40 - Suction Head and Suction Pressure

Suction Line Type: For suction lines, hard tubing is highly recommended over flexible hoses. However, hoses may still be used to take misalignment between the pump and reservoir such as in mobile machines. If a hose is used, make sure a suction hose (NOT a pressure hose) is used. The reason is suction hoses are specially constructed with a spiral wire or supporting tube on the inner layers to prevent collapsing the inner layers due to negative pressure. As shown in Fig. 2.41, if instead a pressure hose is used in suction, internal layers may be separated blocking the line.

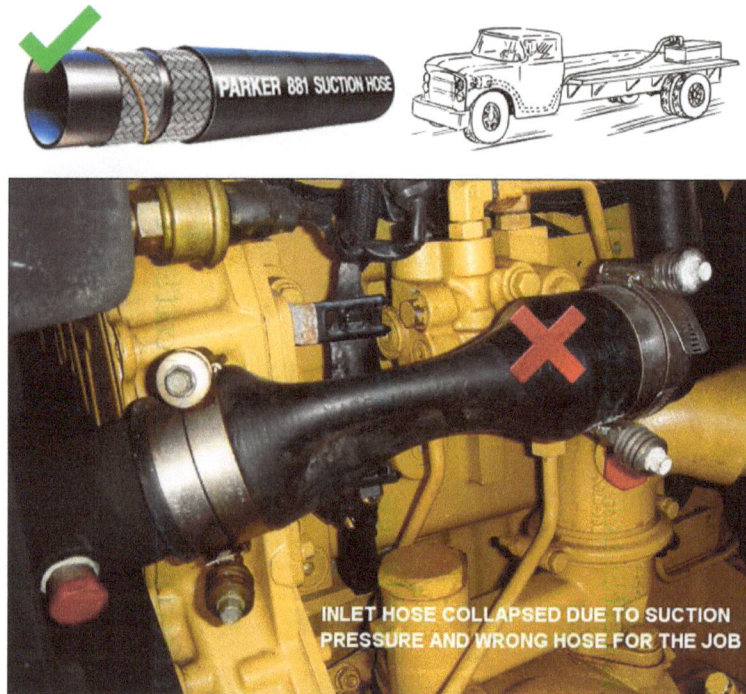

INLET HOSE COLLAPSED DUE TO SUCTION PRESSURE AND WRONG HOSE FOR THE JOB

Fig. 2.41 - Example of Suction Hose Collapse

Suction Line Routing: As shown in Fig. 2.42, if possible, use a straight intake line with minimum possible number of bends to minimize the line losses.

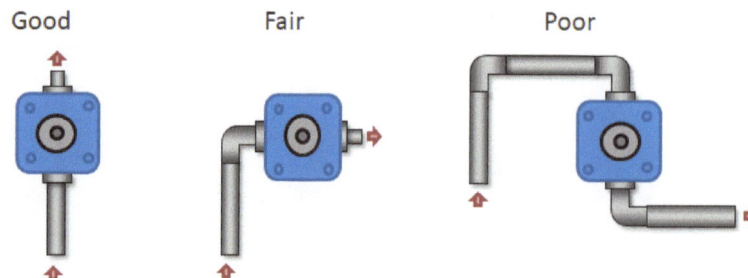

Good   Fair   Poor

Fig. 2.42 - Example of Proper Intake line Routing (Courtesy of Womack)

Suction Strainer: Generally speaking, because of the cavitation concerns, strainers are not recommended for large flow pumps. Installing a *suction strainer* on the suction line subject to approval from the pump manufacturer. If the construction permits and pump manufacturer allows, a great attention should be paid to the following:

- **Micron Size vs. Mesh Size:** Micron size is the size of the largest particles (in microns) that can pass through the screen. Mesh size is the number of holes in one squared inch. Large mesh size means finer filter.

- **Mesh Size:** If no mesh size is reported by the pump manufacturer, 250-500 mesh size is recommended.

- **Surface Area:** A strainer is sized based on the pump flow. Size of the strainer should offer surface area to minimize the pressure drop across the strainer so that pump cavitation is avoided. Review the pump data sheet if found. Otherwise, consult the pump manufacturer regarding recommended surface area. If no information is found, the following rule of thumb is applicable as a guideline. Surface area shouldn't be less than 2 square inches for every GPM of the pump flow (≈ 3 cm² for every liters/min of the pump flow). For example, a 20 GPM pump should have a minimum strainer surface area of 40 square inches. Equations 2.24 A and 2.24B are used to determine the surface area of the strainer in metric and English system of units, respectively.

Suction Strainer Surface Area (cm²) >= 3 x Qp (lit/min)   2.25A

Suction Strainer Surface Area (in²) >= 2 x Qp (gpm)   2.25B

- **Connection with Suction Line:** As shown in Fig. 2.43, a strainer could be aligned with the suction line or a 90° elbow at the bottom of the suction will serve.

- **Bypass Valve:** Suction strainers mounted at the beginning of the suction line inside the reservoir without a bypass valve or pressure drop indicator are not recommended. They are hidden from sight and there is no easy access to remove and clean them. Overtime, without proper cleaning, they will plug restricting flow and cause pump cavitation leading to pump failure.

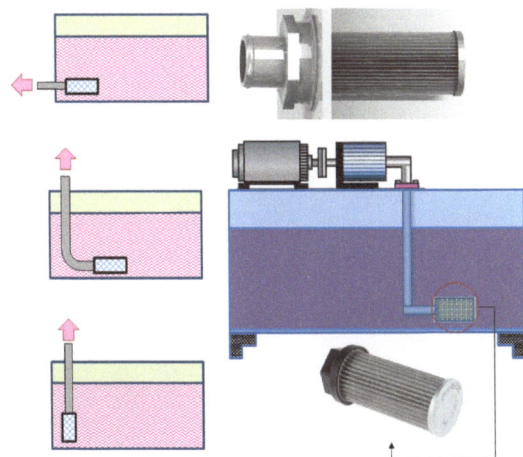

Fig. 2.43 – Suction Strainers

Shutoff Valve: If a pump isn't immersed and installed below the fluid surface, a *shutoff valve* is required to isolate the reservoir in case of disassembling the pump for any reason. Such a valve MUST BE lockable to avoid immediate pump failure if it is accidently closed during the pump operation. As shown in Fig. 2.44, it is very important to protect this valve from accidental closure. Lockable shutoff valves are recommended in such cases.

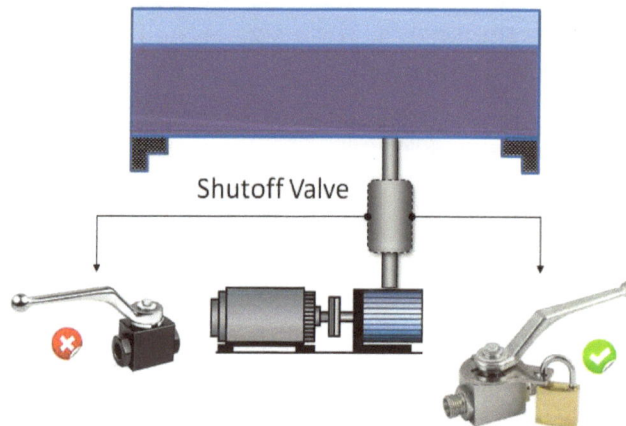

Fig. 2.44 – Shutoff Valve on Suction Line

Suction Pressure Measurement: As shown in Fig. 2.45, adding a vacuum gauge near the pump intake port is an extra helpful solution to monitor pump suction condition. For critical applications, it may be wise to install a vacuum sensor that provides an alarm signal or control signal to shut down the system in case if cavitation occurs. The figure shows an example of a given instruction for the allowable suction pressure of a pump.

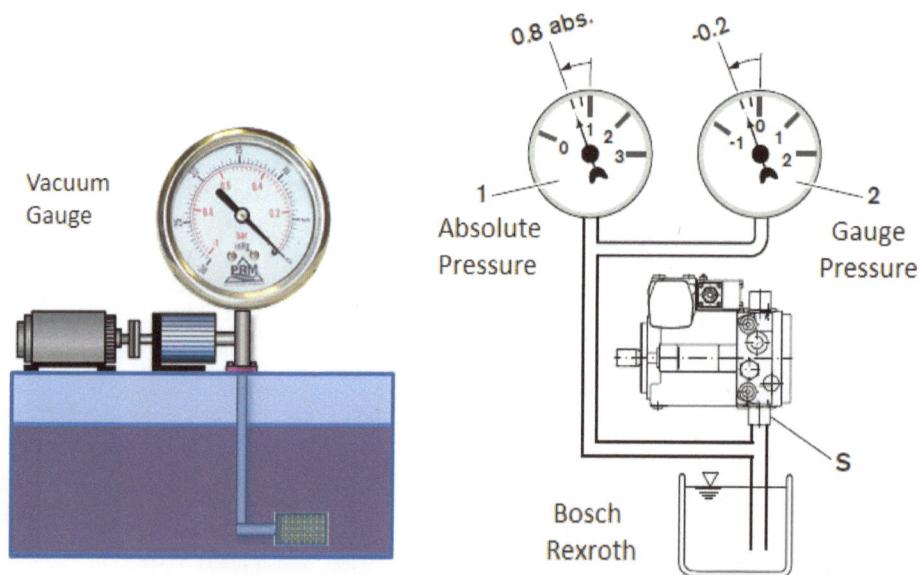

Fig. 2.45 – Pump Suction Pressure

Suction Line Pressure Boosting: For pumps that have large displacement and/or are driven at high speed, to avoid cavitation, it is necessary to boost the pressure of the intake line. As shown in Fig. 2.46, a super-charge circuit is mandatory part of the design of a hydrostatic transmission. However, as shown in the figure, it is applicable for open circuit in industrial applications if needed for large and high-speed pumps.

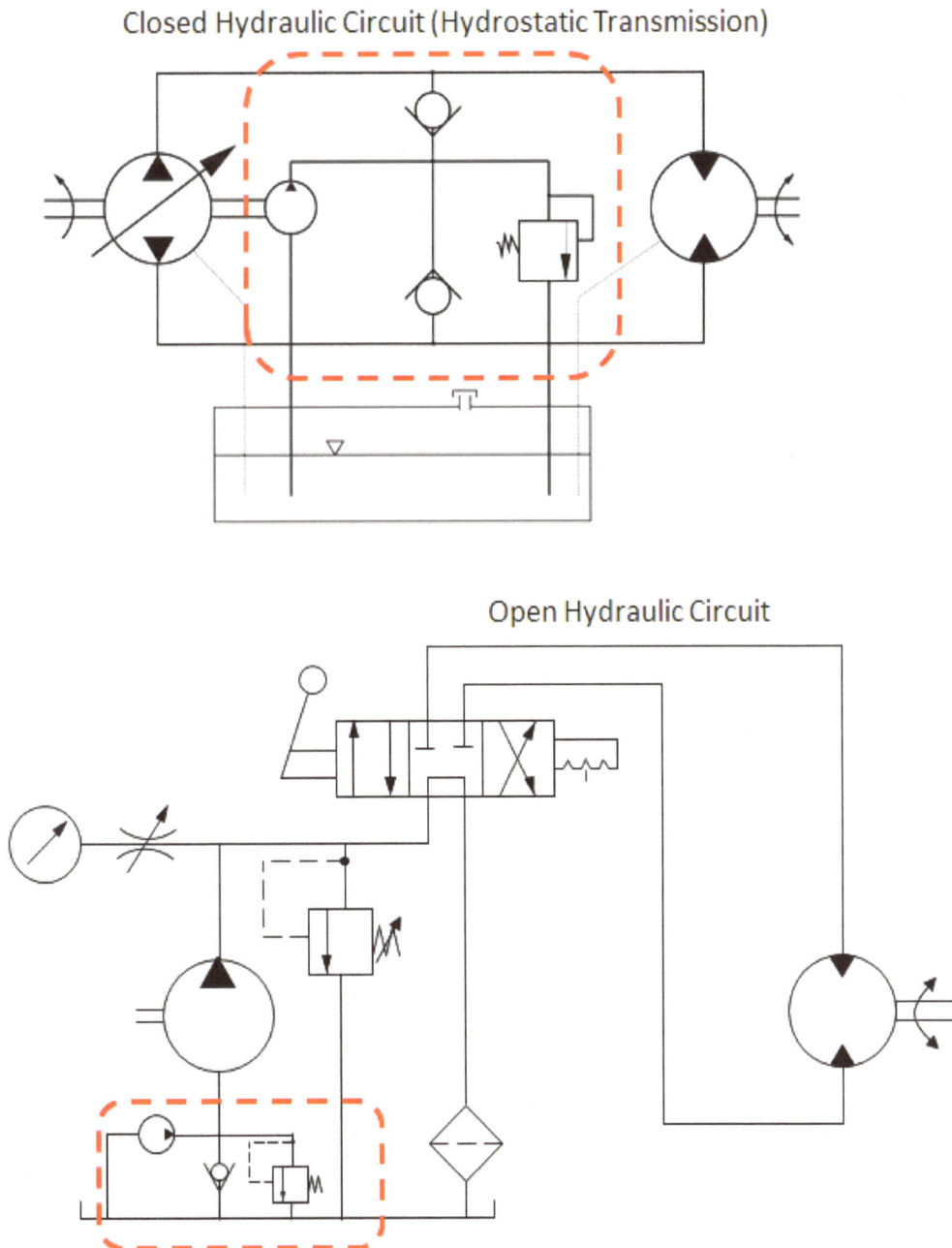

Closed Hydraulic Circuit (Hydrostatic Transmission)

Open Hydraulic Circuit

Fig. 2.46 - Intake Line Pressure Boosting in Closed and Open Hydraulic Circuits

2.4.10- Return Lines

Return Line Placement: First, a *return line* should be placed in the reservoir in accordance with the placement of the suction line and the baffle plate (if used) as discussed in the previous section. Additionally, as shown in Fig. 2.47, the return flow should be directed against the reservoir wall to maximize oil cooling,

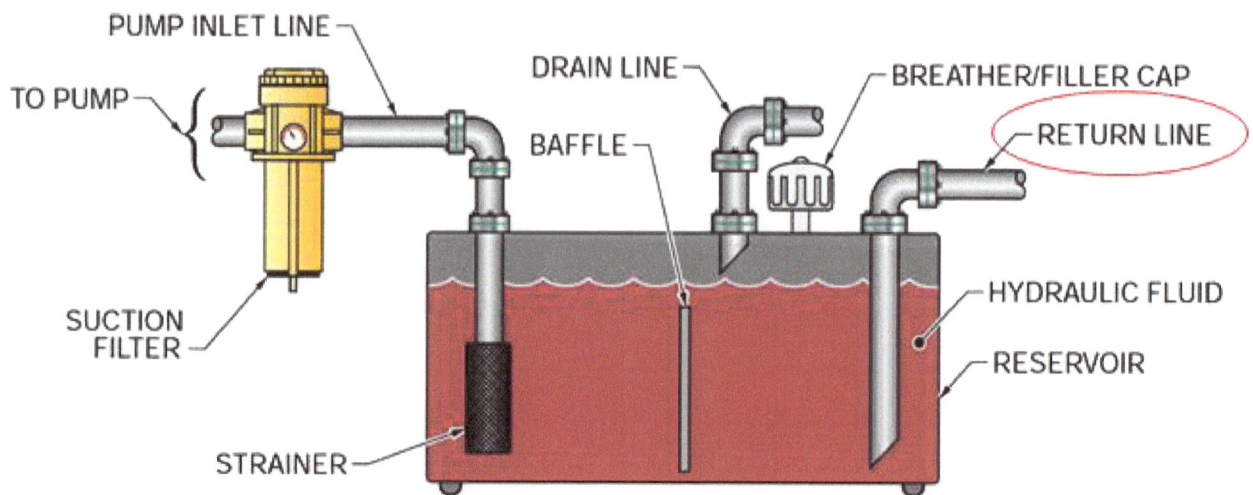

Fig. 2.47 –Return Line (Courtesy of American Technical Publisher)

Return Line Diameter DL: Considering a return line is a line that is used just to bring oil back to the reservoir isn't a wise thinking. It plays a vital role in fluid conditioning. Undersizing the return line may cause back pressure and brings turbulence to the reservoir. Equations 2.19 and 2.20, previously used for suction lines, are also used to determine **DL** with the following considerations.

The rule of thumb is to keep the flow speed within the range of (1.2-2.1) m/s = (4-7) ft/s. Previous equation 2.19 is used to calculate the diameter of the return line based on the given recommended flow speed. However, a common mistake design engineers make is based on a wrong assumption that the return flow equal pump flow. Return flow may be larger than the pump flow if differential cylinders are used or rapidly discharging accumulators. Therefore, prior to sizing a return line, flow distribution analysis must be conducted for all phases of machine operation to determine the maximum return flow based on which the return line is sized.

Return Line Bottom Clearance (RLBC) and Covered Bottom (RLCB): Same rule of thumb that applied to suction line is applied for *return line*. Equations 2.21 and 2.22 are used to determine these parameters.

Diffusers: As it was noted on previous figures, like a suction line, a return line is edged on 45 degrees. This edge acts like a diffuser to help the fluid to get rid of the air. However, as shown in Fig. 2.48 installing a *diffuser* in a hydraulic reservoir is a simple addition that make a big difference in system performance. With special concentric tubes designed with discharge holes 180° opposed, fluid aeration, foaming and reservoir noise are reduced. Pump life is also extended by reducing cavitation to the pump inlet. Figure 2.49 shows the typical stream of flow around the baffle plate between the return line through a diffuser and the suction line through a suction strainer.

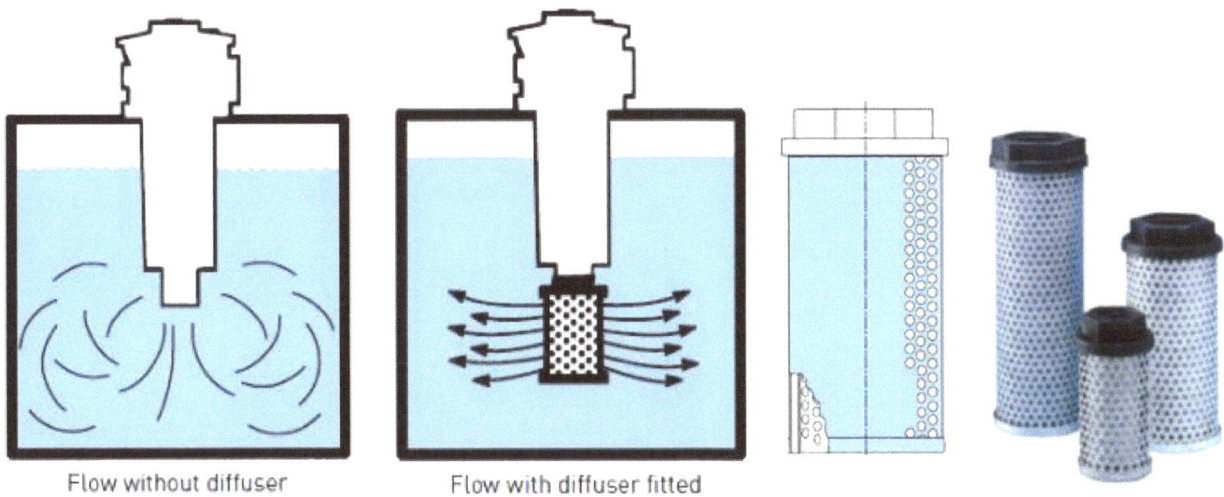

Flow without diffuser   Flow with diffuser fitted

Fig. 2.48- Diffusers on a return line (Courtesy of Parker)

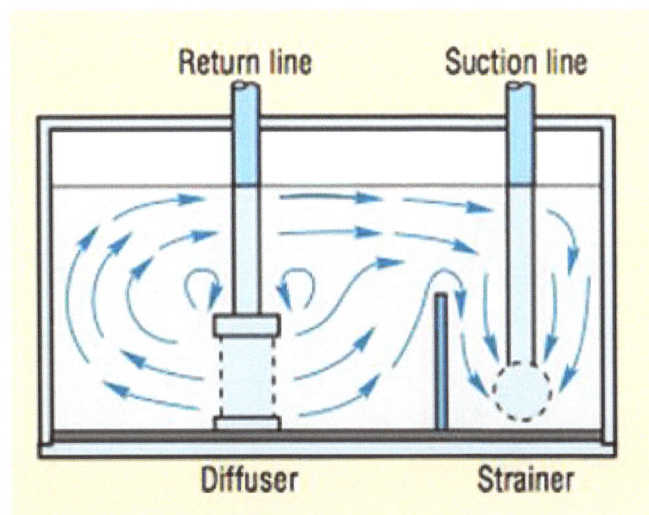

Fig. 2.49- Flow Stream from the Return Line to Suction Line (Courtesy of Parker)

2.4.11- Drain Lines

Some of hydraulic valves (such as sequence valves, over centering valves and some reducing valves) and some pumps (such as bidirectional pumps) have *drain line.* Such drain lines can be combined externally and run via a separate drain line to the reservoir without being combined with the main tank return flow. This is to prevent pressure spikes generated in the main return line from backing up into components and causing false operation. As shown in Fig. 2.50, the drop line for drain flow can discharge either on top or underneath the oil level. Discharging on top of the oil level does not cause excessive turbulence because the oil volume and velocity are low, and it does prevent siphoning of oil out of the reservoir if an external drain connection on a component is opened.

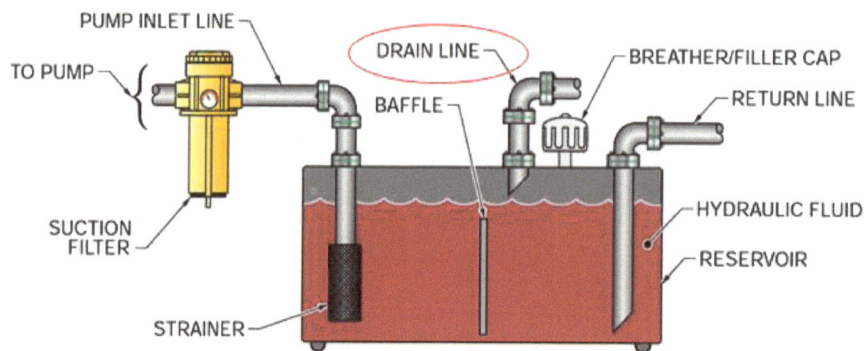

Fig. 2.50 –Drain Line (Courtesy of American Technical Publisher)

2.4.12- Filling Caps

Filling Caps: Figure 2.51 shows the least expensive traditional filling caps. A *filling cap* should be fitted with a sealed cover to prevent the ingress of contaminants when closed. It may contain a filling screen catch relatively large contaminants during filling. Filler caps should be chained to the reservoir to keep them captive. For more protection, the cover should be lockable.

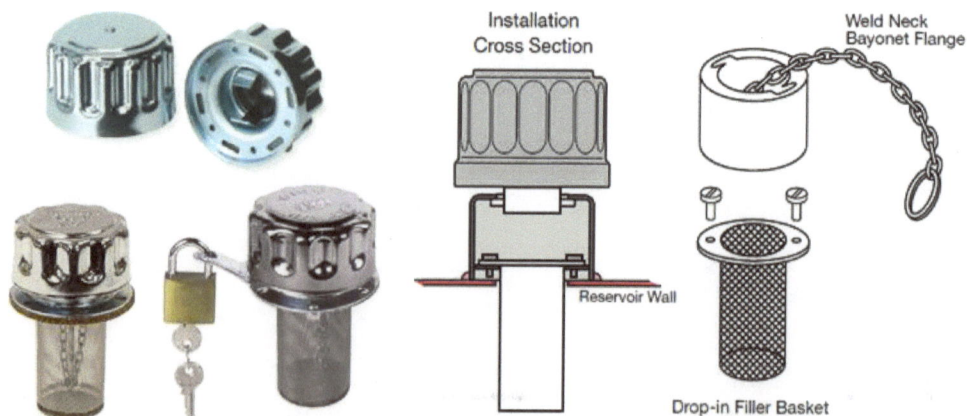

Fig. 2.51 –Filling Caps

2.4.13- Air Breathers

In all systems that use single-acting cylinders, double-acting differential cylinders, and accumulators, the reservoir oil level is dynamically changing. Oil level falls as the cylinders extend and raises up as they retract. So, an open reservoir breathes during the machine operation like a human and hence a very large air volume passes in and out to the air space in the reservoir. If the air is allowed to freely move in and out of the reservoir, everything contained within that air is exchanged with the oil. This can include contaminants and moisture, that both have severe effects on the oil lifetime and the hydraulic system reliability. Therefore, to prevent the dust from getting into the tank, *air breathers* are used. Breathers are available in various configurations, sizes, and working features.

Standard Filler Breathers: As shown in Fig. 2.52, *Standard Filler Breathers* are very similar in shape to the filler caps. They are usually screwed onto a threaded pipe that provides air exchange through the top of the reservoir. Other styles can look like a spin-on oil filter. They are available in various forms such as metallic or non-metallic, flange-mount or thread-mount, and with 10 microns size of a conventional or telescopic strainer. As shown in the figure, manufacturers report the differential pressure across the breather versus the air flow.

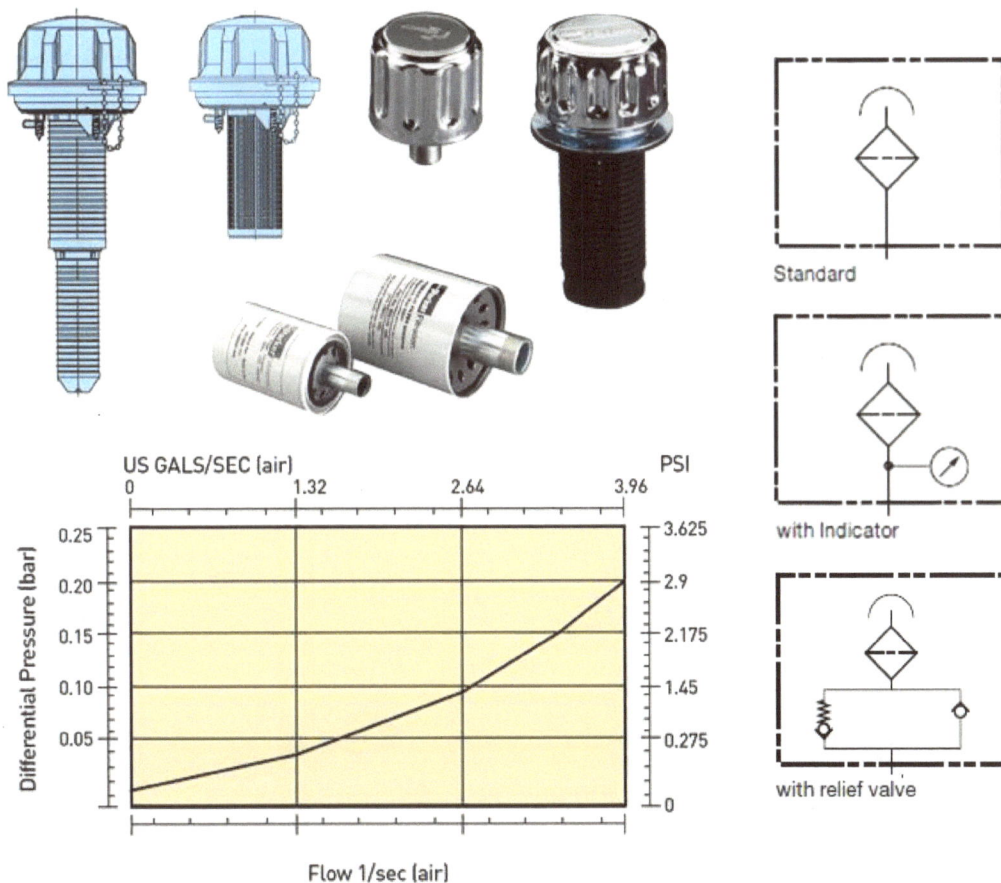

Fig. 2.52- Standard Filler Breathers (Courtesy of Parker)

Desiccant Breathers: In hydraulic systems that use petroleum-based hydraulic fluids, water ingress into the tank is a familiar problem. In such systems, if water content in oil increases above the allowable limit, system faces frequent breakdown and high maintenance costs. For detailed information about contamination by water, review Volume 3 of this textbook series. Therefore, using *Desiccant Breathers* is a must for applications that work in very humid environments such as marine and offshore applications.

Figure 2.53 shows a typical example from industry for a desiccant breather and its hydraulic symbol. As shown in the figure, desiccant breathers are designed with a transparent body filled with silica-gel that are designed to absorb as much as 40% of its weight. The gel's color changes from a blue to light pink color when saturated. The unit also contains regular filter element to capture contaminants as small as 3 microns. As an option, a check valve can be assembled on the air inlet to prevent the saturation of the desiccants during the system shutdown. Another check valve can be assembled on the air outlet to prevent exhaust air from the tank from flowing back through the desiccant so that the desiccant is protected from oil mist. Replace breather when desiccant color changes or when a built-in clogging indicator shows expiration of the drying material.

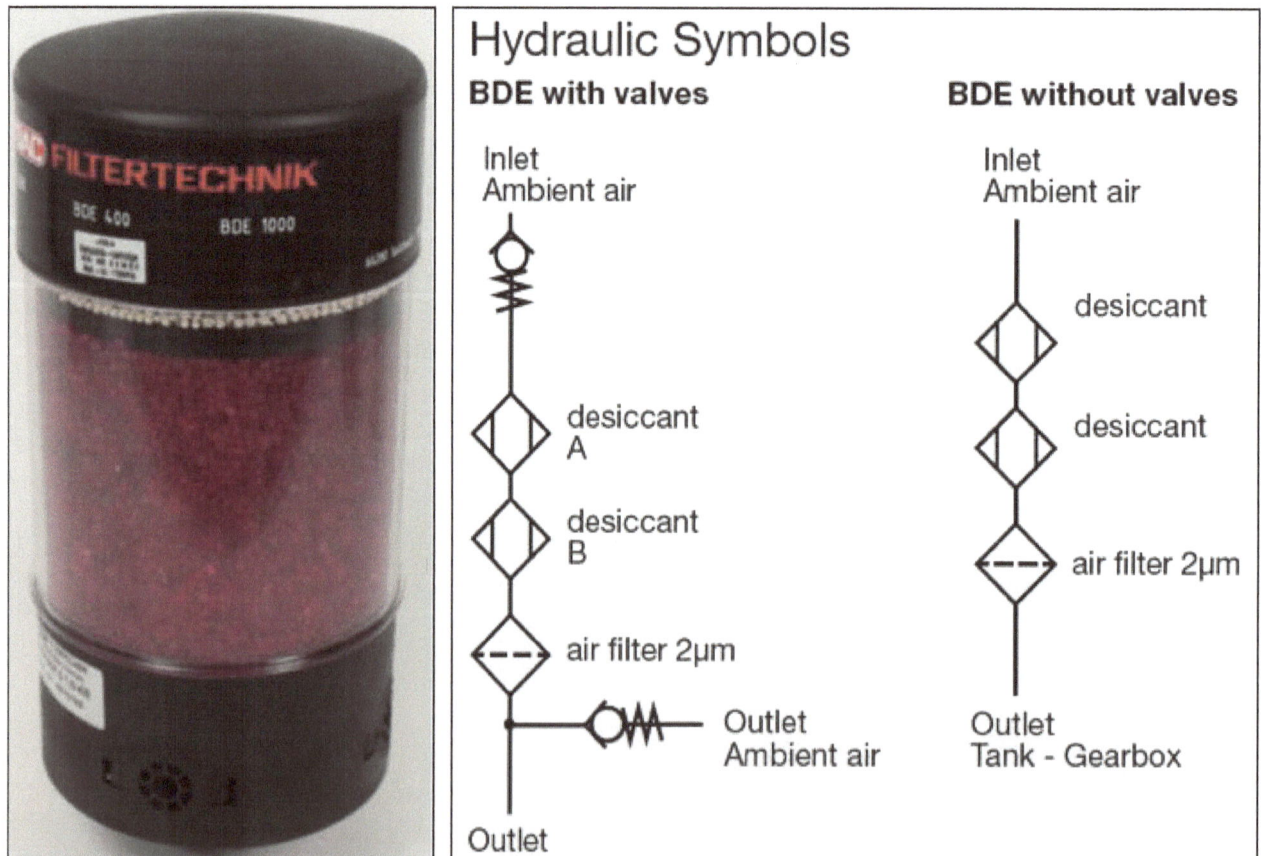

Fig. 2.53 - Breather Dryer (Courtesy of HYDAC)

Construction and Operation of Desiccant Breather: As shown in Fig. 2.54, the desiccant breather consists of two separate chambers which can be filled with two desiccants, which in combination increase total water retention because of the two-stage dewatering. The figure shows a built-in pleated air filter element (absolute filtration of particles > 2 μm) provides the filter with a very high contamination retention capacity (26 g). Such breather dryers can work in temperatures range -30 $^{\circ}$C to 100 $^{\circ}$C (-22 $^{\circ}$F to 212 $^{\circ}$F).

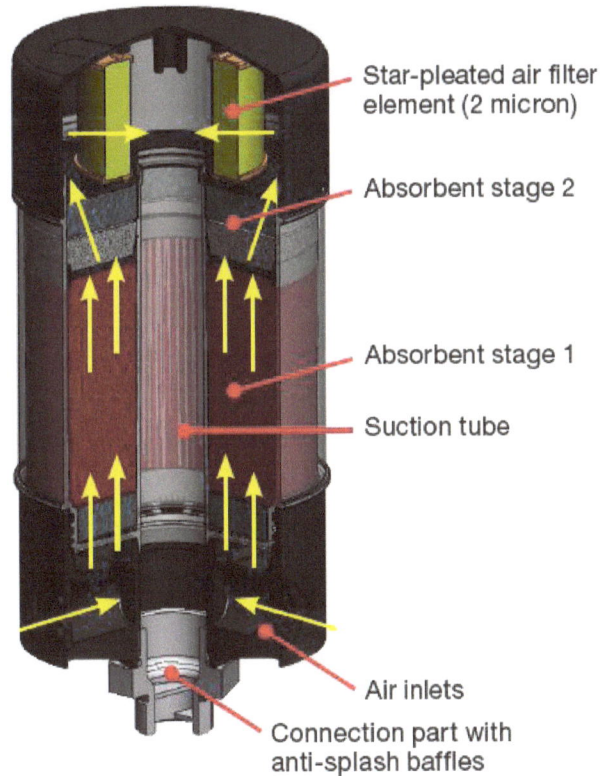

Star-pleated air filter element (2 micron)

Absorbent stage 2

Absorbent stage 1

Suction tube

Air inlets

Connection part with anti-splash baffles

Fig. 2.54 – Construction and Operation of the Desiccant Breather Dryer (Courtesy of HYDAC)

Breathers Dryers: Alternative to using desiccant breathers, *Breather Dryer* can be used. They are breathers with water absorption cartridges. Breather dryers collect and expel moisture out of reservoirs. Unlike desiccant filters, breather dryers cartridge won't change when saturated. Figure 2.55 shows how a breather dryers work.

If a water separator is provided, an indicator that signals when maintenance is required shall be installed.

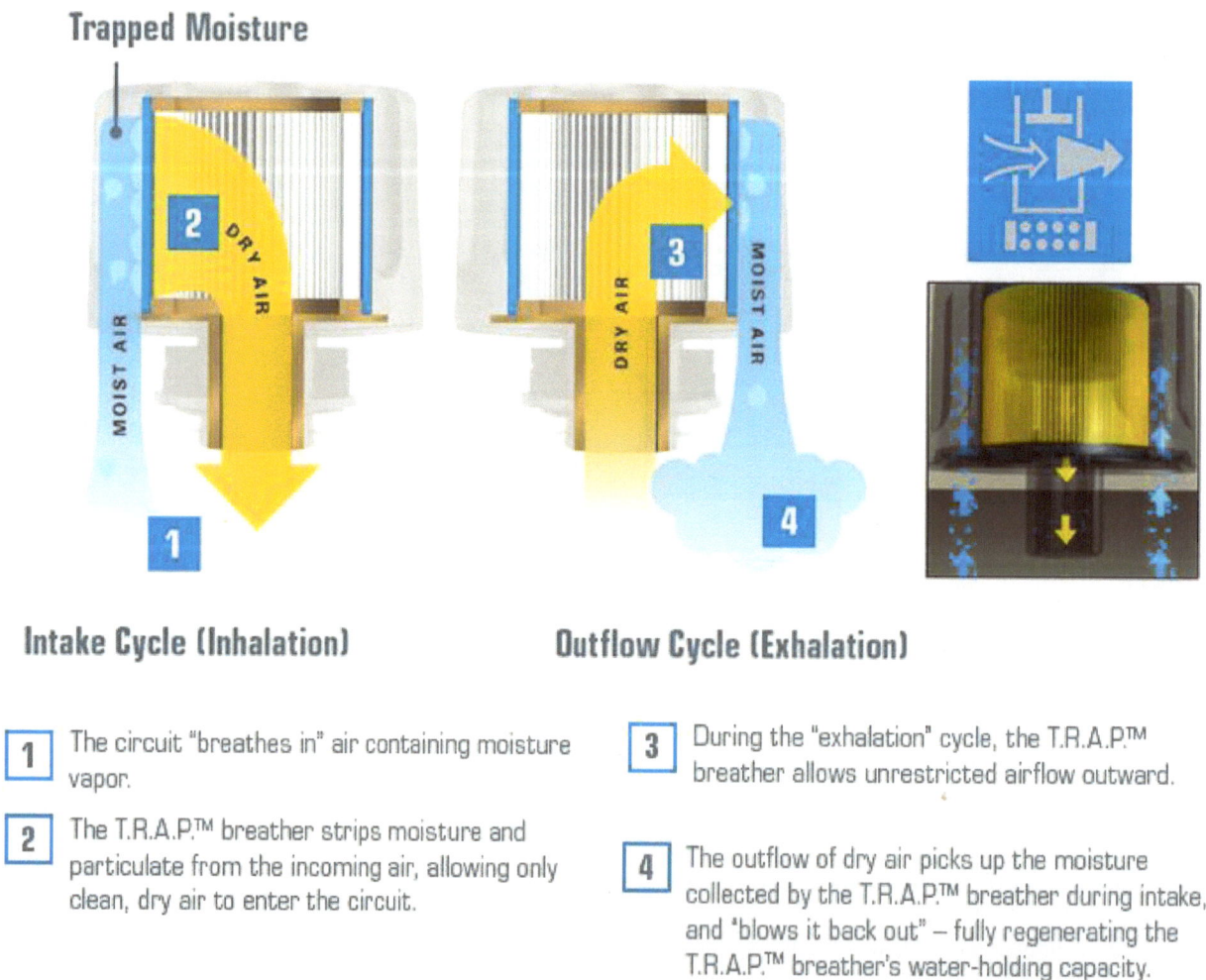

Intake Cycle (Inhalation)   **Outflow Cycle (Exhalation)**

| 1 | The circuit "breathes in" air containing moisture vapor. |

| 2 | The T.R.A.P.™ breather strips moisture and particulate from the incoming air, allowing only clean, dry air to enter the circuit. |

| 3 | During the "exhalation" cycle, the T.R.A.P.™ breather allows unrestricted airflow outward. |

| 4 | The outflow of dry air picks up the moisture collected by the T.R.A.P.™ breather during intake, and "blows it back out" – fully regenerating the T.R.A.P.™ breather's water-holding capacity. |

Fig. 2.55 – Construction and Operation of Breather Dryers (Courtesy of Donaldson)

Sizing of Breather Dryer: Breather dryers are provided in different sizes. Incorrectly sized tank breather filters can place additional strain on the system and reduce the service life of hydraulic filter elements. As shown in Fig. 2.56, larger size has a better water retention, but a relatively larger pressure drops.

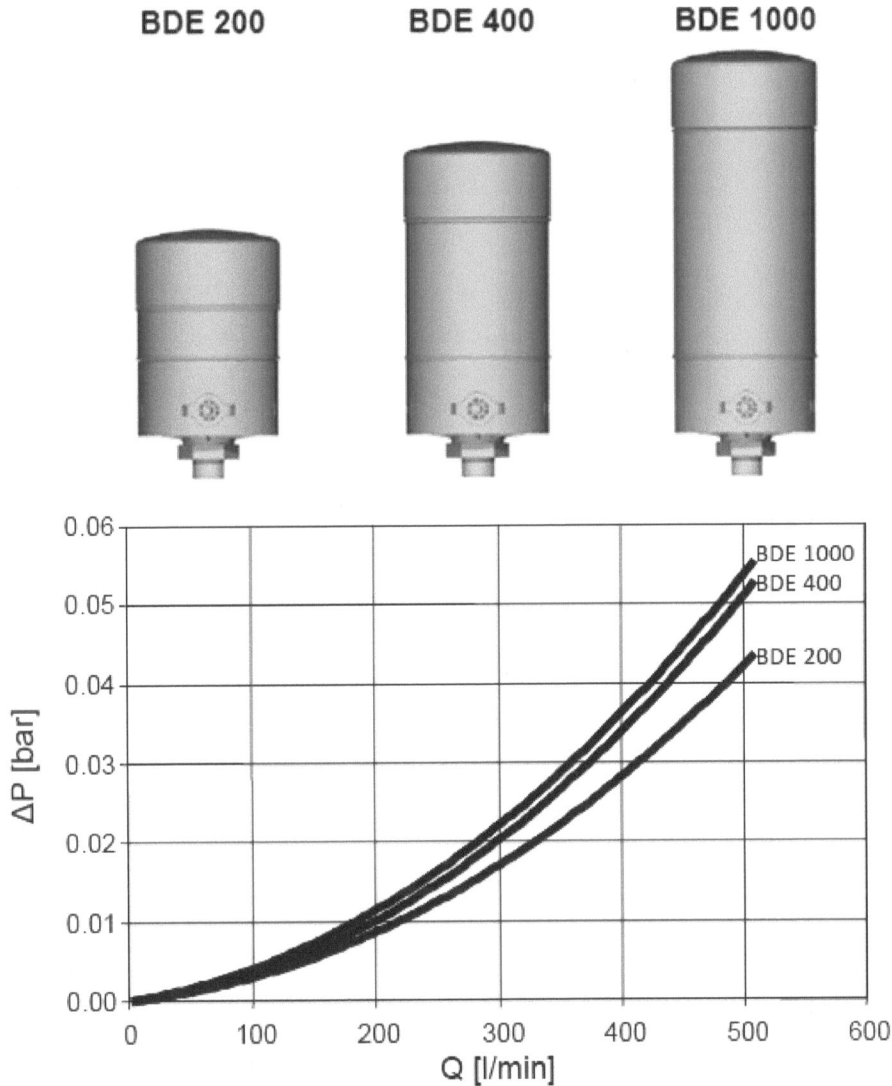

Size	Maximum water retention capacity
200	0.25 l
400	0.50 l
1000	0.75 l

Fig. 2.56 – Sizing of the Breather Dryer (Courtesy of HYDAC)

Air Breathers for Closed Tanks: In some dirty applications, such as mills and foundries, closed reservoirs (pressure-sealed) are recommended. In such reservoirs, the inside pressure is increased above atmospheric by gas bladders and no conventional air breather can be used because the pressure inside the tank is always higher than the atmospheric.

As shown in Fig. 2.57, in some cases, a tank may be completely sealed from the atmosphere but still air space is left on top of the oil surface. Any air returns with the oil back to the tank will be accumulated on top of the oil surface. An air check valve is used to protect the tank against accidental over-pressure. Such a valve should be connected to any point on the tank above oil level. The valve allows free one air flow direction toward the atmosphere. Cracking pressure of 1 to 3 PSI is usual.

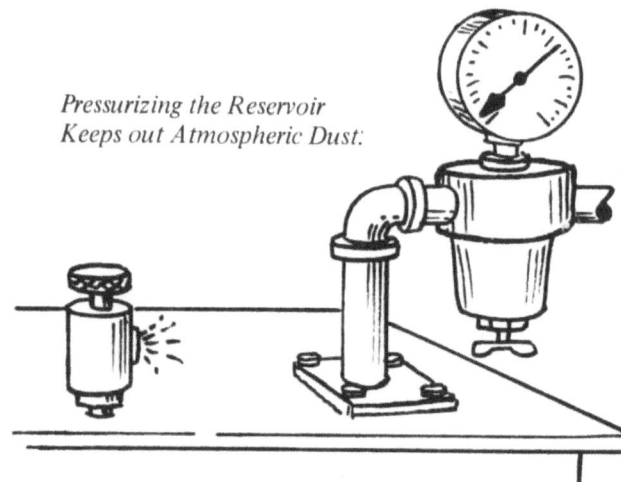

*Pressurizing the Reservoir
Keeps out Atmospheric Dust:*

Fig. 2.57 – Breathers for Pressurized Reservoirs (Courtesy of Womack)

2.4.14- Oil Level and Temperature Indicators

Monitoring the fluid level in hydraulic reservoirs is an essential maintenance action on daily basis. To do so, hydraulic reservoirs must be equipped at least with *Oil Level Indicator (Sight Glass)* that:

- Shall be permanently marked with system fluid high and low levels.
- Shall be fitted so that they are clearly visible during filling.

Optionally, a fully equipped Oil Level Indicator, may have

- Additional marks as appropriate for specific systems.
- *Level Switch* that provides *digital* signal for alarming when oil level is low.
- *Level Sensor* that provides *analog* signal for instantaneous and remote monitoring.
- *Temperature Gauge, Switch, or Sensor* may also be built-in with oil level indicators.

Sight Glasses: As shown in Fig. 2.58, a *Sight Glass* can be as simple as just a rounded pod (1) assembled on the side of the reservoir to show the oil level. However, the upgraded sight glasses can do more than that. They can be used to quick and qualitative machine faults detection. As shown in the figure, they can be used to observe the oil color and clarity (2), oil aeration and foaming (3), varnish formation (4), corrosion (5), wear debris (6) and to help sampling hydraulic fluid (7) for fluid analysis.

Fig. 2.58 – Multifunction Sight Glass (www.luneta.com)

Oil Level and Temperature Indicators: As shown in Fig. 2.59, *Oil Level Indicators* are available in various lengths to show both high and low levels. It is installed and sealed in drilled holes. Some of these gauges also contain a thermometer. On large tanks, it may be necessary to use two small level indicators for observing high and low oil levels. Reservoirs for mobile equipment often use a dipstick to check fluid level because sight gages, though preferred, might be inaccessible or subject to damage.

Oil Level and Temperature Switches: For critical systems, switches are used for the sake of machine protection. Usually, two switch settings are set up; a warning-level setting and a shutdown-level setting.

Most fluid manufacturers specify optimum range of working temperature for their products, typically from 38°C to 54°C (100°F to 130°F) even though many fluids are operated above this temperature range. The critical working temperature for a typical *Petroleum-Based* hydraulic fluid is 70°C (158°F). Every incremental increase of 10°C (18°F) above the critical temperature doubles the oxidation rate of the hydraulic fluid. Thus, cutting its useful life in half. For example, running a system at a consistent 80°C (176°F) would reduce the fluid life by 75%. Set the high-temperature switch at 70°C (158°F) to shut off the pump and prevent oil breakdown.

Fig. 2.59 – Oil Level and Temperature Gauges and Switches (Courtesy of Hydac)

2.5- Hydraulic Reservoir Design Case Study

Statement of the Problem: An open circuit system in an industrial application has a pump discharging 100 liter/min of mineral fluid of 32 cSt viscosity and specific gravity of 0.9. All components of the system are prefilled. However, system operation has been investigated. It was found that a minim of 100 liters and a maximum of 200 liters are still required from the reservoir to satisfy the system flow demand during the machine operation. A steel reservoir is used to host the required oil volume. A ventilation system within the work area guarantees 30 °C temperature difference between the reservoir walls and the surrounding air. Pump suction port is 1.5 in (3.81 mm).

Summary of the given data:
- Pump Flow **Qp** = 100 lit/min & pump suction port = 1.5 in (3.81 mm).
- The pump withdraws the oil through a strainer at the beginning of the suction line.
- Fluid Viscosity **v** = 32 cSt and Specific Gravity **SG** = 0.9.
- Max Fluid Volume required by the system **Vsmax** = 200 liters.
- Reservoir Material: **Steel**
- Temperature Difference (between reservoir walls and surrounding Air) **ΔT** = 30 °C.

Required (referring to Fig. 2.60): Properly design the reservoir including:
❖ Reservoir Sizing:
- Reservoir filling volume of oil **(Vfill)** in liters.
- Oil volume due to thermal expansion **(Vth)** in liters.
- Maximum oil volume in the reservoir **(Vmax)** in liters.
- Air space **(Vas)** in liters.
- Overall Size of the Reservoir **(Vo)** in liters.
- Minimum Absolute Volume **(Vmin)** in liters.

❖ Cooling: Determine the reservoir dimensions for better cooling and calculate the reservoir cooling capacity including:
- Reservoir Length **L** (cm).
- Reservoir Width **W** (cm).
- Area of the reservoir **AR** (cm²).
- Min oil height **Hmin** (cm).
- Maximum oil height **Hmax** (cm).
- Reservoir height **H** (cm).
- Reservoir Ground Clearance **GCR** (cm).
- Contact area with the oil **AC** (m²).
- Cooling Capacity **CC** (kW).

❖ Baffle Plate Dimensioning: Properly design the baffle including:
- Baffle Plate Height **BPH**, Baffle Plate Connecting Area **BPA**, and Width of Connecting Area **BPW**.

❖ Suction Line:
  ▪ Properly size the suction line finding suction line internal diameter **DS**.
  ▪ Suction line placement, bottom clearance **SLBC**, and line covered bottom **SLCB**.
  ▪ Suction line length **LS** (assuming maximum pressure losses = 0.01barg).
  ▪ Check if negative suction pressure NSP isn't falling below -3 psig.
  ▪ Suction strainer surface area.

❖ Return Line:
  ▪ Properly size the return line finding return line internal diameter **DR** based on maximum return flow = 200 lit/min.
  ▪ Return line bottom clearance **RLBC** and return line covered bottom **RLCB**.

Solution (referring to Fig. 2.60):
❖ Reservoir Sizing:
  ▪ Eq. 2.1 → Reservoir filling volume of oil **(Vfill)** = 3 x Qp = 3 x 100 = 300 liters.
  ▪ Eq. 2.2 → Oil volume due to thermal expansion **(Vth)** = 0.1 Vfill = 0.1 x 300 = 33 liters
  ▪ Eq. 2.3 → Maximum oil volume in the reservoir **(Vmax)** = Vfill + Vth = 300 + 33 = 333 liters. This is worst condition when the machine stopped.
  ▪ Eq. 2.4 → Air space **(Vas)** = 0.1 Vmax = 0.1 x 333 = 33.3 liters.
  ▪ Eq. 2.5 → Overall Size of the Reservoir **(Vo)** = Vmax + Vas = 333 + 33.3 = 366.3 liters
  ▪ Eq. 2.6 → Absolute Minimum Volume **(Vmin)** = Vfill – Vsmax = 300 – 200 = 100 liters. It is to be noted that the reservoir must be dimensioned so that this volume is above the lower point of the suction line by at least 5 cm (2 inches).

❖ Cooling:
  ▪ Eq. 2.7 → Reservoir Length **L** = 84 cm.
  ▪ Then Reservoir Length **L** was selected optionally as **L** = 100 cm.
  ▪ Eq. 2.8 → Reservoir Width **W** = 0.85 L = 0.85 x 100 = 85 cm.
  ▪ Eq. 2.9 → Footprint area of the reservoir **AR** = 100 x 85 = 8500 cm^2
  ▪ Eq. 2.10 → Min oil height **Hmin** = Vmin/AR = (100 x 1000) / 8500 = 11.7 cm.
  ▪ Eq. 2.11 → Maximum oil height **Hmax** = Vmax/AR = (333 x 1000) / 8500 = 39 cm.
  ▪ Eq. 2.12 → Reservoir height **H** = Vo / AR = = (366 x 1000) / 8500 = 43 cm.
  ▪ Eq. 2.13 → Reservoir Ground Clearance **(GCR)** in cm = 0.15 x H = 0.15 x 43 = 6.45 cm.
  ▪ Eq. 2.14 → Contact area with the oil **AC** = Hmin x 2 (L + W) = 11.7 x 2 (100 + 85) /10000 = 0.4329 m^2
  ▪ Eq. 2.15A → Cooling Capacity (kW) = AC (m^2) x ΔT (^{o}C) x K
        = 0.4329 x 30 x 0.0058 = 0.753 kW = 0.753 x 3413 = 2570 BTU/HR

❖ Baffle Plate:
  ▪ Eq. 2.16 → Baffle Plate Height **BPH**= Hmax + 1 cm = 39 + 1 = 40 cm.
  ▪ Eq. 2.17 → Baffle Plate Connecting Area **BPCA** = Qp (lit/min) / 2 = 100 / 2 = 50 cm^2
  ▪ Eq. 2.18 → Width of Connecting Area **BPCAW** = BPCA / Hmin (worst condition) = 50 / 11.7 = 4.3 cm.

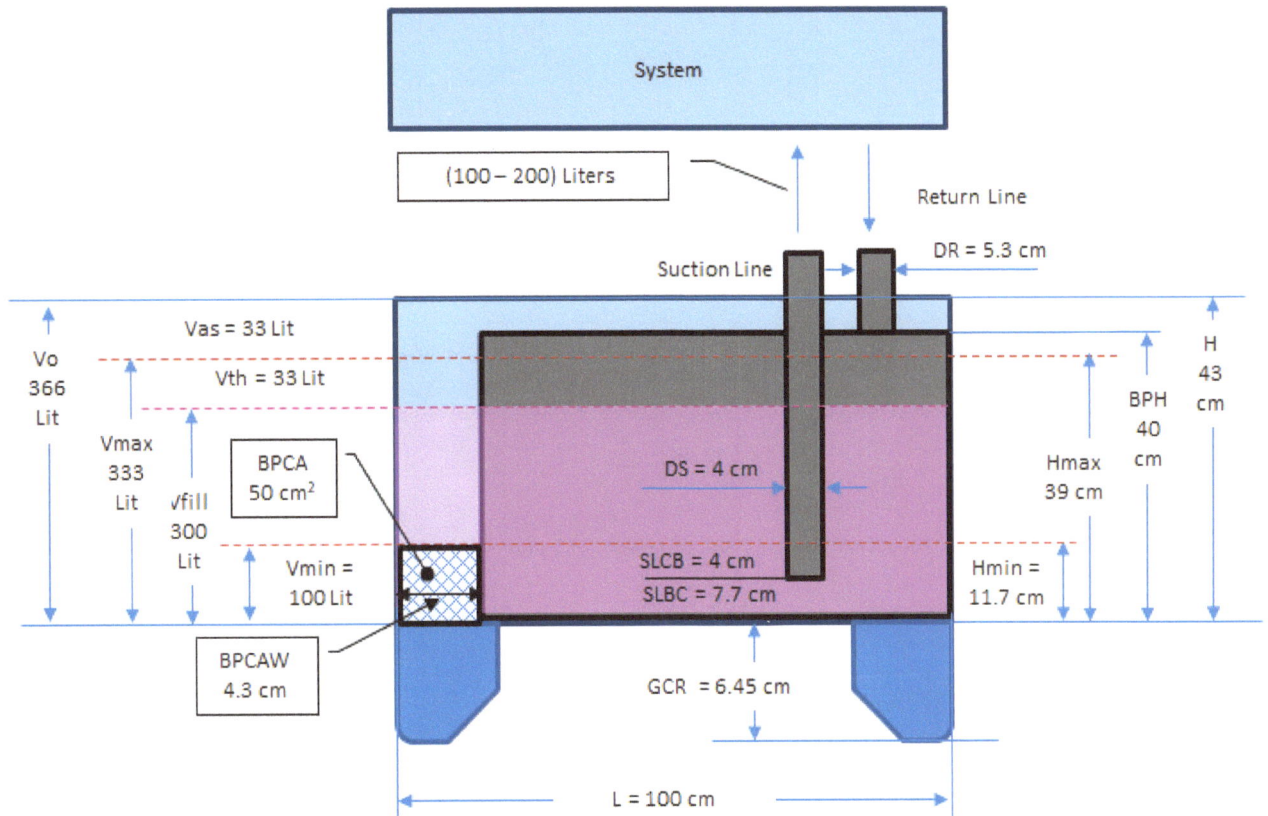

Fig. 2.60 - Case study for Hydraulic Reservoir Design

❖ Suction Line Diameter **DS**:
  ▪ Solution 1: based on the rule of thumb, assuming flow speed = 1 m/s, Eq 2.19A →

$$Q\left[\frac{lit}{min}\right] = \frac{v\,[cm/s] \times A[cm^2] \times 60}{1000} \rightarrow A[cm^2] = \frac{1000 \times Q\left[\frac{l}{min}\right]}{v\,[cm/s] \times 60}$$

$$\rightarrow A[cm^2] = \frac{1000 \times 100}{100\,[cm/s] \times 60} = 16.66 \rightarrow DS[cm] = \sqrt{\frac{4 \times A}{\pi}} = \sqrt{\frac{4 \times 16.66}{\pi}}$$
$$= 4.6\ cm$$

  ▪ Solution 2: Minimum inner diameter based on Reynold's No. = 2000, Eq 2.20A →

$$DS[mm] = \frac{21231\,Q\left[\frac{l}{min}\right]}{v\,[Cst] \times R_e} = \frac{21231 \times 100}{32 \times 2000} = 33.17\ mm = 3.3\ cm$$

- **Solution 3:** Figure 2.61 shows that HCSC validates the calculation resulted from Eq. 2.20A

- **Final Decision** after checking the pump suction port: Suction line diameter is decided to be **DS = 1.5 in = 3.81 cm**.

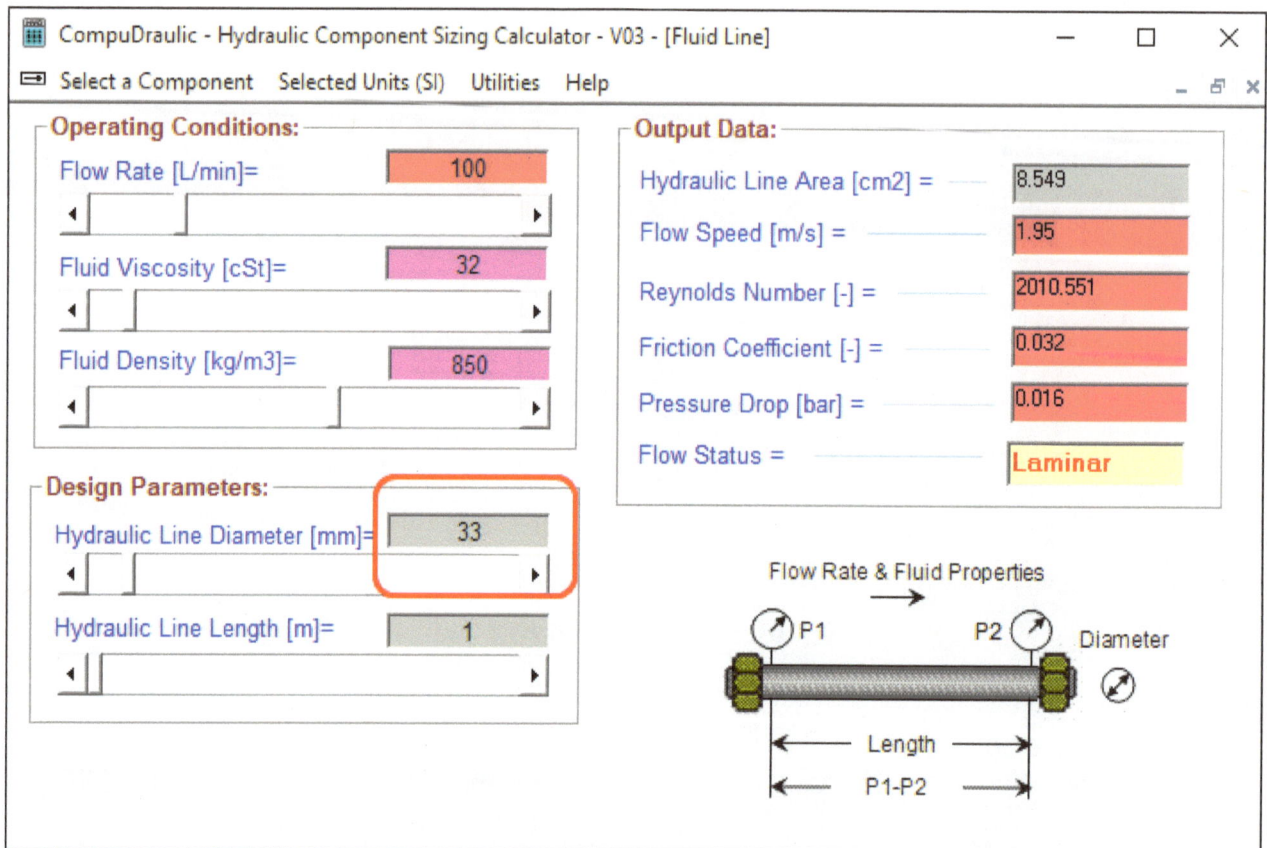

Fig. 2.61 - Validation of Suction Line Diameter using HCSC (www.compudraulic.com)

❖ Suction Line Placement, Bottom Clearance, and Covered Bottom:
  - Since a baffle plate is used, suction and return lines are placed one on each side of the baffle plate and on the same side of the reservoir.

  - Eq. 2.21 → Suction Line Bottom Clearance **(SLBC)** = 2 **DS** = 2 x 3.81 ≈ 7.7 cm.

  - Eq. 2.22 → **SLCB = Hmin − SLBC** = 11.7 − 7.7 = 4 cm.

  - Since **SLCB** < 5 cm, then minimum absolute volume of oil **Vmin** can be increased a bit to make the covered bottom **SLCB** equals at least 5 cm.

❖ Suction Line Length **LS**:
  ▪ **First:** Eq. 2.23 → **LS** = 10 **DS** = 10 x 3.81 = 381 mm.
  ▪ **Second:** Suction line minimum length is at least 25 cm (10 inches) from last turbulent point → **LS** >= 250 mm from the strainer.
  ▪ Then, so far, **LS** >= 381 mm because that satisfies both rules.
  ▪ **Third:** Figure 2.62 shows that the suction line length that develops maximum pressure losses of 0.01 bar can be elongated to 1.1 meter.
  ▪ Then, finally **381 mm >= LS <= 1100 mm. Decision made to use LS = 500 mm.**

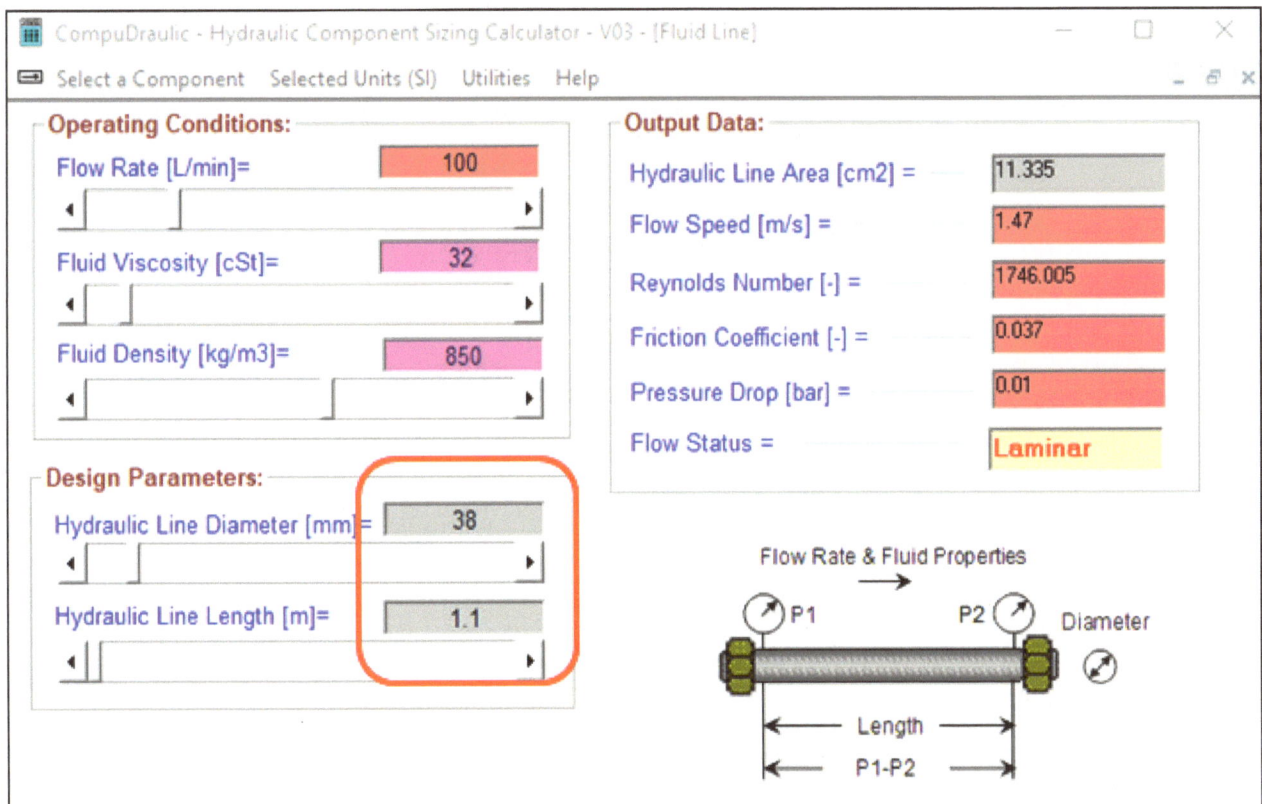

Fig. 2.62- Validation of Suction Line Length using HCSC (www.compudraulic.com)

❖ Suction Pressure:
  ▪ Eq. 2.24A (Hs = Ls = 500 mm & SG = 0.9) →

$$\mathbf{NSP\ (bar)} = -\frac{\mathbf{SG} \times \mathbf{9.81} \times \mathbf{Hs\ (m)}}{\mathbf{100}} = -\frac{0.9 \times 9.81 \times 0.5}{100} = -\mathbf{0.044\ barg}$$
$$= -\mathbf{0.6\ psig}$$

  ▪ Negative Suction Pressure (NSP) = -0.6 psig > -3psig, then it meets the requirement.

❖ Suction Strainer Surface Area:
  ▪ Eq. 2.25A → Suction Strainer Surface Area = 3 x Qp (lit/min) = 3 x 100 = 300 cm².

❖ Return Line Diameter **DR**:
  ▪ Solution 1: A flow distribution has been conducted. Because a differential cylinder is used that has area ratio 2:1, it was found that maximum return flow is 200 lit/min. Therefore, return line size based on the rule of thumb assuming flow speed = 1.5 m/s, Eq 2.19A →

$$Q\left[\frac{lit}{min}\right] = \frac{v\,[cm/s] \times A[cm^2] \times 60}{1000} \rightarrow A[cm^2] = \frac{1000 \times Q\left[\frac{l}{min}\right]}{v\,[cm/s] \times 60}$$

$$\rightarrow A[cm^2] = \frac{1000 \times 200}{150\,[cm/s] \times 60} = 22.21 \rightarrow DR[cm] = \sqrt{\frac{4 \times A}{\pi}} = \sqrt{\frac{4 \times 22.21}{\pi}}$$

$$= 5.3\ cm$$

  ▪ Solution 2: Minimum inner diameter based on Reynold's No. = 2000, Eq 2.20A →

$$DR[mm] = \frac{21231\,Q\left[\frac{l}{min}\right]}{v\,[Cst] \times R_e} = \frac{21231 \times 200}{32 \times 2000} = 66.34\ mm = 6.4\ cm$$

  ▪ Solution 3: Figure 2.63 shows that HCSC validates the calculation resulted from Eq. 2.20A

  ▪ Final Decision: **DR = 7 cm** = 70 mm.

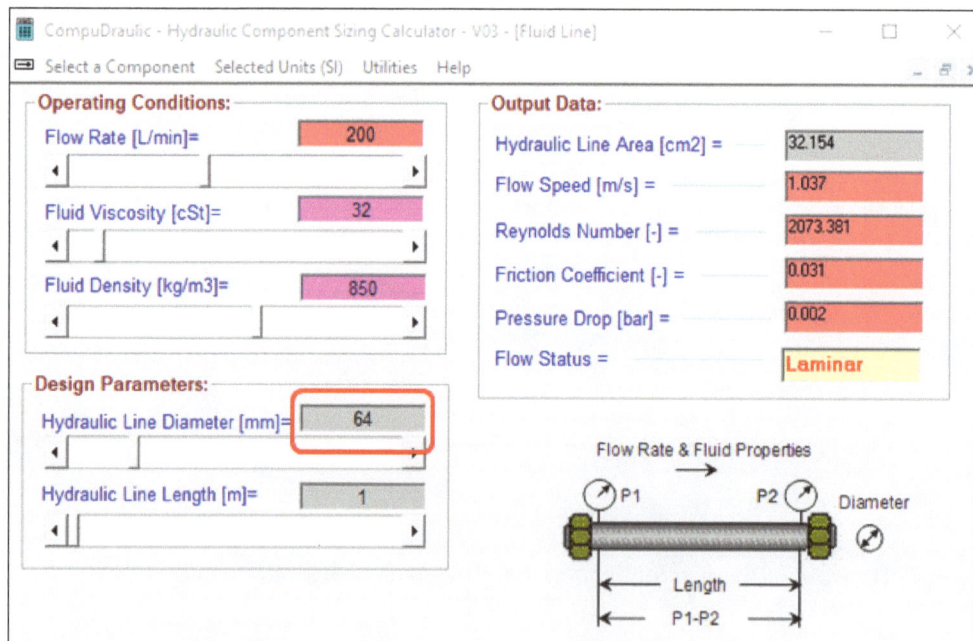

Fig. 2.63 - Validation of Return Line Diameter using HCSC (www.compudraulic.com)

❖ <u>Return Bottom Clearance, and Covered Bottom:</u>

- Eq. 2.21 → Return Line Bottom Clearance **(RLBC)** = 2 **DS** = 2 x 7 = 14 cm.

- Applying Eq. 2.22 → **RLCB = Hmin – RLBC**= 11.7 – 14 = - 2.3 cm, i.e. above **Hmin**!

- Since **RLBC** > **Hmin**, then the choices here are as follows:

- <u>Choice 1:</u> Minimum absolute volume of oil **Vmin** can be increased a bit to make the covered bottom **RLCB** equals at least 5 cm.

- <u>Choice 2:</u> Reducing the bottom clearance to be 5 cm with the consideration that this may result in some back pressure.

- <u>Choice 3: Repeat reservoir sizing in this backward sequence:</u>
  - Eq. 2.22 → **Hmin** = RLCB + RLBC = 5 cm (recommended) + 14 (calculated) = 19 cm
  - Eq. 2.10 → **Vmin** = Hmin x AR = 19 x 8500 = 161500 cm³ = 161.5 liters
  - Eq. 2.6 → Vfill = Vmin + Vsmax = 161.5 + 200 = 361.1 Liter.
  - Eq. 2.2 → Oil volume due to thermal expansion **(Vth)**
  - Eq. 2.3 → Maximum oil volume in the reservoir **(Vmax)**
  - Eq. 2.4 → Air space **(Vas)**
  - Eq. 2.5 → Overall Size of the Reservoir **(Vo)**
  - Eq. 2.7 → Reservoir Length **L** = 84 cm.
  - Eq. 2.8 → Reservoir Width **W**
  - Eq. 2.9 → Footprint area of the reservoir **AR**
  - Eq. 2.10 → Min oil height **Hmin**
  - Eq. 2.11 → Maximum oil height **Hmax**
  - Eq. 2.12 → Reservoir height **H**
  - Eq. 2.13 → Reservoir Ground Clearance
  - Eq. 2.14 → Contact area with the oil **AC**
  - Eq. 2.15A → Cooling Capacity (kW)
  - Eq. 2.16 → Baffle Plate Height **BPH**
  - Eq. 2.17 → Baffle Plate Connecting Area **BPCA**
  - Eq. 2.18 → Width of Connecting Area **BPCAW**

Chapter 3

Hydraulic Transmission Lines

Objectives

This chapter focuses on browsing the construction and the features of the three main hydraulic transmission lines, Pipes, Tubes, and Hoses. For each transmission line, the following topics are presented: sizing, material, construction, and pressure rating. This chapter also presents information about fittings and manifolds.

The following topics are discussed in Chapter 10 in Volume 5 "Safety and Maintenance" of this series of textbooks:
- Transmission Lines Selection and Replacement.
- Transmission Lines Maintenance Scheduling.
- Transmission Lines Installation and Maintenance.
- Transmission Lines Standard Tests and Calibration.
- Transmission Lines Transportation and Storage.

The following topics are discussed in Chapter 10 in Volume 6 "Troubleshooting and Failure Analysis" of this series of textbooks:
- Hydraulic Transmission Lines Inspection.
- Hydraulic Transmission Lines Troubleshooting.
- Hydraulic Transmission Lines Failure Analysis.

Brief Contents

3.1. Basic Types and Contribution of Hydraulic Transmission Lines
3.2. Sizing of Hydraulic Transmission Lines
3.3- Rated Pressures for Hydraulic Lines
3.4- Hydraulic Pipes
3.5- Hydraulic Tubes
3.6- Hydraulic Hoses
3.7- Flanges for Transmission Line Connections
3.8- Rubber Expansion Fittings
3.9- Test Points
3.10-Pressure Measurement Hoses
3.11 Manifolds

Chapter 3: Hydraulic Transmission Lines

3.1. Basic Types and Contribution of Hydraulic Transmission Lines

Hydraulic transmission lines are used to transmit the energy between system components. Modern hydraulic system design trends are to minimize the use of transmission lines, and rather use manifolds to avoid energy losses and external leakage. However, transmission lines are still needed at least to connect control valves to actuators. As shown in Fig. 3.1, there are three types of transmission lines, pipes, tubes, and hoses.

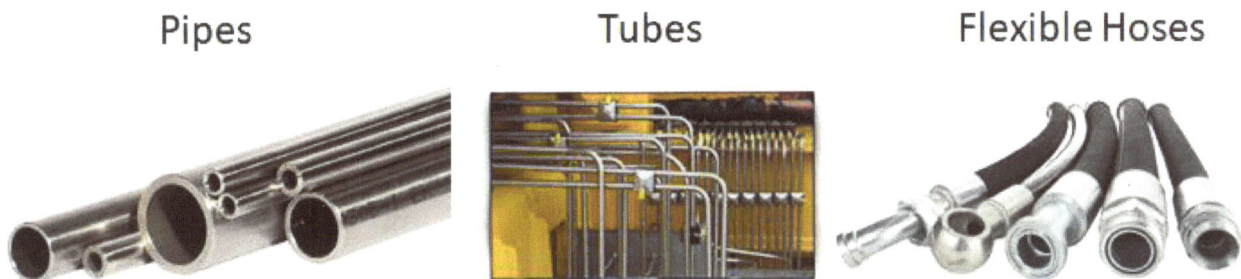

Fig. 3.1 - Types of Hydraulic Transmission Lines

May be the first question is, which type of transmission line is better to use? The following set of bullets answers that question:

- **Interchangeability:** Hydraulic hoses, tubes and pipes are limitedly interchangeable. This means that pipes, tubes, and hoses can't replace each other in every case. Each type has its own advantages that make it more adequate for specific applications.
- **Flexibility:** Pipes are not intended to bend. Tubes can be bend but offers no flexibility. Hoses are intended to flex. However, hoses are shortened under pressure and have limited working temperature range.
- **Stiffness:** Pipes and tubes are best because they have less wall elasticity. Because of a hose wall elasticity, the overall bulk modulus of the system is reduced; and so, the system stiffness, response, and bandwidth.
- **Service Life:** Pipes and tubes have the longest service life because they overcome the environmental conditions. Hoses have the shortest service life they can fail without warning, that is why they require routine replacement because.
- **Cost:** Pipes are the most cost effective and hoses are the most expensive.
- **Applications:** Hoses are more recommended for mobile applications because of limited spaces requirements and relative motion between machine components. Tubes are recommended for industrial application because they are cost effective.
- **Noise and Vibration:** Hoses are better for absorbing vibration and suppressing noise.
- **Heat Transfer:** Pipes and tubes dissipate more heat than hoses can dissipate.

3.2. Sizing of Hydraulic Transmission Lines

Improper sizing, selection, installation, and maintaining of *Hydraulic Transmission Lines* can lead to one or more of the following problems:
- Pump cavitation.
- Turbulent flow.
- System heating-up.
- External leakage.
- Line accidental breakage, system damage and loss of life risks.
- Increased maintenance costs both in replacement parts and labor.
- Increased machine downtime.
- Increased environmental cleanup costs.

Therefore, a hydraulic transmission line must be precisely sized. Sizing a transmission line means calculation of the *Inner Diameter* (ID). Size of transmission lines can be determined mathematically or by charts, software, or tables.

Sizing transmission lines mathematically: that satisfies the following conditions:
- Comply with the recommended flow speed:
  - Equation 2.19 (in Chapter 2) is used for such calculations taking into consideration the following flow speeds:
  - Pressure Line (industrial machines) = 7 - 15 fps. = 2.1- 4.6 m/s.
  - Drain Line = 4 - 7 fps. = 1.2 - 2.1 m/s.
  - Suction Line = 2 - 4 fps. = 0.6 - 1.2 m/s.

- Securing laminar flow:
  - Equation 2.20 (in Chapter 2) is used to size a line based on securing laminar flow in the line (Reynolds Number < 2000).

- Final Solution: the larger size out of the two equations is considered.

Sizing Hydraulic Lines Using Charts: As shown in Fig. 3.2, using the chart, the following steps are required to calculate the inner diameter of the transmission line:
- **Step 1:** Define the flow rate on the upper scale.
- **Step 2:** Define the recommended flow speed based on the transmission line type.
- **Step 3:** Draw a dashed line between the flow rate and the flow speed.
- **Step 4:** Intersection of the dashed line with the middle scale defines the inside area that can be used to calculate the inner diameter.

Example 1: As shown in the figure, assume flow rate of 30 liter/min is flowing in a pressure line with a maximum allowable flow speed is 5.3 m/s. the chart shows inside area = 0.95 cm² and an inner diameter ID = 11 mm.

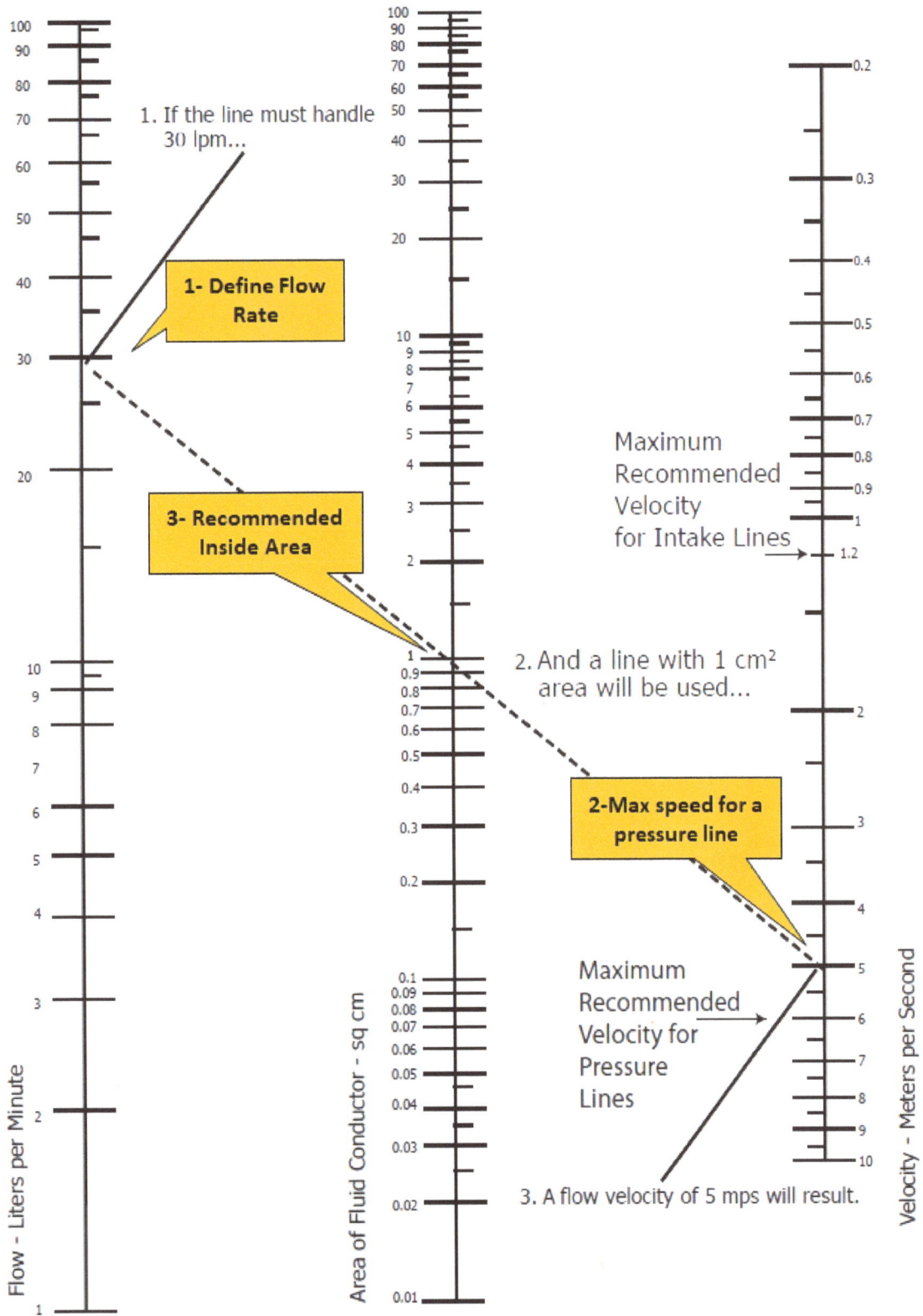

Fig. 3.2 - Example 1 of using Charts for Sizing a Transmission Lines

Example 2: Figure 3.3 shows an example where the flow requirements are increased from 20 to 25 GPM and a decision must be made whether to keep the same line or replace it by another larger size line to cope with an increased flow requirement.

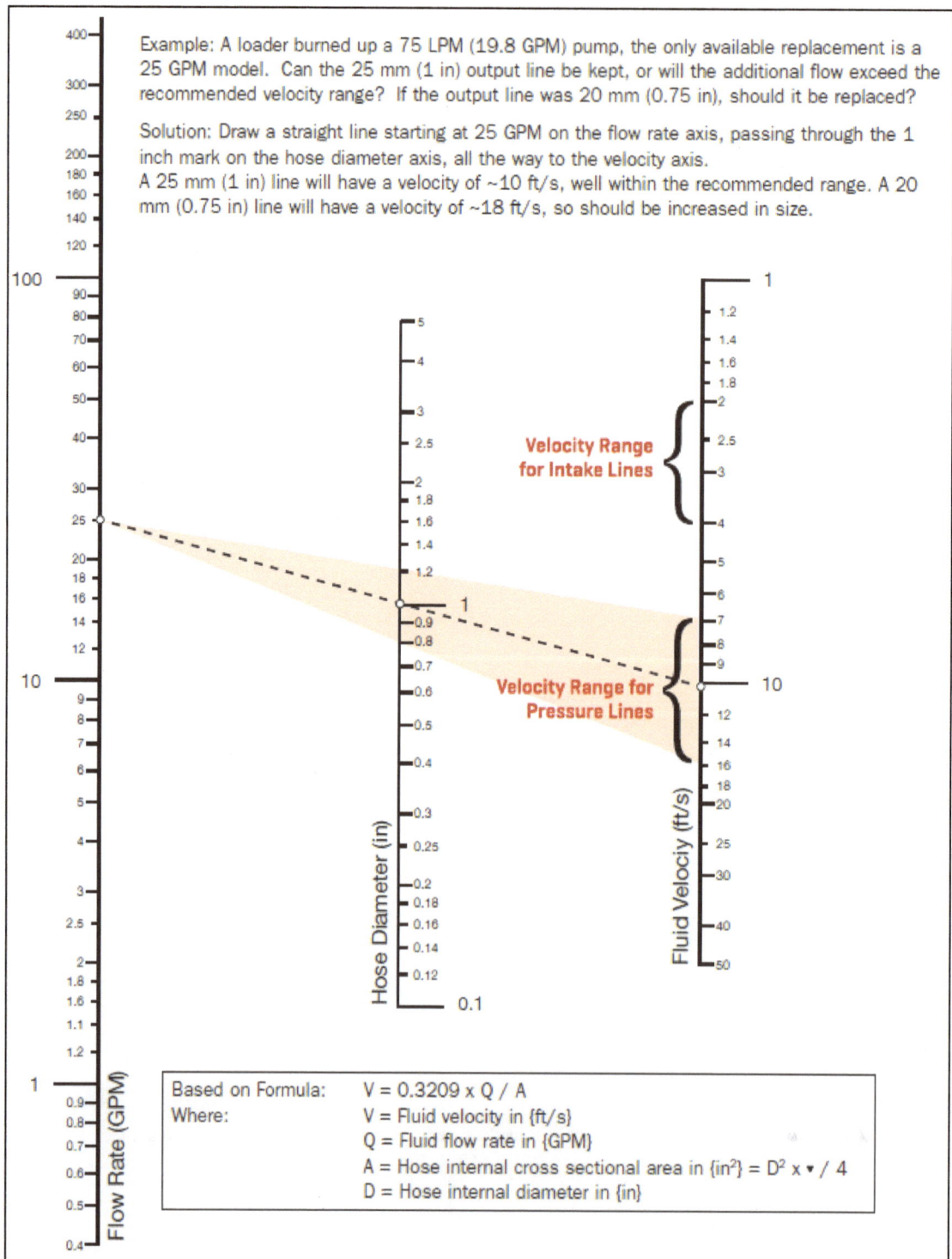

Example: A loader burned up a 75 LPM (19.8 GPM) pump, the only available replacement is a 25 GPM model. Can the 25 mm (1 in) output line be kept, or will the additional flow exceed the recommended velocity range? If the output line was 20 mm (0.75 in), should it be replaced?

Solution: Draw a straight line starting at 25 GPM on the flow rate axis, passing through the 1 inch mark on the hose diameter axis, all the way to the velocity axis.
A 25 mm (1 in) line will have a velocity of ~10 ft/s, well within the recommended range. A 20 mm (0.75 in) line will have a velocity of ~18 ft/s, so should be increased in size.

Velocity Range for Intake Lines

Velocity Range for Pressure Lines

Hose Diameter (in)

Fluid Velociy (ft/s)

Flow Rate (GPM)

Based on Formula: $V = 0.3209 \times Q / A$
Where:
V = Fluid velocity in {ft/s}
Q = Fluid flow rate in {GPM}
A = Hose internal cross sectional area in {in^2} = $D^2 \times \pi / 4$
D = Hose internal diameter in {in}

Fig. 3.3 - Example 2 of using Charts for Sizing a Transmission Lines (Courtesy of Gates)

Sizing Hydraulic Lines Using Software: Figure 3.4, shows the use of *Hydraulic Components Sizing Calculator* (HCSC) to size a hydraulic line. The figure shows that the software validates the results from the chart in example 1 (upper part) and in example 2 (lower part).

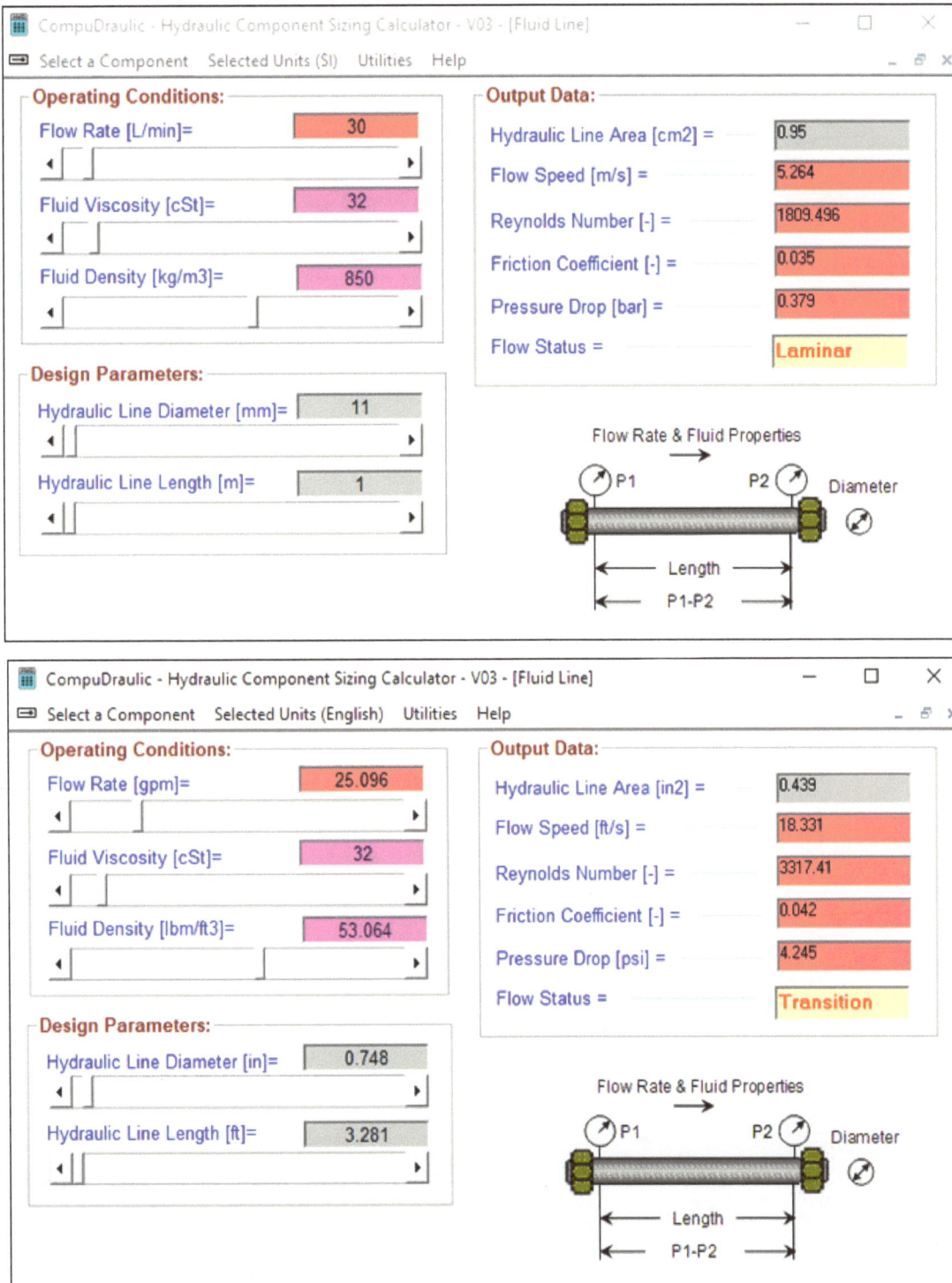

Fig. 3.4 - Using Software for Sizing a Transmission Lines (www.compudraulic.com)

Sizing Hydraulic Lines Using Tables: Using the table shown in Table 3.1, the following steps are required to calculate the inner diameter of the transmission line:
1. Locate the vertical column that shows the recommended flow speed at the top.
2. Go down in this column until the flow rate through the line is met.
3. Go left to find the corresponding inner diameter.

Example 3: The figure shows that, for flow speed 15 fps (4.5 m/s) and flow rate of 8.9 gpm (40 lit/min), inner diameter = 0.493 inches (12.5 mm). Interpolating the results validates the results from the chart and the software for example 1.

STANDARD PIPE – SCHEDULE 40

PIPE SIZE	OD (INCHES)	WALL (INCHES)	ID (INCHES)	INT AREA (SQ. INCHES)	WT/FT (POUNDS)	GPM @ 2 FPS	GPM @ 5 FPS	GPM @ 10 FPS	GPM @ 15 FPS	GPM @ 20 FPS	GPM @ 25 FPS
1/8	0.405	0.068	0.269	0.057	0.244	0.4	0.9	1.8	2.7	3.5	4.4
1/4	0.540	0.088	0.364	0.104	0.424	0.6	1.6	3.2	4.9	6.5	8.1
3/8	0.675	0.091	0.493	0.191	0.567	1.2	3.0	6.0	8.9	11.9	14.9
1/2	0.840	0.109	0.622	0.304	0.850	1.9	4.7	9.5	14.2	18.9	23.7
3/4	1.050	0.113	0.824	0.533	1.130	3.3	8.3	16.6	24.9	33.2	41.6
1	1.315	0.133	1.049	0.864	1.677	5.4	13.5	26.9	40.4	53.9	67.4
1 1/4	1.660	0.140	1.380	1.496	2.270	9.3	23.3	46.6	69.9	93.2	116.6
1 1/2	1.900	0.145	1.610	2.036	2.715	12.7	31.7	63.5	95.2	126.9	158.7
2	2.375	0.154	2.067	3.356	3.649	20.9	52.3	104.6	156.9	209.2	261.5
2 1/2	2.875	0.203	2.469	4.788	5.787	29.8	74.6	149.2	223.9	298.5	373.1
3	3.500	0.216	3.068	7.393	7.568	46.1	115.2	230.4	345.7	460.9	576.1
3 1/2	4.000	0.226	3.548	9.887	9.100	61.6	154.1	308.2	462.3	616.4	770.5
4	4.500	0.237	4.026	12.730	10.779	79.4	198.4	396.8	595.2	793.7	992.1
5	5.563	0.258	5.047	20.006	14.602	124.7	311.8	623.6	935.4	1,247.2	1,559.1
6	6.625	0.280	6.065	28.890	18.954	180.1	450.3	900.6	1,350.9	1,801.1	2,251.4
8	8.625	0.322	7.981	50.027	28.524	311.9	779.7	1,559.4	2,339.2	3,118.9	3,898.6
10	10.750	0.365	10.020	78.854	40.441	491.6	1,229.0	2,458.1	3,687.1	4,916.1	6,145.1
12①	12.750	0.406	11.938	111.932	53.469	697.8	1,744.6	3,489.1	5,233.7	6,978.3	8,722.9
12②	12.750	0.375	12.000	113.097	49.510	705.1	1,762.7	3,525.5	5,288.2	7,051.0	8,813.7

Table 3.1 - Using Tables for Sizing a Transmission Lines

3.3- Rated Pressures for Hydraulic Lines

Pressure carrying capacity is the second most important selection factor of a transmission line.

❖ **Burst Pressure:**
Burst Pressure is the maximum static pressure above which a lines assembly fails. Burst pressure of hoses is based on the structure of the hose. Burst pressure for hard lines (i.e. tubes and pipes) depends on the material and wall thickness. Burst pressure for hydraulic transmission lines is determined by standard test procedure. However, burst pressure for hard lines (i.e., pipes and tubes) can be estimated using *Barlow's formula* as shown in Equation 3.1.

BP = (2 x t x s)/OD   **3.1**

Where:
- BP = Burst Pressure (psi)
- **t** = Wall Thickness [in]
- **s** = Ultimate Strength of Material (psi)
- **OD** = Outer Diameter of Pipe or Tube

❖ **Maximum Allowable Pressure:**
Equation 3.2 is used to calculate the *Maximum Allowable Working Pressure*.

WP = (BP/SF)   **3.2**

Where:
- WP = Allowable Working Pressure.
- BP = Burst Pressure.
- SF = Safety Factor = 4-6 (Typically, but may be different in some applications)

3.4- Hydraulic Pipes

3.4.1- Features of Hydraulic Pipes

Hydraulic *pipes* are rigid, not flexible or intend to bend. They have highest *Tensile Strength* and least expensive among all hydraulic transmission lines. They are used generally to transmit large hydraulic power (large flow and large pressure) for long distances. The weight per unit length is the largest among the hydraulic transmission lines. So, they must be clamped to a fixed frame to carry the weight of the pipe. As shown in Fig. 3.5, many industrial and mobile applications rely on hydraulic pipes in transmitting high hydraulic power.

Fig. 3.5 - Use of Hydraulic Pipes in Industrial (Left) and Mobile (right) Applications

3.4.2- Material of Hydraulic Pipes

Hydraulic pipes are produced from different grades of *Carbon Steel* depending on the pressure rating. Pipes that are used in hydraulic systems are known as black iron pipes and may be either hot or cold drawn. Galvanized pipe should never be used in a hydraulic system because some hydraulic fluids additives react with the zinc coating.

3.4.3- Sizes of Hydraulic Pipes

Sizing of pipes is identified by a *schedule number* established by **ANSI Standard**. As shown in Table 3.2, *Nominal Size* of hydraulic pipes are the *outer diameter* "**OD**" on which the fitting will be assembled. Pipe sizes are ranging between (0.5 - 12) inches (10-300 mm). Hydraulic pipes are produced for different pressure ratings. Obviously, a pipe that works at high pressure has thicker walls. A pipe with a specific OD has different *inner diameters* "**ID**" depending on the standard schedule it belongs to. So, a common mistake is that pipes are sized based on flow only. When pressure increases, moving to a higher schedule may require larger pipe size to maintain same ID, keep flow to be laminar and maintain reasonable pressure drop.

For example, a pipe of nominal size = 1 inch, it has an actual OD = 1.315 inch. This pipe has an ID = 1.049 inch in schedule 40, ID = 0.957 in schedule 80, ID = 0.815 in schedule 160, and ID = 0.599 in double extra heavy schedule. As a result, pipes that belong to large schedules have smaller ID, larger wall thickness, work at higher pressure, handle less flow, and have more weight per unit length. The table shows following schedules:

- Schedule 40, formerly called standard, is for low pressure applications.
- Schedule 80, formerly called extra strong, is for medium pressure applications.
- Schedule 160 or "XS", is for high pressure applications.
- Schedule Double Extra Strong or "XXS", is for extremely high-pressure applications.

NOMINAL SIZE	PIPE O.D.	INSIDE DIAMETER			
		SCHED. 40	SCHED. 80	SCHED. 160	DOUBLE EXTRA HEAVY
1/8	.405	.269	.215	--	--
1/4	.540	.364	.302	--	--
3/8	.675	.493	.423	--	--
1/2	.840	.622	.546	.466	.252
3/4	1.050	.824	.742	.614	.434
1	1.315	1.049	.957	.815	.599
1 1/4	1.660	1.380	1.278	1.160	.896
1 1/2	1.900	1.610	1.500	1.338	1.100
2	2.375	2.067	1.939	1.689	1.503
2 1/2	2.875	2.469	2.323	2.125	1.771
3	3.500	3.068	2.900	2.624	--
3 1/2	4.000	3.548	3.364	--	--
4	4.500	4.026	3.826	3.438	--
5	5.563	5.047	4.813	4.313	4.063

Schedule 40
Schedule 80
Schedule 160
Double Extra Heavy

Table 3.2- ANSI Standard Schedules for Hydraulic Pipes

To get a better sense of ANSI standard for pipes, Fig. 3.6 shows three pipes that have the same OD of one inch, and they belong to different schedules according to ANSI.

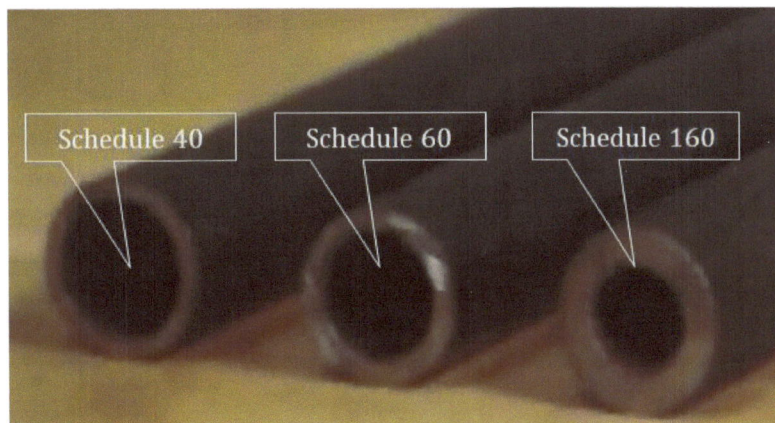

Schedule 40   Schedule 60   Schedule 160

Fig. 3.6 - Hydraulic Pipes According to ANSI Standard

Example of Pipe Size Selection:

Problem Definition: A pump drives a hydraulic motor. Flow distribution analysis is conducted and shows that maximum flow in suction, pressure and return line is 40 GPM.

Required: Referring to Table 2.3 and 3.3, find the nominal and inside diameter of each line. Both the suction and return pipes shall be selected from schedule 40 and pressure pipe from schedule 80.

Suction Line (Purple): Allowable flow speed = 2 FPS, closest upper flow = 46.1 GPM → Pipe Nominal Size = 3 → ID = 3.068 in (Schedule 40).

Return Line (Blue): Allowable flow speed = 5 FPS, closest upper flow = 52.3 GPM → Pipe Nominal Size = 2 → ID = 2.067 in (Schedule 40).

Pressure Line (Red): Allowable flow speed = 10 FPS, flow = 40 GPM → Pipe Nominal Size = 1.25 → ID = 1.278 in (Schedule 80).

PIPE SIZE	OD (INCHES)	WALL (INCHES)	ID (INCHES)	INT AREA (SQ. INCHES)	WT/FT (POUNDS)	GPM @ 2 FPS	GPM @ 5 FPS	GPM @ 10 FPS	GPM @ 15 FPS	GPM @ 20 FPS	GPM @ 25 FPS	WALL (INCHES)	ID (INCHES)	INT AREA (SQ. INCHES)	WT/FT (POUNDS)	GPM @ 2 FPS	GPM @ 5 FPS	GPM @ 10 FPS	GPM @ 15 FPS	GPM @ 20 FPS	GPM @ 25 FPS	PIPE SIZE
				STANDARD PIPE – SCHEDULE 40									EXTRA STRONG PIPE – XS – SCHEDULE 80									
1/8	0.405	0.068	0.269	0.057	0.244	0.4	0.9	1.8	2.7	3.5	4.4	0.095	0.215	0.036	0.314	0.2	0.6	1.1	1.7	2.3	2.8	1/8
1/4	0.540	0.088	0.364	0.104	0.424	0.6	1.6	3.2	4.9	6.5	8.1	0.119	0.302	0.072	0.534	0.4	1.1	2.2	3.3	4.5	5.6	1/4
3/8	0.675	0.091	0.493	0.191	0.567	1.2	3.0	6.0	8.9	11.9	14.9	0.126	0.423	0.141	0.738	0.9	2.2	4.4	6.6	8.8	11.0	3/8
1/2	0.840	0.109	0.622	0.304	0.850	1.9	4.7	9.5	14.2	18.9	23.7	0.147	0.546	0.234	1.087	1.5	3.6	7.3	10.9	14.6	18.2	1/2
3/4	1.050	0.113	0.824	0.533	1.130	3.3	8.3	16.6	24.9	33.2	41.6	0.154	0.742	0.432	1.472	2.7	6.7	13.5	20.2	27.0	33.7	3/4
1	1.315	0.133	1.049	0.864	1.677	5.4	13.5	26.9	40.4	53.9	67.4	0.179	0.957	0.719	2.169	4.5	11.2	22.4	33.6	44.8	56.1	1
1 1/4	1.660	0.140	1.380	1.496	2.270	9.3	23.3	46.6	69.9	93.2	116.6	0.191	1.278	1.283	2.993	8.0	20.0	40.0	60.0	80.0	100.0	1 1/4
1 1/2	1.900	0.145	1.610	2.036	2.715	12.7	31.7	63.5	95.2	126.9	158.7	0.200	1.500	1.767	3.627	11.0	27.5	55.1	82.6	110.2	137.7	1 1/2
2	2.375	0.154	2.067	3.356	3.649	20.9	52.3	104.6	156.9	209.2	261.5	0.218	1.939	2.953	5.017	18.4	46.0	92.0	138.1	184.1	230.1	2
2 1/2	2.875	0.203	2.469	4.788	5.787	29.8	74.6	149.2	223.9	298.5	373.1	0.276	2.323	4.238	7.653	26.4	66.1	132.1	198.2	264.2	330.3	2 1/2
3	3.500	0.216	3.068	7.393	7.568	46.1	115.2	230.4	345.7	460.9	576.1	0.300	2.900	6.605	10.242	41.2	102.9	205.9	308.8	411.8	514.7	3
3 1/2	4.000	0.226	3.548	9.887	9.100	61.6	154.1	308.2	462.3	616.4	770.5	0.318	3.364	8.888	12.492	55.4	138.5	277.1	415.6	554.1	692.6	3 1/2
4	4.500	0.237	4.026	12.730	10.779	79.4	198.4	396.8	595.2	793.7	992.1	0.337	3.826	11.497	14.968	71.7	179.2	358.4	537.6	716.8	896.0	4
5	5.563	0.258	5.047	20.006	14.602	124.7	311.8	623.6	935.4	1,247.2	1,559.1	0.375	4.813	18.194	20.756	113.4	283.6	567.1	850.7	1,134.3	1,417.8	5
6	6.625	0.280	6.065	28.890	18.954	180.1	450.3	900.6	1,350.9	1,801.1	2,251.4	0.432	5.761	26.067	28.543	162.5	406.3	812.6	1,218.8	1,625.1	2,031.4	6
8	8.625	0.322	7.981	50.027	28.524	311.9	779.7	1,559.4	2,339.2	3,118.9	3,898.6	0.500	7.625	45.664	43.342	284.7	711.7	1,423.4	2,135.1	2,846.9	3,558.6	8
10	10.750	0.365	10.020	78.854	40.441	491.6	1,229.0	2,458.1	3,687.1	4,916.1	6,145.1	0.594	9.562	71.810	64.362	447.7	1,119.2	2,238.5	3,357.7	4,477.0	5,596.2	10[3]
12[1]	12.750	0.406	11.938	111.932	53.469	697.8	1,744.6	3,489.1	5,233.7	6,978.3	8,722.9	0.688	11.374	101.605	88.537	633.4	1,583.6	3,167.2	4,750.9	6,334.5	7,918.1	12[3]
12[2]	12.750	0.375	12.000	113.097	49.510	705.1	1,762.7	3,525.5	5,288.2	7,051.0	8,813.7											12[2]

Table 3.3 - Examples of Sizing Hydraulic Pipes

3.4.4- Pressure Rating of Hydraulic Pipes

Table 3.4 presents allowable working pressure and estimated burst pressure for seamless steel pipe (material is ASTM A53 or A106 Grade B steel with *tensile strength* of 60k psi). Presented data are tabulated by ANSI B 31.3 standard based on Barlow's formula. The standard considers lower safety factor for higher schedule number.

Example: As shown in the figure, for a pipe size of nominal size =1, approximate estimated burst pressure = 6k psi in schedule 40, approximate estimated burst pressure = 10k psi in schedule 80, and approximate estimated burst pressure = 16k psi in schedule 160. The reported allowable working pressure for that pipe size indicates that safety factor for schedule 40 ≈ 9, for schedule 80 ≈ 6, and for schedule 160 ≈ 5.

Threaded Connections Pressure Ratings of Seamless Steel

1 inch OD Different schedule

PIPE SIZE & SCHEDULE	ALLOWABLE WORKING PRESSURE (PSIG)		ESTIMATED BURST VALUE P_B (PSIG)	WATER HAMMER FACTOR
	ANSI B31.1	ANSI B 31.3		
1/8-40	994	746	11,369	327.6
1/8-80	3,509	2,632	19,369	512.9
1/4-40	948	711	9,680	178.9
1/4-80	3,103	2,327	16,569	260.0
3/8-40	916	687	8,277	97.55
3/8-80	2,864	2,148	14,500	132.5
1/2-40	884	663	7,409	61.28
1/2-80	2,576	1,932	12,837	79.53
1/2-160	4,445	3,334	18,551	109.2
1/2-XXS	8,577	6,433	33,837	373.3
3/4-40	842	631	6,384	34.92
3/4-80	2,295	1,722	11,070	43.06
3/4-160	4,704	3,528	18,384	62.89
3/4-XXS	7,328	5,496	28,670	125.9
1-40	829	622	5,788	21.55
1-80	2,126	1,595	9,986	25.89
1-160	4,249	3,187	16,465	35.69
1-XXS	6,804	5,103	26,321	66.08
1 1/4-40	806	605	5,091	12.45
1 1/4-80	1,941	1,456	8,778	14.52
1 1/4-160	3,319	2,490	13,043	17.62
1 1/4-XXS	6,575	4,932	22,585	29.53
1 1/2-40	798	598	4,764	9.147
1 1/2-80	1,866	1,399	8,238	10.54
1 1/2-160	3,522	2,642	13,353	13.24
1 1/2-XXS	6,095	4,572	20,869	19.59
2-40	773	580	4,266	5.549
2-80	1,764	1,323	7,500	6.306
2-160	3,834	2,875	13,866	8.331
2-XXS	5,435	4,076	18,514	10.50
2 1/2-40	815	611	4,299	3.889
2 1/2-80	1,750	1,312	7,346	4.394
2 1/2-160	3,074	2,305	11,478	5.250
2 1/2-XXS	5,588	4,191	18,866	7.559
3-40	801	601	3,977	2.519
3-80	1,683	1,263	6,857	2.819
3-160	3,201	2,401	11,589	3.443
3-XXS	5,096	3,822	17,143	4.482
3 1/2-40	790	592	3,780	1.883
3 1/2-80	1,634	1,226	6,540	2.095
3 1/2-XXS	4,783	3,587	16,080	3.186
4-40	789	592	3,653	1.463
4-80	1,604	1,203	6,320	1.620
4-160	3,263	2,447	11,493	2.006
4-XXS	4,556	3,417	15,307	2.386
5-40	772	579	3,408	0.9308
5-80	1,543	1,157	5,932	1.023
5-160	3,270	2,453	11,325	1.275
5-XXS	4,178	3,134	14,021	1.436
6-40	766	575	3,260	0.6445
6-80	1,608	1,206	6,014	0.7144
6-160	3,275	2,456	11,212	0.8812
6-XXS	4,159	3,120	13,838	0.9887

(DISCONTINUED FORMER STANDARD)

Table 3.4 - ANSI B 31.3 Standard for Hydraulic Pipes Pressure Rating

3.4.5- Hydraulic Pipe Assembly

3.4.5.1- Hydraulic Pipes Threads

Pipe Connection: Pipes used in hydraulic systems are not intended to bend and should be seamless, As shown in Fig. 3.7, pipes are traditionally assembled using tapered threads. A pipe thread connection includes external threads cut on the pipe ends and internal threads cut into the openings of the pipe fittings. This thread is tapered 3/4" per foot to assure a positive seal when it is properly tightened.

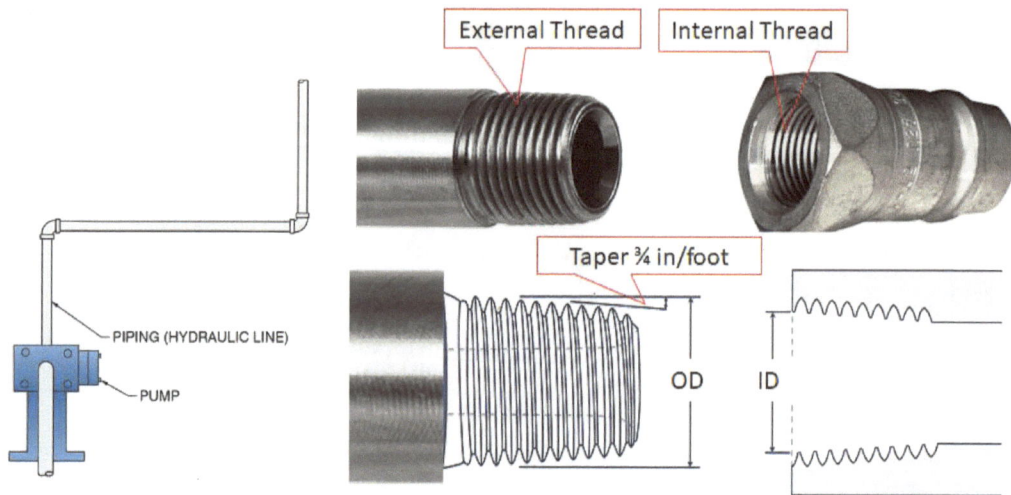

Fig. 3.7 - Pipe Connection by National Thread

Spiral vs. Dryseal Thread: Figure 3.8 shows the standard (*Spiral*) National Pipe Tapered (*NPT*) thread versus the *Dryseal* National Pipe Tapered (*NPTF*) thread. The NPT threads tend to leak under high system pressure because of a continuous internal *spiral clearance* that exists in the threads even when it is tightened. Therefore, NPTF threads are recommended over NPT threads for high pressure hydraulic lines and fuel lines. The NPT/NPTF designations are explained in the tube section.

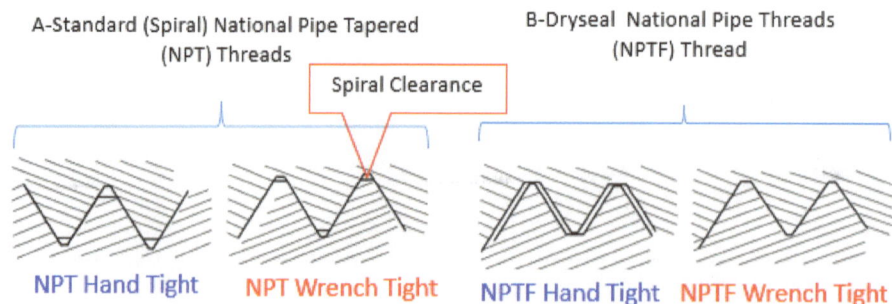

Fig. 3.8 - Standard Spiral Thread versus Dryseal Thread

3.4.5.2- Fittings for Hydraulic Pipes

As shown in Fig. 3.9, a wide variety of threaded connections are available to connect a pipe to hydraulic system. However, Fittings aren't recommended for assembling pipes. Alternatively, Butt & Socket Weld fittings and 4 Bolts 61/62 SAE Flanges are recommended for pipe connections.

Hex Head Stainless Steel Pipe Plug

Hex Stainless Steel Pipe Reducer

Hex Stainless Steel Pipe bushing

Stainless Steel Pipe T-Fitting

Stainless Steel Pipe Cross Fitting

Stainless Steel Pipe Adaptor

Stainless Steel Pipe 90° Elbow Fitting

Stainless Steel Pipe 45° Elbow Fitting

Stainless Steel Pipe Coupling

Fig. 3.9- Hydraulic Pipe Fittings

3.5- Hydraulic Tubes

3.5.1- Features of Hydraulic Tubes

Hydraulic *Tubes* are semirigid fluid conductor that have relatively thinner walls and less weight per unit length as compared to hydraulic pipes. Hydraulic tubes can be bent with the minimum bend radius respected. Tubes are easier to assemble than pipes since no welding is needed. Tubes are hooked to the components by various types of fittings and flanges. Tubes have lower pressure ratings than pipes. As shown in Fig. 3.10, many industrial and mobile applications rely on hydraulic tubes in transmitting hydraulic power.

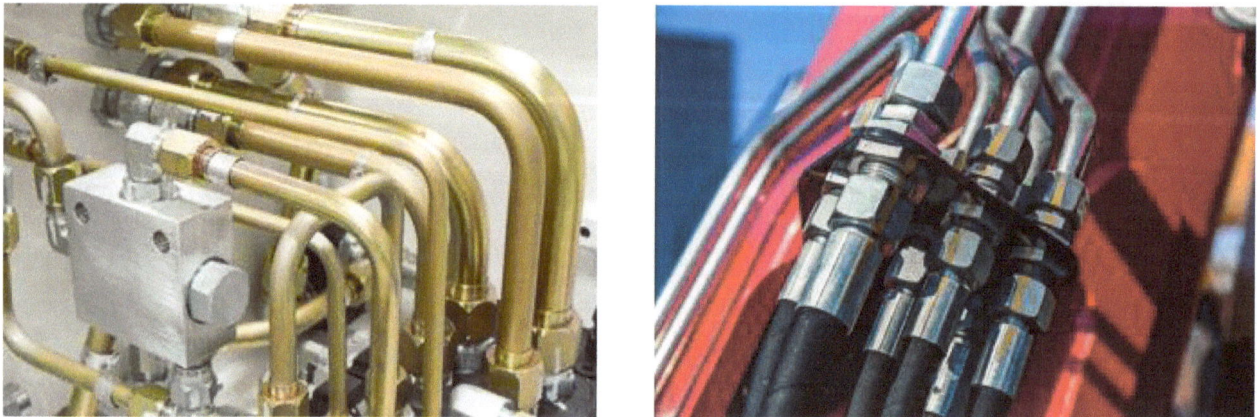

Fig. 3.10 - Use of Hydraulic Tubes in Industrial (Left) and Mobile (right) Applications

3.5.2- Material of Hydraulic Tubes

As shown in Fig. 3.11, hydraulic tubes are produced from various material as follows:
- Stainless Steel: Ultimate Strength = 60/75/100 thousands of psi.
- Low Carbon Steel Ultimate Strength = 55,000 psi.
- Cooper: Ultimate Strength = 32,000 psi.

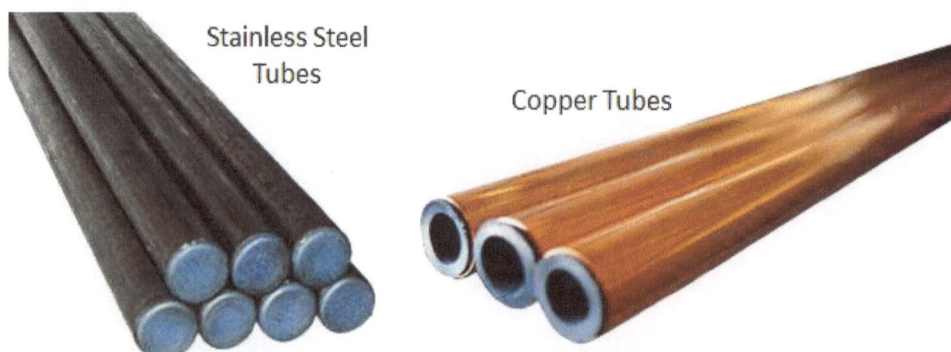

Fig. 3.11 - Material of Hydraulic Tubes

3.5.3- Sizes of Hydraulic Tubes

Like hydraulic pipes, nominal size of hydraulic tubes is the outer diameter "OD". Tube sizes are ranging between (0.125 – 2.25) inches (4-42 mm).

3.5.4- Pressure Rating of Hydraulic Tubes

As shown in Table 3.5, three values for the maximum allowable working pressure are presented based on three formulas to estimate Burt Pressure: *Barlow's Formula "1"*, *Boardman's Formula "2"*, and *Lame's Formula "3"*. In this textbook, like hydraulic pipes, Barlow's formula will only be considered in this textbook because it results in the lowest burst pressure. The values shown in the table are for *Carbon Steel* and based on design *safety factor* of approximately 4:1. Maximum allowable pressure is calculated based on an allowable fiber stress of 12,500 psi which is equivalent to 50% of the minimum yield point and approximately 25% of the *Ultimate Strength* (55000 psi) of the tube material.

Nominal Tube OD, in.		See Note*	Nominal Tube Wall Thickness, in.									
			0.028	0.035	0.049	0.065	0.083	0.095	0.109	0.120	0.134	0.148
1/8	0.125	1	5,600	7,000								
		2	6,800	9,000								
		3	6,650	8,450								
3/16	0.188	1	3,750	4,650								
		2	4,250	5,500								
		3	4,250	5,450								
1/4	0.250	1	2,800	3,500	4,900	6,500						
		2	3,100	3,950	5,800	8,200						
		3	3,100	3,950	5,750	7,800						
5/16	0.312	1	2,250	2,800	3,900	5,200						
		2	2,400	3,100	4,500	6,250						
		3	2,450	3,100	4,500	6,150						
3/8	0.375	1	1,850	2,350	3,250	4,350	5,550	6,350				
		2	2,000	2,500	3,650	5,050	6,700	7,950				
		3	2,000	2,550	3,650	5,000	6,550	7,600				
1/2	0.500	1		1,750	2,450	3,250	4,150	4,750	5,450	6,000		
		2		1,850	2,650	3,650	4,800	5,600	6,600	7,450		
		3		1,850	2,700	3,650	4,800	5,550	6,450	7,200		
5/8	0.625	1		1,400	1,950	2,600	3,300	3,800	4,350	4,800		
		2		1,450	2,100	2,850	3,700	4,350	5,050	5,650		
		3		1,500	2,100	2,850	3,750	4,350	5,050	5,600		
3/4	0.750	1		1,150	1,650	2,150	2,750	3,150	3,650	4,000		
		2		1,200	1,700	2,350	3,050	3,500	4,100	4,600		
		3		1,200	1,750	2,350	3,050	3,550	4,150	4,600		
7/8	0.875	1		1,000	1,400	1,850	2,350	2,700	3,100	3,400		
		2		1,050	1,450	1,950	2,550	2,950	3,450	3,850		
		3		1,050	1,500	2,000	2,600	3,000	3,500	3,900		
1	1.000	1		875	1,200	1,600	2,050	2,350	2,700	3,000	3,350	3,700
		2		900	1,250	1,700	2,200	2,550	3,000	3,300	3,750	4,200
		3		900	1,300	1,750	2,250	2,600	3,000	3,350	3,800	4,200

Note: Wall thicknesses having values shown to the right of the bold line are not normally considered suitable for 37° single flaring to J533.

Table 3.5 – Maximum Allowable Working Pressure (psi) for Carbon Steel Tubes

3.5.5- Hydraulic Tubes Assembly

Unlike hydraulic pipes, tubes are bent to minimize fittings and connected by fitting, or flanges.

3.5.5.1- Tube Bending

During tube bending, wall thickness of the outer surface is reduced and accordingly pressure rating too. Therefore, tube bending must respect *minimum bend radius* reported by tube manufacturer. Figure 3.12 shows that minimum bend radius is based on the inner surface or based on the center line of the tube. The figure also shows manual mandrills of different sizes for tube bending. Table 3.6 shows minimum bend radius based on the size of the tube.

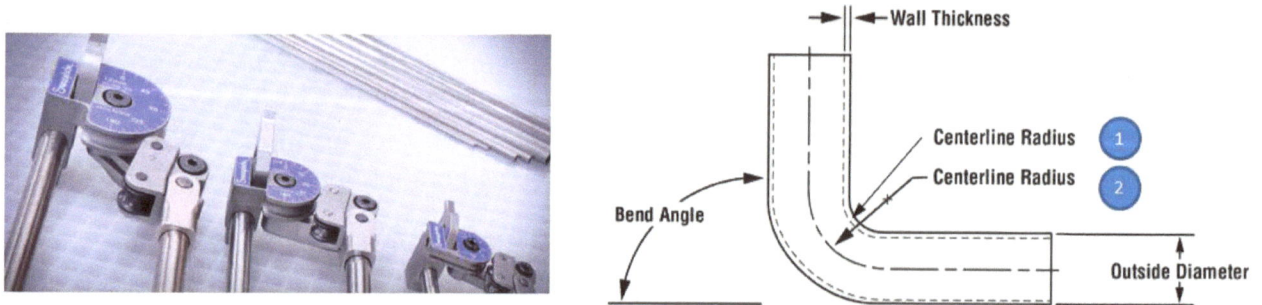

Fig. 3.12 - Manual Mandrills of Different Sizes for Tube Bending

TUBE OD (INCHES)	CLOSE① BEND (INCHES)	NORMAL BEND (INCHES)	MS33611 STANDARD	METRIC SIZES		
				TUBE OD (MM)	CLOSE① BEND (MM)	NORMAL BEND (MM)
0.125	—	0.281③	—	4	—	9
0.188	—	0.422③	—	6	—	13
0.250	0.562	0.750	—	8	—	18
0.312	0.687	1.000	—	10	32	32
0.375	0.937	1.250	1.125	12	—	32
0.500	1.250	2.000	1.500	15	38	38
0.625	1.500	2.500	1.875	16	38	38
0.750	1.750	3.000	2.250	18	44	44
0.875	2.000	3.500	2.625	20	44	44
1.000	3.000	4.000	3.000	22	—	89
1.125	3.500	4.500	3.375	25	—	100
1.250	3.750	5.000	3.750	28	—	112
1.500	5.000	6.000	4.500	30	—	128
1.750	—	7.000	—	35	—	140
2.000	—	8.000	—	38	—	152
2.250	—	9.000③	—	42	—	168

Table 3.6 - Minimum Bend Radius for Hydraulic Tubes of Different Sizes

3.5.5.2- Tube Flaring

The standard **SAE J533B** covers specifications for 37° and 45° for *tube flaring.* 37° single and double flares shall conform to the dimensions specified in Fig. 3.13 and Table 3.7. 45° Single and double flares shall conform to the dimensions specified in Fig. 3.14 and Table 3.8.

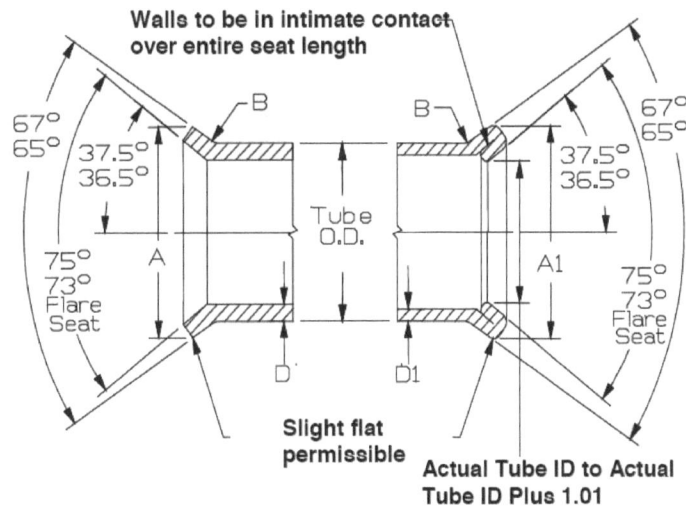

Fig. 3.13 - 37º Single and Double Flaring According to SAE J533B

Nominal Tube OD	A Single Flare Diameter		A₁ Double Flare Diameter		B Radius	Dᵇ Single Flare Wall Thickness	D₁ᵇ Double Flare Wall Thickness
	in		in		in	in	in
in	Max	Min	Max	Min	±0.02	Max	Max
1/8	0.200	0.180	0.200	0.180	0.03	0.035	0.025
3/16	0.280	0.260	0.280	0.260	0.03	0.035	0.028
1/4	0.360	0.340	0.360	0.340	0.03	0.065	0.035
5/16	0.430	0.400	0.430	0.400	0.03	0.065	0.035
3/8	0.490	0.460	0.490	0.460	0.04	0.065	0.049
1/2	0.660	0.630	0.660	0.630	0.06	0.04	0.049
5/8	0.790	0.760	0.790	0.760	0.06	0.04	0.049
3/4	0.950	0.920	0.950	0.920	0.08	0.04	0.049
7/8	0.748	1.040	1.070	1.040	0.08	0.109	0.065
1	0.916	1.170	1.200	1.170	0.09	0.120	0.065
1-1/8	1.041	1.350	1.380	1.350	0.09	0.120	0.065
1-1/4	1.157	1.480	1.510	1.480	0.09	0.120	0.065
1-1/2	1.730	1.700	1.730	1.700	0.11	0.120	0.065
1-3/4	2.110	2.080	2.110	2.080	0.11	0.120	0.065
2	2.360	2.330	2.360	2.330	0.11	0.134	0.065

Table 3.7 - 37º Single and Double Flaring Standard Dimensions According to SAE J533B

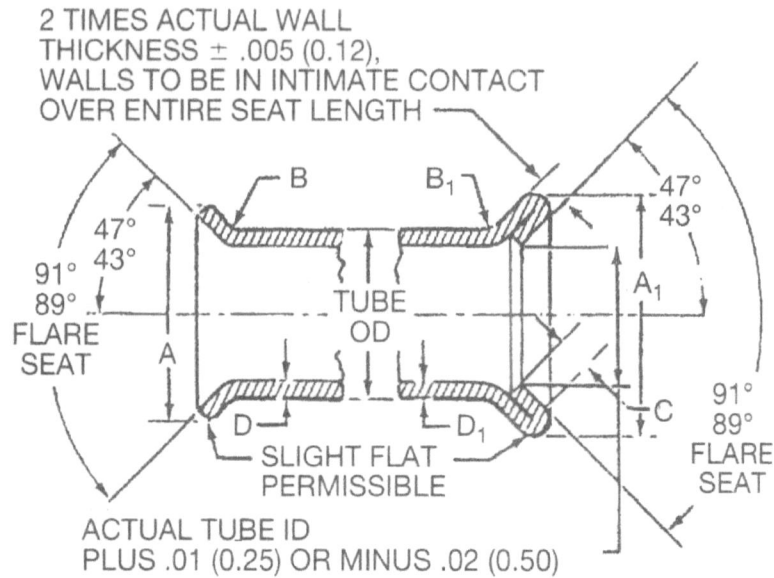

Fig. 3.14 - 45⁰ Single and Double Flaring According to SAE J533B

Nominal Tube OD	A Single Flare Diameter		A$_1$ Double Flare Diameter		B Single Flare Radius	B$_1$ Double Flare Radius	C Double Flare Coined Seat Length	Db Single Flare Wall Thickness	D$_1$[b] Double Flare Wall Thickness
in	in		in		in	in	in	in	in
	Max	Min	Max	Min	±0.01	±0.01	Min	Max	Max
1/8	0.181	0.171	0.213	0.198	0.02	0.04	0.040	0.035	0.025
3/16	0.249	0.239	0.280	0.265	0.02	0.04	0.040	0.035	0.028
1/4	0.325	0.315	0.360	0.345	0.02	0.04	0.040	0.049	0.035
5/16	0.404	0.388	0.425	0.410	0.02	0.04	0.062	0.049	0.035
3/8	0.487	0.471	0.500	0.485	0.02	0.04	0.062	0.065	0.049
7/16	0.561	0.545	0.570	0.555	0.02	0.04	0.062	0.065	0.049
1/2	0.623	0.607	0.640	0.625	0.02	0.04	0.062	0.083	0.049
9/16	0.676	0.660	0.712	0.697	0.02	0.04	0.062	0.083	0.049
5/8	0.748	0.732	0.772	0.757	0.02	0.04	0.062	0.095	0.049
3/4	0.916	0.900	0.912	0.897	0.02	0.04	0.062	0.109	0.049
7/8	1.041	1.025	—	—	0.02	—	—	0.109	—
1	1.157	1.141	—	—	0.02	—	—	0.120	—

Table 3.8 - 45⁰ Single and Double Flaring Standard Dimensions According to SAE J533B

3.5.5.3- Tube Fittings

What is a Tube Fitting? A fitting is a connection between a transmission line and a hydraulic component.

How to Identify a Fitting? Selecting proper fitting is a challenging task. Fittings are identified by Standards, Body Configuration, Seal Configuration, Thread Characteristics, Size, Material, and Interchangeability.

Fitting Body General Configurations: Fittings are provided in various configurations such as male, female, male to male, female to female, and male to female. Figure 3.15 shows that fittings could be fixed or rotating. Rotating "*Swivel*" fittings are used to connect stationary and rotating machine assemblies.

Fixed Joins:
1. Straight Male Stud Fitting
2. Elbow Banjo Fitting
3. Equal Elbow Fitting
4. Equal T Fitting
5. Straight Male Fitting
6. Equal Cross Fitting

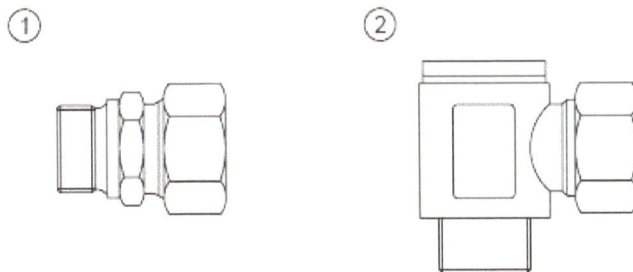

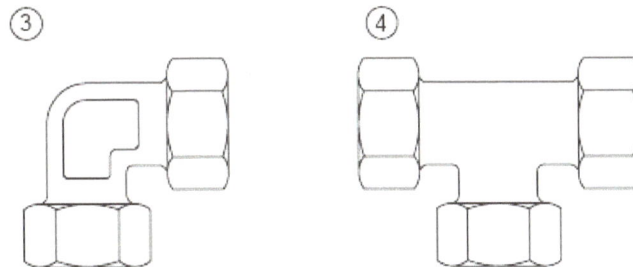

Rotating "Swivel" Joints:
1. Seal
2. Ball Race

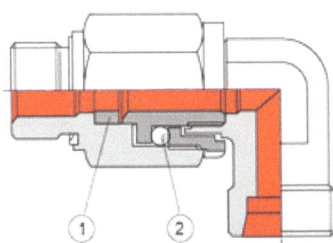

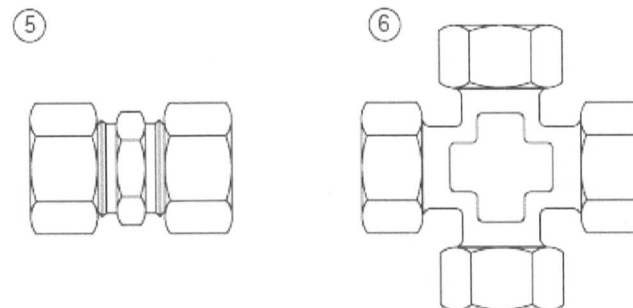

Fig. 3.15 - Fitting Body Configurations (Courtesy of Bosch Rexroth)

Fitting Seal General Configuration: As shown in Fig. 3.16, sealing between the fitting and the tube is provided by several methods as follows:

- **O-Ring Sealing:** This type of seal is excellent for high-pressure applications. The O-ring is being compressed makes the seal. In both *O-Ring Face Seal* and O-*Ring Cone Seal*, the nipple is welded to the tube.
- **Mated Surfaces Sealing using Flareless Fitting:** In *Flareless* fittings, the seal takes place using a *Compression Ring*, also referred to as a *Cutting Ring*.
- **Mated Surfaces Sealing using Flared Fitting:** In *Flared* fittings, the seal takes place between flared tube and fitting with coned seat. Different seat angles are available.
- **Thread Interference Sealing:** This is the easiest type of fitting to use. A characteristic of this thread is that the male is thinner at the front than it is at the back. As the male is threaded into the female, the edges of the thread distort by flattening out. This distortion creates the seal.

Fitting Standards: There are several standards for hydraulic fittings such as: International Standard "ISO", North American "JIC/SAE/NPT", British "B", Metric French, "GAZ" German "DIN", Japanese "JIS".

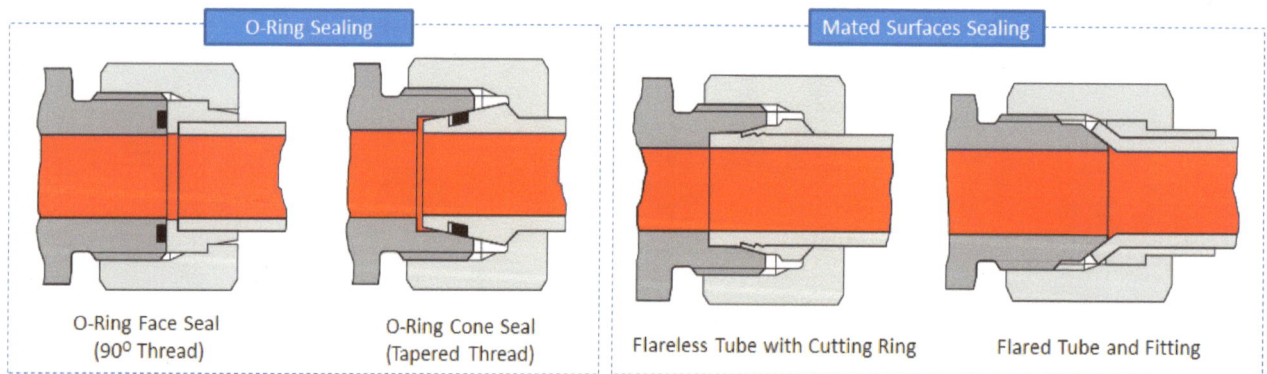

Fig. 3.16 - Fitting Seal Configurations (Courtesy of Bosch)

3.5.5.4- Threaded Sealing Fittings

North American Standards Thread Types:

The *North American Standard* for threads is widely used for over 100 years. They are used to effectively seal pipes for fluid and gas transfer. They are available in iron or brass for low-pressure applications and carbon steel and stainless steel for high-pressure

The following designations are used to identify thread type.

N = National, **P** = Pipe, **S** = Straight, **T** = Tapered, **F** = Fuel, and **M** = Swivel Mechanical Joints.

Figure 3.17 shows the common thread sealing fittings:

- **National Pipe Tapered (NPT):** *National Pipe Tapered* thread is a *Spiral tapered* thread used commonly for low pressure application because of the spiral clearance. As shown in Fig. 3.17, the spiral clearance increases with the large size fittings and with the frequent use of the fitting.

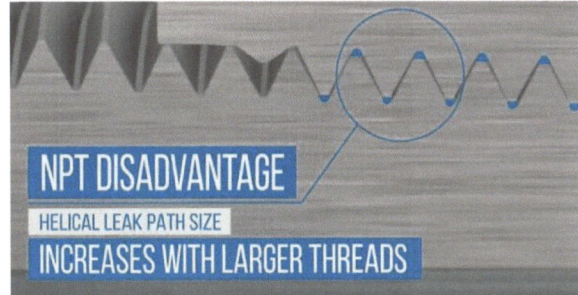

Fig. 3.17 – Spiral Clearance on NPT Fittings (Courtesy of Brennan Industries)

- **National Pipe Tapered for Fuels (NPTF):** As shown in Fig. 3.18, *National Pipe Tapered Fuel* thread is a *Dryseal* thread for Fuel and made to specifications ASME 1.20.3. It is used for both male and female ends. The NPTF male will mate with the NPTF, NPSF, or NPSM females. The NPTF male has tapered threads and a 30° inverted seat. The NPTF female has tapered threads and no seat. <u>The seal takes place by deformation of the threads.</u>

- **National Pipe Straight for Fuels (NPSF):** *National Pipe Straight* thread for Fuels is used for female ends and properly mates with the NPTF male end.

- **National Pipe Straight for Swivel Mechanical Joints (NPSM):** *National Pipe Straight* thread for Mechanical Joint has a straight thread and a 30° internal chamfer. The female half of this connection has a straight thread and an inverted 30° seat. <u>The seal takes place on the 30° seat.</u>

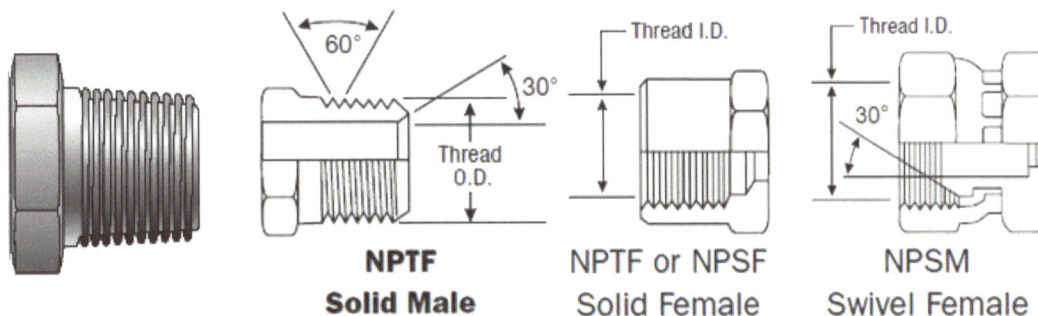

Fig. 3.18 - National Pipe Thread (Courtesy of Gates)

Figure 3.19 shows crossectional views of NPTF and NPSM fittings.

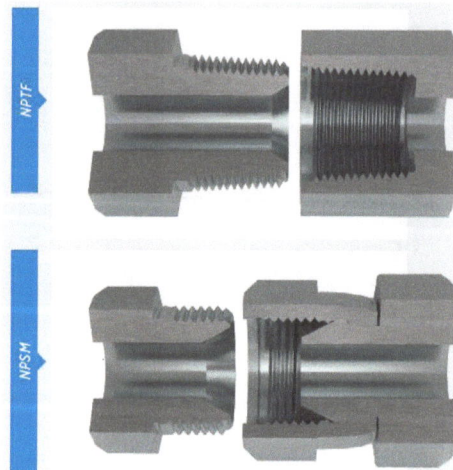

Fig. 3.19 – Crossectional Views of NPTF and NPSM Fittings (Courtesy of Brennan Industries)

North American Standards Thread Types (SAE):

- As shown in Fig. 3.20, the SAE 45° inverted flare male will mate with an SAE 42° inverted flare female only. The male has straight threads and a 45° inverted flare. The female has straight threads and a 42° inverted flare. <u>The seal is made between the 45° flare seat on the male and the 42° flare seat on the female.</u>

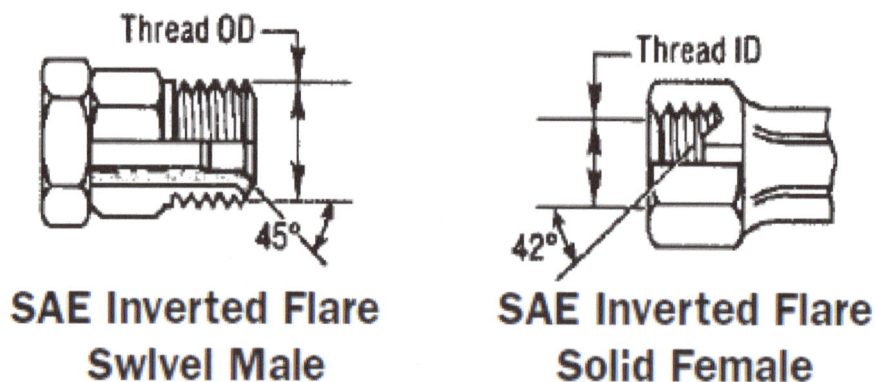

SAE Inverted Flare Swivel Male

SAE Inverted Flare Solid Female

Fig. 3.20 - SAE Inverted Flare Thread (Courtesy of Gates)

British Standards Thread Types (BSP):

- **British Standard Pipe Parallel (BSPP) (Fig 3.21):** Popular couplings have *British Standard Pipe* (BSP) threads, also known as *Whitworth* threads. The BSPP male has straight threads and a 30° seat. The BSPP (parallel) male will mate with a BSPOR (parallel) female or a female port. The BSPP (parallel) connector is similar to, but not interchangeable with the NPSM connector. The thread pitch is different in most sizes, and the thread angle is 55° instead of the 60° angle on NPSM threads.

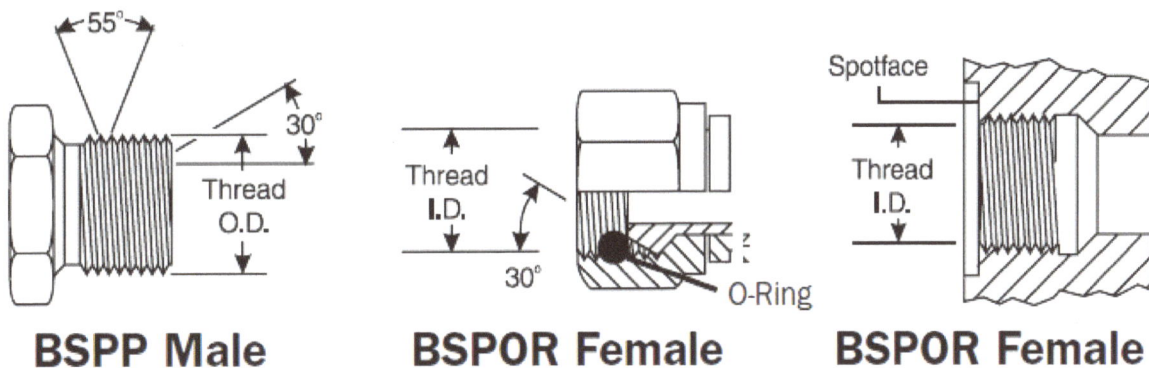

Fig. 3.21 - British Standard Pipe Parallel Thread (Courtesy of Gates)

- **British Standard Pipe Tapered (BSPT) (Fig 3.22):** It is a popular coupling. The BSPT (tapered) male will mate with a BSPT (tapered) female, or a BSPOR (parallel) female. The BSPT (parallel) connector is similar to, but not interchangeable with the NPTF connector. The thread pitch is different in most sizes, and the thread angle is 55° instead of the 60° angle on NPSM threads.

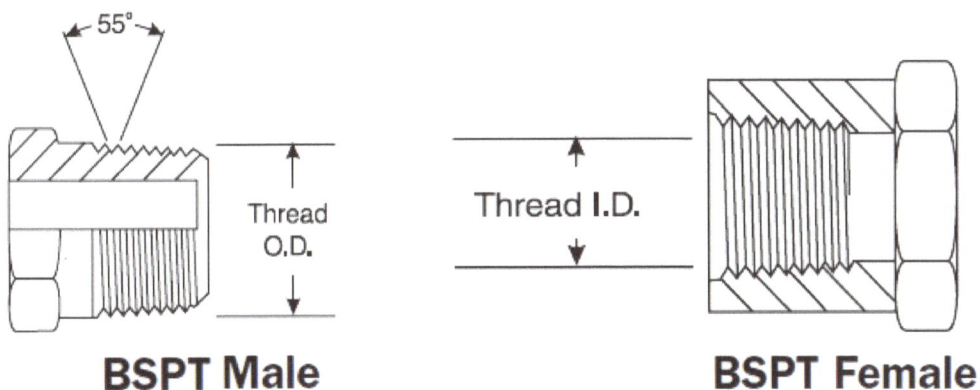

Fig. 3.22 - British Standard Pipe Tapered Thread (Courtesy of Gates)

Figure 3.23 shows sectional views for BSPT and PSPP fittings.

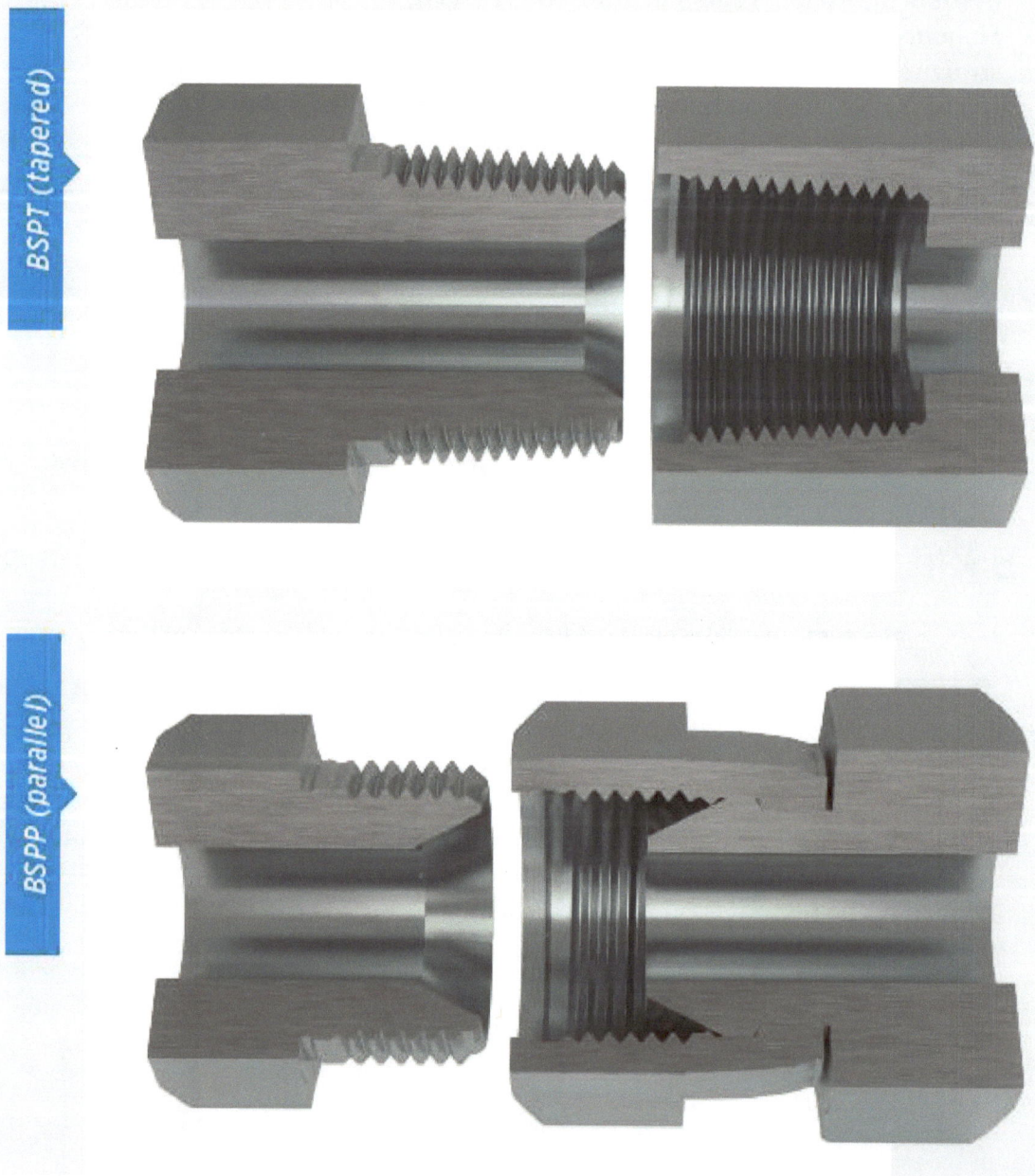

Fig. 3.23 - Sectional Views for BSPT and PSPP Fittings (Courtesy of Brennan Industries)

Metric Standards for Thread Type Couplings:

Popular couplings have metric threads. Both French "GAZ" and German "DIN" are considered metric. However, they are not the same because the threads are different in some sizes. The French GAZ couplings use fine threads in all sizes, and German DIN couplings use coarse threads in larger sizes.

Metric French Standard Thread Type (GAZ):

GAZ 24° Metric French (GAZ) (Fig. 3.24): The GAZ 24° cone male will mate with the female 24° cone or the female tube fitting. The male has a 24° seat and straight metric threads. The female has a 24° seat or a tubing sleeve. When measuring the flare angle with the seat angle gauge, use the 12° gauge. The seat angle gauge measures the angle from the connector centerline.

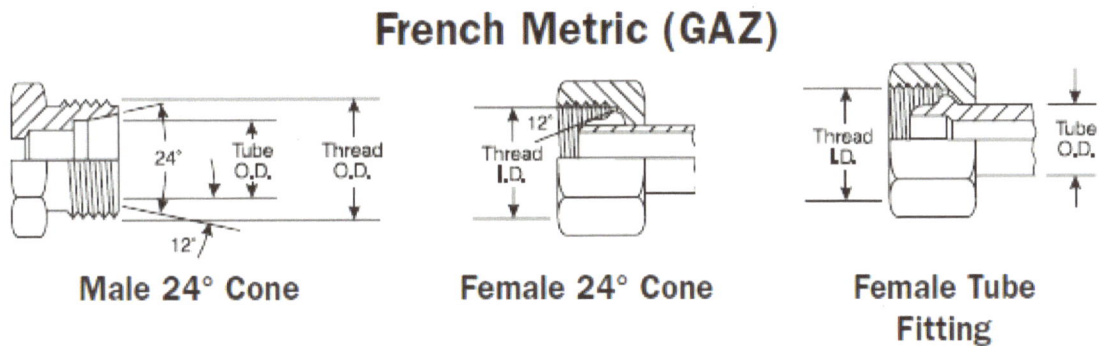

Fig. 3.24 - GAZ 24° Metric French Thread (Courtesy of Gates)

Metric German Standard Thread Type (DIN):

DIN 2353 24° Cone Metric German (Fig. 3.25): The DIN 24° cone male will mate with any of the females shown. The male has a 24° seat, straight metric threads, and a recessed counterbore which matches the tube O.D. of the coupling used with it. The mating female is a 24° cone with O-ring, a metric tube fitting or a universal 24° cone.

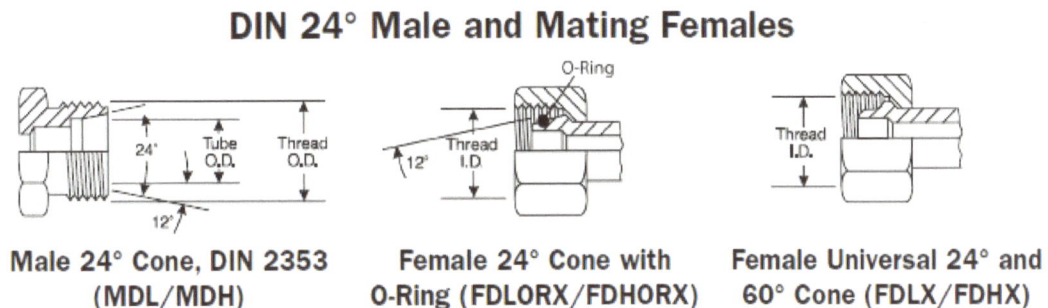

Fig. 3.25 - DIN 24° Cone Metric German Thread (Courtesy of Gates)

- **Example of DIN 2353 - Innovative Metric Fitting (EO-2) (Fig. 3.26):**
  o <u>Description:</u> The *EO-2* is a high-pressure tube fitting generation with elastomeric seals on all joints. EO-2 is a true metric straight thread design according to 24° bite-type standards such as: ISO 8434-1, DIN 2353, or DIN 3861. It consists of a body, a functional nut, and an elastomeric seal.

  o <u>Operation:</u> The elastomeric seal assures a hermetically sealed tube joint. It is located in between the inner cone of the fitting body and the tube surface, thus blocking the only possible leak path. Due to its large cross-section, the seal effectively compensates for all manufacturing tolerances on tube and fitting cone. The sealing effect is pressure supported which makes the EO-2 fitting suitable for high pressure applications. The static compression also eliminates air-ingress into the fluid system below atmospheric pressure conditions.

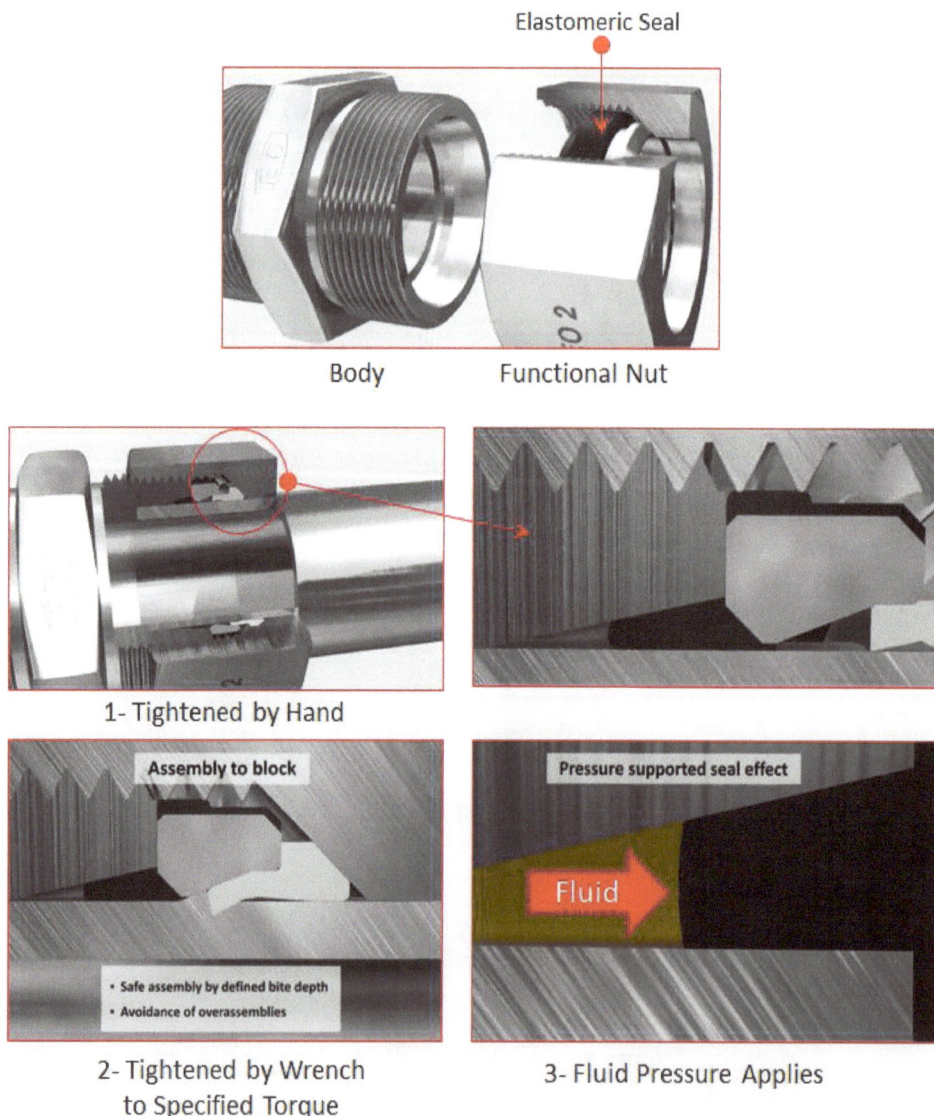

Fig. 3.26 - EO-2 Fitting (Courtesy of Parker)

- **DIN 6711 60° Cone Metric German (Fig. 3.27):** The DIN 60° cone male will mate with the female universal 24° or 60° cone connector only. The male has a 60° seat and straight metric threads. The female has a 60° universal seat and straight metric threads.

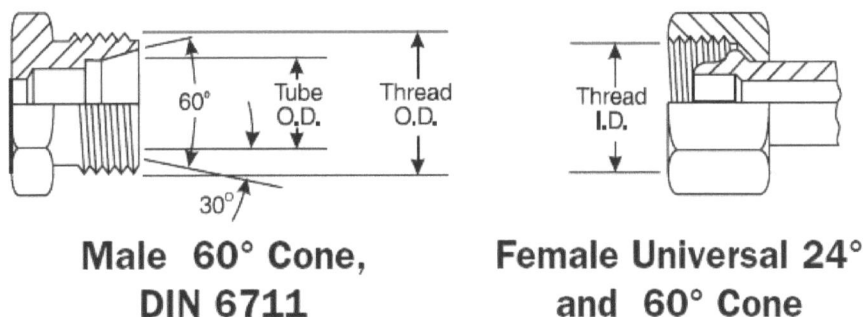

Fig. 3.27 - DIN 60° Cone Metric German Thread (Courtesy of Gates)

- **DIN 3852 Couplings Type A & B (Fig. 3.28):** The male DIN 3852 Type A & B couplings will mate with the female DIN coupling shown below. The male and female type A & B couplings have straight threads. The seal occurs when the ring seal (Type A) or the face seal (Type B) mates with the face of the female port.

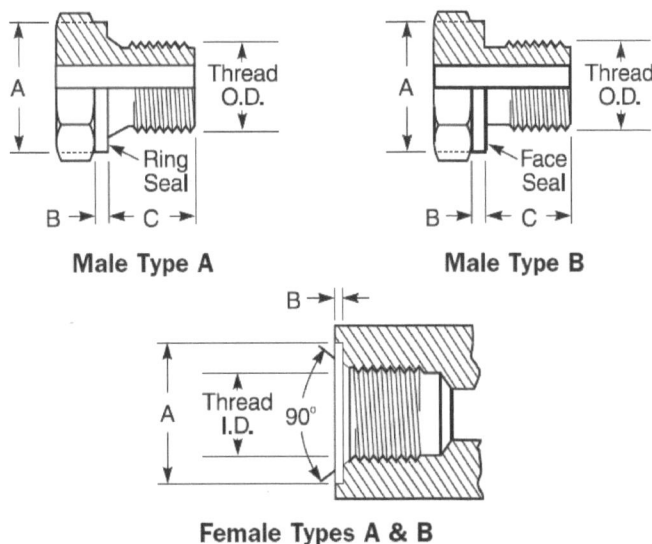

Fig. 3.28 - DIN 3852 Thread (Courtesy of Gates)

Japanese Standards Thread Types (JIS):

There are two popular types of coupling styles in Japan, Japanese Industrial Standard "JIS" and Komatsu. These couplings look like Male JIC and Female JIC Swivel couplings. However, there are two major differences: The threads are BSP and the seat angle is only 30° instead of 37° for JIC.

- **Japanese JIS-30° Flare Parallel Threads (Fig. 3.29):** The Japanese 30° flare male connector will mate with a Japanese 30° flare female only. The male and female have straight threads and a 30° seat. The seal is made on the 30° seat. The threads on the Japanese 30° flare connector conform to JIS B 0202, which are the same as the BSPOR threads. Both the British and Japanese connectors have a 30° seat, but they are not interchangeable because the British seat is inverted.

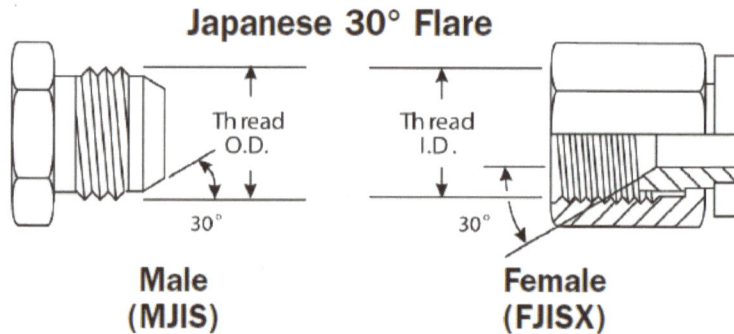

Fig. 3.29 - Japanese 30° Flare Thread (Courtesy of Gates)

- **Japanese Tapered Pipe Thread (Fig. 3.30):** The Japanese tapered pipe thread connector is identical to and fully interchangeable with the BSPT (tapered) connector. The Japanese connector does not have a 30° flare and will not mate with the BSPOR female. The threads conform to JIS B 0203, which are the same as BSPT threads. The seal on the Japanese tapered pipe thread connector is made on the threads.

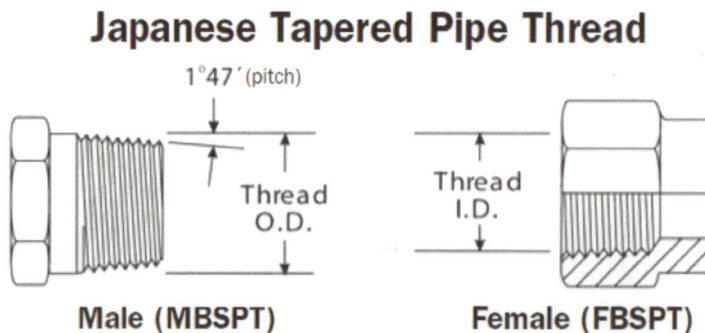

Fig. 3.30 - Japanese Tapered Pipe Thread (Courtesy of Gates)

3.5.5.5- O-Ring Sealing Fittings

North American Standards (SAE):

- **SAE-JI926 Straight Thread O-Ring Boss (ORB):** Figure 3.31 shows the *O-Ring Boss* male type. The male will only mate with an O-ring boss female, and the female is generally found on ports. The male has straight threads and an O-ring. The female has straight threads and a sealing face. The seal is made at the O-ring on the male and the sealing chamfer on the female. It can be configured as a straight or in elbow coupling to ease routing a line. Figure 3.32 shows typical O-Ring Boss fittings of different configurations.

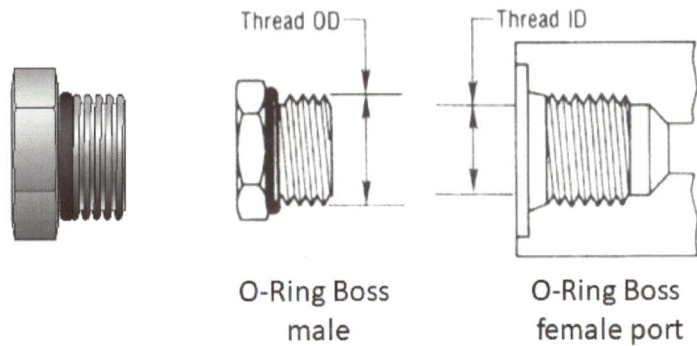

Fig. 3.31 - SAE JI926 Straight Thread O-Ring Boss (ORB)

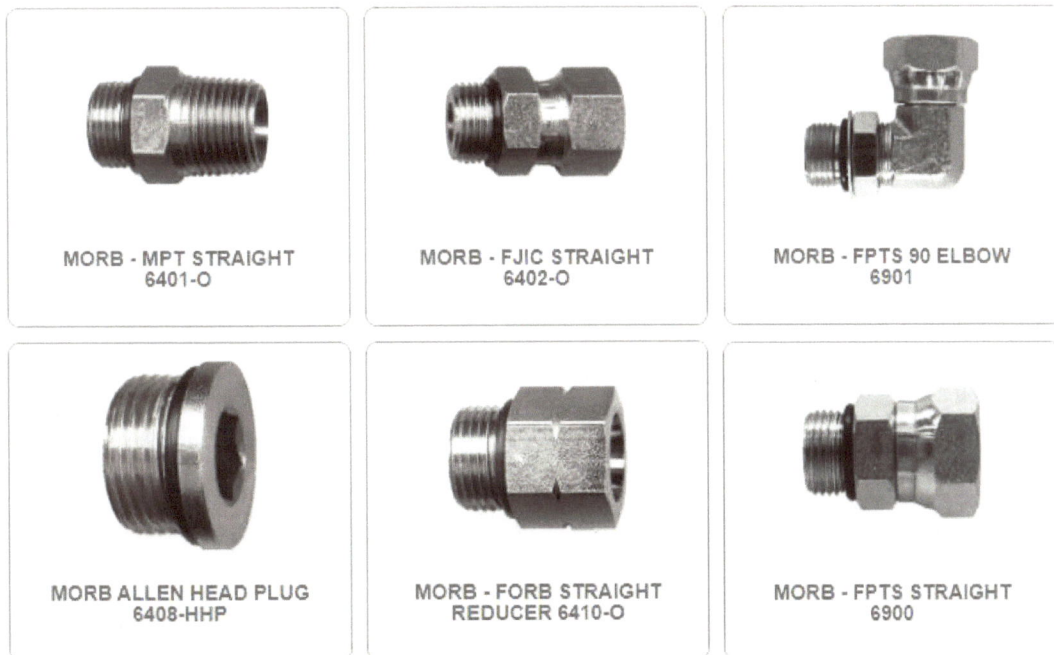

Fig. 3.32 - Typical O-Ring Boss Fittings of Different Configurations (www.redl.com)

ORB is recommended by the *National Fire Protection Association* (N.F.P.A.) for leak prevention in medium and high-pressure hydraulic systems. Figure 3.33 shows a sectional view of an ORB fitting.

Fig. 3.33 – Sectional View of an ORB Fitting (Curtesy of Brennan Industries)

- **O-Ring Face Seal (ORFS) – SAE-J1453 (Fig. 3.34):** *In the O-Ring Face Seal,* the O-Ring is placed in a captive groove on the face of the fitting. The solid male O-ring face seal will mate with a swivel female O-ring face seal SAE J1453 fitting only. An O-ring rests in the O-ring groove in the male coupling. The seal is made when the O-ring in the male contacts the flat face on the female coupling. The limitation on using such fittings comes from the maximum operating temperature under which the O-Ring performs properly. Figure 3.35 shows typical O-Ring Face Seal fittings of different configurations.

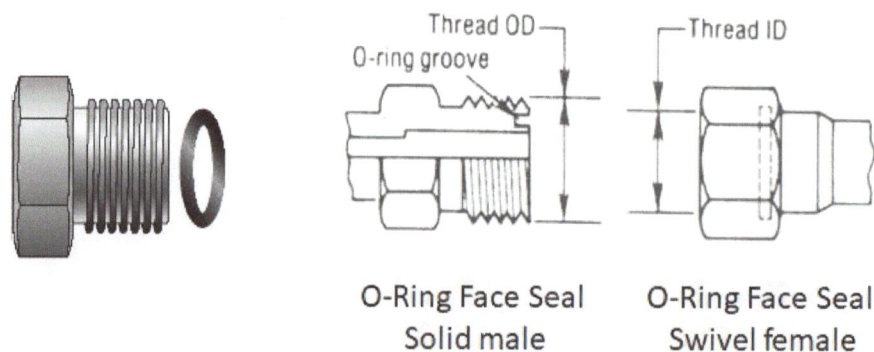

Fig. 3.34 - O-Ring Face Seal SAE J1453 (Courtesy of Gates)

**Fig. 3.35 - Typical O-Ring Face Seal Fittings of Different Configurations
(Courtesy of Parker)**

British Standards:
- Same as the SAE (ORB and ORFS) but the thread characteristic is different.

Metric Standard:
- ISO/DIS 8434-3 same as SAE (ORB and ORFS) but the thread characteristic is different.

Example of O-Ring Face Seal (O-Lok) (Fig. 3.36):

Description: *O-Lok* fittings are O-ring face seal (ORFS) fittings developed to eliminate leakage in high pressure hydraulic systems. The O-Lok system is designed to meet the needs of mobile equipment, mining, agriculture and other heavy equipment.

Features:
- Working pressure up to 630 bar.
- Available as standard in steel and stainless steel.
- Available in sizes from 1/4" to 2".
- Can be connected to both tubes and hoses.
- Can be used with tubes in different materials.
- Corrosion resistant.
- Easy to assemble.
- Best choice if you work with construction machinery.
- Meets international standards SAE J1453 and ISO8434-3.

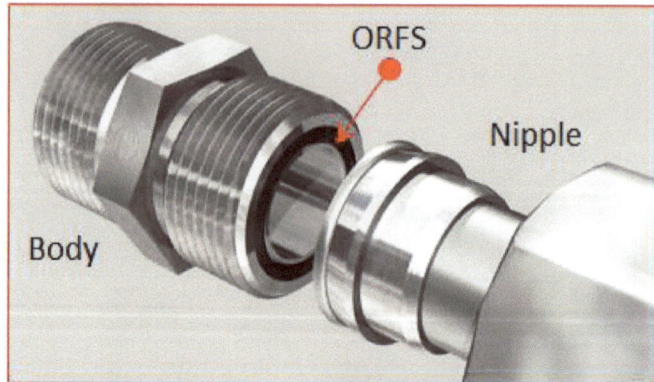

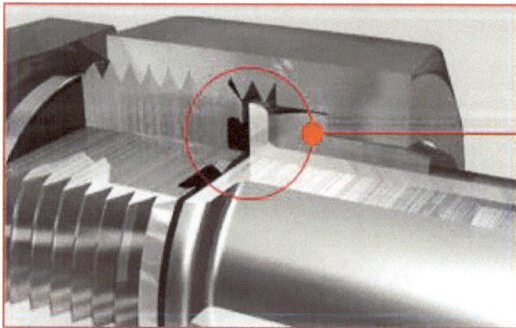

1- Tightened by Hand

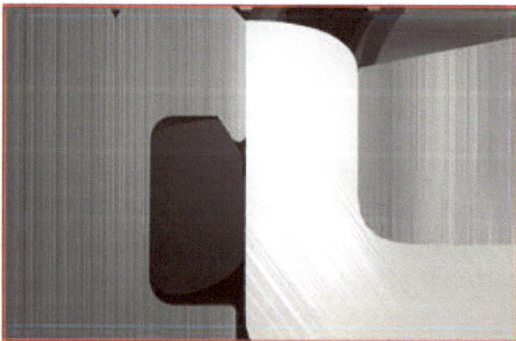

2- Tightened by Wrench
to Specified Torque

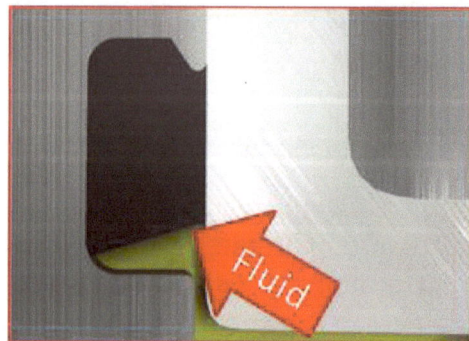

3- Fluid Pressure Applies

Fig. 3.36 - O-Lok Fitting (Courtesy of Parker)

3.5.5.6- Flareless Fittings with Cutting Rings

Flareless Tube Fittings (Fig. 3.37): Flareless pip assembly is also referred to as the *Standpipe* assembly. It could be North American (NASP) or Metric (MSP). This assembly consists of flareless solid male that mates only with a female flareless nut and cutting ring (compression sleeve). The male has straight threads and a 24° seat. The female has straight threads and a cutting ring for a sealing surface. <u>The seal is made between the cutting ring and the 24° seat on the male, and between the cutting ring and the tube on the female.</u> Figure 3.38 shows typical standpipe flareless assemblies.

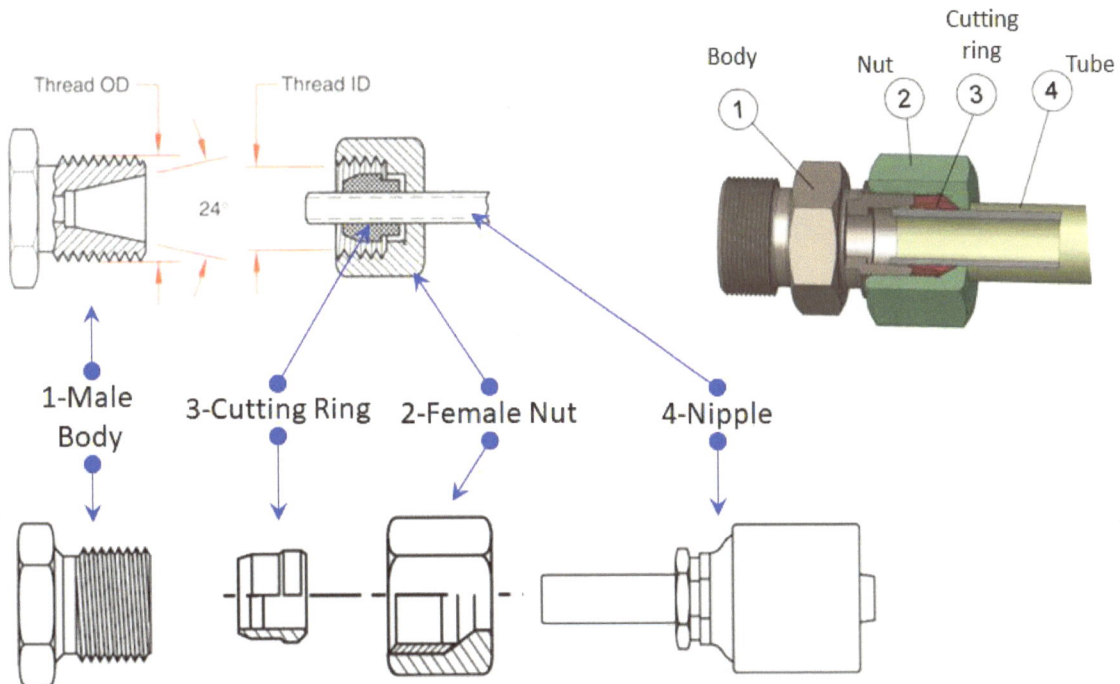

Fig. 3.37 - North American or Metric Flareless Assembly (Courtesy of Gates)

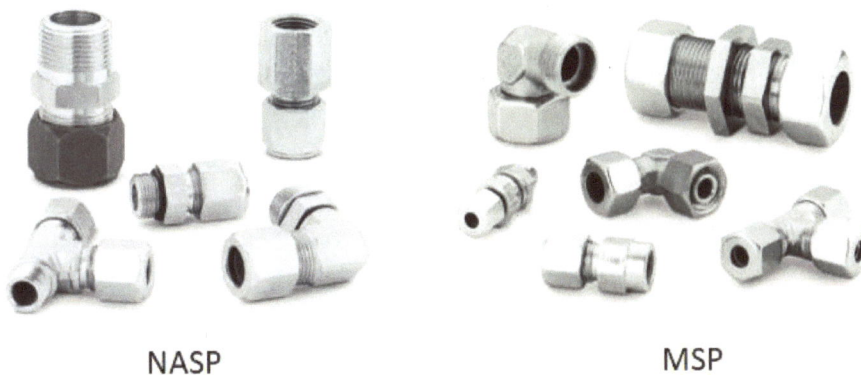

NASP

MSP

Fig. 3.38 - Typical North American or Metric Flareless Assembly (Courtesy of Parker)

How Cutting Rings Work (Fig. 3.39): As shown in the figure, after tightening to specified torque, the two cutting edges on the inner surface of the cutting ring carve into the outer surface of the tube, hence ensuring the necessary holding power and sealing for high operating pressures. Simultaneously, the outer surface of the cutting ring mates with the inner surface of the body.

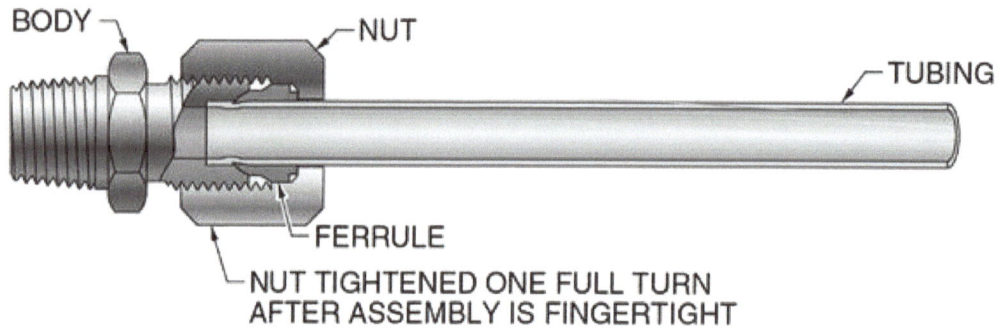

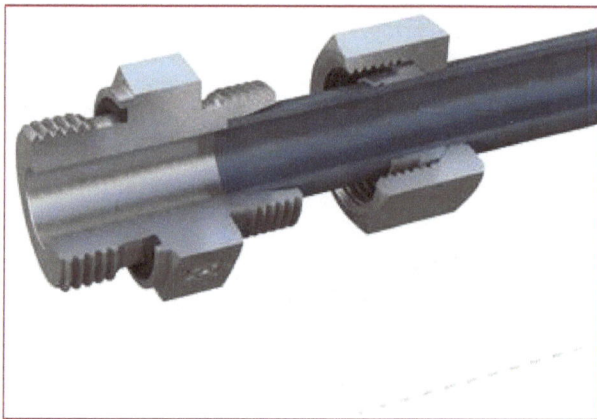

BODY — NUT — TUBING — FERRULE — NUT TIGHTENED ONE FULL TURN AFTER ASSEMBLY IS FINGERTIGHT

1- Untightened

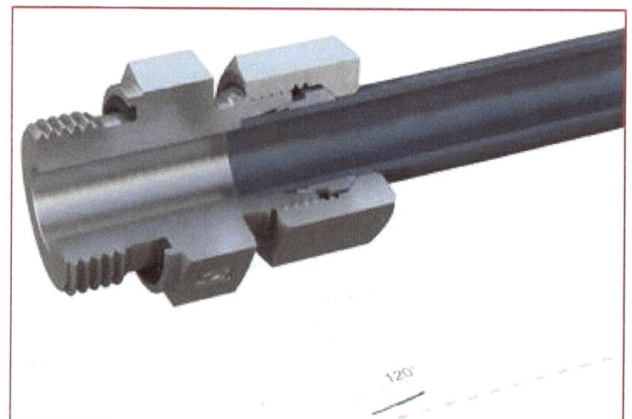

2- Tightened by Hand

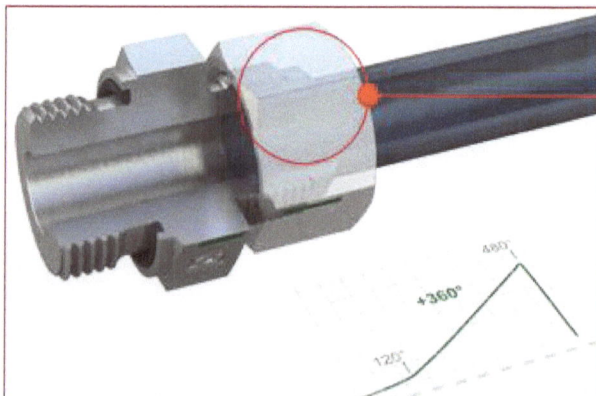

3- Tightened by Wrench to Specified Torque

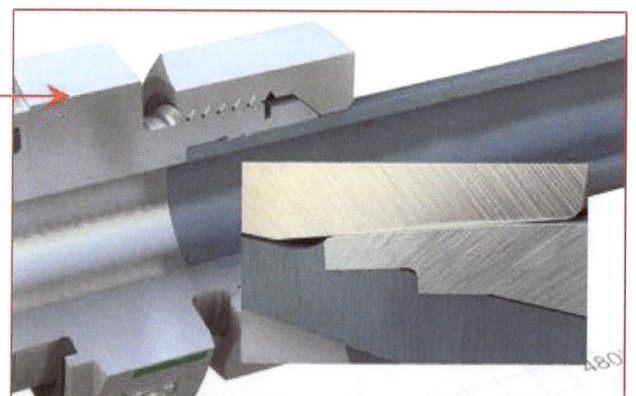

Fig. 3.39 - How Cutting Rings Work (www.stauffusa.com)

3.5.5.7- Flared Fittings for Flared Tubes

North American Standards:

- **SAE-J514 37⁰ Flared Fittings (Fig. 3.40):** The *Joint Industrial Council* (JIC) is now defunct, and this standard is included as a part of SAE J5I4. Both male and female have straight threads and a 37⁰ flare seat. Seat angle must be carefully measured to differentiate between the two 37⁰ and 45⁰. Figure 3.41 shows typical JIC 37⁰ fittings. They are common in most fluid power systems. They are available in various material such as Nickel Alloys, Brass, Carbon Steel, and Stainless Steel. The absence of O-Ring makes these fittings recommended for high-temperature applications.

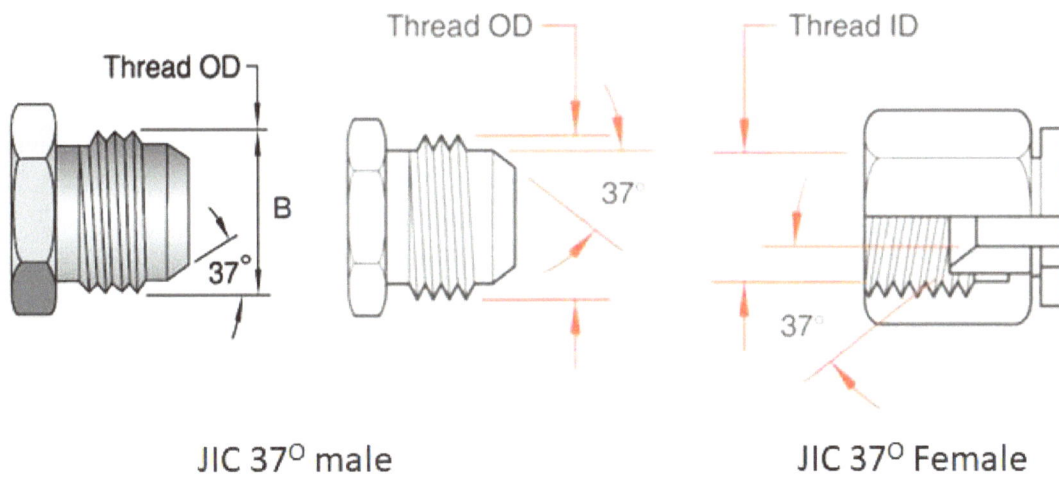

Fig. 3.40 – Characteristics of 37⁰ Flare Fitting (Courtesy of Gates)

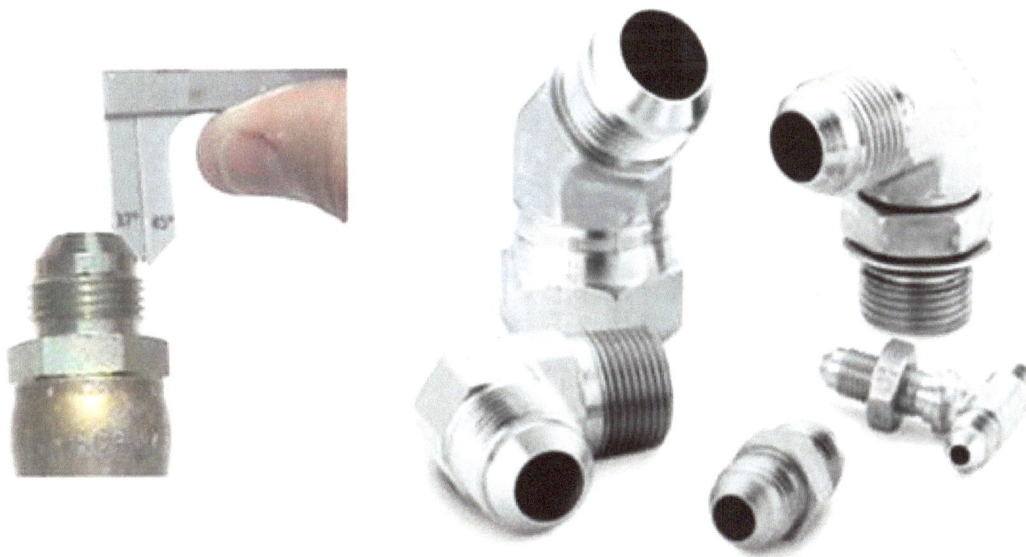

Fig. 3.41 – Typical 37⁰ Flared Fitting Configurations (Courtesy of Parker)

Figure 3.42 shows a sectional view of a sectional view of a 37⁰ flared fitting.

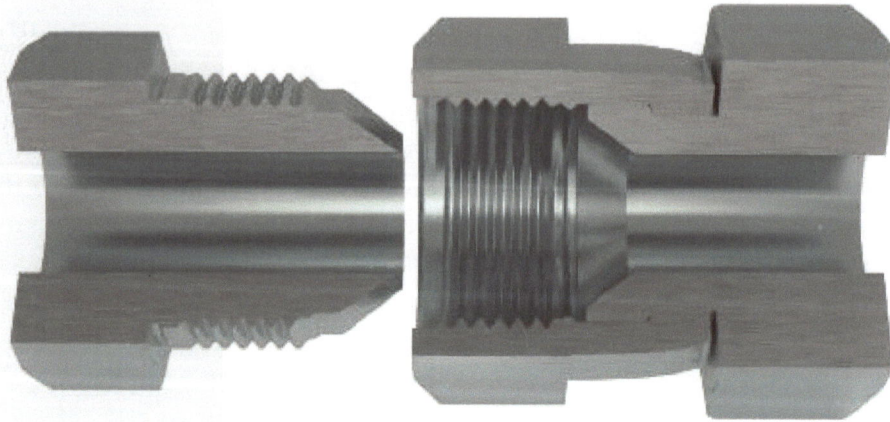

**Fig. 3.42 – Sectional View of a 37⁰ Flare Fitting Configurations
(Courtesy of Brennan Industries)**

Example of SAE-514 37⁰ Flare (Triple-Lok) (Fig. 3.43):

Description: *Triple-Lok* 37° flare fittings are the widely known and used.

Features as reported by manufacturer:
- Reliable and compact design. Easy to assembly.
- Meet the strict requirements of SAE-J514 and ISO 8434-2 industry standards.
- Adaptable to inch or metric tubes, worldwide available and accepted.
- Used for both tube and hose adapters.
- Working pressure up to 600 bar (9000 psi) depends on size.
- Available in sizes   from 1/4" size to 2" size.

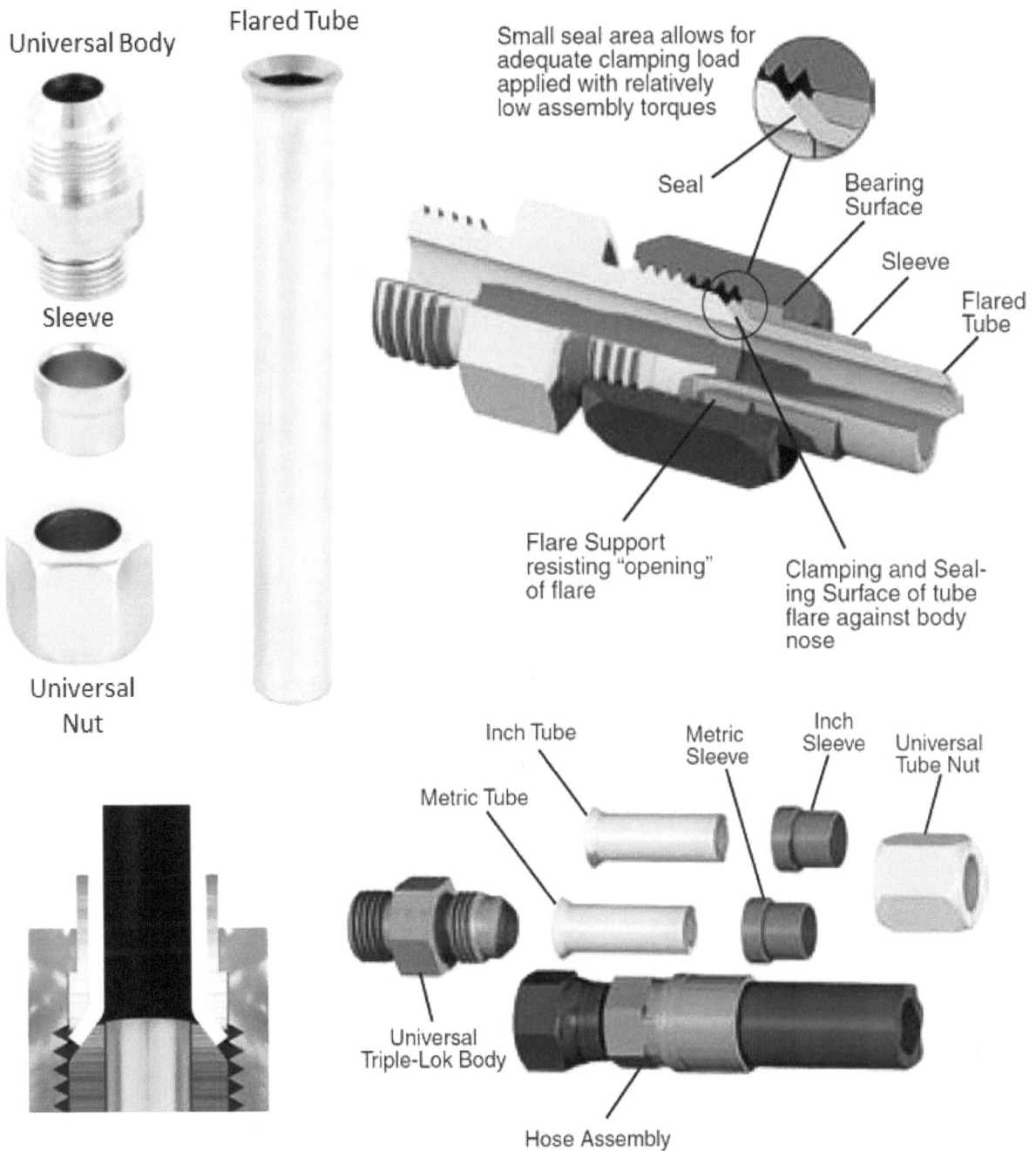

Fig. 3.43 - Triple-Lok Fitting (Courtesy of Parker)

- **SAE-J512 45⁰ Flare (Fig. 3.44):** The SAE 45° flare will only mate with a SAE 45° flare female. Both male and female couplings have straight threads and a 45° flare seat. The seal is made on the 45° flare seat. Once again, because some sizes of this coupling have the same threads as the JIC 37° flare, carefully measure the seat angle before use. Figure 3.45 shows typical JIC 45⁰ fittings. Such fittings are typically brass material and are not used in high-pressure hydraulic systems. They rather used in low-pressure fluid systems such as fuel systems, brakes, air conditioning systems, power steering, transmission, refrigerators, and oil coolers.

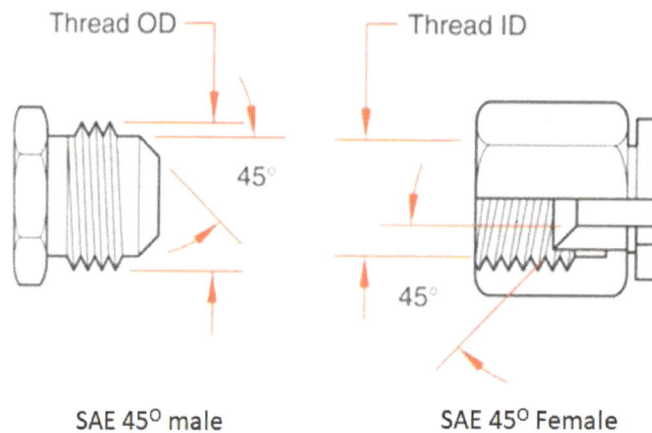

Fig. 3.44 - SAE 45⁰ Flare (Courtesy of Gates)

Fig. 3.45 - Typical JIC 45⁰ Flare (Courtesy of Parker)

Metric Standard:

- 037° flare fitting ISO 8434-2 ® 90° thread fitting with O-ring.

Japanese Standard:

- **Japanese 30° Flare Parallel Threads (Fig. 3.46):** The Japanese 30° flare male connector will mate with a Japanese 30° flare female only. The male and female have straight threads and a 30° seat. The seal is made on the 30° seat. The threads on the Japanese 30° flare connector conform to JIS B 0202, which are the same as the BSPOR threads. Both the British and Japanese connectors have a 30° seat, but they are not interchangeable because the British seat is inverted.

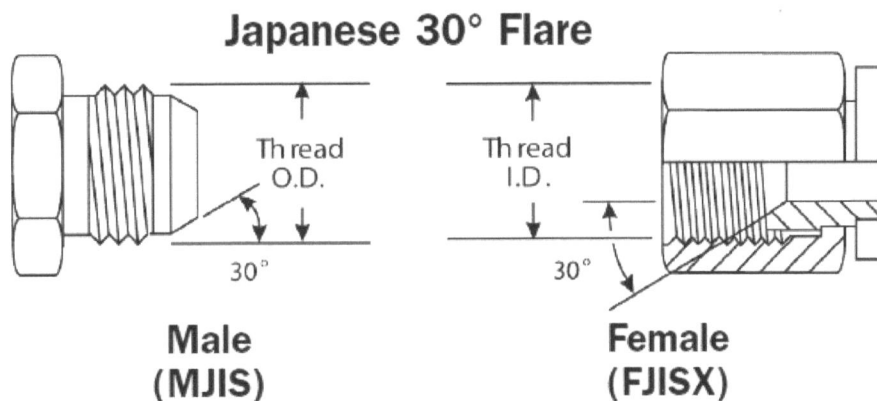

Fig. 3.46 - Japanese 30⁰ Flare Thread (Courtesy of Gates)

3.5.5.8- Adaptors

To connect fittings from foreign (British or Metric) standards to a fitting from North American National Standard, *Adaptors* are used. Figure 3.47 shows examples of available adaptors:
1. Male Foreign Standard Pipe Tapered Thread to Male JIC 37° Flare.
2. Male Foreign Standard Pipe Parallel to Male JIC 37° Flare.
3. Male Foreign Standard Pipe Parallel to Male Pipe NPTF.
4. Male Foreign Standard Pipe Parallel with O-Ring Boss to Male JIC 37° Flare.
5. Male Foreign Standard Pipe Parallel with O-Ring Boss to Male Flat-Face O-Ring.
6. Female Foreign Standard Pipe Parallel to Male Pipe NPTF.

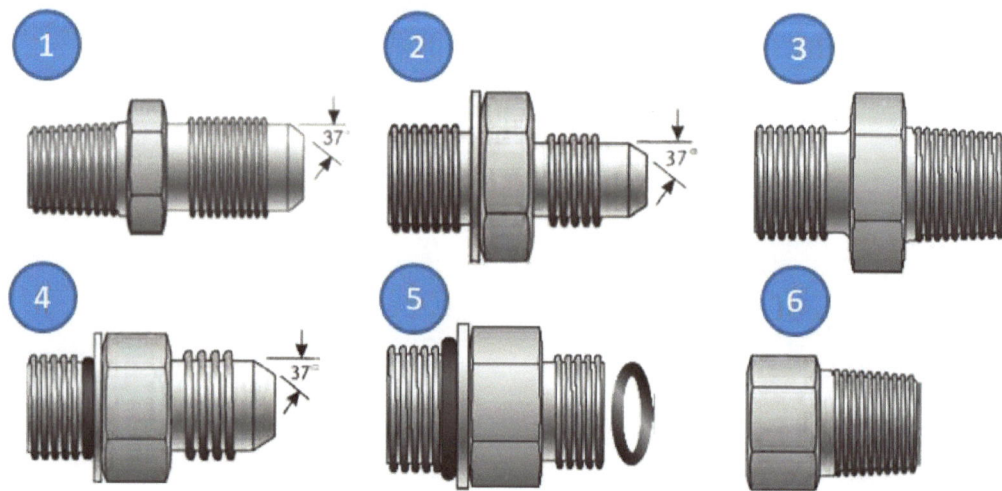

**Fig. 3.47 - Adaptors to Connect Fittings from Foreign Standards to SAE Standards
(Courtesy of Gates)**

Figure 3.48 shows typical adapters.

Fig. 3.48 - Typical Adapters (Courtesy of Parker)

3.5.5.9- Thread Characteristics Identification

❖ **Identifying Fittings by Measurements (Fig. 3.49):**

Step 1- measure the Thread Diameter: use a combination O.D./I.D. *Caliper* to measure the thread diameter. It is to be noted that the threads of a used fitting can become worn and distorted, so the measurements may not be exact.

Step 2- Measure the Thread Pitch: use a *Thread Pitch Gauge* to identify the threads. Place the gauge on the threads until it fits snugly. For British and other European threads, the thread pitch gauge measures the threads per inch. For metric connections, measure the distance between threads.

Step 3: Measure the Seat Angle: If the port is angled, determine the seat angle by using a *Seat Angle Gauge* on the sealing surface. The centerline of the fitting and the gauge must be parallel.

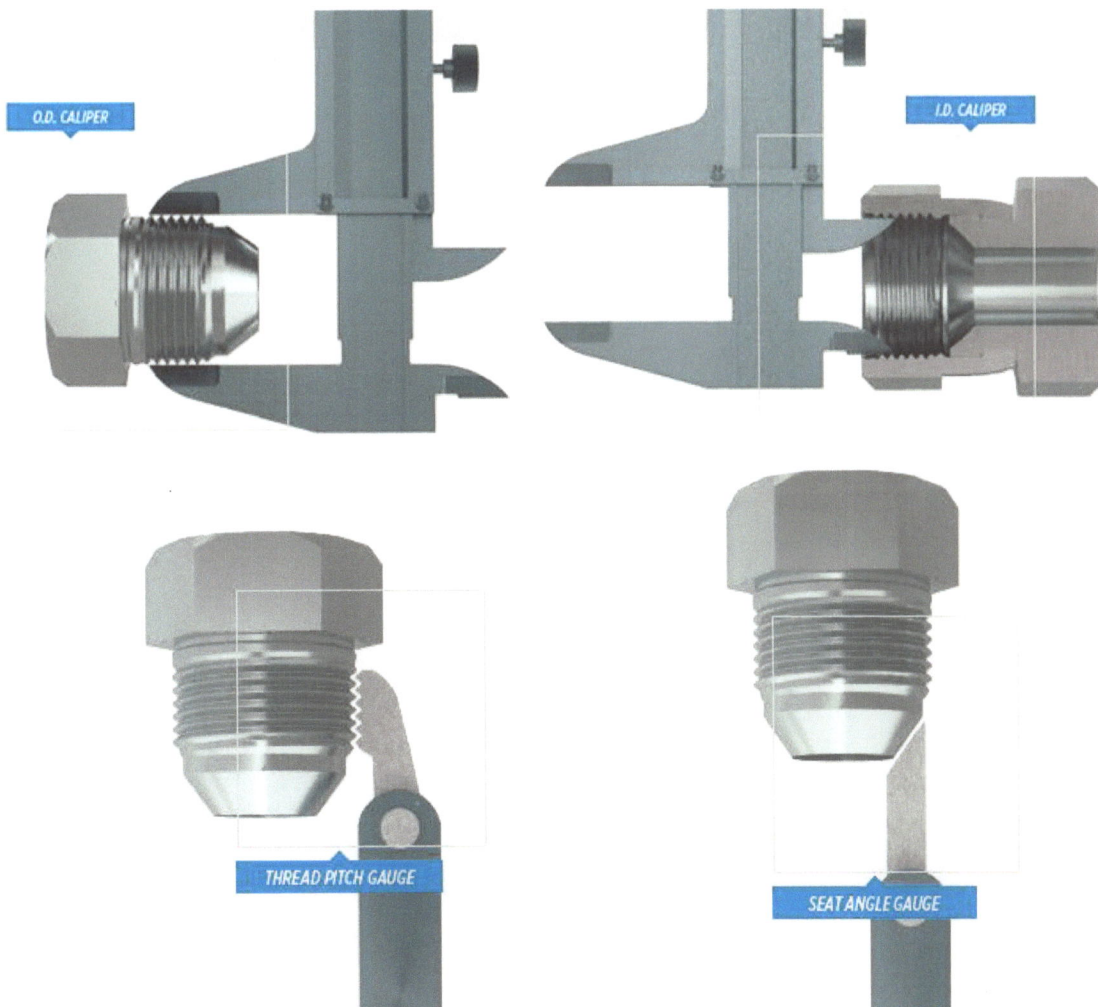

Fig. 3.49 – Identifying Fittings by Measurements (Courtesy of Brennan Industries)

❖ **Identifying Fittings by Thread ID Kit (Fig. 3.50):**

A *Thread ID Kit* is used for faster and easier identification of fittings. For ease of use, each fitting is color coded, and the size of the fitting is imprinted on the side. Each fitting has a female connector in one end and a male connector in the other end.

Fig. 3.50 – Identifying Fittings by Thread ID Kit (Courtesy of Brennan Industries)

3.5.5.10- Swivel Joints

Conventional Swivel Joints: In some systems, there might be a need for a component to rotate through an angular displacement, up to 360⁰ in some cases. In such systems, *Swivel Joints* are usable with hoses and rigid lines. As shown in Fig. 3.51, straight swivel joints provide rotation around one axis only, and elbow swivel joints provide rotation around two axes. The figure shows the construction of the swivel joint that guarantees rotary motion, sealing, and protection from introduced contamination.

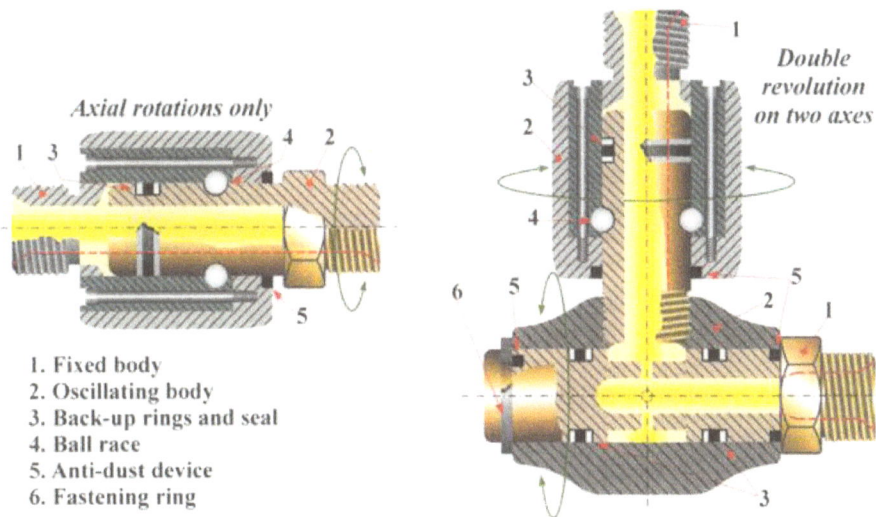

Fig. 3.51 - Construction of Conventional Swivel Joints (Courtesy of Assufluid)

Figure 3.52 shows various styles of swivel joints with the following technical characteristics:
- Pressure capabilities up to 350 bar (5000 psi).
- Variety of port options with size range is 1/4" through 2".
- Full flow design minimizes pressure drop for optimum system performance.
- Nickel plating and a wide range of seal options.
- Hardened bearing races for extended service life.
- Sealed bearing design isolates bearing race from media and environment.

Fig. 3.52 - Typical Conventional Swivel Joints (Courtesy of Parker)

Innovative Swivel Joints - WEO Plug-In hose fittings (Fig. 3.53):

Description: Over the time, more applications industry-wide are being converted from threaded to thread-less connectors, *WEO Plug-In* hose fittings are good choice for leak-free innovative swivel joints. As shown in the figure, such fittings are available in various configurations and sizes.

Features: WEO Plug-In quick swivel connection has the following features:
- Easy to install where no follow-up tightening needed.
- Minimum space requirement enables new system designs.
- Easy to service leak-free connection.
- Work injuries associated with connection/disconnection are eliminated.
- Longer hose life lower overall cost.

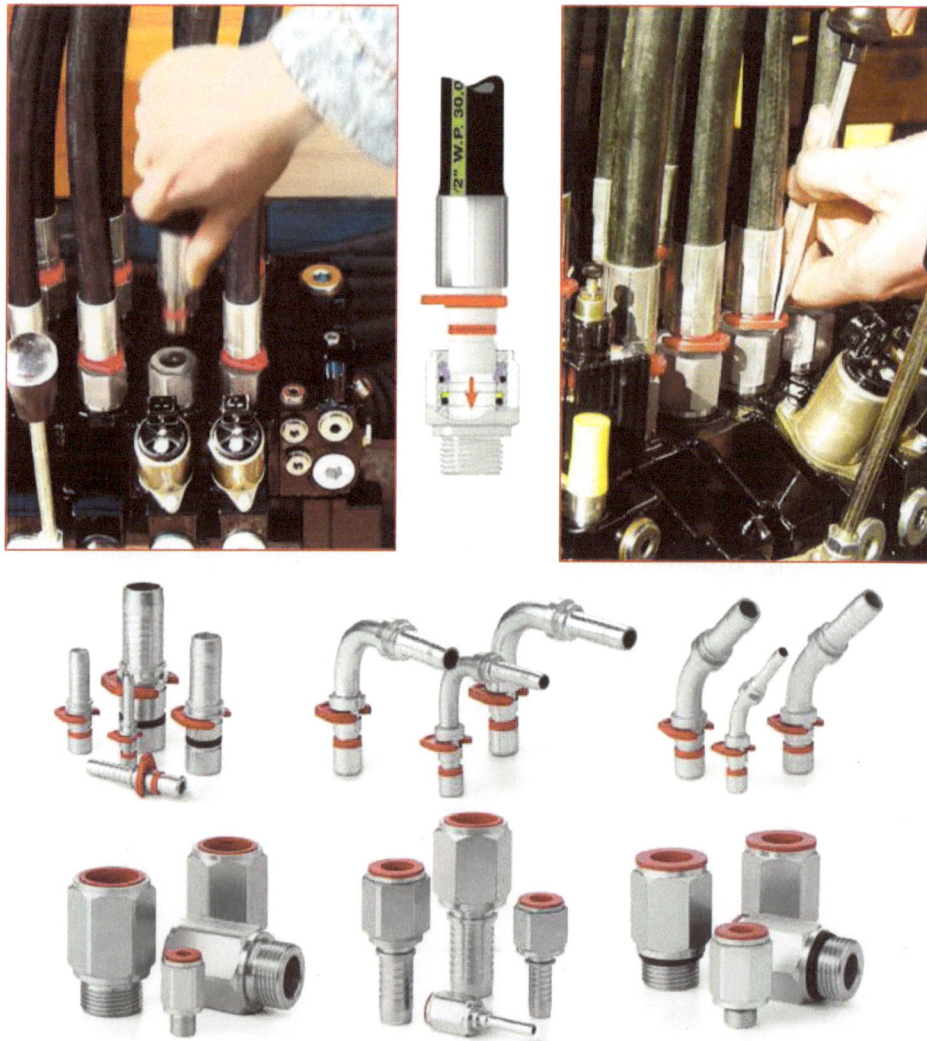

Fig. 3.53 - WEO Plug-In Swivel Joints (Courtesy of CEJN)

Technical Characteristics:

- Working Pressure:
  - Sizes (1/4 – 3/4) inch have a maximum working pressure of 350 bar (5075 psi).
  - Size 1-inch has a maximum working pressure of 250 bar (3625 psi).
- Burst Pressure: A minimum safety factor of 4:1. So, burst pressure is a minimum of four times the working pressure.
- Temperature: -30°C to +100°C (-22°F to +212°F).

Construction: Figure 3.54 shows the construction of the WEO Plug-in swivel joints.

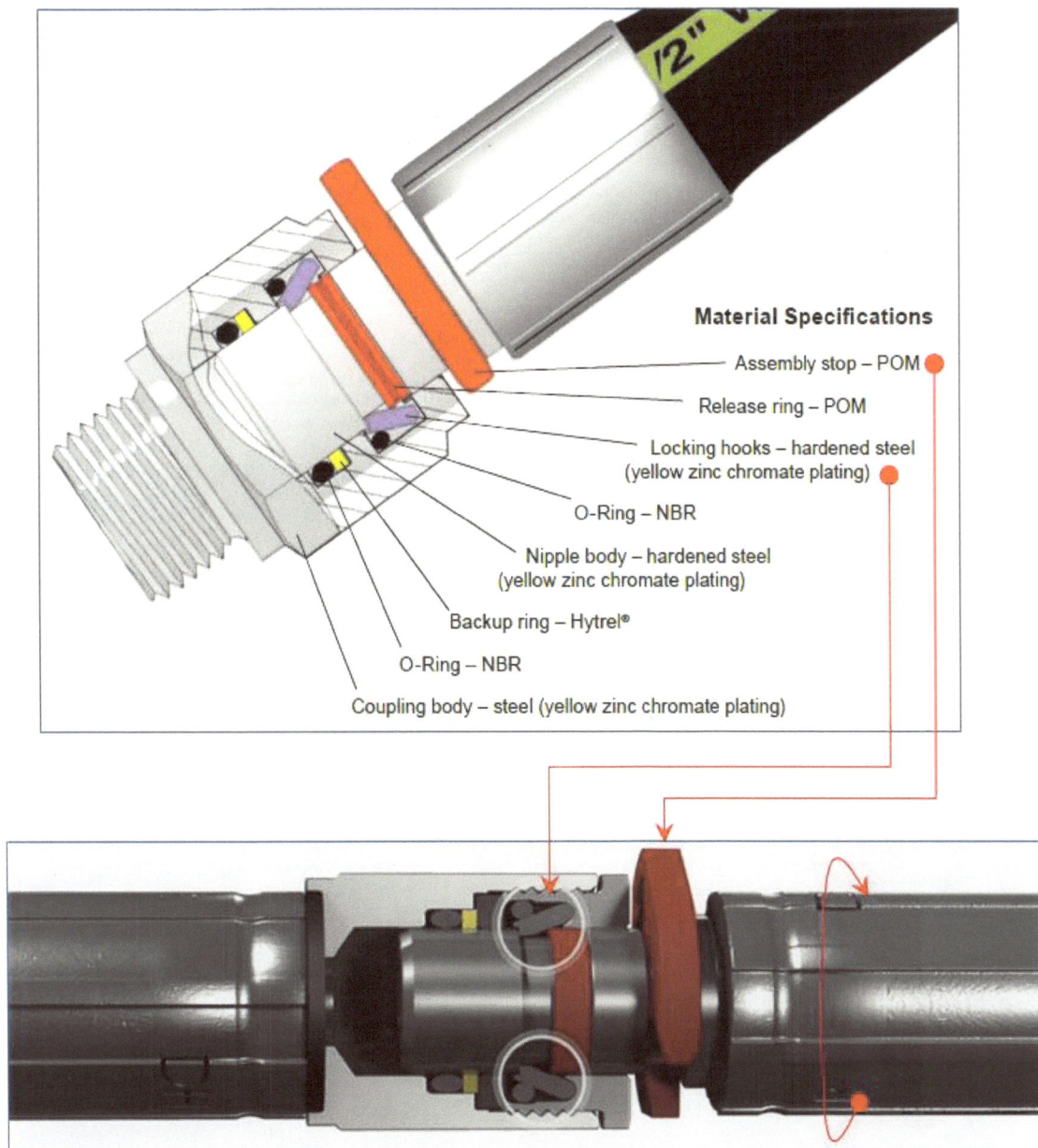

Fig. 3.54 - Construction of WEO Plug-In Swivel Joints (Courtesy of CEJN)

<u>Operation:</u> As shown in Fig. 3.55, WEO swivel joint fittings are designed with an innovative click-to-connect feature that provides easy engagement of the product's male and female halves, without the aid of tools or wrenches.

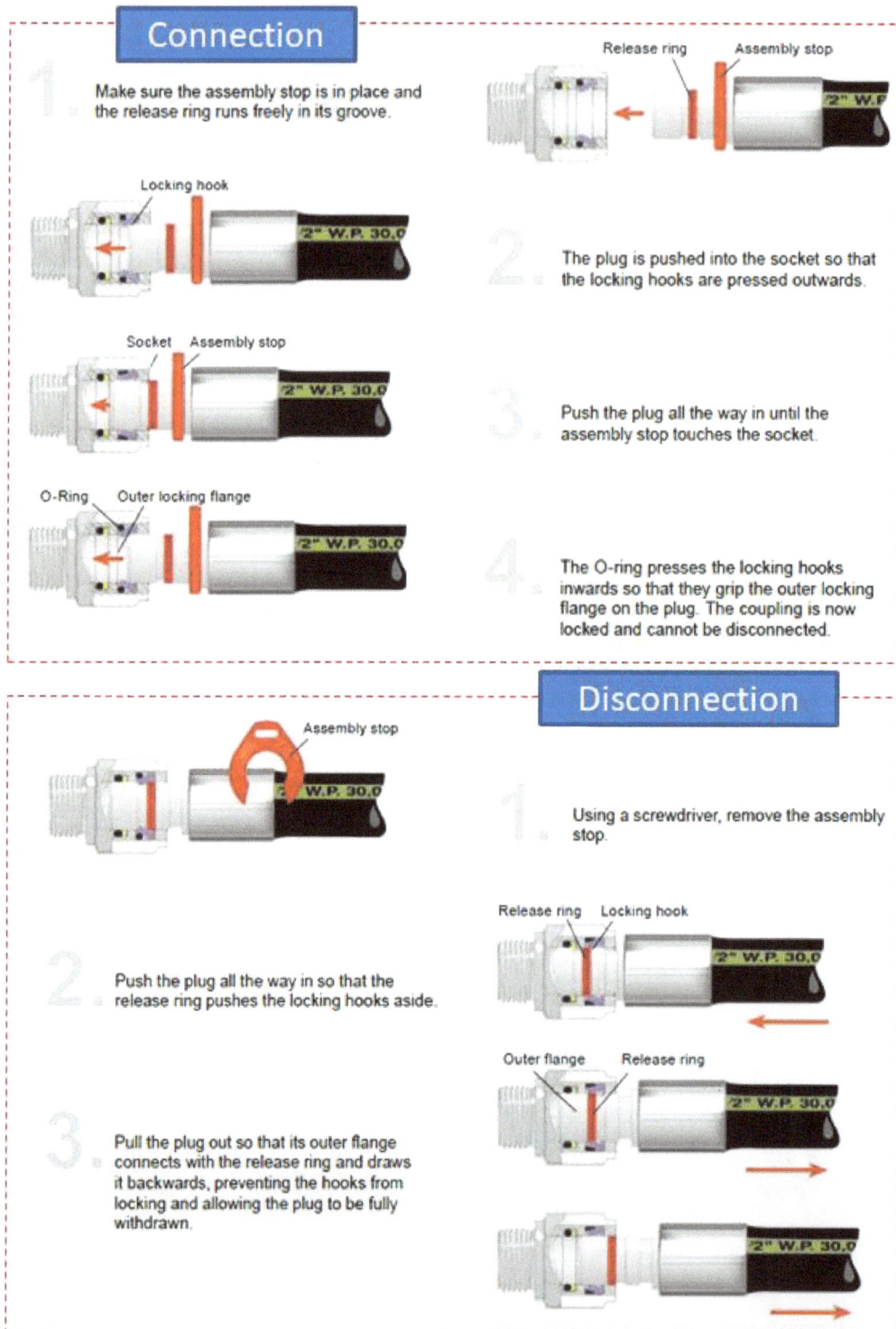

Fig. 3.55 - Operation of WEO Plug-In Swivel Joints (Courtesy of CEJN)

3.6- Hydraulic Hoses

3.6.1- Features of Hydraulic Hoses

Hydraulic hoses have the following features:

- Easy installation and flexible routing.
- Permit connections to components on moving parts.
- Permit misalignment between components.
- Isolate noise, shocks, and vibrations.
- Require compatibility with system hydraulic fluid.
- Low equivalent bulk modulus due to the flexible walls.
- Reduce hydraulic stiffness and consequently system response.
- Usually more expensive than tubes or pipes.

Countless number of applications both in mobile and industrial side rely on hydraulic hoses. Figure 3.56 shows hydraulic hoses connecting hydraulic cylinders to the hydraulic system. The lower photo shows hydraulic hoses in stationary hydraulic power unit.

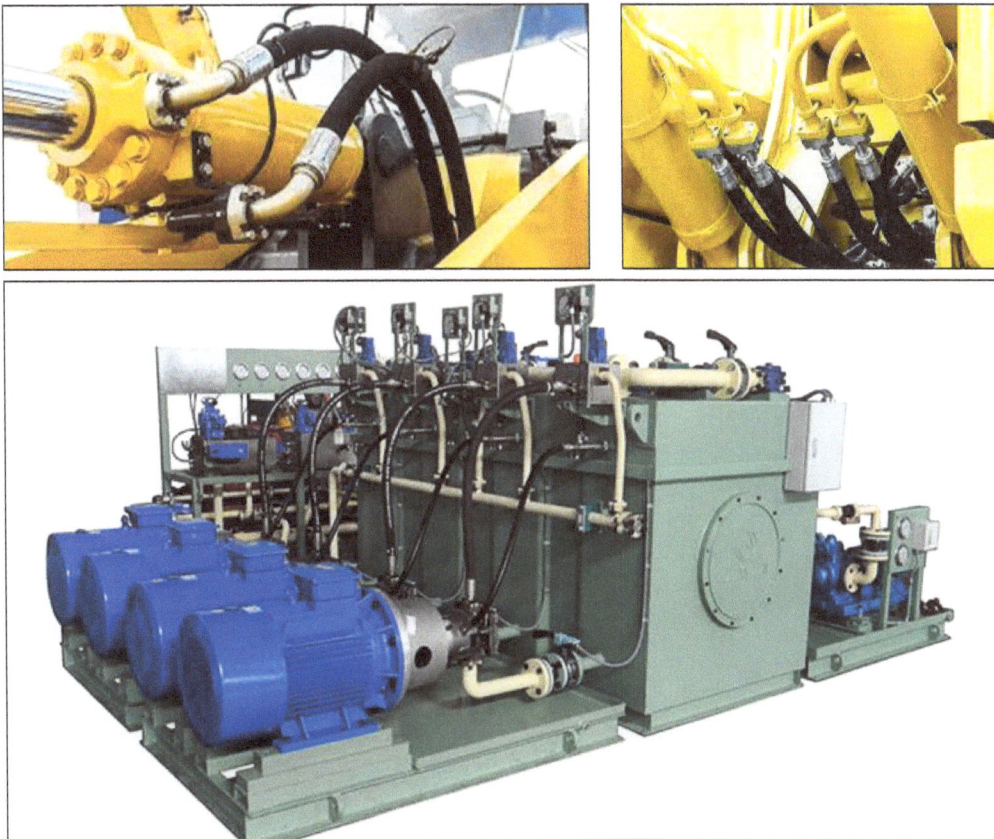

Fig. 3.56 - Examples of using Hydraulic Hoses in Mobile and Industrial Applications

3.6.2- Material and Construction of Hydraulic Hoses

As shown in Fig. 3.57, hydraulic hoses are generally constructed from the following main three parts:

Central Tube: The *Central Tube* is a seamless tube of synthetic rubber or plastic that must be compatible with the hydraulic fluid. This layer works as leakage barrier and must withstand the operating temperature range (low to high).

Reinforcement Layers: *Reinforcement Layers* consist of braided or spiraled textile strings or metal wires. The number of reinforcement layers is proportional to the maximum allowable pressure and inversely proportional to the hose flexibility. Figure 3.58 shows braded versus spiraled reinforcement layers. Hoses with braded reinforcement layers are for low to medium pressures. Hoses with spiraled reinforcement layers are for medium to high pressure. Braided hose is generally more flexible than spiral hose.

Outer Protective Cover: The *Outer Protective Cover* is to protect the hose from environmental conditions, hose abrasion, etc. Various materials are used such as plastic, cloth, etc. Hoses for special applications are covered by abrasion resistant covers.

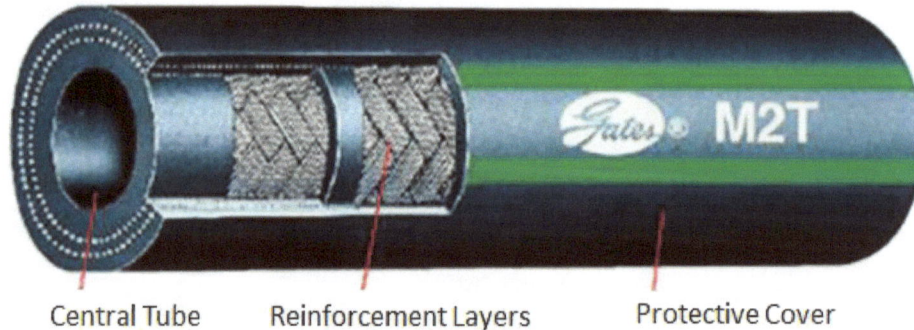

Central Tube   Reinforcement Layers   Protective Cover

Fig. 3.57 - Basic Construction of Hydraulic Hoes (Courtesy of Gates)

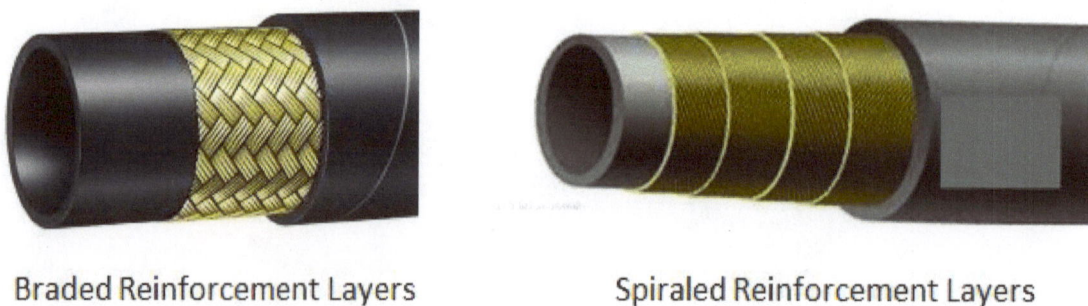

Braded Reinforcement Layers   Spiraled Reinforcement Layers

Fig. 3.58 - Spiraled versus Braded Reinforcement Layers in Hydraulic Hoses

3.6.3- Sizes of Hydraulic Hoses

As shown in Fig. 3.59, unlike hydraulic pipes and tubes, nominal size of hydraulic hose is based on the inner diameter "ID". That is because the outer diameter depends on the hose structure and its pressure rating. As shown in Table 3.9, hoses have size range (3/16 – 4) inches (4 – 100) mm. The table shows that the dimension of hydraulic hoses is designated in 16th of an inch using the (-#) format. An example is -8 means 8/16, i.e. half inch.

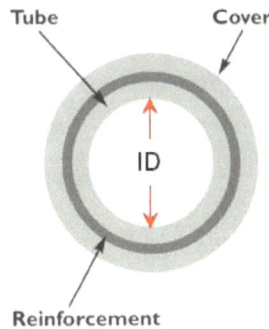

Fig. 3.59 - Nominal Size of Hydraulic Hoses (Courtesy of Gates)

Dash No.	Hose I.D.	
	Inches	Millimeters
-3	3/16	4.8
-4	1/4	6.4
-5	5/16	7.9
-6	3/8	9.5
-8	1/2	12.7
-10	5/8	15.9
-12	3/4	19.0
-14	7/8	22.2
-16	1	25.4
-20	1-1/4	31.8
-24	1-1/2	38.1
-32	2	50.8
-40	2-1/2	63.5
-48	3	76.2
-56	3-1/2	88.9
-64	4	101.6

Table 3.9- Size Range of Hydraulic Hoses (Courtesy of Gates)

3.6.4- Pressure Rating of Hydraulic Hoses

Hydraulic hose construction and performance is covered by various national and international standards including ISO, SAE, DIN, etc. The SAE standards being the most widely followed in the U.S. SAE standards provide general, dimensional, and performance specifications for the most common hoses used in hydraulic systems in mobile and stationary equipment.

SAE J517 Standard: It provides guidelines for a series of hydraulic hoses. These manufacturer-driven SAE standards have been based on design, construction, and pressure ratings to ensure that hydraulic hoses meet minimum construction and performance requirements.

Designations of hydraulic hoses follows **SAE-J517 (May 1989)** industry standard. An example of a hydraulic hose designation based on SAE standard is as **SAE100R2AT-16.** The interpretation of this designation is as follows:

- **SAE100RX:** This part describes basic construction and pressure rating.
- **A or B:** A is for braided wire and B is for spiral wire reinforcement layers.
- **T:** Appears only if the cover does not need to be removed to attach the fitting.
- **Dash number:** is the inside diameter in sixteenths of an inch. (e.g. -16 means ID = 1 in).

Table 3.10 presents SAE hose data. As shown in the table, a safety factor of 4 is used to calculate the maximum allowable working pressure as function of the burst pressure. The table also shows that, for a specific nominal size, with the increase of the maximum operating pressure, both outer diameter and minimum bend radius increases. The following examples show interpretation for the presented data.

Referring to Table 3.10:
- **SAE 100R1A:** A hose with single layer of braided wire.
- **SAE 100R2A:** A hose with double layers of braided wire.
- **SAE 100R2B:** Two spiral plies and one braid of wire.
- **SAE 100R4:** Wire Inserted Hydraulic Suction Hose.
- **SAE 100R5:** Single Wire Braid, Textile Covered Hydraulic Hose.
- **SAE 100R6:** Single Fiber Braid, (Nonmetallic), Rubber Covered Hydraulic Hose.
- **SAE 100R7:** Thermoplastic Hydraulic Hose.
- **SAE 100R8:** High Pressure Thermoplastic Hydraulic Hose.
- **SAE 100R9:** High Pressure, 4-Spiral Steel Wire, Rubber Covered Hydraulic Hose.
- **SAE 100R10:** Heavy Duty, 4-Spiral Steel Wire, Rubber Covered Hydraulic Hose.
- **SAE 100R11:** Heavy Duty, 6-Spiral Steel Wire Rubber Covered Hydraulic Hose.
- **SAE lOOR12:** Heavy Duty, High Impulse, 4-Spiral Wire, Rubber Cover Hydraulic Hose.
- **SAE 100R13:** Heavy Duty, High Impulse, Multiple Spiral Wire, Rubber Covered Hose.
- **SAE lOOR14:** This specification covers hose for use with petroleum, synthetic, and water base hydraulic fluids.

¹Minimum burst pressure is 4 times maximum operating pressure.

I.D. INCHES	DASH NO. REF.	SAE NO. & TYPE SPEC.	O.D. MAX. INCHES	MIN. BEND (INTERNAL) RAD. IN. AT MAX. OPERATING PRESSURE	¹MAX. OPERATING PRESSURE PSIG
1/8	-3	100R14	0.268	1.5	1,500
3/16	-3	100R1-A	0.531	3.5	3,000
3/16	-3	100R1-AT	0.494	3.5	3,000
3/16	-3	100R2-A&B	0.656	3.5	5,000
3/16	-3	100R2-AT&BT	0.557	3.5	5,000
3/16	-3	100R3	0.531	3	1,500
3/16	-4	100R5	0.539	3	3,000
3/16	-3	100R6	0.469	2	500
3/16	-3	100R7	0.450	3.5	3,000
3/16	-3	100R8	0.575	3.5	5,000
3/16	-3	100R10-A	0.781	4	10,000
3/16	-3	100R11	0.906	4	12,500
3/16	-4	100R14	0.324	2	1,500
1/4	-4	100R1-A	0.656	4	2,750
1/4	-4	100R1-AT	0.557	4	2,750
1/4	-4	100R2-A&B	0.719	4	5,000
1/4	-4	100R2-AT&BT	0.619	4	5,000
1/4	-4	100R3	0.594	3	1,250
1/4	-5	100R5	0.601	3.38	3,000
1/4	-4	100R6	0.531	2.5	400
1/4	-4	100R7	0.538	4	2,750
1/4	-4	100R8	0.660	4	5,000
1/4	-4	100R10-A	0.844	5	8,750
1/4	-4	100R11	0.969	5	11,250
1/4	-5	100R14	0.397	3	1,500
5/16	-5	100R1-A	0.719	4.5	2,500
5/16	-5	100R1-AT	0.619	4.5	2,500
5/16	-5	100R2-A&B	0.781	4.5	4,250
5/16	-5	100R2-AT&BT	0.682	4.5	4,250
5/16	-5	100R3	0.719	4	1,200
5/16	-6	100R5	0.695	4	2,250
5/16	-5	100R6	0.594	3	400
5/16	-5	100R7	0.615	4.5	2,500
5/16	-6	100R14	0.458	4	1,500
3/8	-6	100R1-A	0.812	5	2,250
3/8	-6	100R1-AT	0.713	5	2,250
3/8	-6	100R2-A&B	0.875	5	4,000
3/8	-6	100R2-AT&BT	0.777	5	4,000
3/8	-6	100R3	0.781	4	1,125
3/8	-6	100R6	0.656	3	400
3/8	-6	100R7	0.725	5	2,250
3/8	-6	100R8	0.800	5	4,000
3/8	-6	100R9-A	0.875	5	4,500
3/8	-6	100R9-AT	0.831	5	4,500
3/8	-6	100R10-A	0.969	6	7,500
3/8	-6	100R11	1.094	6	10,000
3/8	-6	100R12	0.828	5	4,000
3/8	-7	100R14	0.526	5	1,500
13/32	-6.5	100R1-A	0.844	5.5	2,250
13/32	-6.5	100R1-AT	0.744	5.5	2,250
13/32	-8	100R5	0.789	4.62	2,000
13/32	-8	100R14	0.562	5.25	1,000
1/2	-8	100R1-A	0.938	7	2,000
1/2	-8	100R1-AT	0.846	7	2,000
1/2	-8	100R2-A&B	1.000	7	3,500
1/2	-8	100R2-AT&BT	0.908	7	3,500
1/2	-8	100R3	0.969	5	1,000
1/2	-10	100R5	0.945	5.5	1,750
1/2	-8	100R6	0.812	4	400
1/2	-8	100R7	0.885	7	2,000
1/2	-8	100R8	0.970	7	3,500
1/2	-8	100R9-A	1.000	7	4,000
1/2	-8	100R9-AT	0.958	7	4,000
1/2	-8	100R10-A	1.125	8	6,250
1/2	-8	100R11	1.250	8	7,500
1/2	-8	100R12	0.966	7	4,000
1/2	-10	100R14	0.663	6.5	800
5/8	-10	100R1-A	1.062	8	1,500
5/8	-10	100R1-AT	0.971	8	1,500
5/8	-10	100R2-A&B	1.125	8	2,750
5/8	-10	100R2-AT&BT	1.034	8	2,750
5/8	-10	100R3	1.094	5.5	875
5/8	-12	100R5	1.101	6.5	1,500
5/8	-10	100R6	0.938	5	350
5/8	-10	100R7	1.015	8	1,500
5/8	-10	100R8	1.175	8	2,750
5/8	-10	100R12	1.111	8	4,000
5/8	-12	100R14	0.793	7.75	800
3/4	-12	100R1-A	1.219	9.5	1,250
3/4	-12	100R1-AT	1.127	9.5	1,250
3/4	-12	100R2-A&B	1.281	9.5	2,250
3/4	-12	100R2-AT&BT	1.190	9.5	2,250
3/4	-12	100R3	1.281	6	750
3/4	-12	100R4	1.375	5	300
3/4	-12	100R6	1.000	6	300
3/4	-12	100R7	1.125	9.5	1,250
3/4	-12	100R8	1.300	9.5	2,250
3/4	-12	100R9-A	1.266	9.5	3,000
3/4	-12	100R9-AT	1.255	9.5	3,000
3/4	-12	100R10-A	1.469	11	5,000
3/4	-12	100R10-AT	1.450	11	5,000
3/4	-12	100R11	1.594	11	6,250
3/4	-12	100R12	1.241	9.5	4,000
3/4	-12	100R13	1.306	9.5	5,000
3/4	-14	100R14	0.917	9	800
7/8	-14	100R1-A	1.344	11	1,125
7/8	-14	100R1-AT	1.252	11	1,125
7/8	-14	100R2-A&B	1.406	11	2,000
7/8	-14	100R2-AT&AB	1.315	11	2,000
7/8	-16	100R5	1.266	7.38	800
7/8	-16	100R14	1.061	9	800
1	-16	100R1-A	1.547	12	1,000
1	-16	100R1-AT	1.440	12	1,000
1	-16	100R2-A&B	1.609	12	2,000
1	-16	100R2-AT&BT	1.531	12	2,000
1	-16	100R3	1.547	8	565
1	-16	100R4	1.625	6	250
1	-16	100R7	1.445	12	1,000
1	-16	100R8	1.520	12	2,000
1	-16	100R9-A	1.609	12	3,000
1	-16	100R9-AT	1.594	12	3,000
1	-16	100R10-A	1.797	14	4,000
1	-16	100R10-AT	1.790	14	4,000
1	-16	100R11	1.953	14	5,000
1	-16	100R12	1.542	12	4,000
1	-16	100R13	1.567	12	5,000
1	-18	100R14	1.175	12	800
1 1/8	-20	100R5	1.531	9	625
1 1/8	-20	100R14	1.320	16	600
1 1/4	-20	100R1-A	1.875	16.5	625
1 1/4	-20	100R1-AT	1.766	16.5	625
1 1/4	-20	100R2-A&B	2.062	16.5	1,625
1 1/4	-20	100R2-AT&BT	1.953	16.5	1,625
1 1/4	-20	100R3	1.812	10	375
1 1/4	-20	100R4	2.000	8	200
1 1/4	-20	100R9-A	2.062	16.5	2,500
1 1/4	-20	100R9-AT	1.997	16.5	2,500
1 1/4	-20	100R10-A	2.062	18	3,000
1 1/4	-20	100R10-AT	2.060	18	3,000
1 1/4	-20	100R11	2.219	18	3,500
1 1/4	-20	100R12	1.912	16.5	3,000
1 1/4	-20	100R13	2.021	16.5	5,000
1 3/8	-24	100R5	1.781	10.5	500
1 1/2	-24	100R1-A	2.125	20	500
1 1/2	-24	100R1-AT	2.047	20	500
1 1/2	-24	100R2-A&B	2.312	20	1,250
1 1/2	-24	100R2-AT&BT	2.203	20	1,250
1 1/2	-24	100R4	2.250	10	150
1 1/2	-24	100R9-A	2.312	20	2,000
1 1/2	-24	100R10-A	2.312	22	2,500
1 1/2	-24	100R10-AT	2.310	22	2500
1 1/2	-24	100R11	2.469	22	3,000
1 1/2	-24	100R12	2.167	20	2,500
1 1/2	-24	100R13	2.315	20	5,000
1 13/16	-32	100R5	2.266	13.25	350
2	-32	100R1-A	2.688	25	375
2	-32	100R1-AT	2.594	25	375
2	-32	100R2-A&B	2.812	25	1,125
2	-32	100R2-AT&BT	2.703	25	1,125
2	-32	100R4	2.750	12	100
2	-32	100R9-A	2.875	26	2,000
2	-32	100R10-A	2.844	28	2,500
2	-32	100R10-AT	2.840	28	2,500
2	-32	100R11	3.031	28	3,000
2	-32	100R12	2.688	25	2,500
2	-32	100R13	2.862	25	5,000
2 3/8	-40	100R5	2.922	24	350
2 1/2	-40	100R2-A&B	3.312	30	1000
2 1/2	-40	100R11	3.656	36	2500
2 1/2	-40	100R4	3.250	14	62
3	-48	100R4	3.750	18	56
3	-48	100R5	3.609	33	200
3 1/2	-56	100R4	4.250	21	45
4	-64	100R4	4.750	24	35

Table 3.10 – Hoses Specifications in accordance with SAE Standard

Table 3.11 presents more data about fluid compatibility and temperature range in accordance with the SAE standard.

SAE DIMENSIONAL AND PERFORMANCE STANDARDS FOR HYDRAULIC HOSE							
SAE standard hydraulic hose type/application	Compatible hydraulic fluids	Temperature range, °F	ID range, in.	Maximum operating pressure, psi	Proof pressure range,psi	Minimum burst pressure range, psi	Minimum bend radius, in.
100R1—Steel wire reinforced, rubber coated	Petroleum & water based	-40 to 212	3/16 to 2	575 to 3,250	1,150 to 6,500	2,300to 13,000	3.5 to 25
100R2—High-pressure steel wire, reinforced rubber cover	Petroleum & water based	-40 to 212	3/16 to 2	1,150 to 6,000	2250to 12,000	4,500to 24,000	3.5 to 25
100R3—Double fiber, braid rubber cover – High-temp, low-pressure	Petroleum & water based	-40 to 212	3/16 to 1-1	375 to 1,500	750 to 3,000	1,500 to 6,000	3 to 10
100R4—Wire inserted, hydraulic suction and return	Petroleum & water based	-40 to 212	3 to 4	35 to 300	70 to 600	140 to 1,200	5 to 24
100R5—Single wire braid, textile cover; Transportation/DOT	Petroleum & water based	-40 to 212	3/16 to 3-1/16	200 to 3,000	400 to 6,000	800to 12,000	3 to 33
100R6—Single fiber braid, rubber cover—Transportation	Petroleum & water based	-40 to 212	1/16 to 3	300 to 500	600 to 1,000	1,200 to 2,000	2 to 6
100R7—Single fiber braid, thermoplastic–Hydraulic	Petroleum, water based & synthetic	-40 to +212	1/8 to 1	1,000 to 3,000	2,000 to 6,000	4,000to 12,000	1 to 12
100R8—High pressure, thermoplastic-Hydraulic	Petroleum, water based & synthetic	-40 to 212	1/8 to 1	2,000 to 6,000	4,000to 12,000	8,000to 24,000	1 to 12
100R9	No longer part of the SAE standard.						
100R10	No longer part of the SAE standard.						
100R11	No longer part of the SAE standard.						
100R12—Heavy duty, high impulse, four-spiral wire reinforced, rubber cover–Hydraulic	Petroleum & water based	-40 to 250	3/8 to 2	2,500 to 4,000	5,000 to 8,000	10,000 to 16,000	5 to 25
100R13—Heavy duty, high impulse, 4- & 6-spiral steel wire reinforced, rubber cover; Hydraulic	Petroleum & water based	-40 to 250	3 to 2	5,000	10,000	20,000	9.5 to 25
100R14—High temperature, corrosive fluids, PTFE-lined hydraulic hose, single-stainless steel braid	Petroleum, water based & synthetic	-65 to +400	3/16 to 1-1	600 to 1,500	1,200 to 6,000	2,500to 12,000	1.5 to 16
100R15—Heavy duty, ultra-high-pressure, 6-spiral steel wire reinforced, rubber cover–Hydraulic	Petroleum based	-40 to 250	3/8 to 1-1	6,000	12,000	24,000	6 to 21
100R16—Compact, high-pressure, 2-braided wire reinforced rubber cover; Hydraulic	Petroleum & water based	-40 to 212	1 to 1-1	1,800 to 5,800	3,600to 11,600	7,200to 23,200	2 to 8
100R17—Compact, maximum operating pressure, 1- and 2-steel braided wire reinforced rubber cover; Hydraulic	Petroleum & water based	-40 to 212	3/16 to 1	3,000	6,000	12,000	2 to 6
100R18—Thermoplastic, synthetic-fiber reinforcement, thermoplastic cover; Hydraulic	Petroleum, water-based, synthetic	-40 to 212	1/8 to 1	3,000	6,000	12,000	1 to 10
100R19—Compact, maximum operating pressure, 1- and 2-braided steel wire, reinforced rubber cover; Hydraulic	Petroleum & water based	-40 to 212	3/16 to 1	4,000	8,000	16,000	2 to 6

Table 3.11 - SAE Standard for Hydraulic Hoses (Hydraulic & Pneumatic Magazine)

ISO 18752 Standard Pressure Classes: More recently, many large OEMs switched to ISO standards in their design and manufacturing process to ensure the sale and service of their equipment globally. ISO Standard 18752, released in 2006, takes a different approach centered around the design practices of users who typically design hydraulic systems based on performance and pressure requirements. Table 3.12 shows the ISO pressure class specifications.

Class		35	70	140	210	250	280	350	420	560
MWP[a] (bar)		35	70	140	210	250	280	350	420	560
MWP[a] (MPa)		3.5	7	14	21	25	28	35	42	56
MWP[a] (psi)		500	1000	2000	3000	3500	4000	5000	6000	8000
Nominal Size ISO	Inch									
5	-3	●	●	●	●	●	●	●	●	N/A
6.3	-4	●	●	●	●	●	●	●	●	N/A
8	-5	●	●	●	●	●	●	●	●	N/A
10	-6	●	●	●	●	●	●	●	●	N/A
12.5	-8	●	●	●	●	●	●	●	●	N/A
16	-10	●	●	●	●	●	●	●	●	●
19	-12	●	●	●	●	●	●	●	●	●
25	-16	●	●	●	●	●	●	●	●	●
31.5	-20	●	●	●	●	●	●	●	●	●
38	-24	●	●	●	●	●	●	●	●	N/A
51	-32	●	●	●	●	●	●	●	●	N/A
63	-40	●	●	N/A	N/A	N/A	N/A	N/A	N/A	N/A
76	-48	●	●	N/A	N/A	N/A	N/A	N/A	N/A	N/A
102	-64	●	N/A	N/A	N/A	N/A	N/A	N/A	N/A	N/A

Note: ● = Applicable   N/A = Not Applicable   [a] = Maximum Working Pressure

Table 3.12 - ISO 18752 Standard Pressure Classes for Hydraulic Hoses (Courtesy of Parker)

ISO 18752 Standard Resistance to Impulse: As shown in Table 3.13, in ISO 18752, hydraulic hoses specifications are classified according to their resistance to impulse pressure in four grades: A, B, C, and D. Each grade requires a specific number of impulse cycles at a certain temperature and impulse pressure in order to meet the standard. In addition, each grade is then classified by the OD of the hose into standard types (AS, BS, CS) or compact types (AC, BC, CC, DC). Compact types have a smaller OD and bend radius than the standard types.

Grades and Types				
Grade	Type[a]	Resistance to Impulse		
		Temperature °C	Impulse Pressure (% of MWP[b])	Minimum Number of Cycles
A	AS	100	133%	200,000
	AC			
B	BS	100	133%	500,000
	BC			
C	CS	120	133% and 120%[c]	500,000
	CC			
D	DC	120	133%	1,000,000

[a] Standard or compact, e.g. CS is grade C and standard type.

Standard types have larger outside diameters and larger bend radii and compact types have smaller outside diameters and smaller bend radii.

[b] Maximum working pressure.

[c] 120% of the MWP shall be used for classes 350, 420 and 560 instead of 133%.

Table 3.13 - ISO 18752 Standard Pressure Classes for Hydraulic Hoses (Courtesy of Parker)

3.6.5- Hydraulic Hose Bend Radius

When application space is tight, bend radius is another important consideration. Otherwise, Installation of a hose at less than the minimum listed bend radius kinks the hose, reduce the hose life, and may result in premature failure.

Minimum bend radius of a hose depends on hose construction and is reported by the hose manufacturer. Table 3.14 shows an example of typical minimum bend radius reported by the hose manufacture

387
Hydraulic – Constant Working Pressure
ISO 18752

# Part Number 387	Standard Cover 387	Tough Cover 387TC	Super Tough 387ST	Hose I.D. inch	Hose I.D. mm	Hose O.D. inch	Hose O.D. mm	Working Pressure psi	Working Pressure MPa	Minimum Bend Radius inch	Minimum Bend Radius mm	Weight bs/ft	Weight kg/m	Vacuum Rating inches of Hg	Vacuum Rating kPa	Parkrimp 43 Series	Parkrimp 77 Series
387	ISO 18752 Performance																
387-4	AC	AC	AC	1/4	6,3	0.53	13,4	3000	21,0	2	50	0.16	0,24	24	80	●	
387-6	AC	AC	AC	3/8	10	0.69	17,4	3000	21,0	2-1/2	65	0.23	0,34	24	80	●	
387-8	AC	AC	AC	1/2	12,5	0.82	20,7	3000	21,0	3-1/2	90	0.29	0,43	24	80	●	
387-10	AC	AC	AC	5/8	16	0.94	23,9	3000	21,0	4	100	0.33	0,49	24	80	●	
387-12	AC	AC	AC	3/4	19	1.10	27,8	3000	21,0	4-3/4	120	0.58	0,86	24	80	●	
387-16	AC	AC	AC	1	25	1.40	35,4	3000	21,0	6	150	0.79	1,17	24	80	●	
387-20	BC	CC	CC	1-1/4	31,5	1.82	46,3	3000	21,0	8-1/4	210	1.74	2,59	18	60	●	●
387-24	BC	CC	CC	1-1/2	38	2.08	52,8	3000	21,0	10	250	2.01	2,99	18	60		●
387-32	BC	CC	CC	2	51	2.61	66,2	3000	21,0	12-1/2	320	2.75	4,09	18	60		●

Table 3.14 - Example of Minimum Bend Radius Reported by the Hose Manufacturer
(Courtesy of Parker)

As shown in Fig. 3.60, when considering the bend radius of a hose assembly, a minimum straight length of twice the hose's outside diameter should be allowed between the hose fitting and the point at which the bend starts.

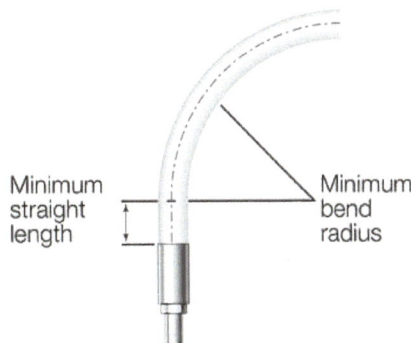

Fig. 3.60 - Minimum Straight Length for Bent Hoses

3.6.6- Hydraulic Hose Assembly

Before cutting a hose, make sure you understand the difference between "*Cut Hose Length*" and "*Assembly Overall Length*". As shown in Fig. 3.61. the overall assembly length (OAL) for a straight hose includes the hose cut length (HCL) of the hose plus the length of the two hose couplings C1 and C2. The overall assembly length of bended hoses is calculated as shown in Fig. 3.62.

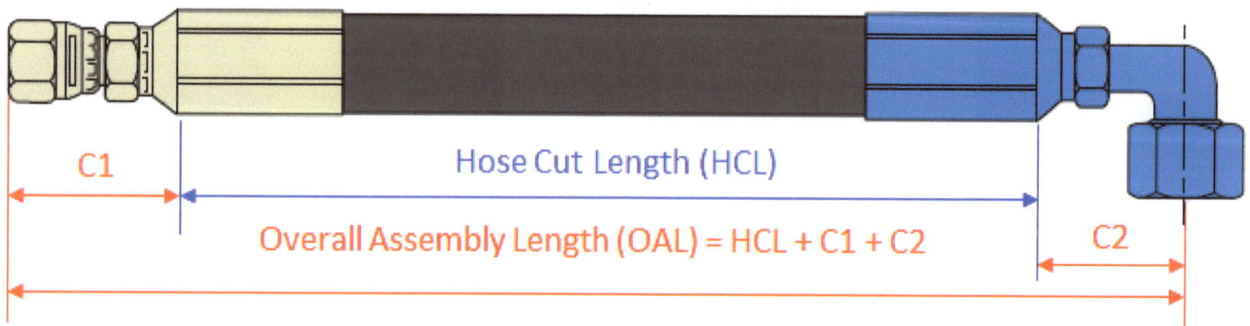

C1

Hose Cut Length (HCL)

Overall Assembly Length (OAL) = HCL + C1 + C2

C2

Fig. 3.61 - Straight Hose Assembly Length

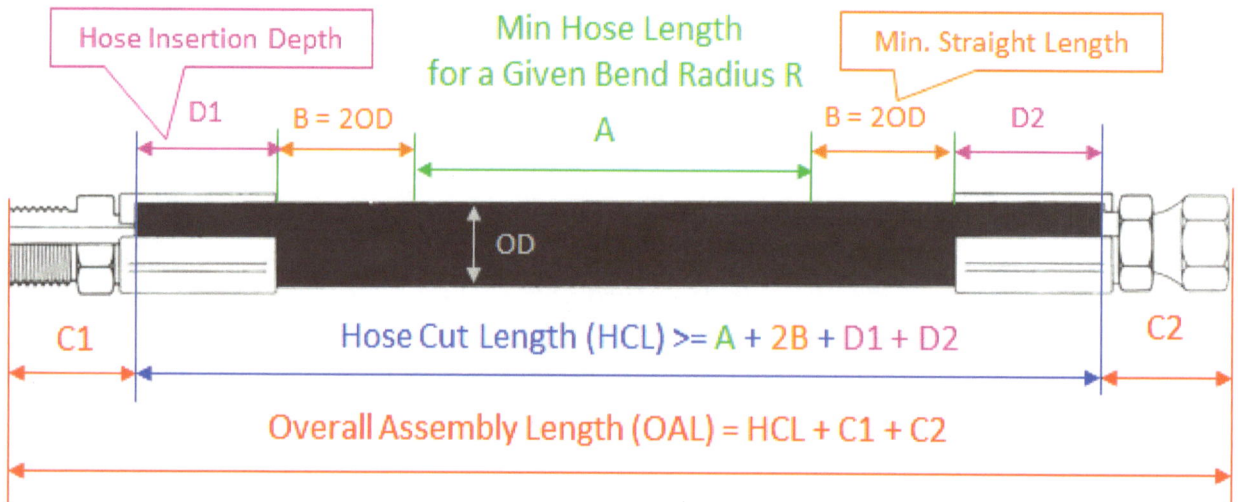

Hose Insertion Depth

Min Hose Length
for a Given Bend Radius R

Min. Straight Length

D1   B = 2OD   A   B = 2OD   D2

OD

C1

Hose Cut Length (HCL) >= A + 2B + D1 + D2

Overall Assembly Length (OAL) = HCL + C1 + C2

C2

Fig. 3.62 - Bent Hose Assembly Length

Equation 3.3 shows a general Formula for Minimum Hose Cut Length for a Given Bend Radius:

A = (Angle of Bend/360) x 2 π R   3.3

Example: to make a 90° bend for hose that has a minimum bend radius r = 4.5 inches, the minimum hose length is **A = (90/360) x 2 x π x 4.5 = 7 inches.**

Figure 3.63 shows the overall assembly length according to DIN 20066 for various hose configurations.

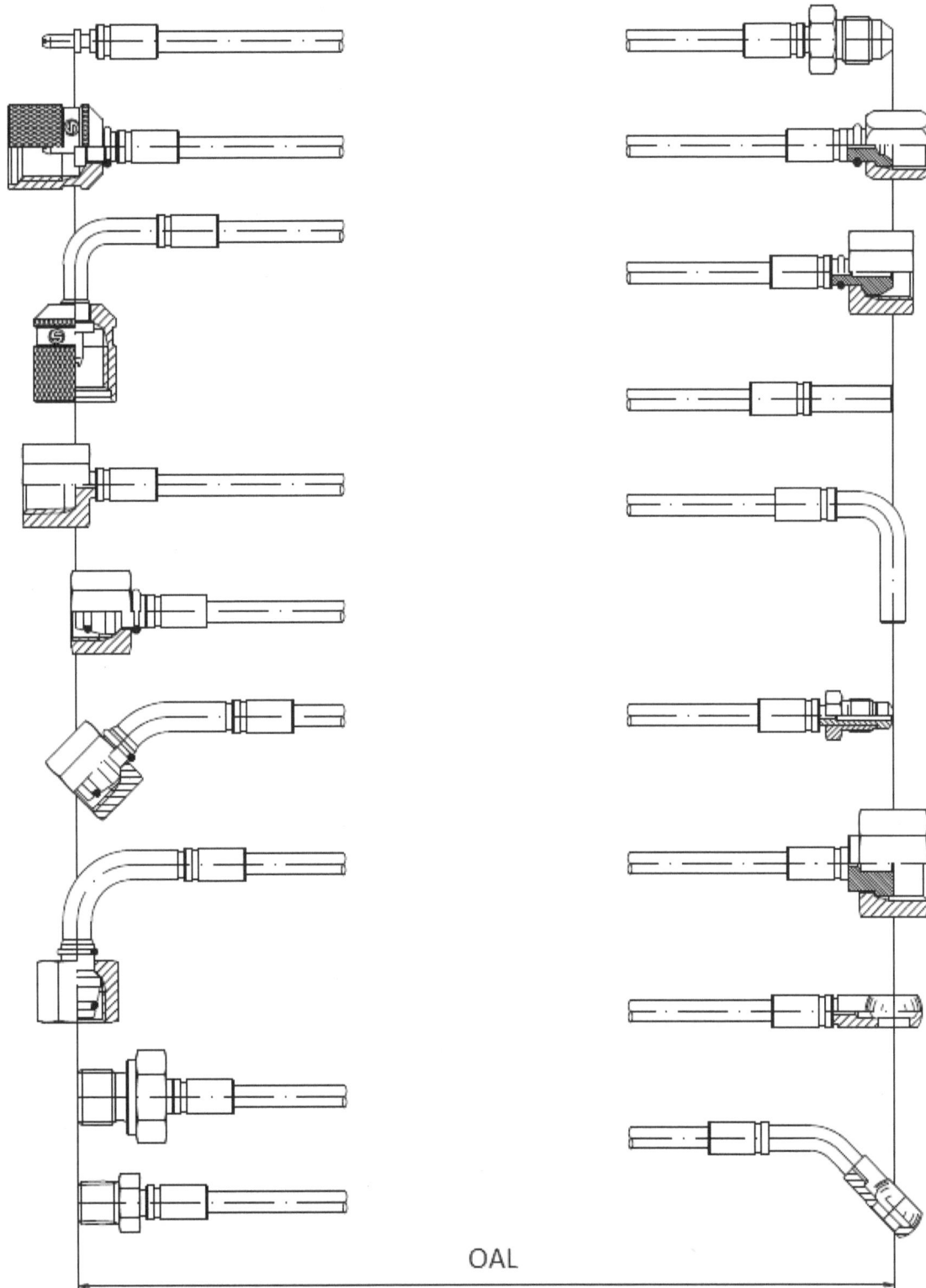

OAL

Fig. 3.63 - Hose Overall Assembly Length according to DIN 20066

3.6.7- Hose Couplings

3.6.7.1- Identification of Hose Couplings

A hydraulic hose requires two end joints to connect the hose from both ends. These end joints are referred to *Hose Couplings.* As shown in Fig. 3.64, a coupling is identified as follows:
- The *hose end* for hose attachment. The hose end is identified by the hose size and type to which it is attached.
- The *thread end* (or adaptor) for port attachment. The thread end, as shown in the figure, are available in different configurations with sizes depending on the port to which it will be attached.

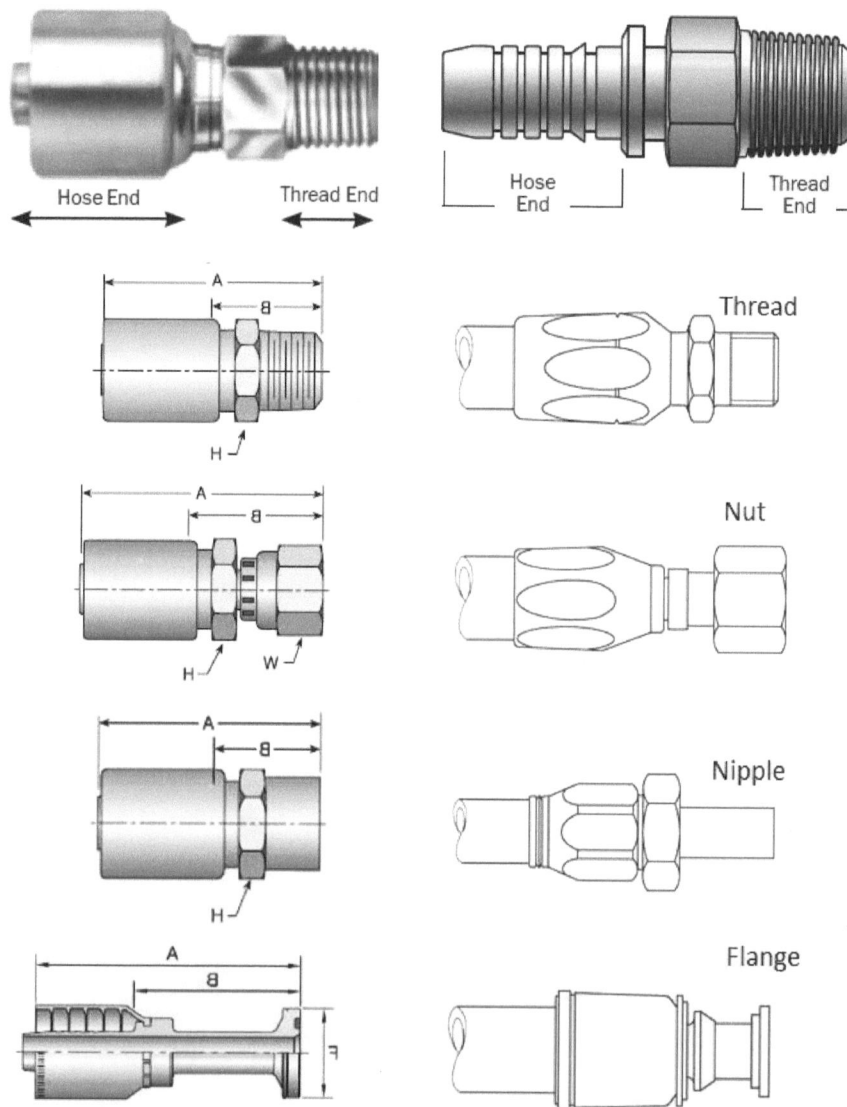

Fig. 3.64 - Identification of Hose Couplings (Courtesy of Gates)

3.6.7.2- Reusable versus Permanent Couplings

As shown in Fig. 3.65, there are two types of hydraulic couplings *Reusable Couplings* and *Permanent Couplings* for one-time use.

Reusable Couplings: As shown in Fig. 3.66, reusable (*field Attachments or Screw in Connectors*) hose fittings are an excellent choice for emergency repairs. Assembly can be accomplished with hand wrenches; there is no crimper required. However, these connectors aren't recommended if there are high pulsating pressure.

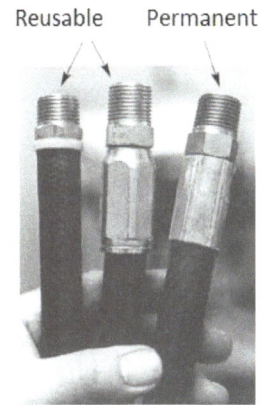

Fig. 3.65 - Permanent versus Reusable Hose Couplings (Courtesy of Gates)

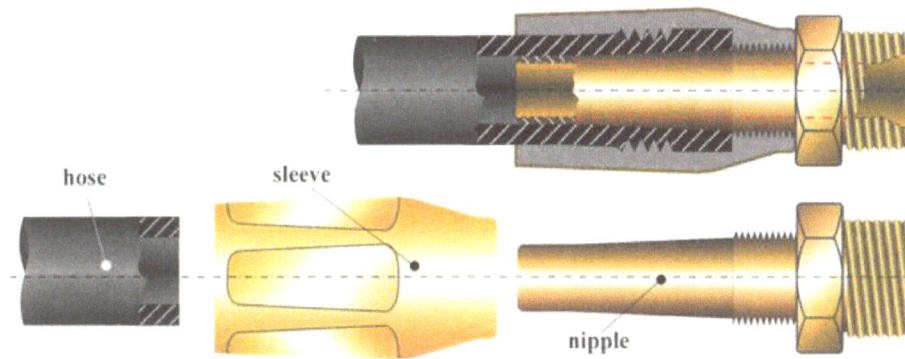

Fig. 3.66 - Reusable Hose Couplings (Courtesy of Assufluid)

Permanent Couplings: As shown in Fig. 3.67, the nipple ensures seizing and prevent hose narrowing. They require *crimping equipment* to assemble to a hose.

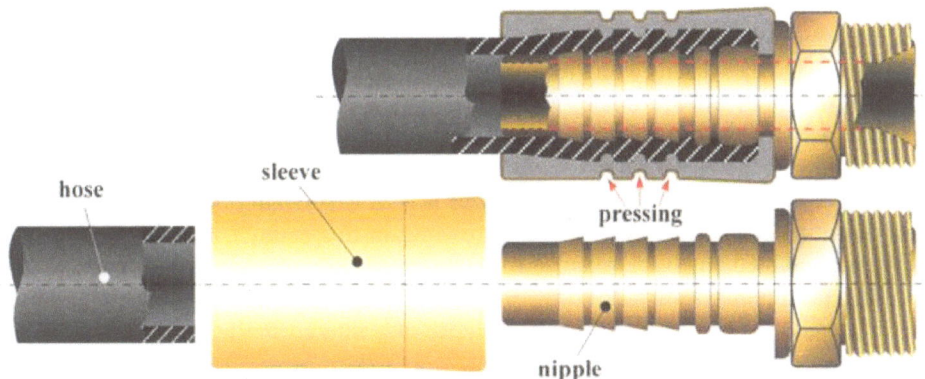

Fig. 3.67 - Permanent Hose Couplings (Courtesy of Assufluid)

As shown in Fig. 3.68, permanent hose couplings are available in either preassembled or two-piece configurations.

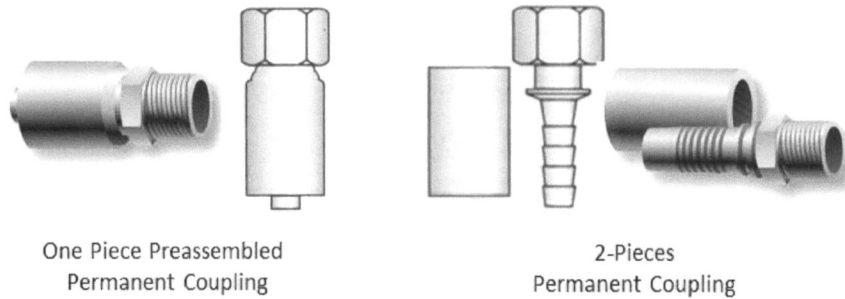

One Piece Preassembled
Permanent Coupling

2-Pieces
Permanent Coupling

Fig. 3.68 - Permanent Hose Couplings (Courtesy of Gates)

3.6.7.3- Hose Couplings Configurations

Figure 3.69 shows the common configuration of hose couplings. However, Fig. 3.70 shows typical configurations for high and medium pressure.

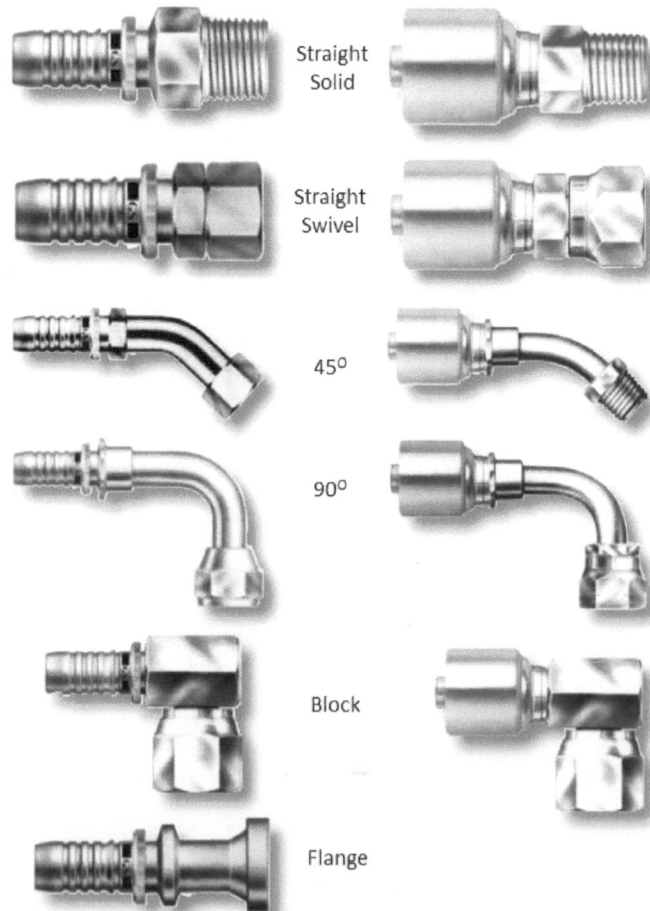

Straight
Solid

Straight
Swivel

45°

90°

Block

Flange

Fig. 3.69 - Hose Coupling Common Configurations (Courtesy of Gates)

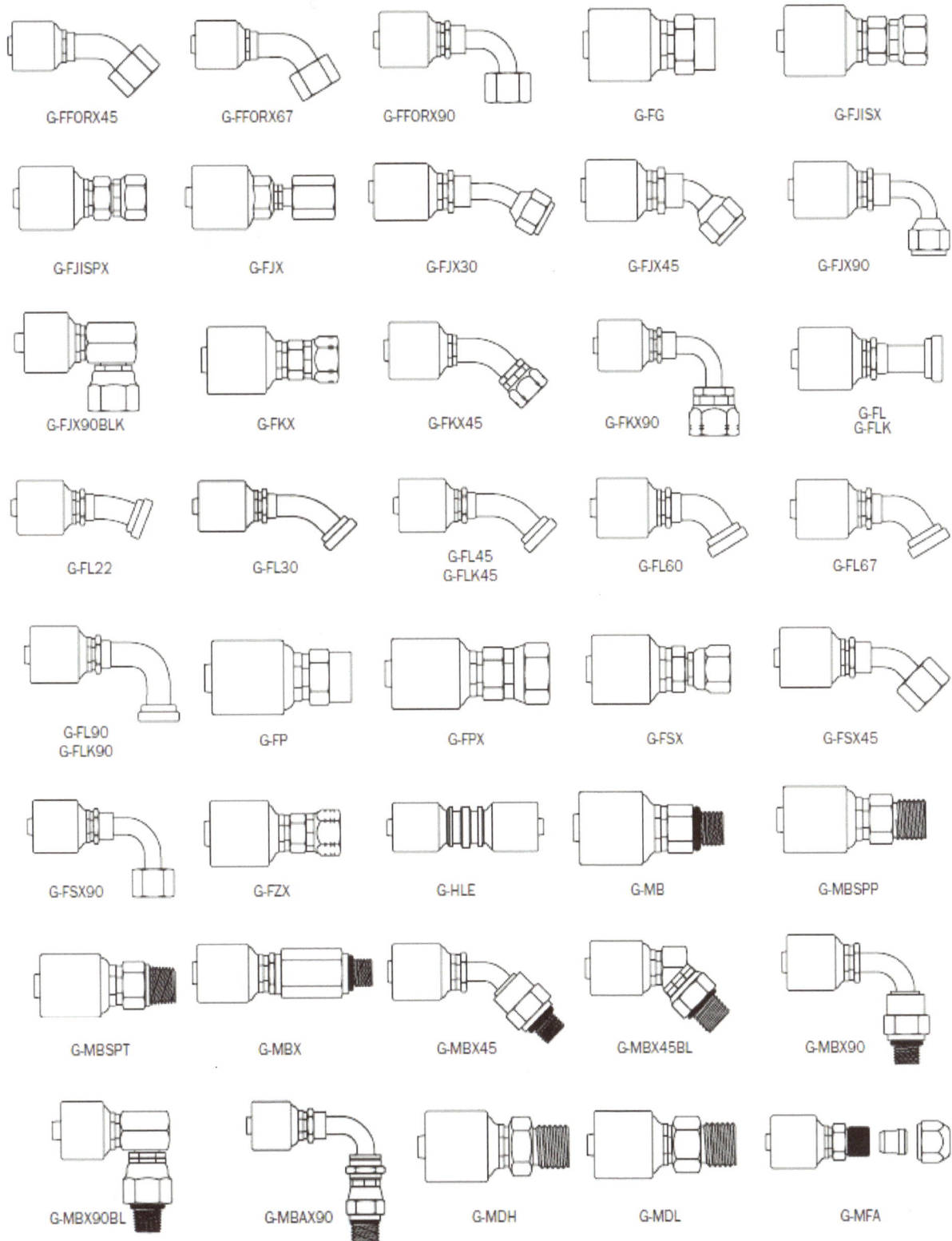

G-FFORX45	G-FFORX67	G-FFORX90	G-FG	G-FJISX
G-FJISPX	G-FJX	G-FJX30	G-FJX45	G-FJX90
G-FJX90BLK	G-FKX	G-FKX45	G-FKX90	G-FL G-FLK
G-FL22	G-FL30	G-FL45 G-FLK45	G-FL60	G-FL67
G-FL90 G-FLK90	G-FP	G-FPX	G-FSX	G-FSX45
G-FSX90	G-FZX	G-HLE	G-MB	G-MBSPP
G-MBSPT	G-MBX	G-MBX45	G-MBX45BL	G-MBX90
G-MBX90BL	G-MBAX90	G-MDH	G-MDL	G-MFA

**Fig. 3.70 - Hose Coupling Configurations for Medium and High Pressure
(Courtesy of Gates)**

3.6.7.4- Hose Crimping

Crimpable Coupling: Figure 3.71 shows a typical design of one piece crimpable hose coupling. The teeth in the crimpable fittings bite down to the hose wire for a metal-to-metal grip with maximum integrity. One-piece fittings can be combined with many types of hose covers.

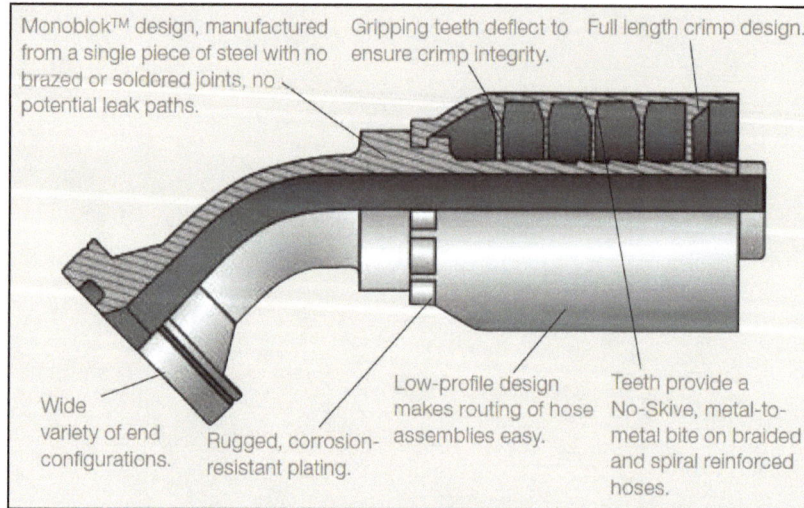

**Fig. 3.71 - Typical Design of Preassembled Cimpable Hose Coupling
(Courtesy of Parker)**

Crimping Machines: As shown in Fig. 3.72, a crimper is a hydraulic ram that uses fluid under pressure to extend the ram and crimp the fittings. They are available in various styles and power capacity.

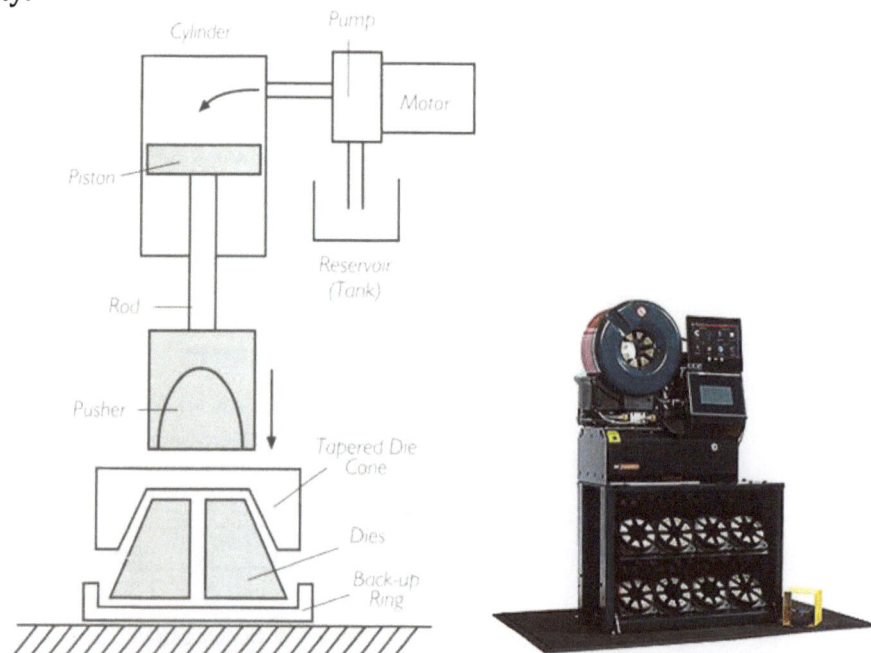

Fig. 3.72 - Hose Crimping Machine (Courtesy of Gates)

Hose Crimping Method:
Before crimping a hose, be aware of the following:

- Never re-crimp used hose with permanent or field attachable couplings.
- As shown in Fig. 3.73, some hose design requires that part of the rubber cover be removed from the end so that the fitting can be installed.

Fig. 3.73 - Prepare a Hose for Crimping

For a specific crimping machine, refer to instructions provided by the appropriate operating manual. However, the following steps shows general guidelines:

1. Setup the crimping machine referring to the manufacturer' instructions identifying the following information:
   A. Ferrule selection (if necessary).
   B. Die selection.
   C. Approximate crimp setting.
   D. Finished crimp diameter.
2. Load the selected die into the crimper.
3. Adjust the machine to the proper crimp setting.
4. Insert the hose assembly and properly align the coupling with the die fingers, as required by the particular model crimper that you're using.
5. Install die cone, if used.
6. Always wear safety glasses and keep hands and clothing away from the moving parts of the crimper.
7. Activate the crimping mechanism.
8. Remove crimped assembly from dies and measure final crimp diameter.

Measuring Crimp Diameters: As shown in Fig. 3.74, *Crimp Diameter* is measured between the ridges, halfway between the top and bottom of the coupling. When using calipers, be sure the caliper fingers do not touch the ridges.

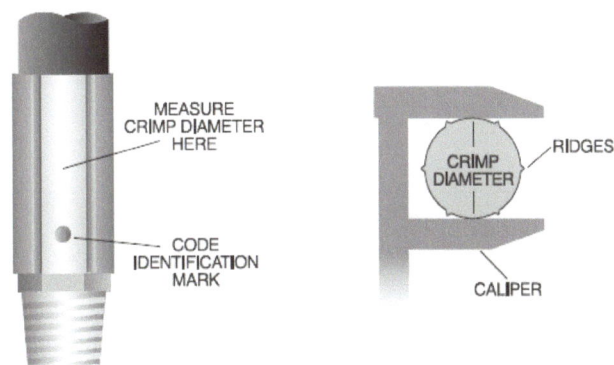

Fig. 3.74 – Measuring Crimp Diameter (Courtesy of Gates)

3.6.8- Quick Connect Couplings

Conventional Quick Connect Coupling: A *Quick* Connect Coupling is a component which can quickly connect and disconnect hoses without the use of tools or special devices. As shown in Fig. 3.75, according to ISO 16028, a *Flat Face* quick connect coupling consists of the following two pieces:
- **Coupler:** Other nomenclatures are "Female", "Socket", "Body".
- **Nipple:** Other nomenclatures are "Male", "Plug", "Adapter".

Connection is performed by pushing the movable part (usually the coupler) towards the fixed part (usually the nipple). Disconnection is performed by pulling the nut (2) in the coupler backward. Such couplings are used for lines where no residual pressure or fluid leakage is found upon connection or disconnection.

Caution: These types of couplings can exhibit high-pressure drop (5-10 bar) or more if not sized properly. These couplings should be sized based on the internal flow area of the coupling, which is typically significantly less than the internal area of the connection part.

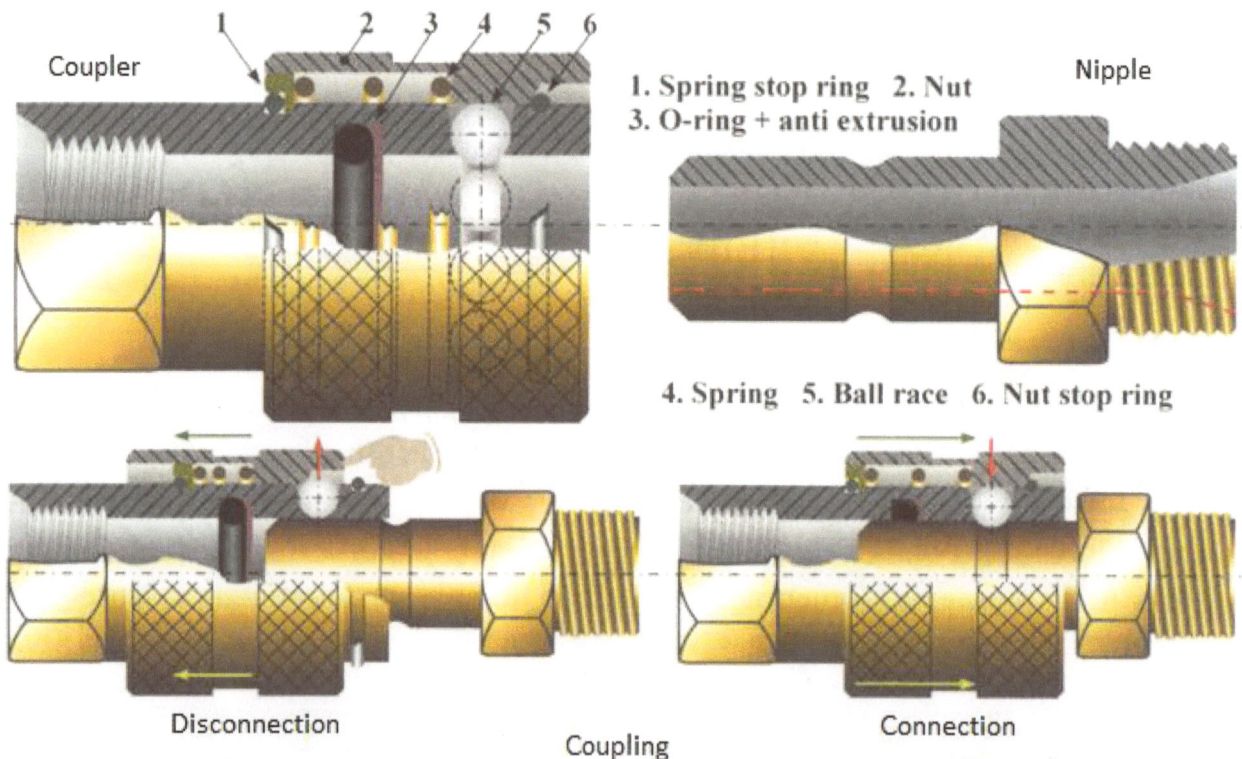

Fig. 3.75 - Flat Face Quick Connect Coupling (Courtesy of Assufluid)

Quick Connect Coupling with Nonreturn Valve: Figure 3.76 shows a single non-return valve (7) is included in the female coupler design. When the coupling is disconnected, if there is pressure, the fluid itself presses on the surface ensuring perfect sealing.

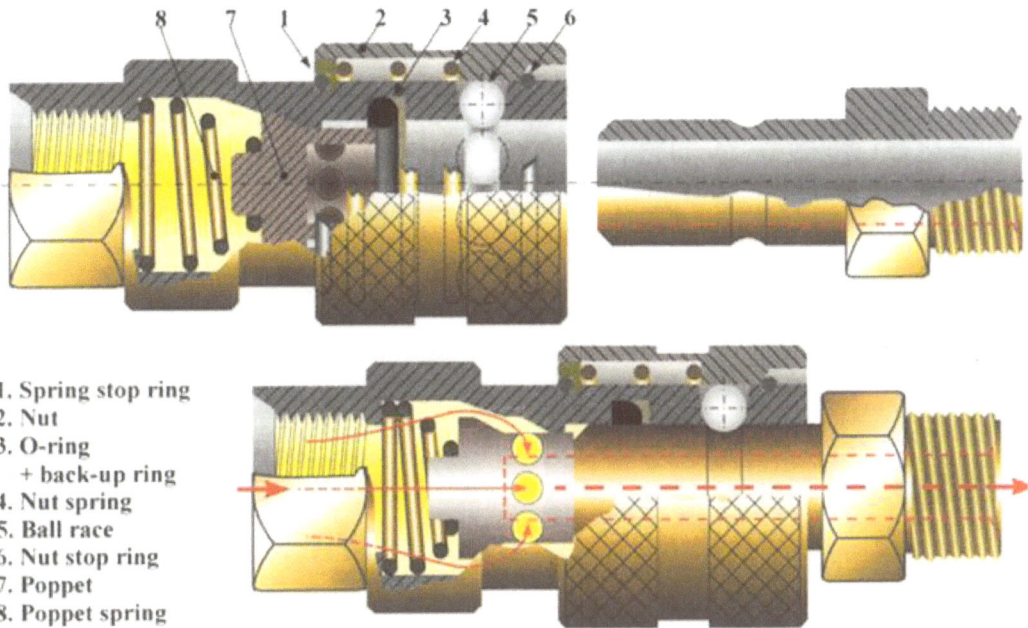

1. Spring stop ring
2. Nut
3. O-ring
   + back-up ring
4. Nut spring
5. Ball race
6. Nut stop ring
7. Poppet
8. Poppet spring

Fig. 3.76 - Quick Connect Coupling with Single Non-Return Valve (Courtesy of Assufluid)

Figure 3.77 shows another configuration where two non-return valves are included, one in the female side (7) and one in the male side (9). Such quick connect coupling is recommended for connecting two pressurized lines. Couplings with non-return valves work under pressure 400-600 bar (6000-9000 psi) depends on the design of the couplings.

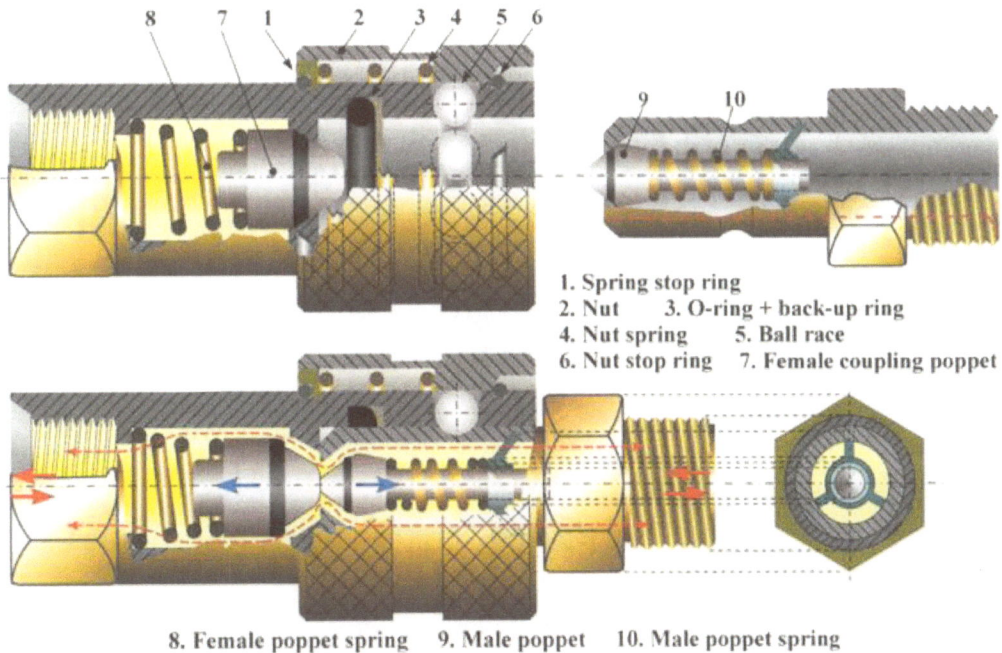

1. Spring stop ring
2. Nut   3. O-ring + back-up ring
4. Nut spring   5. Ball race
6. Nut stop ring   7. Female coupling poppet

8. Female poppet spring   9. Male poppet   10. Male poppet spring

Fig. 3.77 - Quick Connect Coupling with Double Non-Return Valves (Courtesy of Assufluid)

Innovative iLok™ Coupling: In some applications, such as mining machines, workers often are involved in dangerous practices to free the couplings when it's time to move the equipment. They use hammers to beat on the coupling ferrule or the hose itself. The result is often a cracked ferrule or damaged hose, leading to a leak or catastrophic burst when lines are re-pressurized. The *iLok™ Coupling* is an innovative quick connect coupling. Figure 3.78 shows the male and the female side of such a connector that has the following characteristics:

- Easy to connect and disconnect by hand.
- Withstands high impulse applications.
- Use a secure, visible locking system.
- Release residual pressure away from workers during disconnection.
- Compact to fit in tight spaces and smooth to prevent abrasion against adjoining hoses.

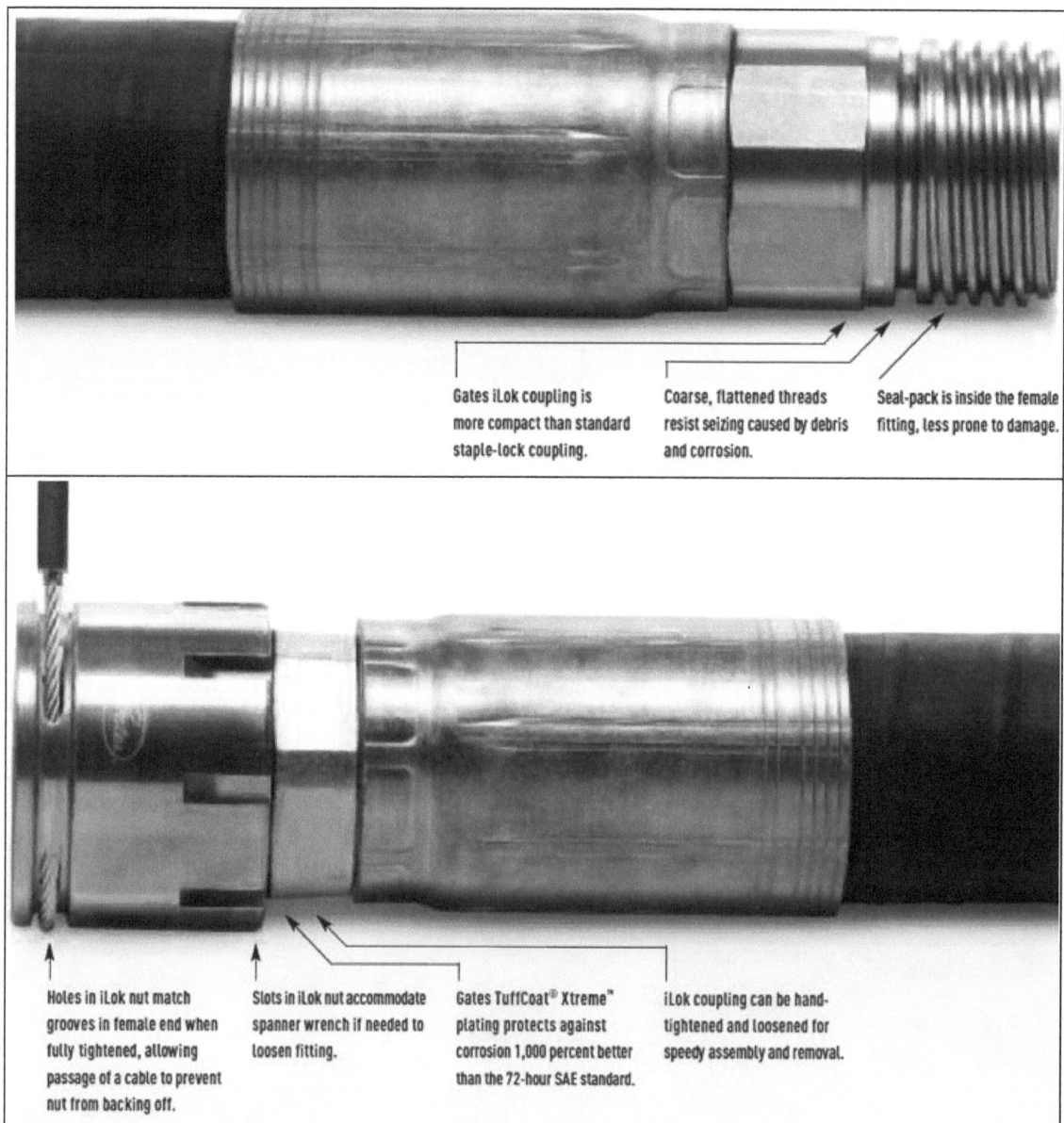

Gates iLok coupling is more compact than standard staple-lock coupling.

Coarse, flattened threads resist seizing caused by debris and corrosion.

Seal-pack is inside the female fitting, less prone to damage.

Holes in iLok nut match grooves in female end when fully tightened, allowing passage of a cable to prevent nut from backing off.

Slots in iLok nut accommodate spanner wrench if needed to loosen fitting.

Gates TuffCoat® Xtreme™ plating protects against corrosion 1,000 percent better than the 72-hour SAE standard.

iLok coupling can be hand-tightened and loosened for speedy assembly and removal.

Fig. 3.78 - Male and Female Parts of iLok™ Quick Connect Coupling (Courtesy of Gates)

As shown in Fig. 3.79, when fully connected, openings in the iLok nut align with a groove in the female fitting to accommodate a cable lock. The cable won't pass through the grooves unless the nut is completely tightened, providing positive proof of a secure connection. The cable lock "flags" indicates that a connection has been made. The iLock has the following technical specifications:

- Impulse tested to 133% of operating pressure at 250°F (+121°C) for 1 million impulse cycles with no leaks or failures.
- Exceeds Code 62 of SAE J518 flange specifications.
- 4:1 design factor (burst pressure to working pressure ratio).
- TuffCoat® Xtreme™ plating provides red rust protection that exceeds the 72-hour SAE standard by 1000 percent.
- MSHA-approved for underground mining applications.

Fig. 3.79 - Locking Mechanism of iLok™ Quick Connect Coupling (Courtesy of Gates)

Innovative TLX Coupling: Figure 3.80 shows same concept has been developed by another manufacturer under a brand name of *TLX Quick Connect Coupling*.

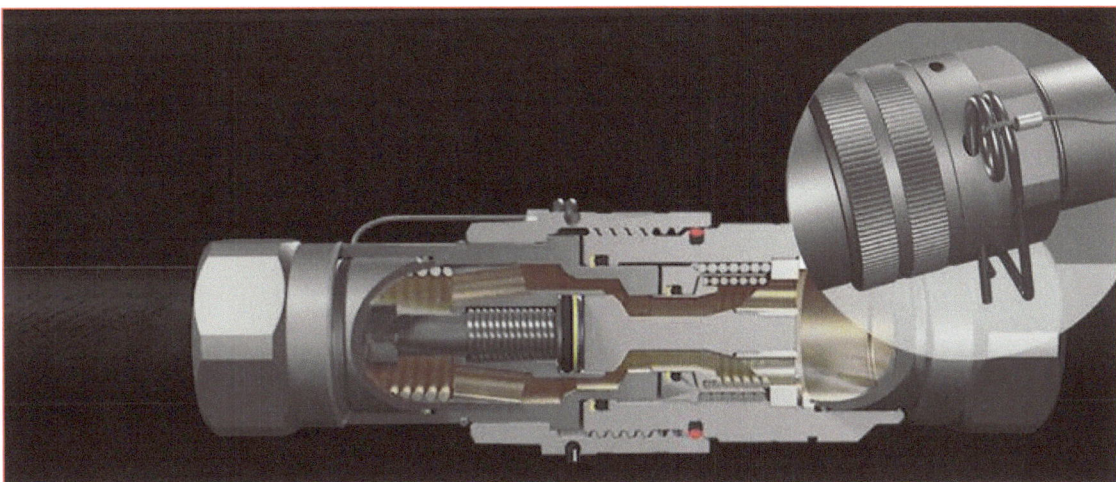

Fig. 3.80 - Locking Mechanism of TLX Quick Connect Coupling (Courtesy of CEJN)

Universal Push-to-Connect (UPTC) Assembly:

<u>Description (Fig. 3.81)</u>: *Universal Push-to-Connect* (*UPTC*) assembly is designed to achieve reliable leak-free connections for hose and tube assembly. UPTC assembly utilizes standard O-Ring Face Seal (ORFS) or EO (24º DIN cone) fitting bodies, and it is suited for hydraulic hose (rubber or thermoplastic) and tube (inch or metric) assemblies.

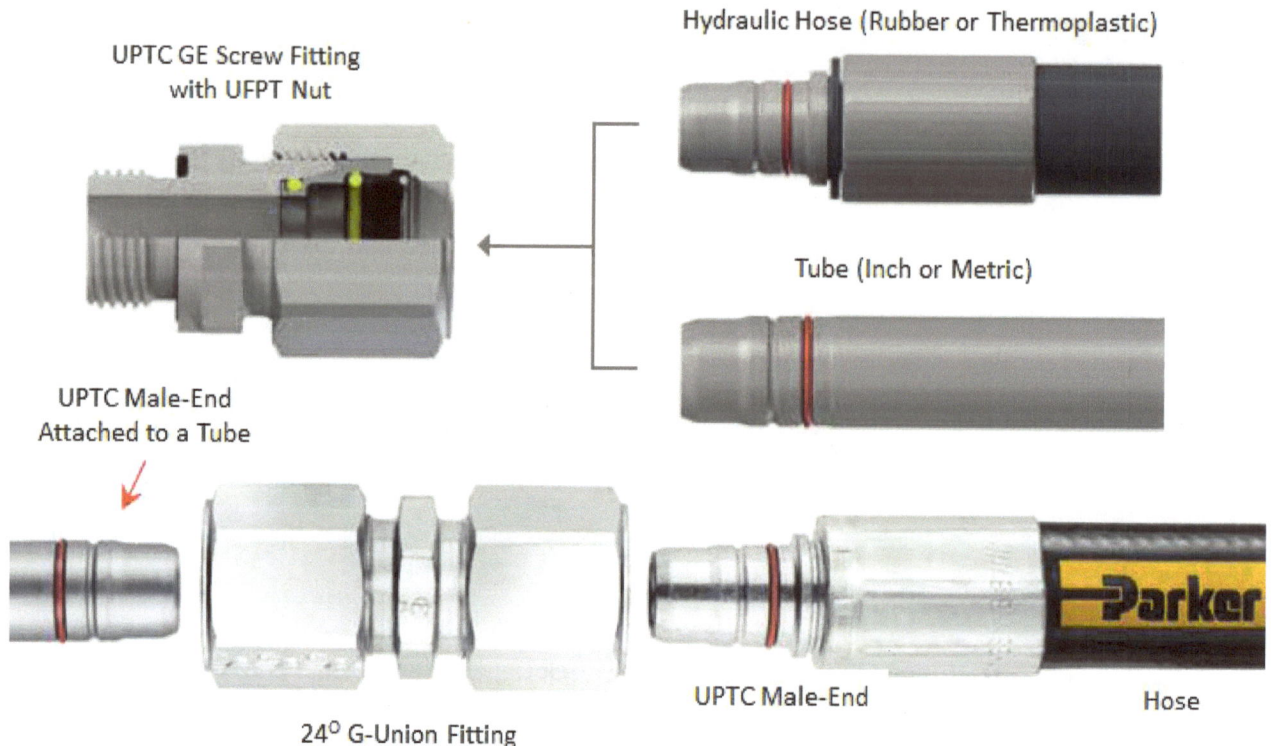

Fig. 3.81 - Universal Push-to-Connect (UPTC) (Courtesy of Parker)

<u>Features:</u>
- <u>Assembly Time Savings:</u> Simple push-to-connect design requires low pushing force with no special assembly tools.
- <u>Space Saving:</u> Easy to use in hard-to-reach areas and allows for a more compact system.
- <u>Assured Proper Connection:</u> Visual and tactile installation indicators easily help to assure proper connection, assembly, inspection, and diagnosis to minimize error and its costly consequences.
- <u>Elimination of Hose Twist:</u> Self-aligning nipple eliminates hose twisting during assembly for longer service life.
- <u>Leak-Free Performance:</u> Leak-free performance is assured by mathematical simulation, rigorous lab tests and significant field-tests with leading equipment manufacturers.
- <u>Easy Design Implementation:</u> UPTC allows for incorporation into any current hydraulic system with virtually no adverse effect on other components, or the system.
- <u>Excellent Field Serviceability:</u> Allowing standard wrench disassembly and damaged hose replacement.

<u>Assembling (Fig. 3.82):</u> As shown in the figure, assembly process is as easy as pushing the push-in male part (1) inside the system nut (2) with low pushing force as low as 60N. When a click is heard, and the red ring (6) isn't visible, this means that the stainless-steel clip ring (3) is snapped correctly, and the connection is done properly.

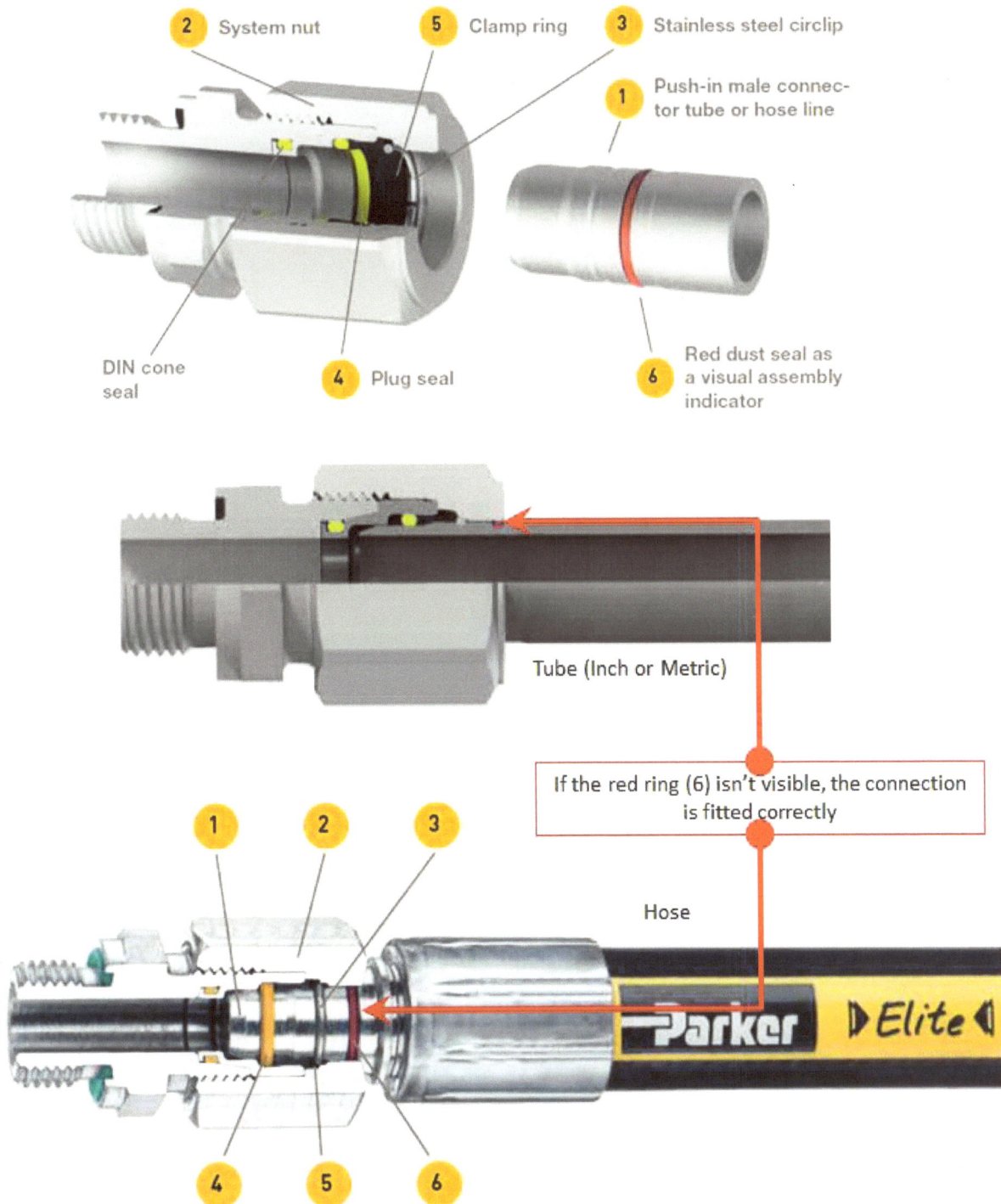

Fig. 3.82 - Assembling of Universal Push-to-Connect (UPTC) (Courtesy of Parker)

Sealing (Fig. 3.83): after proper assembling, sealing is taken place by series of sealing defense lines; the cone seal, the plug seal (4), and the clamp ring (5).

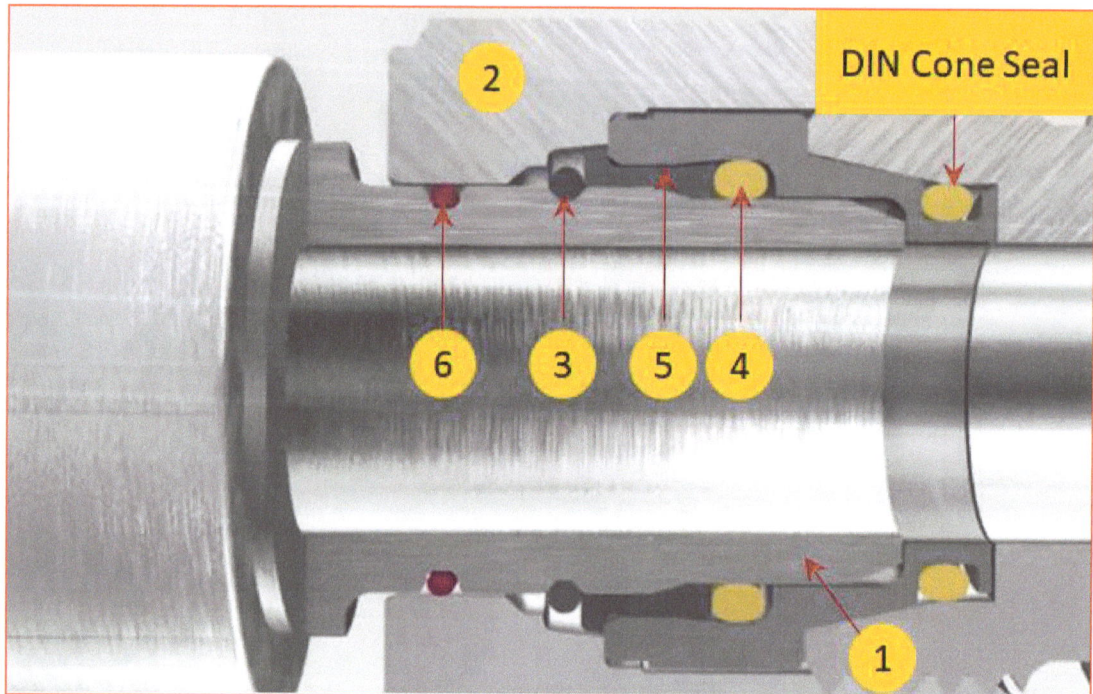

Fig. 3.83- Sealing of Universal Push-to-Connect (UPTC) (Courtesy of Parker)

Disassembling (Fig. 3.84): As shown in the figure, during disassembling, the system nut (2) will be removed from the screw socket. The male end and UPTC system nut remain as a single unit.

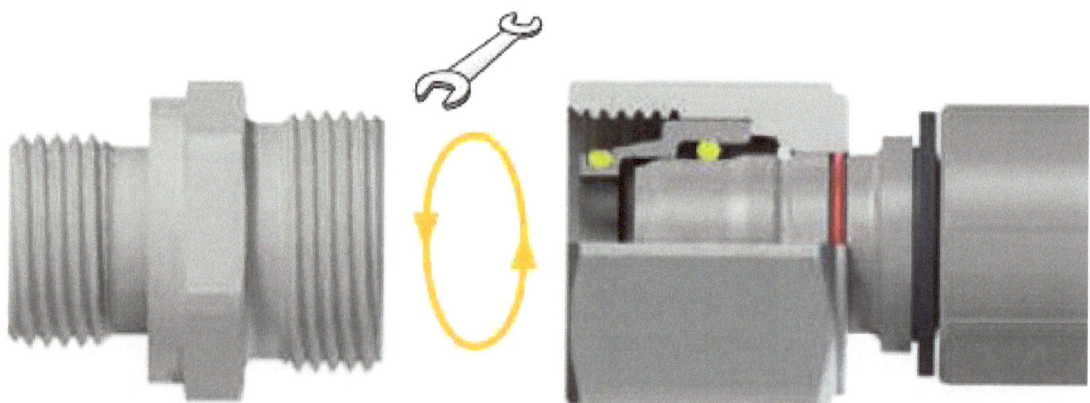

Fig. 3.84 - Disassembling of Universal Push-to-Connect (UPTC) (Courtesy of Parker)

Push-Lok Assembly for Low Pressure:

Description (Fig. 3.85): *Push-Lok* is an easy-to-use complete line of hose and fittings for low pressure applications. They are structured, like high pressure hoses, containing an inner tube, reinforcement layer, and a protective cover. In applications where several hose lines carry different media, Push-Lok colors reduce timely "tracing" of lines, preventing connection/disconnection of wrong line and unnecessary, downtime.

Features:
- Easy assembly with no clamps or special tools is required during installation.
- Time and cost saving.
- Ensures reliable, durable, leak-free service.
- High functional safety with a design factor of 4.
- Wide range of hose and fittings for a wide range of applications.
- Wide range of fluid compatibility that is color-coded.

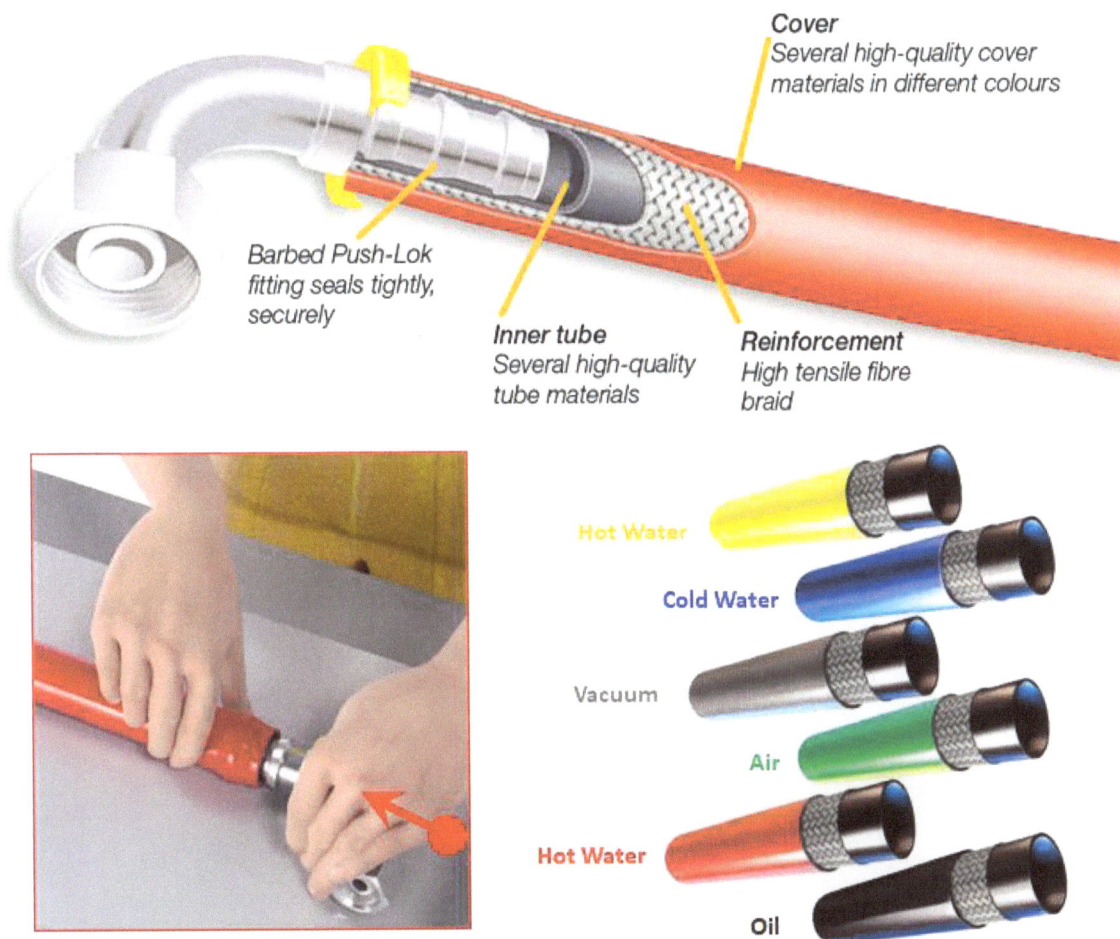

Fig. 3.85 - Push-Lok Hose and Fitting Assembly (Courtesy of Parker)

Assembling and Disassembling (Fig. 3.86): During assembly, Push-Lok fittings will provide an effective grip only when the Push-Lok hose is pushed fully on the insert, where the cropped end of the hose should be fully concealed by the plastic collar.

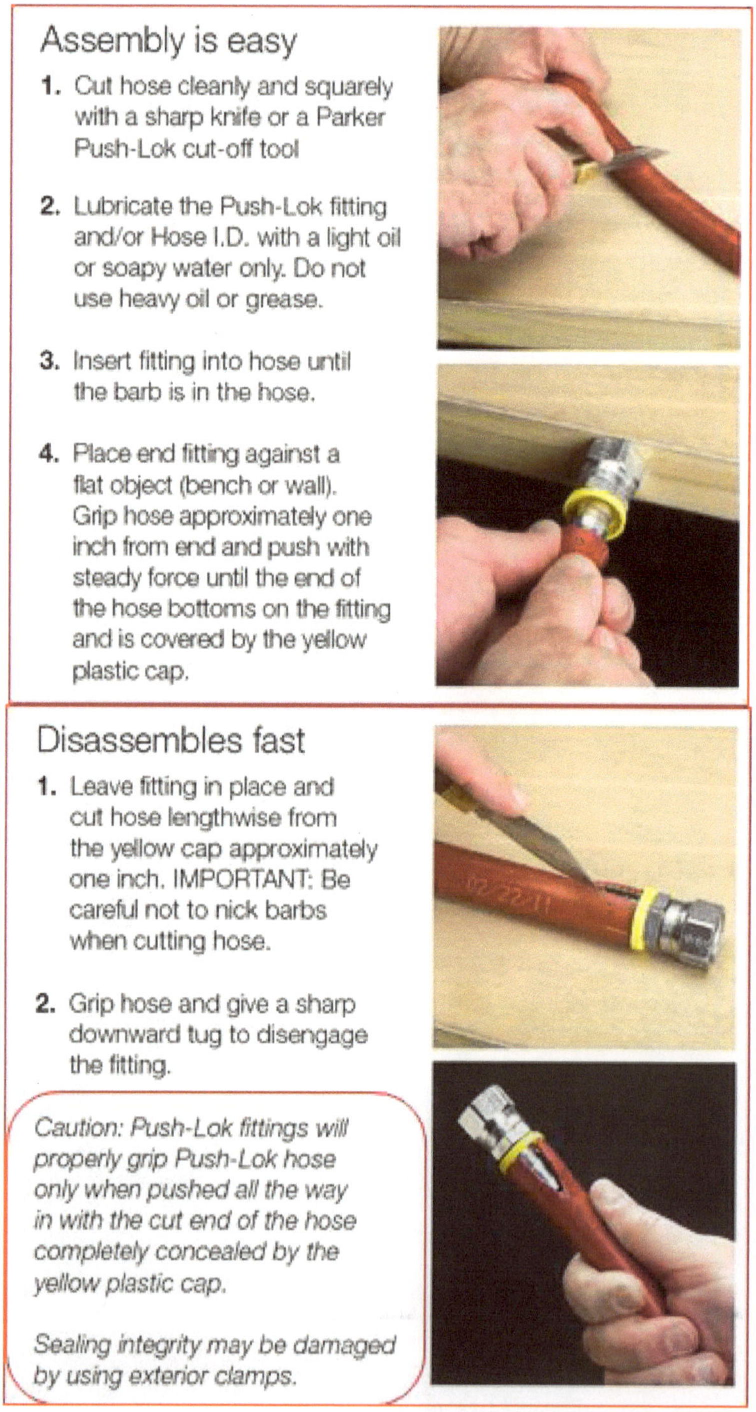

Assembly is easy

1. Cut hose cleanly and squarely with a sharp knife or a Parker Push-Lok cut-off tool

2. Lubricate the Push-Lok fitting and/or Hose I.D. with a light oil or soapy water only. Do not use heavy oil or grease.

3. Insert fitting into hose until the barb is in the hose.

4. Place end fitting against a flat object (bench or wall). Grip hose approximately one inch from end and push with steady force until the end of the hose bottoms on the fitting and is covered by the yellow plastic cap.

Disassembles fast

1. Leave fitting in place and cut hose lengthwise from the yellow cap approximately one inch. IMPORTANT: Be careful not to nick barbs when cutting hose.

2. Grip hose and give a sharp downward tug to disengage the fitting.

Caution: Push-Lok fittings will properly grip Push-Lok hose only when pushed all the way in with the cut end of the hose completely concealed by the yellow plastic cap.

Sealing integrity may be damaged by using exterior clamps.

**Fig. 3.86 - Assembling and Disassembling Push-Lok Hose and Fitting Assembly
(Courtesy of Parker)**

Innovative Quick Connect System for Low Pressure (Fig 3.87): The innovative *Quick-Connect* system is an easy to install and cost-effective alternative to conventional hose connection fittings. The patented system is suitable for both suction and return lines and has already been tested more than 100,000 times in many different applications in the field. Tubes that are equipped with these connections can simply be plugged into the appropriate adapters by the user and fixed with the enclosed locking clips. They work in temperature up to 120 °C (250 °F) and operating pressure up to 10 bar (150 psi).

Plastics and sealing geometry are selected in such a way that the connections can also be reliably installed to metal adaptors. This makes an attractive product range available which offers the right connection for practically every application.

The locking clip prevents the hose from being pulled off the adaptor. However, the line remains rotatable. In many applications this represents a further advantage since the line can thus move into a position as free of stress as possible.

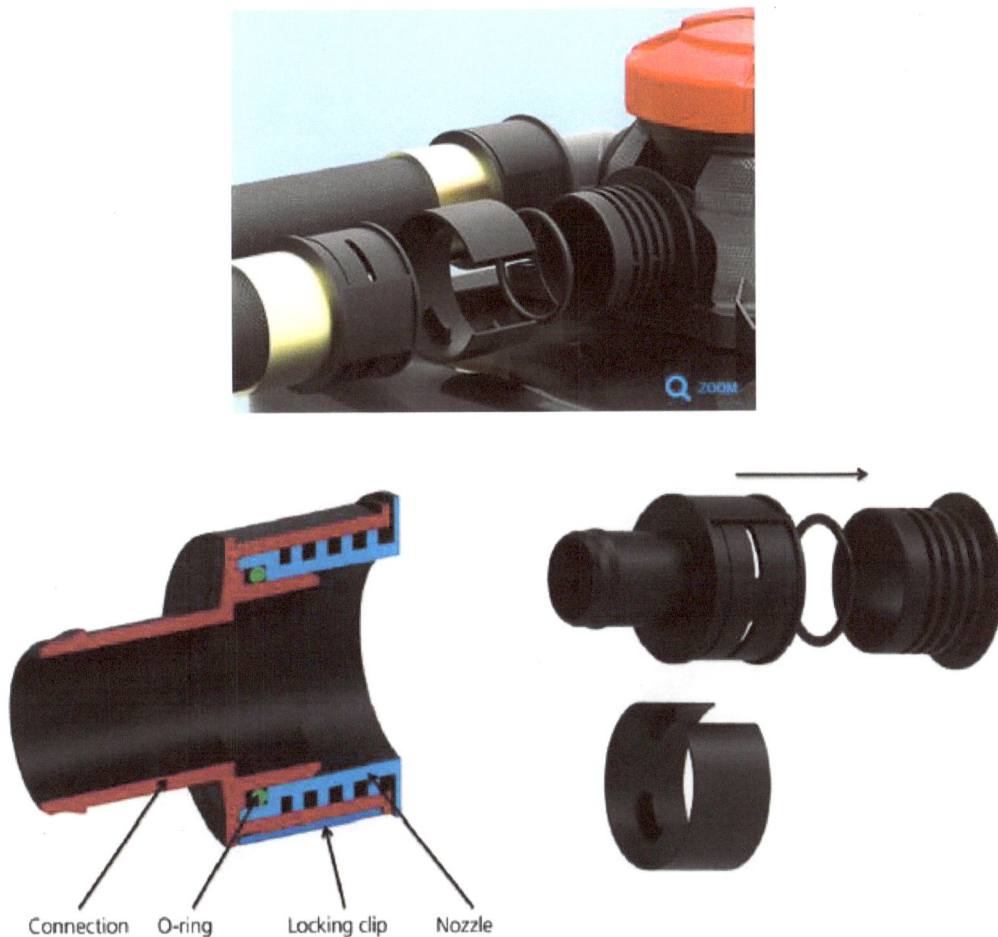

Connection   O-ring   Locking clip   Nozzle

Fig. 3.87 - Innovative Quick Connect Coupling (Courtesy of ArgoHyots)

CEJN Multi-X Quick Connect Couplings (Fig. 3.88): *Multi-X Quick Connect Couplings* are designed to connect multiple lines simultaneously with great flexibility, high performance, easy installation and trouble-free operation. They are designed to meet and exceed the demands of even the most challenging mobile hydraulic application.

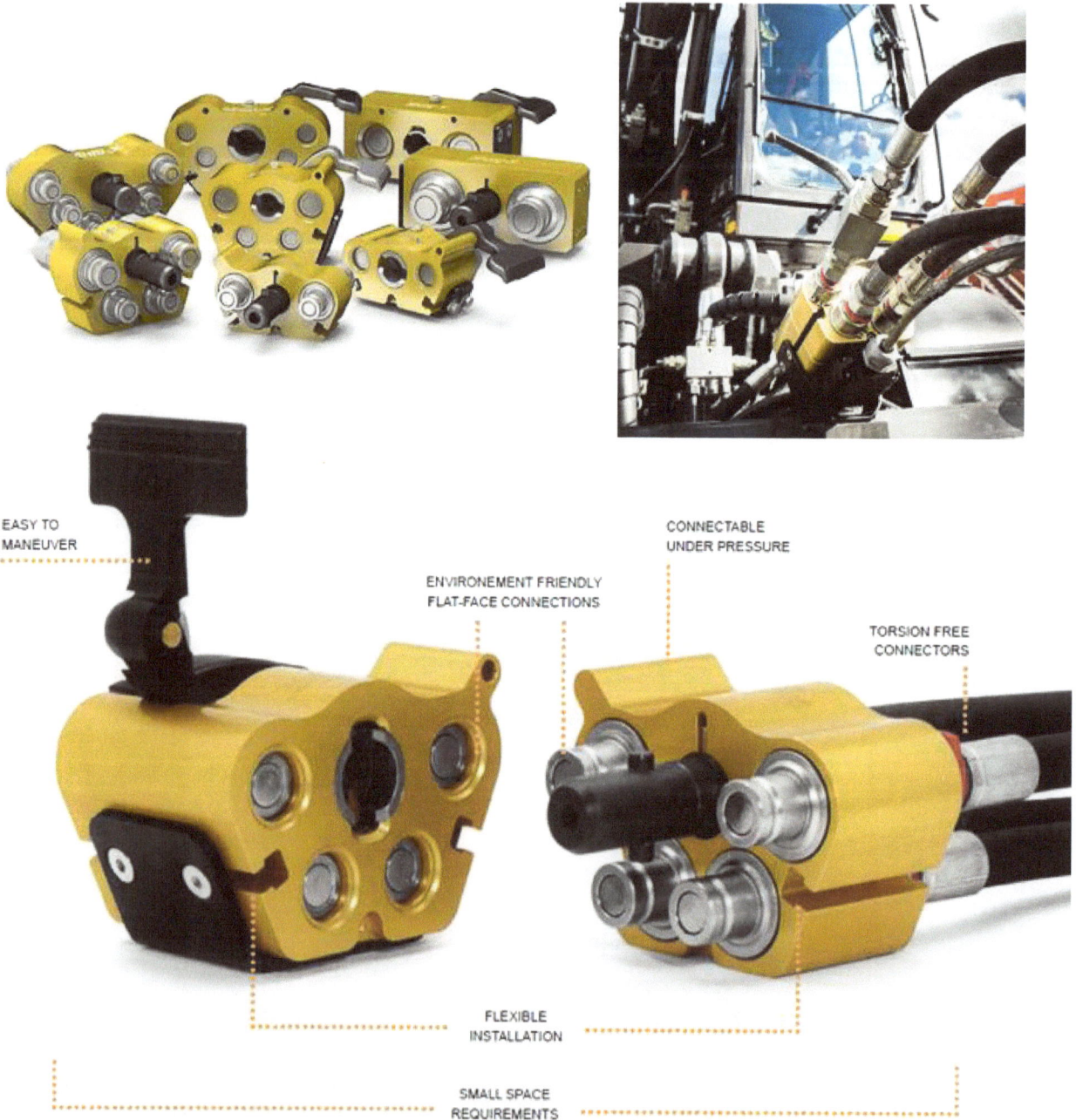

EASY TO
MANEUVER

CONNECTABLE
UNDER PRESSURE

ENVIRONEMENT FRIENDLY
FLAT-FACE CONNECTIONS

TORSION FREE
CONNECTORS

FLEXIBLE
INSTALLATION

SMALL SPACE
REQUIREMENTS

Fig. 3.88 - Multi-X Quick Connect Coupling (Courtesy of CEJN)

Figure 3.89 shows the concept of operation of Multi-X quick connect couplings.

Working pressure of up to 350 bar (5000 psi) can be used on half the port simultaneously with the other half of the port used as return lines with a maximum pressure 50 bar (725 psi).

Both the female and male plate can be used in the fixed part.

Electric connectors can also be easily attached.

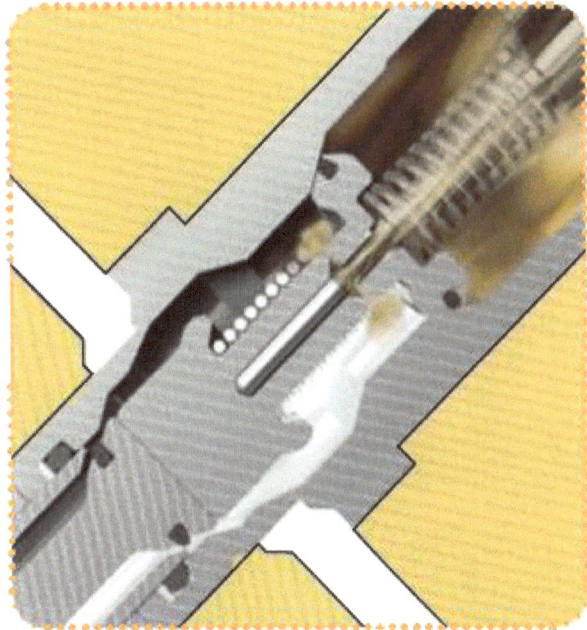

Pressure is connected without spillage making it simple to connect with residual pressure in the system

Fig. 3.89 - Conceptual Operation of Multi-X Quick Connect Coupling (Courtesy of CEJN)

3.6.9- Hose Guards

Using *Hose Guards* is recommended for the following reasons:

- **Hose Protection:** Hose guards prolong the service life of the hose because they offer hose protection against harsh environmental and abrasion conditions.
- **Operator Protection:** Hose guards provide protection for the machine operators against whiplash and oil injection hazards when a hose assembly fails.
- **Machine Protection:** Hose guards prolong the service life of the hose and reduce the costly machine downtime and liabilities.

Hose guards are available in several styles and sizes as follows:

- **Metallic Spring Guards and Armor Guards (Fig. 3.90):** They distribute bending radii to avoid kinking in hose lines and protect hose from abrasion and deep cuts. Guards are constructed of steel wire and plated to resist rust.

Steel Spring Guard

Steel Armor Guard

sapphirehydraulics.com

Fig. 3.90 - Metallic Spring and Armor Guards (Courtesy of Parker)

- **Nonmetallic Spiral Wrap Hose Guards (Fig. 3.91):** They protect against abrasion and extreme physical abuse. They are resistant to oil, lubricants, gasoline, most solvents and can withstand ambient temperatures from -40° to +300° F. They are available in different colors to make it easy for hose identifications via color coding, e.g. red color for high pressure hoses and blue color for low pressure hoses.

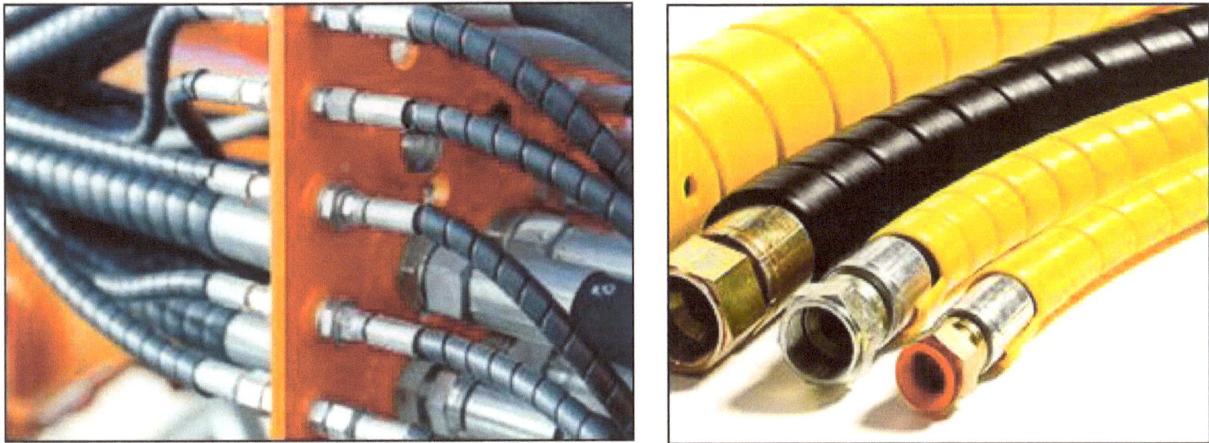

Fig. 3.91- Nonmetallic Spiral Wrap Guards (sapphirehydraulics.com)

- **Hose Shields (Fig. 3.92):** They are durable wear shields made of an abrasion and hydrocarbon resistant material that is impervious to solvents, oils, grease, and gasoline.

- **Hose Protection Sleeves (Fig. 3.93):** They protect hoses from contamination, direct sunlight, and the working environment. They are made of heat and chemical resistant materials.

Fig. 3.92- Hose Shields
(www.epha.com)

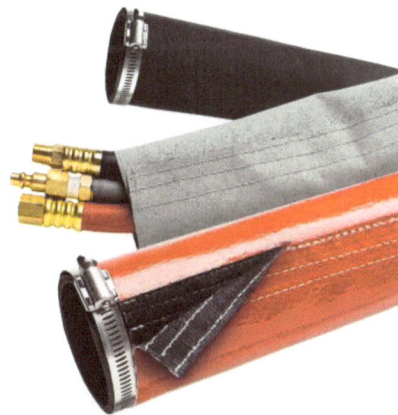

Fig. 3.93- Hose Protection Sleeves
(www.sealsaver.com)

Hose Oil-Injection Protection Sleeves (Fig. 3.94):

A pinhole leak in a pressurized hydraulic hose can inject fluid through human body skin into the muscle, nerve, and tissue that can cause loss of life if not treated immediately. Ordinary nylon protection sleeves do not offer the level of safety needed when lives are at risk. As shown in the figure, a special sleeving system from Gates Corporation (LifeGuard™) protects workers within a three-feet circle around the line. This specially designed sleeve has the following characteristics:

- Working pressure 400 bar (6,000) psi and working temperature up to 100 °C (212°F).
- Compatible with a wide range of fluids, including environmentally friendly.
- Its inner layer is made of tightly woven, extruded filament nylon designed to absorb the energy of a hydraulic hose burst or pinhole leak by stretching up to 20 percent.
- The outer sleeve is resistant to abrasion but not specifically designed for abrasive environments. Nylon material layers contains escaped fluids and redirects them to the clamped ends of the hose.
- The sleeve is secured at either end of the hose with special "channel" clamps. The clamps allow leaking fluid to safely escape, so it will not collect behind the sleeve and cause a burst. Plus, leaked fluid allows for fast hose failure detection.

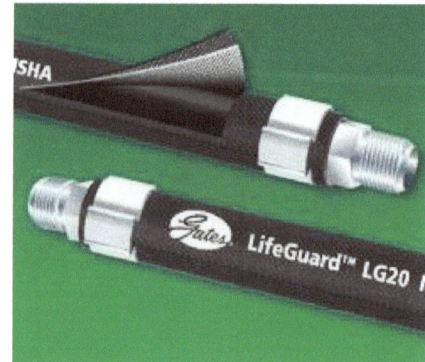

Fig. 3.94- Hose Oil-Injection Protection Sleeve "Lifeguard" (Courtesy of Gates)

3.6.10- Hose Whip Restraints

Major damage or injury can occur from whipping hoses, especially at higher pressures. As shown in Fig. 1.95, *Hose Whip Restraints* are designed to prevent whipping of a pressurized hose in the event of the hose separating from its fitting. So, the *Hose Whip Restraint* provides an additional level of safety by preventing damage to nearby equipment or injury to operators near the failed hose. The system consists of a hose collar and a cable assembly. The hose collar is selected based on the outside diameter of the hose, and the cable assembly is selected based on the type of hose connection.

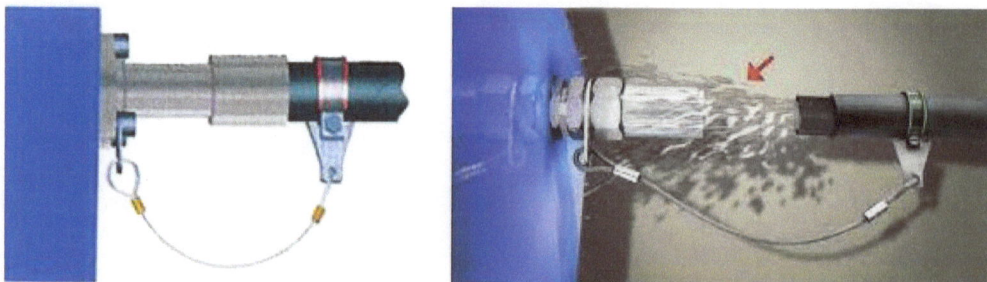

Fig. 3.95- Hose Wipe Restraint (Courtesy of Parker)

3.6.11- Hose Condition Monitoring

As shown in Fig. 3.96, the *LifeSense* from Eaton is a patented technology that intelligently monitors the condition of hose assemblies in real time. LifeSense can detect internal fatigue as well as external abrasion. LifeSense notifies users when a hose assembly is approaching the end of its useful life. Notification can be received on base units and mobile communication devices. The system requires DC power supply.

Fig. 3.96 - Real Time Hoses Condition Monitoring (www.hydrotechnik.com)

It was proved that, as shown in Fig. 3.97, using such a method of real time hose condition monitoring extends the service life of hoses as compared to time-bases traditional replacement method.

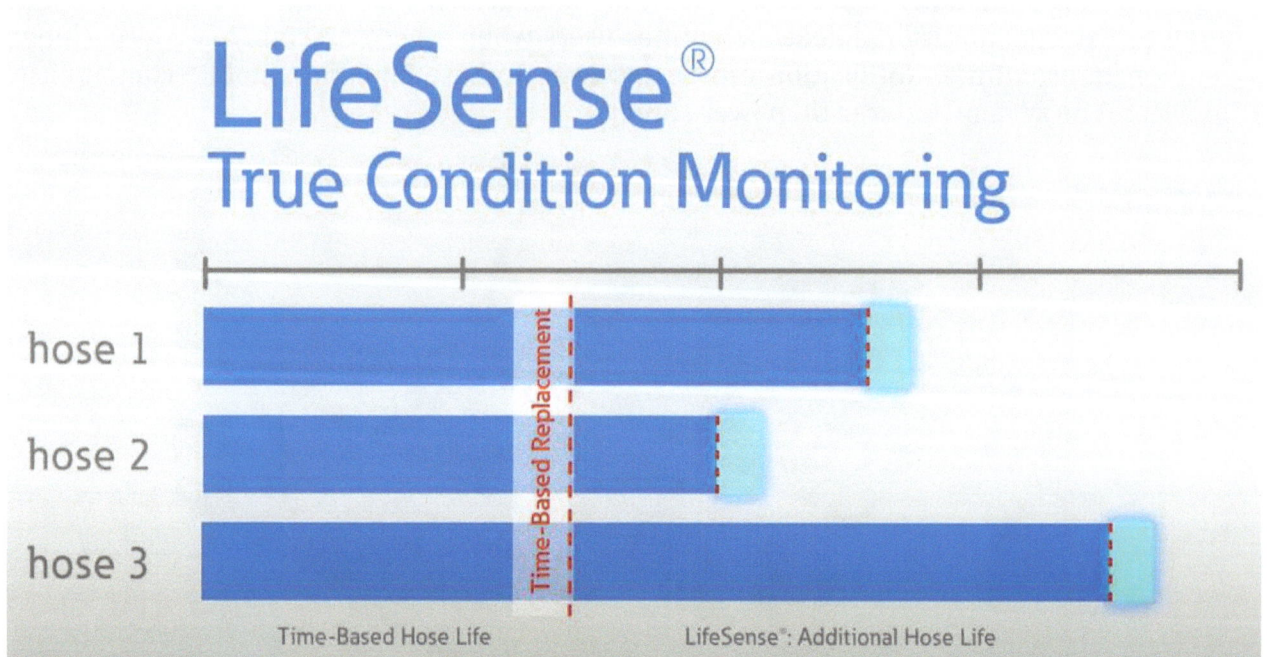

Fig. 3.97- Real Time Hoses Condition Monitoring Maximizes Hose Life (www.hydrotechnik.com)

3.6.12- Selection of Hose Couplings

Size: Make sure coupling size is properly selected in accordance with the size and type of the hose and the port of hydraulic component that the hose is connected to.

Working Pressure: Make sure that maximum allowable pressure for the couplings meets or exceeds the maximum allowable pressure for the hose. Some fittings don't seal well at high pressures and can develop oil leakage. O-ring-type fittings as well as solid port connectors work well at high pressures. Avoid the use of swivel staked nut couplings at extremely high pressures. Also, avoid the use of aluminum ferrules on high impulse applications.

Working Temperature: Mated surfaces type of fittings are recommended under extreme temperature fluctuations. Use of O-rings sealing fittings limits the maximum working temperature. It may be necessary to use O-ring materials that are suitable for high temperatures.

Fluid Compatibility: While hydraulic hose is commonly selected by its compatibility with fluid, couplings usually are not. However, it is advisable to check the chemical resistance charts for compatibility with coupling materials and O-rings.

Corrosion Resistance: In case of hose exposure to corrosive environment, hose couplings should be marked and selected as corrosion resistant and tested under SAE J516 and ASTMB I 17 *Salt-Spray Tests.*

Vibration: Coupling selection should consider if the end connection is exposed to motion and/or vibration, which can potentially weaken or loosen a connection. Use of split flange, or other couplings that use an O-ring for sealing, perform better under vibration. Avoid use of couplings that seal on the threads in such case.

Use of Quick Connect Coupling: check if the application requires the ability to connect and disconnect under pressure.

Use of Adaptors: You may want to select a coupling based on the need of adapters. Some couplings connect directly to a port, while others connect to adapters. Connecting directly to the port eliminates the need for an additional connection but can make installation more difficult. As shown in Fig. 3.98, Adapters can make installation easier; eliminate the need for coupling orientation but introduce an additional connection or possible leak point.

Fig. 3.98 - Use of Adaptors (Courtesy of Gates)

3.6.13- Hydraulic Hoses Selection Criteria

When selecting a hydraulic hose for an application, Fig. 3.99, provides a key word "STAMP" as a reminder for what to check. Interpretation of the key word is as follows:

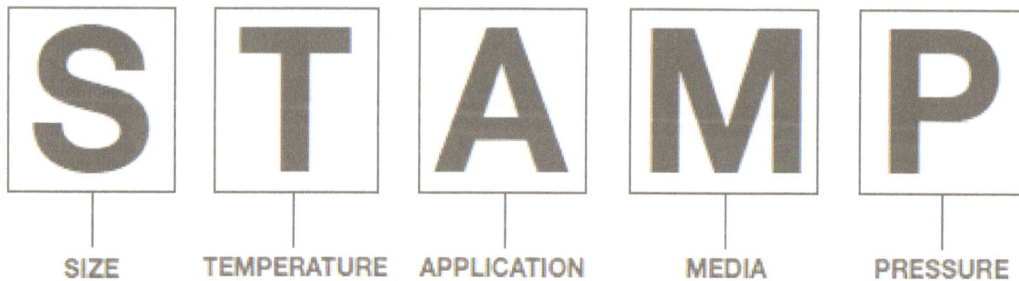

SIZE   TEMPERATURE   APPLICATION   MEDIA   PRESSURE

Fig. 3.99 - Keyword for Selecting Hydraulic Hoses (Courtesy of Parker)

Size: Line size has a direct effect on laminar flow condition and pressure losses. Therefore, a hose should be properly sized as per the rules presented in section 1.2.

Temperature: Working temperature of a hose must at least meet or exceed the application working temperature.

Application: Application in which the hose is going to be used should also be considered. For example, applications where hoses will encounter rubbing or abrasive surfaces, would be best handled by a family of abrasion resistant hoses with special cover.

Media (Fluid Compatibility) Material of a hose must be compatible the working hydraulic fluid. Hose manufacturers should provide hose-fluid compatibility charts.

Pressure: Maximum allowable static and dynamic pressure must be considered when specifying a hose as follows:

- Static Pressure: It is the maximum steady state pressure in the part of the system where this hose will be used. A common misconception that is the maximum pressure in a hydraulic system is the pressure setting of the relief valve. This isn't true in all cases. An example of that is pressure intensification at the rod side of a double acting differential cylinder when a system is designed with meter out speed control. In such a case, depending on the system design, configuration, and operating conditions, rod side pressure could be 3-4 times over the relief valve setting.

- Dynamic Pressure: Pressure pulses, spikes, and shocks are different forms of dynamic pressure. Pressure pulsation results from a cyclic load. Pressure spikes likely happened in transient time, e.g., when changing a direction of an actuator. Pressure shocks occur due to rapid increase in the load. Dynamic pressure causes fatigue of the transmission lines. Dynamic pressure likely isn't captured by pressure gauges, it rather requires special instrumentation, e.g., pressure transducer with appropriate readout device.

- Inside Negative Pressure: Suction hoses are structured with reinforcement layers to work under vacuum. If a pressure hose is used as a suction hose, it may collapse because pressure hoses are structured for inside positive pressure only.

- Outside Positive Pressure: If the environmental conditions aren't atmospheric, such as if a hose is used under water, make sure the hose structure can withstand the outside positive pressure. In this case, consult the manufacturer if necessary.

- Low-Pressure Hoses: *Low-Pressure* hydraulic hoses are designed for applications with operating pressures less than 20 bar (300 psi) such as lubrication systems or drain lines using mineral fluids. Their reinforcement layer is usually textile.

- Medium-Pressure Hose: *Medium-Pressure* hoses are designed for applications with operating pressures 20-200 bar (300 – 3000 psi) such as heavy-duty trucks, fleet vehicle, and aircrafts.

- High-Pressure Hoses: *High-Pressure* hoses are designed for applications with operating pressures up to 400 bar (6000 psi) such as construction equipment. These hoses are often called "two-wire" braid hose because they generally have a reinforcement of two-wire braids of high-tensile strength steel.

- Extremely High-Pressure Hoses: *Extremely High-Pressure* hoses are designed for applications with operating pressures above 400 bar (6000 psi) such as off-highway equipment and heavy-duty machinery where extremely high impulse or pressure surges are encountered. These hoses are often called "four-wire" or "six-wire" hoses.

In addition, the following conditions are also important to be reviewed:

Electrical Conductivity: Certain applications require that the hose be *nonconductive* to prevent electrical current flow or to maintain electrical isolation. For applications near high voltage electric lines, only special nonconductive hose can be used. Other applications require the hose and the fittings and the hose/fitting interface to be sufficiently conductive to drain off static electricity.

Length: Hose length must be wisely specified. Neither short nor long hoses are recommended. Short hoses can be overstretched when pressure is applied resulting in detaching end joints. Unnecessary long hoses result in vibration, rubbing, pressure losses, etc.

Hoses Pictogram: Figure 3.100 shows typical manufacturer's pictogram that are printed on the hose cover to indicate main hose specifications.

#	Standard Cover	Tough Cover	Super Tough	Hose I.D.		Hose O.D.		Working Pressure		Minimum Bend Radius		Weight		Vacuum Rating	
Part Number	387	387TC	387ST	inch	mm	inch	mm	psi	MPa	inch	mm	lbs/ft	kg/m	inches of Hg	kPa
387	ISO 18752 Performance														

Fig. 3.100 - Hydraulic Hoses Pictogram (Courtesy of Parker)

Figure 3.101 shows examples of hoses with pictogram marking.

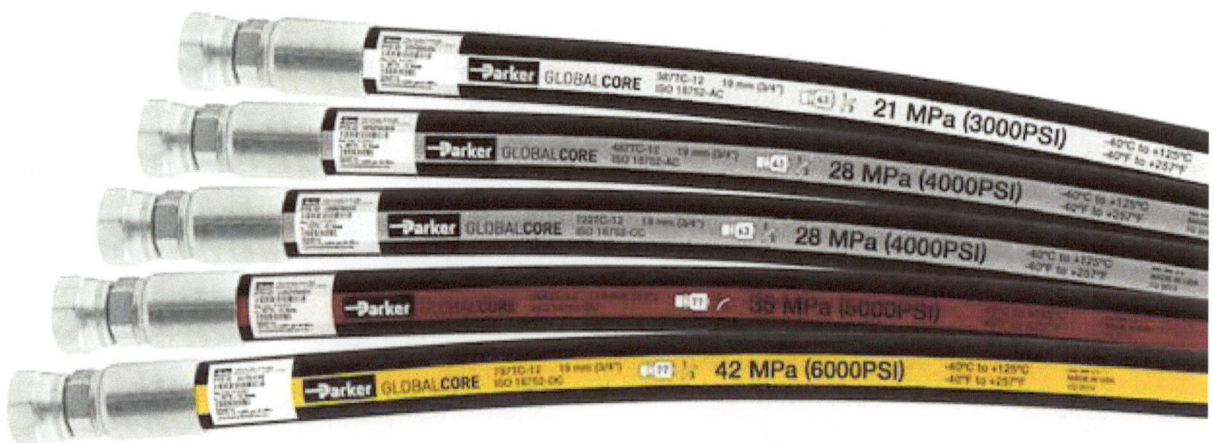

**Fig. 3.101 - Example of Hydraulic Hoses with Pictogram Marked
(Courtesy of Parker)**

Example 1 – High Pressure Hoses: Figure 3.102 shows an example of hoses that are developed for high and ultra-high-pressure applications. They are used in high pressure service tube cleaning applications, such as, heat exchanger tube cleaning in the chemical and refining industries and as waterblast hoses. They can handle pressures above 13,000 psi. Available with a Tough Cover, for improved abrasion resistance in many applications.

PERFORMANCE CHARACTERISTICS	
SPECIFICATIONS MET	DIN EN1829-2 compliant
DESIGN FACTOR	2.5
HOSE INSIDE DIAMETER	1/8, 5/32, 3/16, 1/4, 5/16 inch 3.2, 4, 4.8, 6.4, 7.9 mm
HOSE OUTSIDE DIAMETER	.280, .300, .310, .370, .460, .530 inch 7.1, 7.7, 7.9, 9.5, 11.6, 13.5 mm
MAXIMUM OPERATING PRESSURE	15000, 15950, 17400, 20300, 21750 psi 1034, 1100, 1200, 1400, 1500 bar
MAXIMUM OPERATING TEMPERATURE	158 °F 70 °C
MINIMUM OPERATING TEMPERATURE	14 °F -10 °C
MINIMUM BEND RADIUS	2-3/8, 3, 3-3/4, 4-3/8, 4-3/4 inch 60, 75, 95, 110, 125 mm
COMPATIBLE FITTING	AX, TX (dependent on hose)
MEDIA	Water, Chemicals
APPLICATION	Industrial Cleaning - Waterblast, Shipbuilding, Construction
COVER COLOR	Green, Blue (dependent on hose)
WEIGHT	0.05 - 0.07 lb/ft 0.07 - 0.10 kg/m
HOSE INNER TUBE MATERIAL	Polyoxymethylene
COVER MATERIAL	Polyamide
HOSE REINFORCEMENT MATERIAL	Two spiral layers of maximum tensile steel wire

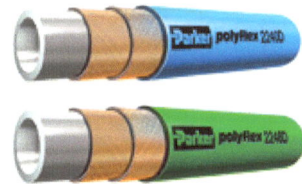

Fig. 3.102 - Examples of Hoses for High Pressure (Courtesy of Parker)

Example 2 – High Pressure Collapse Resistant Hoses: Figure 3.103 shows an example of *High Pressure Collapse Resistant* (HCR) hose is specifically designed to serve the oil & gas market in subsea hydraulics. Its state-of-the-art construction provides a minimum crush resistance of 6,600 psi external pressure.

PERFORMANCE CHARACTERISTICS	
SPECIFICATIONS MET	ISO 13628-5, API 17E (1.5 design factor)
DESIGN FACTOR	4
HOSE INSIDE DIAMETER	1/2, 1 inch 13, 25 mm
HOSE OUTSIDE DIAMETER	1.04, 1.83 inch 26.4, 46.4 mm
MAXIMUM OPERATING PRESSURE	5000 psi 34.5 MPa 345 bar
MAXIMUM OPERATING TEMPERATURE	131 °F 55 °C
MINIMUM OPERATING TEMPERATURE	-40 °F -40 °C
MINIMUM BEND RADIUS	4, 11.8 inch 102, 300 mm
COMPATIBLE FITTING	HV
MEDIA	Hydraulic Fluids
APPLICATION	Oil & Gas, BOP Stack, Subsea
COVER COLOR	Blue, Yellow
WEIGHT	.45 - 1.44 lb/ft .67 - 2.15 kg/m
HOSE INNER TUBE MATERIAL	Polyamide 11 with 316L SS Carcass
COVER MATERIAL	Polyurethane
HOSE REINFORCEMENT MATERIAL	Aramid Fiber Braid
HOSE I.D. (SIZE)	-8, -16
HOSE TYPE	Subsea, BOP Stack, Oil & Gas
NOTES	Parflex offers an unlimited number of hose assembly configurations. Please contact the division for part numbers and/or ordering assistance.
COLLAPSE PRESSURE	6600 psi 45.6 MPa

Fig. 3.103 - Examples of Hoses for High Collapse External Pressure (Courtesy of Parker)

Example 3 - Low Temperature Hoses: Figure 3.104 shows an example of low temperature thermoplastic hydraulic hoses feature optimum performance in cold temperatures, especially for applications requiring flexibility in cold climates with temperatures as low as -70°F. They are designed for construction and agriculture equipment that work in cold climates.

PERFORMANCE CHARACTERISTICS	
SPECIFICATIONS MET	100R17, SAE J517 100R18,SAE 100R18, SAE J517 for less than 50 micro-amps leakage under 75,000 volts per ft.
DESIGN FACTOR	4
HOSE INSIDE DIAMETER	1/4, 3/8, 1/2, 5/8, 3/4 inch 6, 10, 13, 16, 19 mm
HOSE OUTSIDE DIAMETER	.490, .640, .660, .770, .840, .970, 1.080, 1.420 inch 11.9, 16.3, 16.7, 19.5, 21.3, 24.6, 26.2, 28.7 mm
MAXIMUM OPERATING PRESSURE	3000 psi 20.7 MPa 207 bar
MAXIMUM OPERATING TEMPERATURE	250 °F 121 °C
MINIMUM OPERATING TEMPERATURE	-70 °F -57 °C
MINIMUM BEND RADIUS	1-1/4, 2, 2-3/4, 3-1/2, 4, 6-1/2 inch 32, 50.8, 51, 69.9, 88.9, 89, 102, 165 mm
COMPATIBLE FITTING	56 Series, CG Series, 43 Series
MEDIA	Air, Hydraulic Fluid, Oil Water
APPLICATION	Cold Applications, Construction Equipment, Hydraulics, Industrial Equipment, Over-the-Sheave, Pneumatics - dependent on size
COVER COLOR	Black, Orange
WEIGHT	0.07 - 0.40 lb/ft 0.10 - 0.60 kg/m
HOSE INNER TUBE MATERIAL	Copolyester, Non conductive Copolyester
COVER MATERIAL	Copolyester
HOSE REINFORCEMENT MATERIAL	Fiber, High Tensile Steel Wire Braid
HOSE I.D. (SIZE)	-4, -6, -8, -10, -12

Fig. 3.104 - Example of Hoses for Cold Temperature (Courtesy of Parker)

Example 4 - High Temperature Hoses: Figure 3.105 shows an example of high temperature hoses. PTFE hoses feature superior chemical compatibility, resist moisture, perform over a wide temperature range (up to 230 $^{\circ}$C = 450°F) and offer a low coefficient of friction to reduce pressure drop.

PERFORMANCE CHARACTERISTICS	
SPECIFICATIONS MET	SAE 100R14A, SAE 100R14B, FDA CFR 177.1550 (natural)
DESIGN FACTOR	4
HOSE INSIDE DIAMETER	3/16, 1/4, 5/16, 13/32, 1/2, 5/8, 7/8, 1-1/8 inch 5, 6, 8, 10, 13, 16, 19, 22, 29 mm
HOSE OUTSIDE DIAMETER	.330, .400, .460, .560, .580, .660, .790, 1.060 inch 8.2, 10.1, 11.6, 14.3, 16.8, 20.1, 26.9 mm
MAXIMUM OPERATING PRESSURE	1000, 1200, 1500, 2000, 2500, 3000 psi 6.9, 8.3, 10.3, 13.8, 17.2, 20.7 MPa 69, 83, 103, 138, 172, 207 bar
MAXIMUM OPERATING TEMPERATURE	275, 450 °F 135, 232 °C
MINIMUM OPERATING TEMPERATURE	-40 to -73 °C -40 to -100 °F
MINIMUM BEND RADIUS	2, 3, 4, 5, 6-1/2, 7-1/2, 9 inch 50, 75, 100, 127, 165, 191, 229 mm
COMPATIBLE FITTING	90 Series, 91/91N Series
MEDIA	Adhesive, Air, Chemicals, Cooking Oil, Fluid, Water
APPLICATION	Fluid Handling, Chemical Transfer, Manufacturing / Industrial, Medical/Pharmaceutical, Packaging, Instrumentation, Transportation
COVER COLOR	Red, Brown
WEIGHT	0.06 - 0.39 lb/ft 0.09 - 0.58 kg/m
HOSE INNER TUBE MATERIAL	Natural, Static Dissipative PTFE
COVER MATERIAL	Silicone, Polyurethane
HOSE REINFORCEMENT MATERIAL	Stainless Steel Braid
HOSE I.D. (SIZE)	-4, -5, -6, -8, -10, -12, -16, -20
HOSE TYPE	Natural, Static-Dissipative
VACUUM PRESSURE RATING	10, 12, 14, 28 Inch-Hg 254, 305, 355, 711 mm-Hg

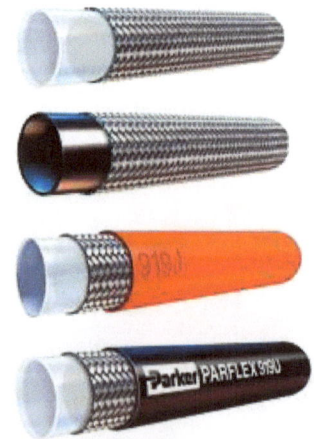

Fig. 3.105 - Examples of Hoses for High Temperature (Courtesy of Parker)

Example 5 – Nonconductive Hoses: Figure 3.106 shows an example of nonconductive hydraulic hose. It is the optimal solution for medium pressure hydraulic service where hydraulic circuit exposure and contact with high voltage may be encountered.

PERFORMANCE CHARACTERISTICS	
SPECIFICATIONS MET	SAE J517 electrical standards (non-conductivity), SAE J517 100R7, DNV-GL Type Approved, ANSI A92.2, ISO 3949-R7-2
DESIGN FACTOR	4
HOSE INSIDE DIAMETER	1/4 inch 6 mm
HOSE OUTSIDE DIAMETER	.470 inch 11.9 mm
MAXIMUM OPERATING PRESSURE	3,000 psi 20.7 MPa 207 bar
MAXIMUM OPERATING TEMPERATURE	212 °F 100 °C
MINIMUM OPERATING TEMPERATURE	-40 °F -40 °C
MINIMUM BEND RADIUS	1-1/2 inch 38 mm
COMPATIBLE FITTING	56 Series, 51R Series
MEDIA	Hydraulic Fluids, Lubricants
APPLICATION	Aerial Lift, Hydraulics, Mobile Equipment
COVER COLOR	Orange
WEIGHT	0.06 lb/ft 0.09 kg/m
HOSE INNER TUBE MATERIAL	Nylon
COVER MATERIAL	Polyurethane
HOSE REINFORCEMENT MATERIAL	Fiber
HOSE I.D. (SIZE)	-4

Fig. 3.106 - Examples of Nonconductive Hydraulic Hoses (Courtesy of Parker)

Example 6 – Fluid Compatibility: Table 3.15. shows an example of fluid compatibility chart provided by a manufacturer.

Chemical Name	Neoprene (Poly-Chloroprene)	Nitrile (Acrylonitrile and Butadiene)	Butyl (Isobutylene and Isoprene)	Hypalon (Chlorosulfonated Polyethylene)	EPDM (Ethylene Propylene Diene)	CPE (Chlorinated Polyethylene)
ASTM-SAE Designation SAE J14 & SAE J200	SC BC	SB BG	R AA	TB CE	R AA	None None
Flame Resistance	Very Good	Poor	Poor	Good	Poor	Good
Petroleum Base Oils	Good	Excellent	Poor	Good	Poor	Very Good
Diesel Fuel	Good to Excellent	Excellent	Poor	Poor	Poor	Very Good
Resistance to Gas Permeation	Good	Good	Outstanding	Good to Excellent	Fair to Good	Good
Weather	Good to Excellent	Poor	Excellent	Very Good	Excellent	Good
Ozone	Good to Excellent	Poor for Tube Good for Cover	Excellent	Very Good	Outstanding	Good
Heat	Good	Good	Excellent	Very Good	Excellent	Excellent
Low Temperature	Fair to Good	Poor to Fair	Very Good	Poor	Good to Excellent	Good
Water-Oil Emulsions	Excellent	Excellent	Good	Good	Poor	Excellent
Water-/Glycol Emulsions	Excellent	Excellent	Excellent	Excellent	Excellent	Excellent
Diesters	Poor	Poor	Excellent	Fair	Excellent	Very Good
Phosphate Esters	Fair (For Cover)	Poor	Good	Fair	Very Good	Very Good
Phosphate Ester Base Emulsions	Fair (For Cover)	Poor	Good	Fair	Very Good	Very Good

Table 3.15 - Hose-Fluid Compatibility Chart (Courtesy of Gates)

Example 7- Hoses for Extreme Conditions: Figure 3.107 shows an example of extreme high-pressure hoses. In such hoses, the central tube is made from synthetic combatable rubber and surrounded by 4-6 reinforcement layers of spiraled high-tensile steel wire. These reinforcement layers are separated by thin adhesive layer of rubber to keep the wires in place. The reinforcement layers are spiraled in opposite direction to balance the pressure force and increase the ability to withstand pressure impulses. As shown in the figure, compared with conventional spiral hose, the *Compact Spiral* design 797 hose offers measurably greater advantages in routing and installation, product size and weight

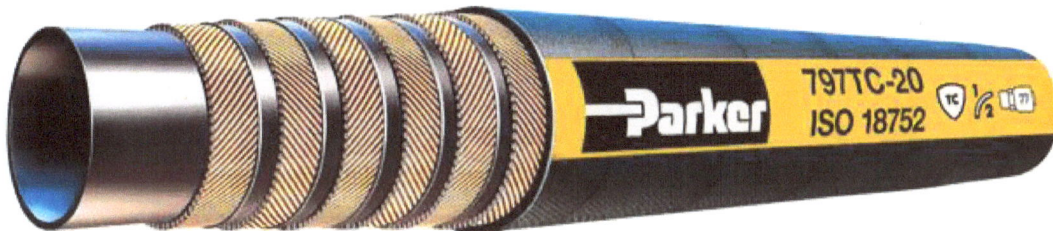

797
Hydraulic – Constant Working Pressure
ISO 18752 - AC/BC/CC/DC

#		TC	Super Tough	Hose I.D.		Hose O.D.		Working Pressure		Minimum Bend Radius		Weight		Parkrimp	Parkrimp
Part Number	Standard Cover	Tough Cover	Tough Cover											43 Series	77 Series
	797	797TC	797ST												
	ISO 18752 Performance			inch	mm	inch	mm	psi	MPa	inch	mm	lbs/ft	kg/m		
797-4	AC	AC	AC	1/4	6,3	0.51	13,0	6000	42,0	2	50	0.21	0,31	●	
797-6	BC	CC	CC	3/8	10	0.66	17,0	6000	42,0	2-1/2	63	0.31	0,46	●	
797-8	BC	DC	DC	1/2	12,5	0.83	21,1	6000	42,0	4	100	0.45	0,67		●
797-10	BC	DC	DC	5/8	16	0.94	23,9	6000	42,0	4-1/2	115	0.54	0,80		●
797-12	BC	DC	DC	3/4	19	1.10	27,9	6000	42,0	5-1/4	135	0.78	1,16		●
797-16	BC	DC	DC	1	25	1.40	35,7	6000	42,0	6-1/2	165	1.17	1,74		●
797-20	BC	DC	DC	1-1/4	31,5	1.77	44,9	6000	42,0	8-3/4	225	1.95	2,89		●
797-24	BC	CC	CC	1-1/2	38	2.08	52,8	6000	42,0	12	305	2.66	3,96		●
797-32	BC	CC	CC	2	51	2.66	67,6	6000	42,0	15	380	4.37	6,50		●

Fig. 3.107 - Hydraulic Hoses for Extreme Conditions (Courtesy of Parker)

3.7- Flanges for Transmission Line Connections

Flanges versus Fittings:
- Connection and Sealing: The major difference between them is how they achieve an effective seal. In fittings, the connection and seal are achieved by tightening threads between the mating halves. In flanges the connection and seal are achieved by bolting together the flanges on two mating halves that have an O-ring in between.

- Line Size: As shown in Fig. 1.108, fittings are most effective and easy to install for small hose diameters (< 1" OD). Large size hoses (> 1" OD) are exposed to severe forces, high pressures, vibration, and shock loads. Therefore, flanges are required to achieve a secure connection. Flanges are also ideal for installation in areas where there is not enough swing clearance for a wrench.

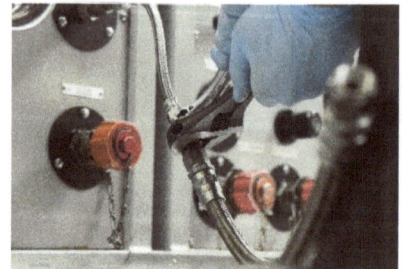

Fig. 3.108 - Use of fittings in Hoses (mac-hyd.com)

Flanges According to SAE-J518 Standard: There are two 4-Bolts flanges known worldwide as "*SAE Flanges*" per the codes shown below. As shown in Fig. 3.109, the SAE standard specifies the dimensions of the flange based on the line size. However, some OEM have their own flange dimensions.

- **SAE Standard Code 61** is for Pmax = (3000 – 5000) psi, depending on the size.
- **SAE Standard Code 62** is for Pmax = 6000 psi regardless of size.

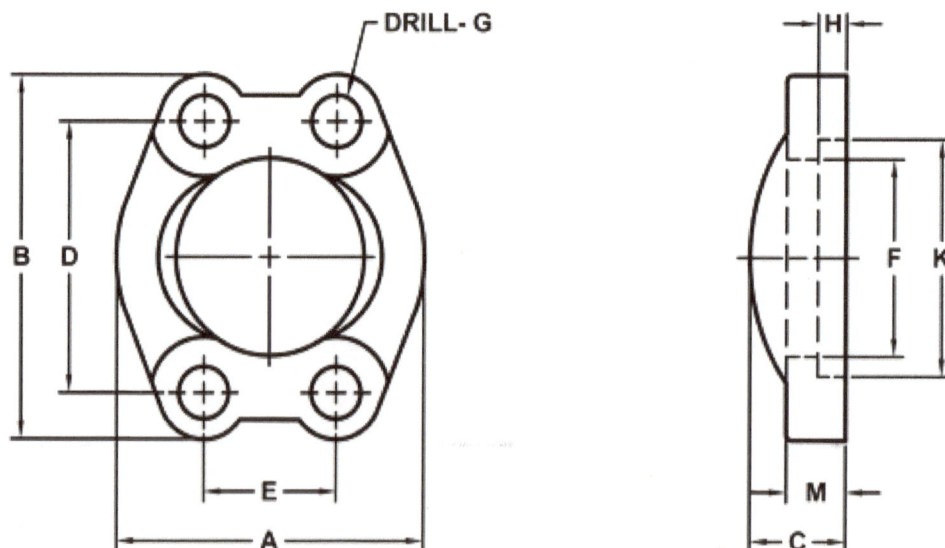

Fig. 3.109 - Standard Flange Dimensions

Components of Flange Assembly: As shown in Fig. 3.110, flanges could be one-piece or Split Flange, i.e. two pieces. Flange connections use a static face seal (a high durometer O-ring), and clamps/bolts for holding power as shown in the figure. The O-ring seal is compressed between the bottom of the groove in the flange head and the flat surface of the port or flange pad, providing a reliable soft seal. The flange, the O-ring, and four bolts are sold as a kit. Having an O-Ring limits using flange connections in high-temperature applications.

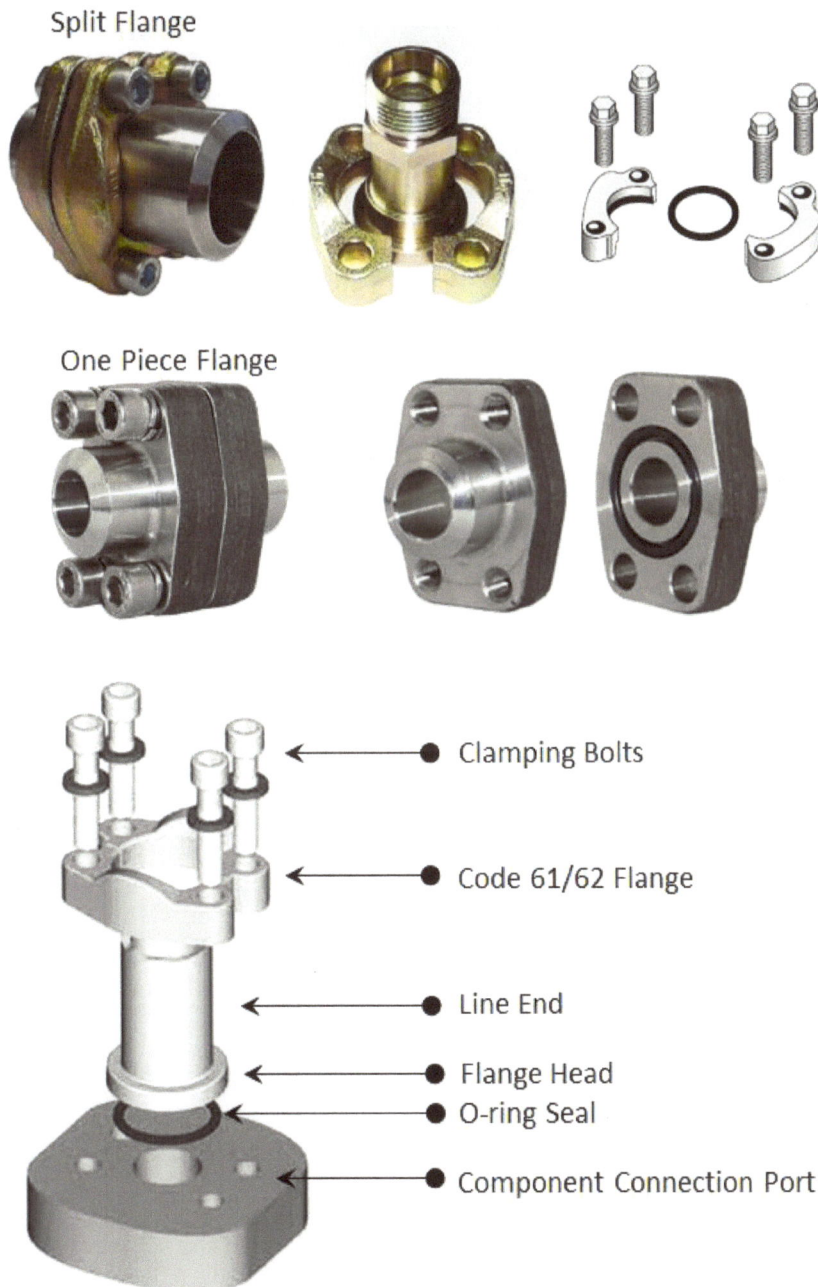

Split Flange

One Piece Flange

Clamping Bolts

Code 61/62 Flange

Line End

Flange Head

O-ring Seal

Component Connection Port

Fig. 3.110- Components of Flange Assembly

One-Piece versus Split Flanges: Flanges are used with all kinds of lines (Pipes, tubes, and hoses) usually as a connection on pumps and motors. As shown in Fig. 3.111, 4-bolt one-piece welded-end flanges could connect pipe sizes up to 5 inches. Figure 3.112 shows that use of split flanges is more adequate for connecting tubes and hoses.

Fig. 3.111 - Hydraulic Pipe Flange Assembly

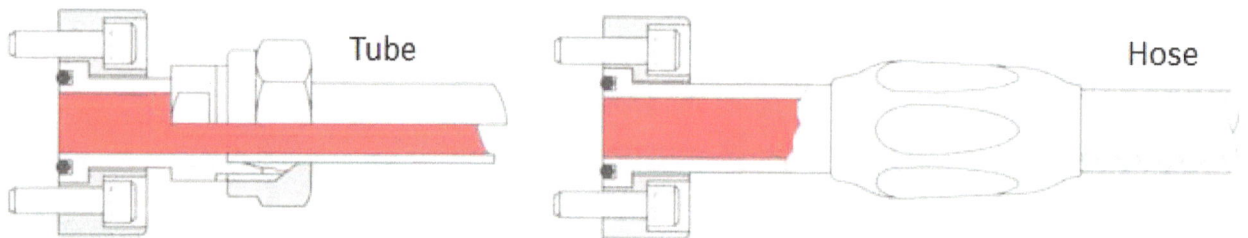

Fig. 3.112 - Hydraulic Tube and Hose Flange Assembly

Square Flanges: As shown in Fig. 3.113, *Square Flanges* are widely used to connect between two tubes or two pipes in hydraulic systems. Square flanges are suitable to connect a tube or a pipe to some hydraulic component, such as a hydraulic valve, hydraulic block, or other component. They are available in full range of dimensions and materials, mainly carbon steel.

Square flange usually is designed and manufactured in different standards in different countries as follows:

- ANSI/ISO 6164/ASME Square flanges.
- JIS B2291, JIS F7806, DIN 3901, BS 16.5.
- Customized square flanges are also available according to customers' drawing and specifications.

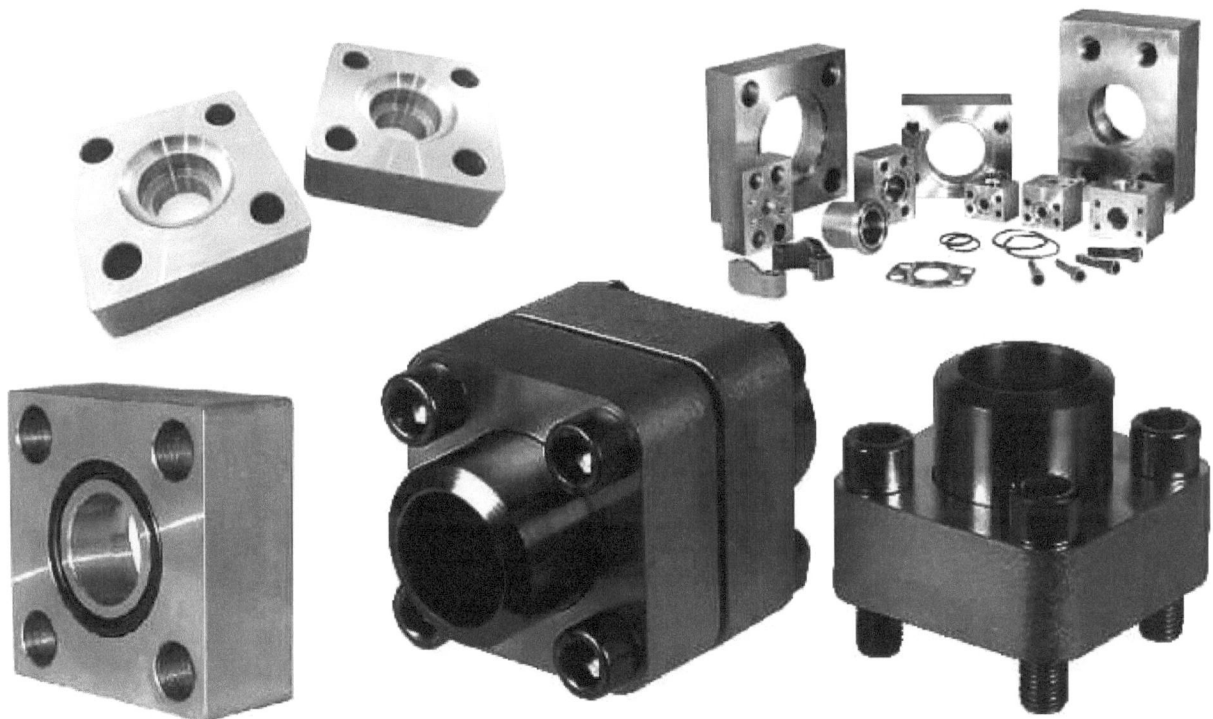

Fig. 3.113 - Square Flanges (https://flanges-pipe.com)

3.8- Rubber Expansion Fittings

Reasons to use Rubber Expansion Fittings: They are used in a pipework system to:
- Absorb expansion and contraction caused by thermal stresses due to temperature fluctuation.
- Absorb expansion and contraction caused by mechanical stresses due to movement of foundations, fluctuating loads or vibration during startup and flow surges.
- Isolate structure-borne noise.
- Accommodating inaccuracy and misalignment in installation.

Material for Rubber Expansion Fittings:
- They are made of natural rubber or synthetic rubber depending on the pressure.

Working Conditions for Rubber Expansion Fittings:
- The pressure strength of rubber expansion fittings depends on the size and design of the fittings. Usually, they are good for pressures below 20 bar (300 psi). The preferred point for locating rubber expansion fittings in hydraulic systems is in the suction and drain lines. It can also connect to isolation valve.
- Usually, expansion fittings with synthetic fiber reinforcement can be used up to +10°C and with steel wire reinforcement up to + 130°C.

Figure 3.114 shows various styles of rubber expansion fitting.

Fig. 3.114 - Rubber Expansion Fittings (www.grainger.com)

3.9- Test Points

Function of Test Points: *Test points* were developed and widely used in the design of hydraulic systems for the following purposes:
- Pressure monitoring and venting.
- Sampling of hydraulic fluids.
- Connection with instruments and sensors.

Features of Test Points (Fig. 3.115): Test points have the following features:
- Ensures 100% leakage free coupling while under system pressure.
- Available sealing materials are NRB, FKM and EPDM.
- Available in various styles and sizes.
- Available standards are ISO, SAE, and international threads.
- Available in steel and stainless steel.

Fig. 3.115 - Series 1620 Test Points (www.hydrotechnik.com)

Innovative Test Point (Fig. 3.116): An innovative test points (ICON™) were developed to allow easy and tight seal connection with various fluid power components. As shown in the figure, they can connect to various tube ends configurations. Moreover, they add sensing capabilities, include instant LED visual feedback, and data output to verify sealing connection accuracy. Hence, they maximize production efficiency, and optimize maintenance.

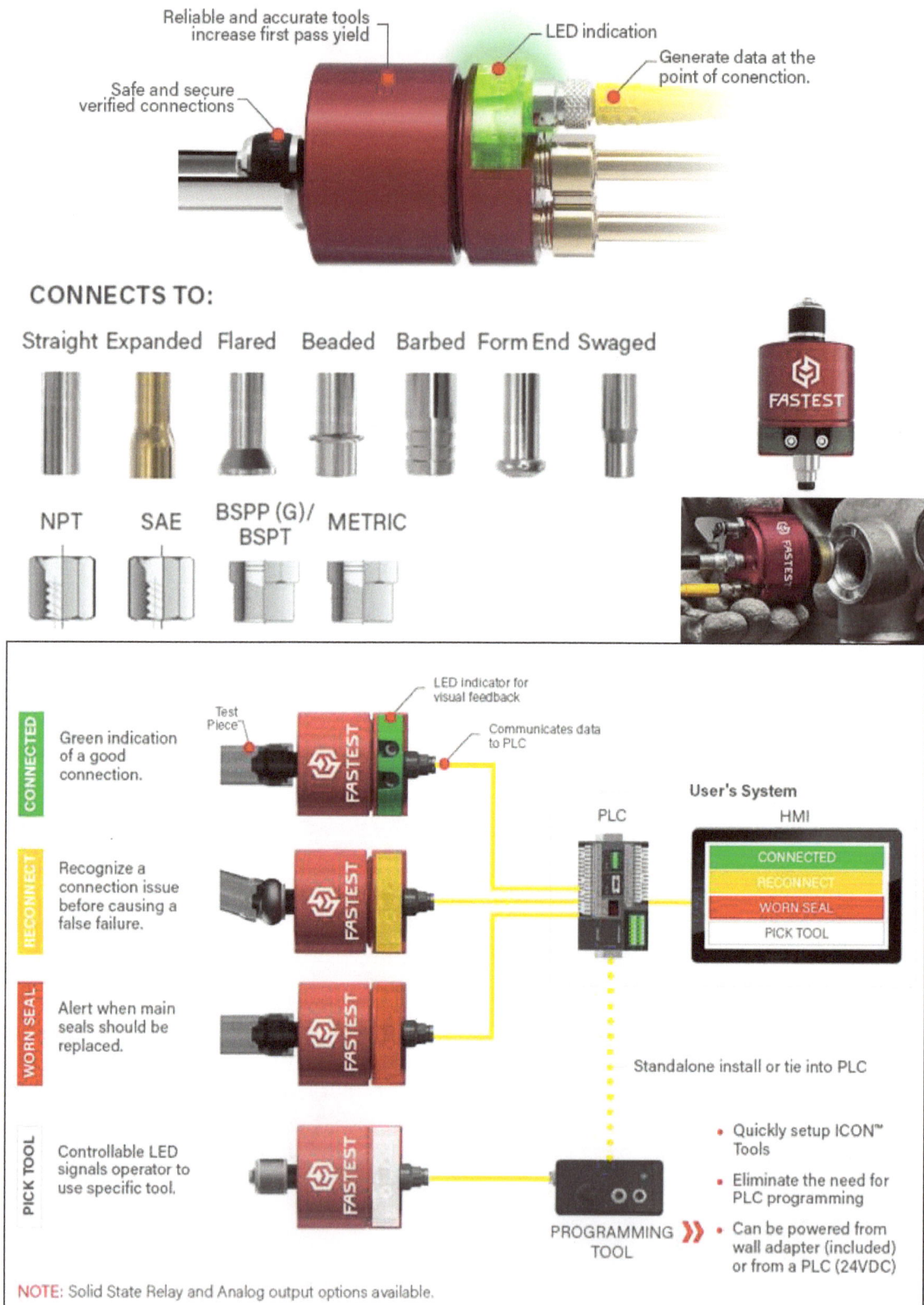

CONNECTS TO:

Straight Expanded Flared Beaded Barbed Form End Swaged

NPT SAE BSPP (G)/BSPT METRIC

CONNECTED — Green indication of a good connection.

Test Piece

LED indicator for visual feedback

Communicates data to PLC

RECONNECT — Recognize a connection issue before causing a false failure.

WORN SEAL — Alert when main seals should be replaced.

PICK TOOL — Controllable LED signals operator to use specific tool.

User's System

PLC

HMI

CONNECTED
RECONNECT
WORN SEAL
PICK TOOL

Standalone install or tie into PLC

PROGRAMMING TOOL

- Quickly setup ICON™ Tools
- Eliminate the need for PLC programming
- Can be powered from wall adapter (included) or from a PLC (24VDC)

NOTE: Solid State Relay and Analog output options available.

Fig. 3.116 - Innovative Test Points (Courtesy of CEJN)

3.10- Pressure Measurement Hoses

Function of Measurement Hoses: *Measurements Hose Lines* are used for purposes of pressure measuring, transmitting control pressure, and system diagnostic.

Features of Measurement Hoses: Measurement hoses have the following features:
- They have been developed for applications with pressures of up to 600 bar (9000 psi).
- They are characterized by their resistance to aggressive fluids.
- As shown in Fig. 3.117, they are constructed from a core, reinforcement layer, and protective cover.
- As shown in Fig. 3.118, they are supported by anti-kinking spirals or Aluminum protective covers.
- As shown in Fig. 3.119, they can be combined with test points.
- As shown in Fig. 3.120, use of these hoses is associated with pressure drop that should be reported by the hose manufacturer.

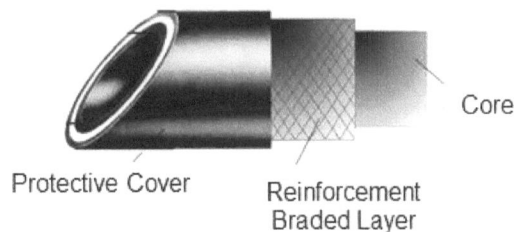

Fig. 3.117 - Construction of Measurement Hoses (www.hydrotechnik.com)

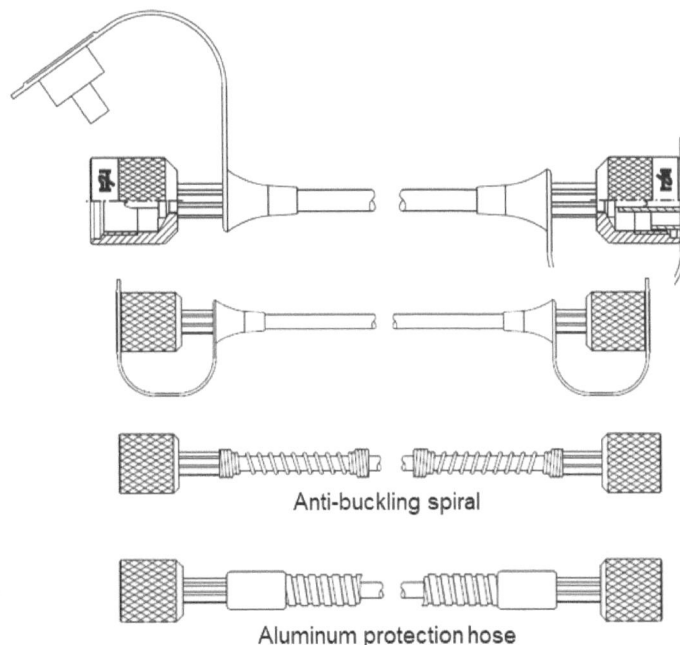

Fig. 3.118 - Measurement Hoses Supported by Anti-Buckling and Protection Covers (www.hydrotechnik.com)

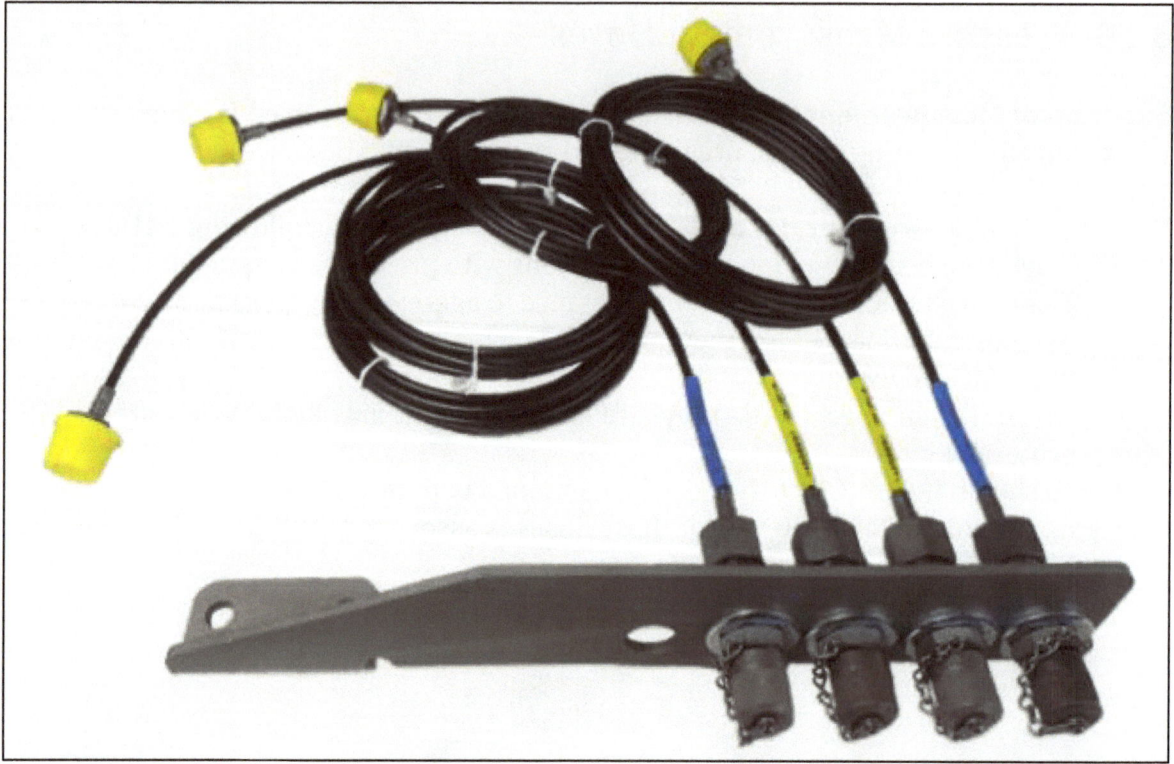

Fig. 3.119 - Measurement Hoses Combined with Test Points (www.hydrotechnik.com)

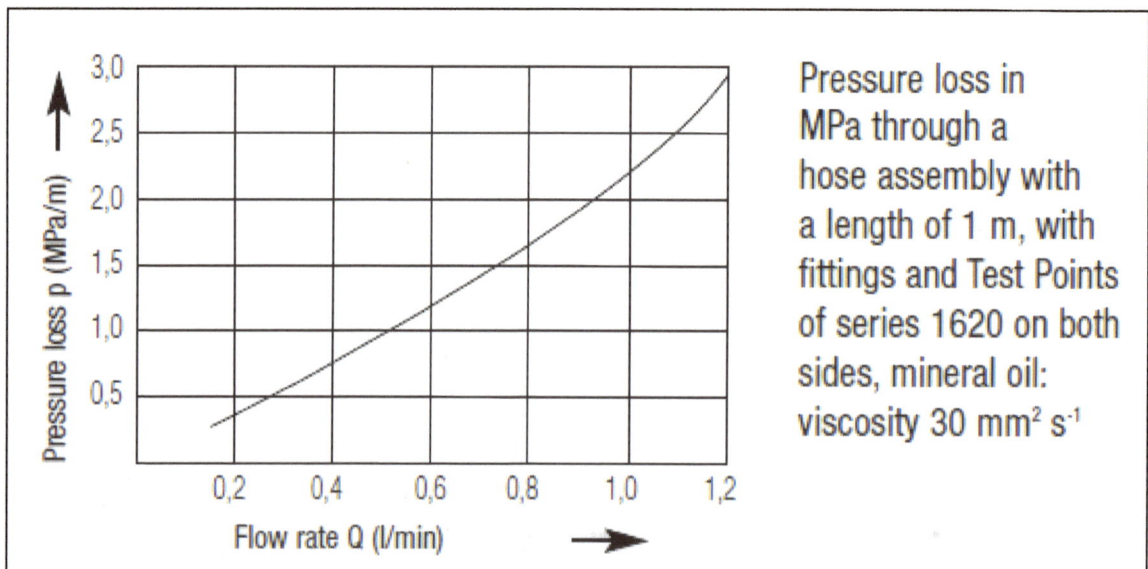

Pressure loss in MPa through a hose assembly with a length of 1 m, with fittings and Test Points of series 1620 on both sides, mineral oil: viscosity 30 mm^2 s^{-1}

Fig. 3.120 - Pressure Drop in Measurement Hoses (www.hydrotechnik.com)

Routing of Measurement Hoses: Reliability of the system and lifetime of the hose assembly are dependent on the correct installation. Measurement hoses can easily be routed to make maintenance and diagnostic activities easier. However, as shown in Fig. 3.121, routing them should follow the best practices to avoid hose failure.

1. Under load, the length of a hose pipe can change. shortening causes an additional tensile stress of the hose and the connections. Therefore, the hose pipe should be in an unpressurised state. Also, union nuts must only be tightened for the torque specified by manufacturer.

2. With curved assemblies, attention must be paid to the bending radius. Sharp bends must be avoided wherever possible.

3. 90° hose fittings are also available to aid in the fitting of hose assemblies to maximize life and operation of the assembly.

4. 90° hose fittings can also aid in installing hose assembly in tight positions.

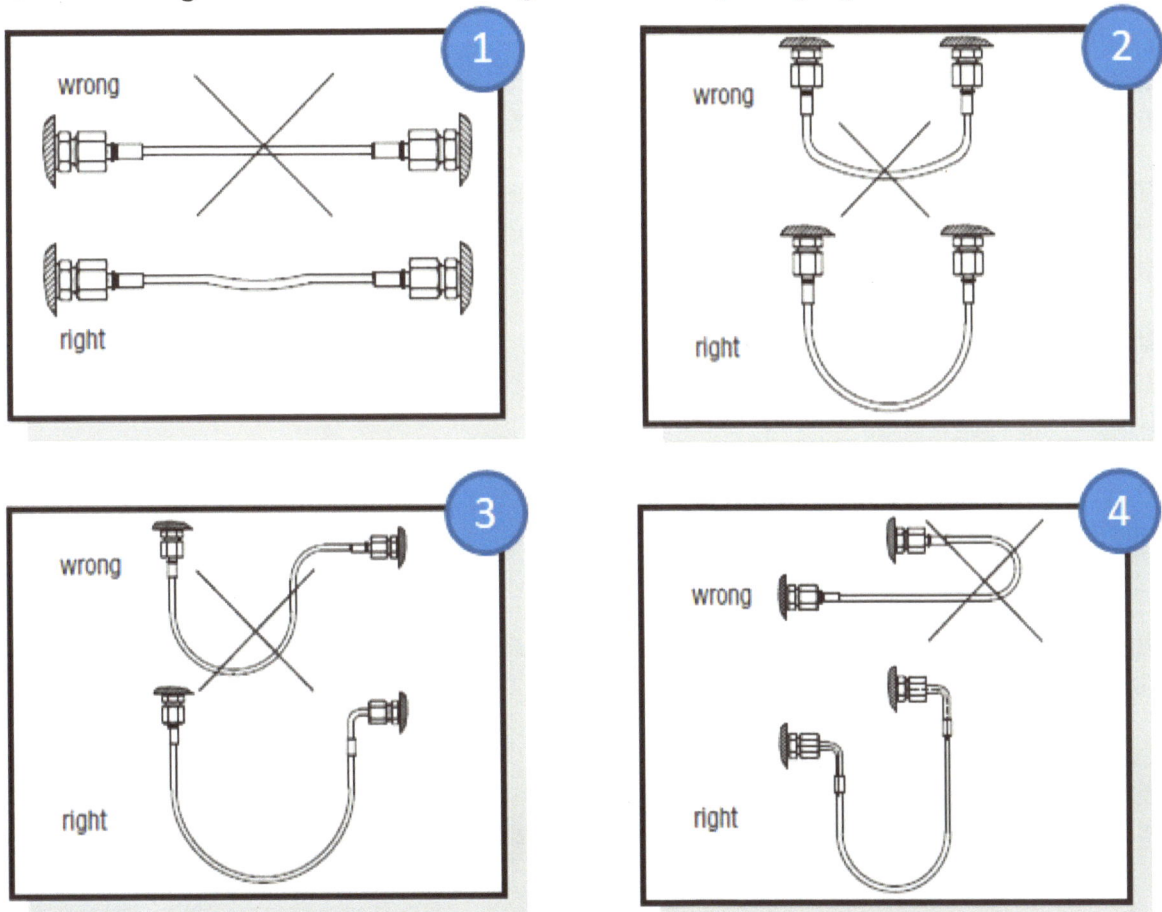

Fig. 3.121 - Best Practices of Routing Measurement Hoses (www.hydrotechnik.com)

3.11 Manifolds

Manifolds are machined blocks with internal passages to connect hydraulic components. The main purpose of using such manifolds is to minimize use of transmission lines for the sake of developing compact, fast response, and energy efficient systems. Subplates are manufactured according to ANSI/NFPA standard for various types and sizes of hydraulic valves. They are also customized for group of components. There are several software platforms, such as AutoCAD and Automation Studio, used to customize a manifold based on hydraulic circuit diagram and components of specific brands. Figure 3.122 shows common type of manifolds.

Subplates (1): *Subplates* allow a simple solution to mount a single valve in a hydraulic system. Several different valve sizes, port sizes, and mounting configurations are available.

Cover Plates (2): Also known as "blanking plates," "blocking plates," or "cap-off plates". A *Cover Plates* is used to cover a subplates after a valve is removed or to cover a valve pattern intended for future use.

Bar Manifolds (3): *Bar Manifolds* reduce plumbing and the number of leak points. Hence, improve maintenance of hydraulic systems. Multi-station manifolds allow several valves to be mounted to a single manifold which has common pressure and tank connections.

Tapping Plates (4): Also referred to as *Sandwich Plates* or *Module Plates*. They provide threaded port access to the A, B, P, T, X and Y passages of an existing valve manifold to add a test port or gauge ports.

Valve Adaptors (5): This *Stack Module* mounts between the existing manifold and a valve. Used to convert the existing manifold valve pattern to an alternative size valve without disrupting the existing manifold configuration.

Din Bodies (6): These are single valve bodies to house DIN slip-in logic valves used in applications that require higher flow than a typical subplate mounted or threaded cartridge valve.

Header & Junction Blocks (7): Also known as *Distribution Manifolds*. Used to connect several threaded connections into one common block to eliminate leaks and reduce plumbing labor.

Custom Engineered Manifolds (8): They are designed for specific applications after the circuit is designed and components are selected. These designs optimize application with a single integrated manifold solution, eliminating leak points and labor costs associated with standard off-the-shelf discrete manifolds.

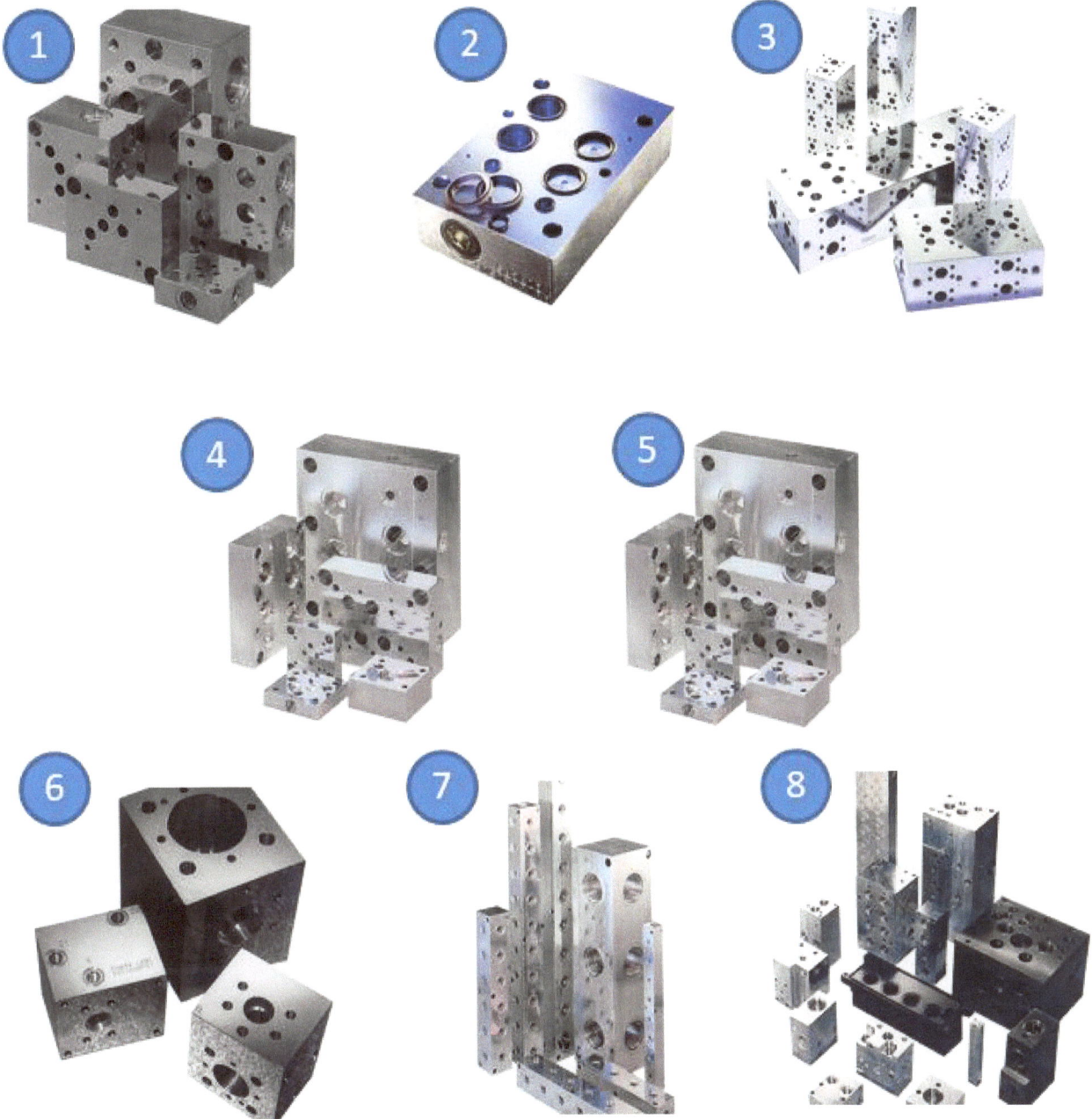

Fig. 3.122 – Hydraulic Manifolds (www.daman.com)

Chapter 4

Hydraulic Sealing Elements

Objectives

This chapter provides a knowledge base for fluid power users to become familiar with the commonly used seals in hydraulic components. This chapter presents an overview of hydraulic sealing elements including seal functions, classifications, and materials. This chapter also presents 15 various properties of hydraulic seals and the relevant standard test methods. This chapter also presents sealing solutions for cylinders and rotational shafts.

Brief Contents

Chapter 4: Hydraulic Sealing Elements

The following topics are discussed in Chapter 3 in Volume 6 "Troubleshooting and Failure Analysis" of this series of textbooks:

- Hydraulic Seals Inspection
- Hydraulic Seals Troubleshooting
- Hydraulic Seals Failure Analysis

This chapter presents an overview of hydraulic sealing elements including seal functions, classifications, materials, physical properties and testing methods.

4.1- Introduction to Hydraulic Sealing Elements

The technology of producing hydraulic sealing elements and the types of seals are very broad. Additionally, sealing solutions vary widely depending on the hydraulic components, application, and working conditions. Another challenge is that manufacturers of hydraulic seals treat the production process as confidential to maintain their competitiveness in the marketplace.

4.1.1- Functions of Hydraulic Sealing Elements

Hydraulic seals are basically used to:

- Prevent external leakage and consequently:
  - Save money on oil. Assuming $10/Gallon is a current price, Figure 4.1 shows significant money savings by having seals working properly.
  - Minimize environmental damage.
  - Improve work zone safety.
  - Reduce accident liability due to fire or slip and falls.

- Prevent internal leakage from a high-pressure chamber to a low-pressure chamber and consequently:
  - Help components and systems function properly and efficiently.
  - Reduce heat generation.

- Provide controlled lubrication for adjacent parts or surfaces.

- Some hydraulic sealing elements have the primary function of preventing dirt, dust, and other contaminants from getting into the hydraulic components.

Leakage Rate	Monthly Losses	Yearly Losses
1 Drop/5Sec.	6.6 Gallon = $66	80 Gallon = $800
1 Drop/Sec.	34 Gallon = $340	409 Gallon = $4090
3 Drop/Sec.	113 Gallon = $1130	1243 Gallon = $12430
Steady Stream	720 Gallon = $7200	8640 Gallon = $86400

Spend tens of dollars
to change the seal on time

Save hundreds of dollars
on Troubleshooting

Save thousands of dollars on
machine shutdown

Fig. 4.1 - Cost of Oil Leakage

4.1.2- Applications of Hydraulic Sealing Elements

On the systems level, hydraulic sealing elements are used in almost every other industry whenever hydraulic systems are used. Figure 4.2 shows application examples where hydraulic seals are used. On the component level, hydraulic sealing elements are used in all components including pumps, actuators, valves, accumulators, etc.

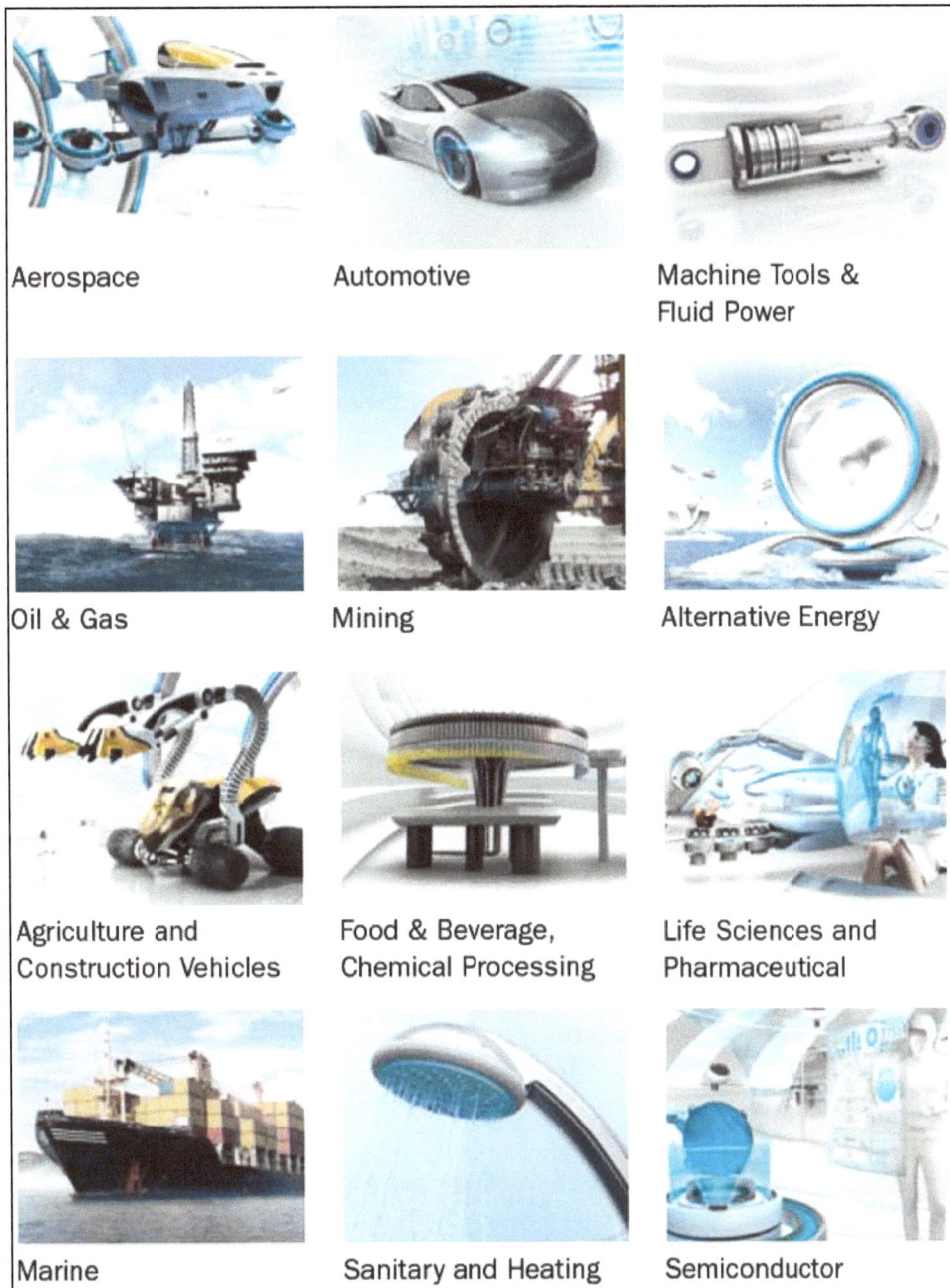

Fig. 4.2 - Examples of Sealing Applications (Courtesy of Trelleborg)

4.1.3- Classifications of Hydraulic Sealing Elements

Figure 4.3 shows the basic classifications of hydraulic sealing elements. Hydraulic sealing elements are broadly classified as *Static Seals* or *Dynamic Seals*.

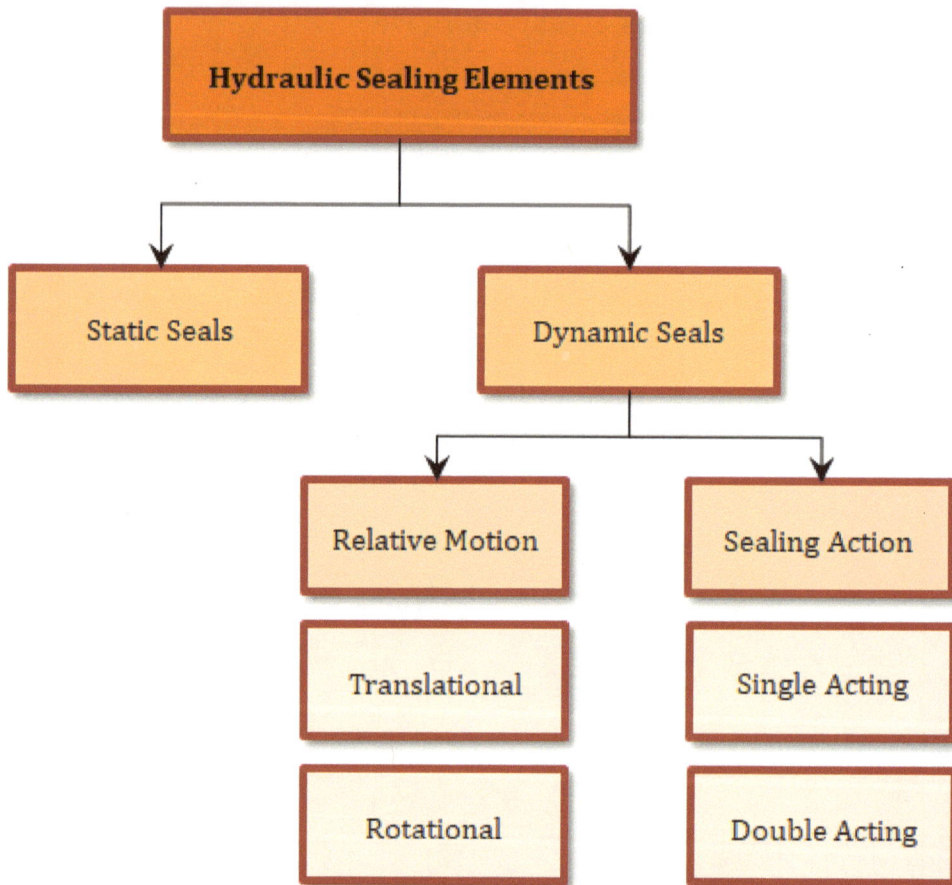

Fig. 4.3 - Classifications of Hydraulic Sealing Elements

Static Seals:
- Used to seal between mating surfaces that have no relative motion between them.
- Used to fill confined or none-confined spaces.
- Basic static seals include Sealing Rings and Gaskets.

Dynamic Seals:
- Used to seal and lubricate between surfaces at least one of them is moving.
- A dynamic seal could be installed on a stationary surface, e.g. Cylinder rod seal.
- A dynamic seal could be installed on a moving surface, e.g. Cylinder piston seal.
- Relative motion between sealed surfaces could be translational as in cylinders and spool valves or could be rotational as in pumps and motors.
- A dynamic seal could seal in one or two directions.

Based on specific construction, the following are common sealing configurations:

- Sealing Rings
- Cup Seals
- U-Cup Seals
- T-Shaped Seals
- V-Packings
- Spring-Energized Seals
- Glands
- Wear-Rings
- Back-up Rings
- Rod Wipers (Scrapers)

4.2- Sealing Rings

4.2.1- Features and Basic Use of Sealing Rings

Sealing Rings are common sealing elements because of they have several advantages:
- They are inexpensive and readily available.
- They are available in all sizes and all type of elastomers.
- They work over a wide range of operating pressure.
- They are easy to assemble and to replace.
- They require small room to fit inside the components.
- Their failure analysis isn't difficult.
- Used for both static and dynamic sealing and for translational and rotational shafts.
- Used limitedly in high temperature application.
- As shown in Fig. 4.4, they can be sold as separate pieces with specific dimensions or as *Cord Stock*.

Fig. 4.4 - Example of O-Rings Cord Stock (Courtesy of MFP Seals)

As shown in Fig. 4.5, *Sealing Rings* are used as static seals against external leakage from between parts in a hydraulic component.

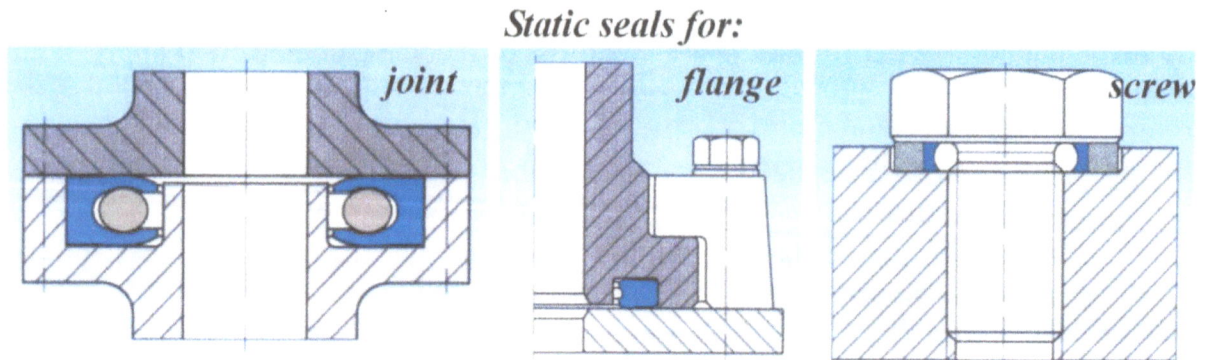

Fig. 4.5 - Use of Sealing Rings as Static Seals (Courtesy of Assofluid)

As shown in Fig. 4.6, Sealing Rings have limited use as dynamic seals under the following conditions:

- For short strokes.
- For relatively small diameter applications.
- For relatively low-pressure applications.

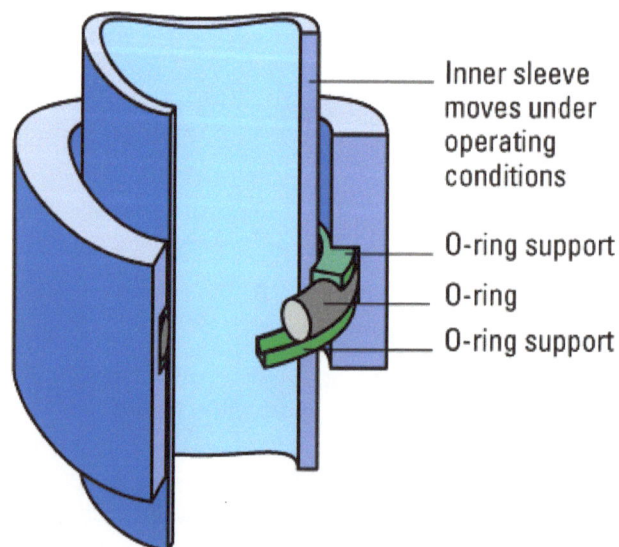

Fig. 4.6 - Use of Sealing Rings as Dynamic Seals

4.2.2- Configurations of Sealing Rings

Sealing Rings have various cross sections as shown in Fig. 4.7. Despite the various cross sections, they are typically called O-Rings. In this chapter, O-Ring refers to the ones with circular cross section.

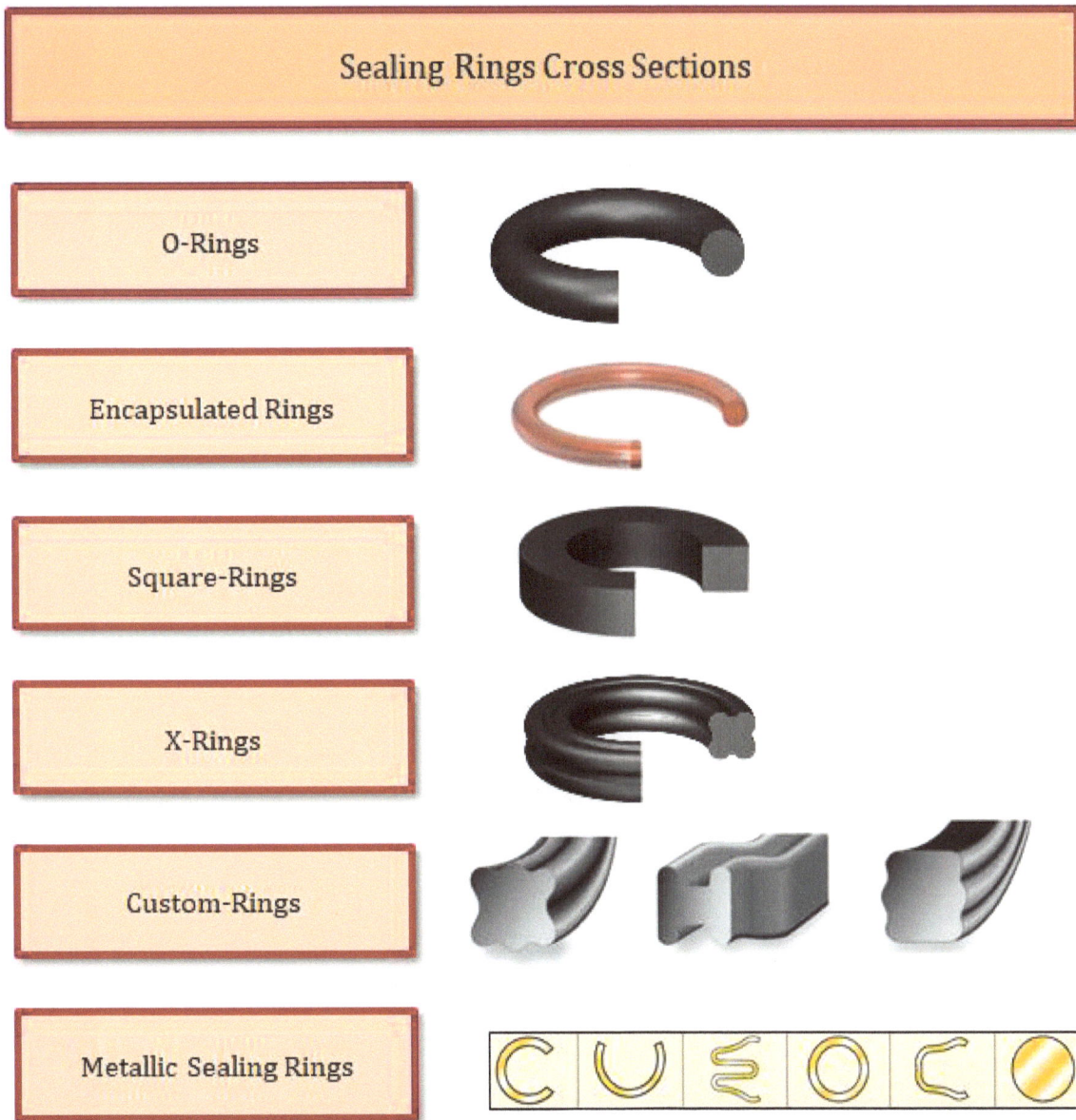

Fig. 4.7 - Use of Sealing Rings as Dynamic Seals

4.2.3- O- Rings

O-Rings were one of the first developed and are the most commonly known type of sealing rings.

4.2.3.1- O-Rings Construction

As shown in Fig. 4.8, an *O-Ring* is a sealing element with solid circular cross section. They are produced in a wide range of materials and dimensions.

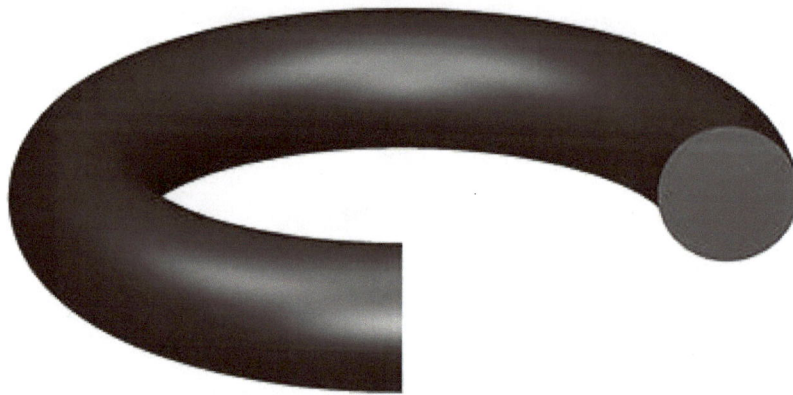

Fig. 4.8 - Basic Construction of Static O-Rings (Courtesy of MFP Seals)

4.2.3.2- O-Rings Sealing Mechanism

As shown in Fig. 4.9, an O-Ring is compressed (15-30) % of its initial volume after assembly. The initial squeeze, which acts in a radial or axial direction depending on the installation, gives the O-Ring its initial sealing condition. When system pressure applies, sealing condition increases proportional to system pressure.

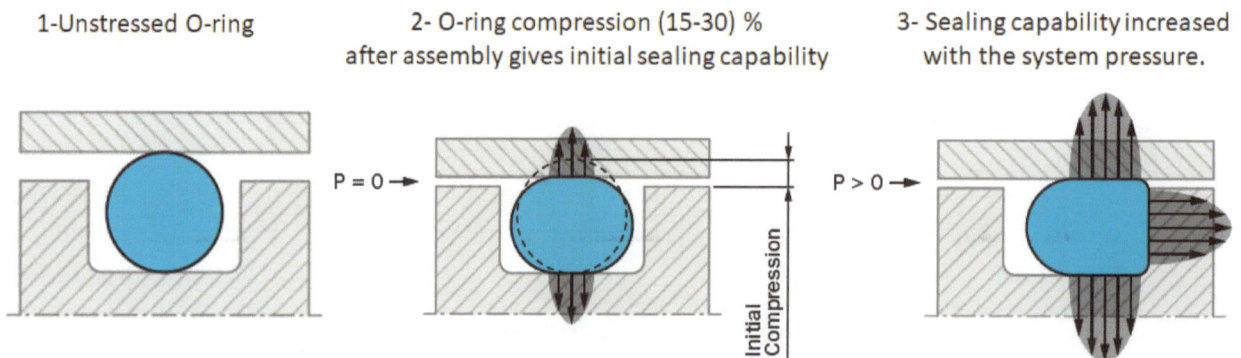

1-Unstressed O-ring

2- O-ring compression (15-30) % after assembly gives initial sealing capability

3- Sealing capability increased with the system pressure.

Fig. 4.9 - O-Ring Sealing Mechanism (Courtesy of Trelleborg)

4.2.3.3- O-Rings Main Dimensions

O-Rings are produced in inch and metric standard sizes and are available from numerous manufacturers. Standard sizes are based on **ISO 3601** and **SAE AS568** standards

Figure 4.10 shows the main dimension of O-Rings as follows:
- Inside Diameter (ID).
- Radial and Axial Crosssection Width (W).
- Radial and axial Flashes due to rubber injection.

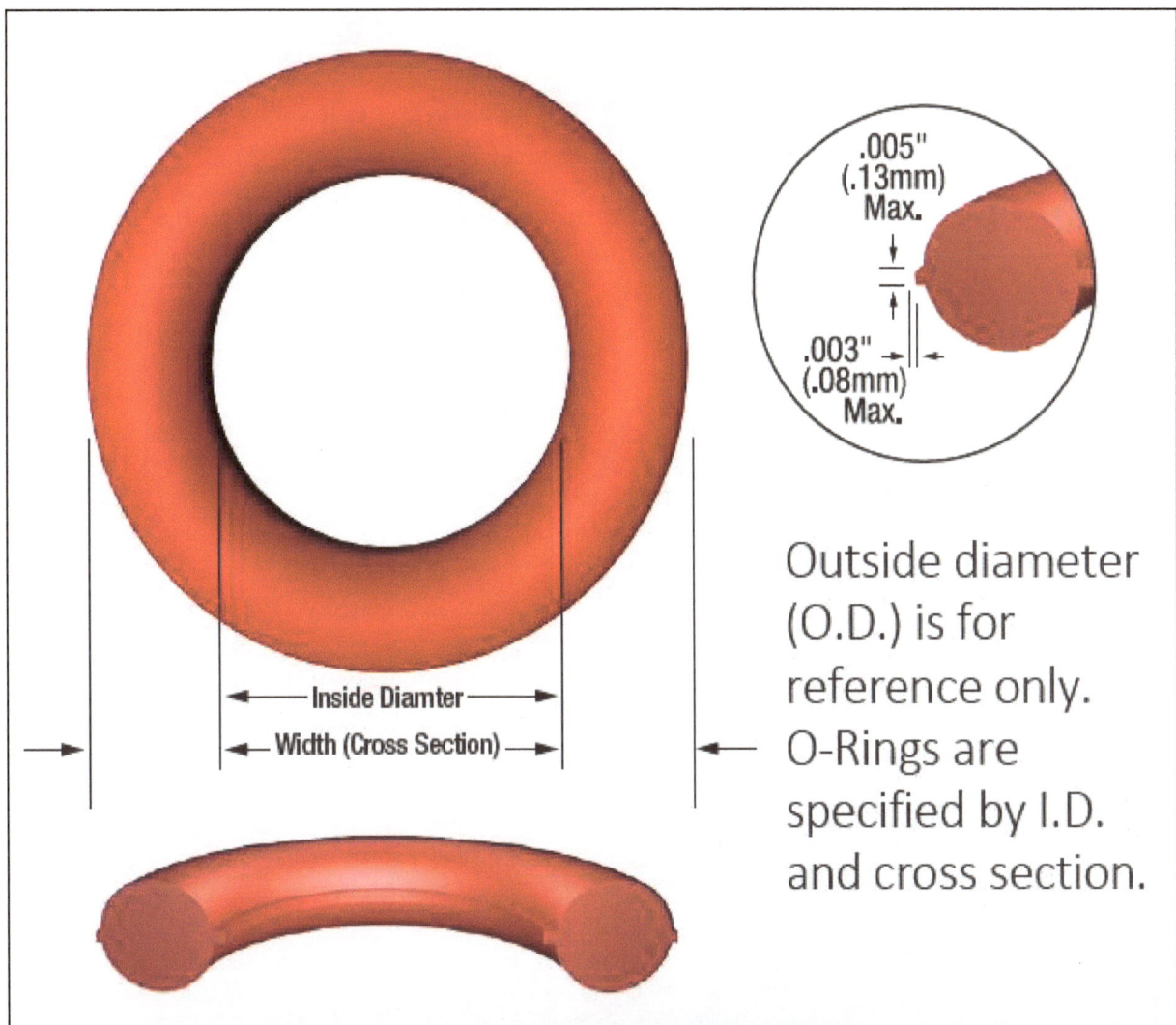

Fig. 4.10 - Main Dimensions of Basic O-Rings (applerubber.com)

4.2.4- Encapsulated O-Rings

Figure 4.11 shows the construction of an *Encapsulated O-Ring.* It consists of an elastomeric base material that is encapsulated with *Teflon* industrial coating (FEP). The FEP coating provides increased chemical, temperature and wear resistance that would otherwise not be achievable with the base compound alone. Figure 4.12 shows typical examples of construction of encapsulated O-Rings.

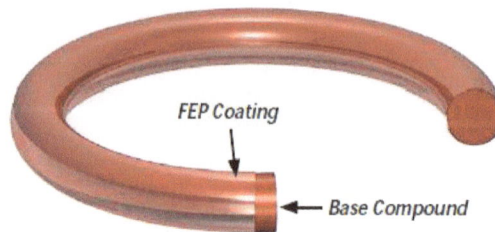

Fig. 4.11 - Construction of Encapsulated O-Rings (Courtesy of MFP Seals)

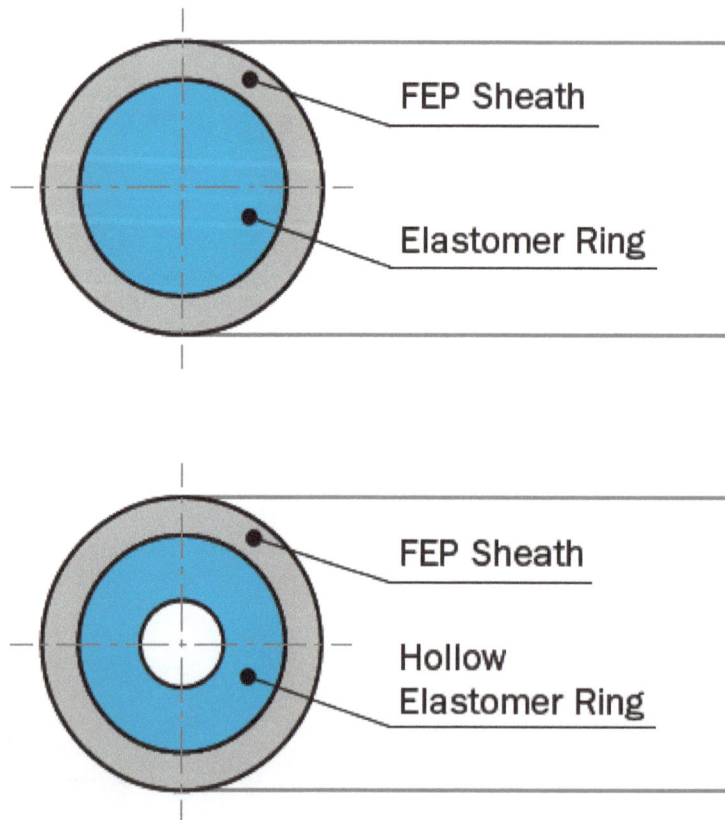

Fig. 4.12 - Typical Construction of Encapsulated O-Rings (Courtesy of Trelleborg)

4.2.5- Square-Rings

Figure 4.13 shows the construction of *Square-Rings.* They are used for the same function as O-Rings. They are directly interchangeable in certain sizes with the O-Rings, using the same groove. Due to their larger sealing surface, Square-Rings can normally handle higher pressures than O-Rings. Square-Rings are commonly used for static applications. Figure 4.14 shows the main dimensions of Square-Rings.

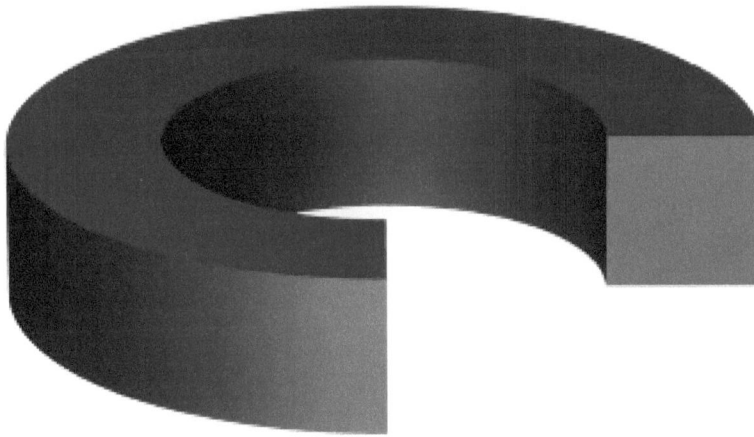

Fig. 4.13 - Construction of Square-Rings (Courtesy of MFP Seals)

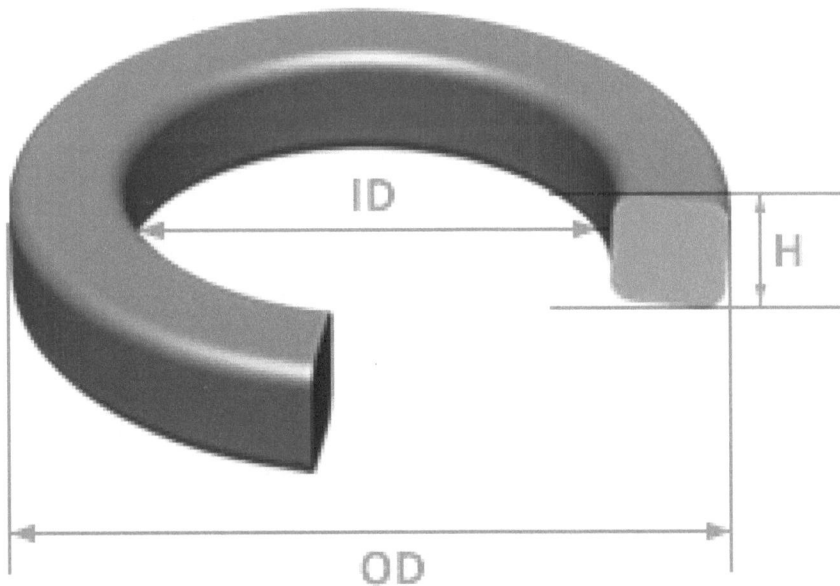

Fig. 4.14 - Main Dimensions of Square-Rings (www.marcorubber.com)

4.2.6- X- Rings

Figure 4.15 shows the construction of *X-Rings* that are also referred to as *Quad-Rings*. They are designed to offer two contact points on each sealing surface versus one contact point in O-Rings. This dual sealing lip feature gives positive sealing with reduced radial squeeze, and thus reduced friction. By reducing the friction, the wear of ring is reduced giving the seal a longer service life over the standard O-Ring design. X-Rings are excellent for use in rotary applications and for providing seal stability due to a design that resists twisting in the groove. Figure 4.16 shows the main dimension of X-Rings.

Fig. 4.15 - Construction of X-Rings (Courtesy of MFP Seals)

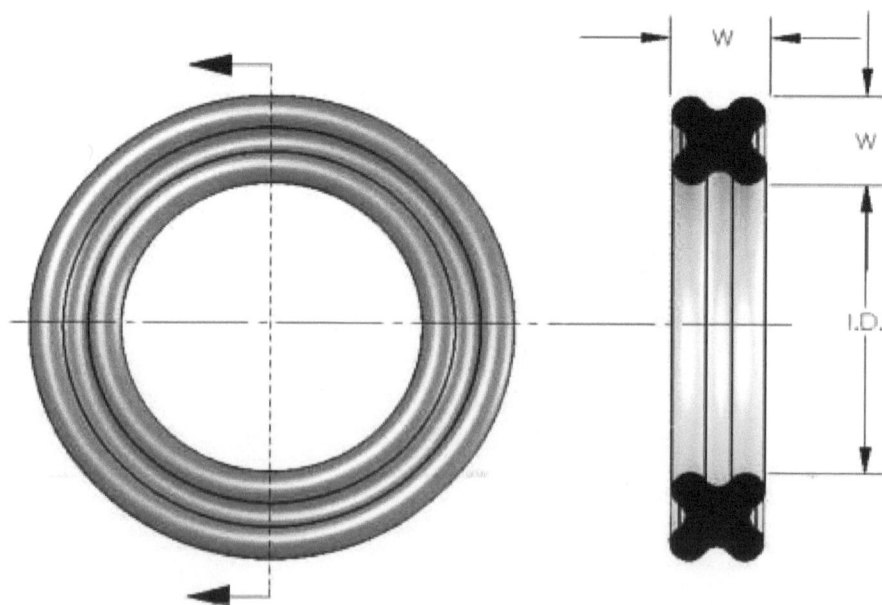

Fig. 4.16 - Main Dimensions of X-Rings (www.marcorubber.com)

4.2.7- Custom-Rings

Figure 4.17 shows the construction of a *Custom-Ring,* example 1. Such a ring is designed for dynamic sealing applications providing near zero leakage at pressures up to 138 bar (2000 psi). This six-lobed configuration, designed with two primary and four backup sealing surfaces, has excellent sealing features in very difficult applications. It can be used with standard O-Ring grooves.

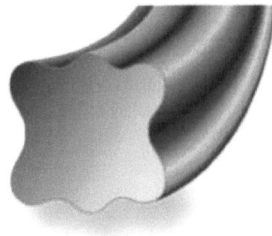

Fig. 4.17 - Quad-O-Dyn® Brand Seals (www.mnrubber.com)

Figure 4.18 shows the construction of a *Custom-Ring,* example 2. Such a ring Ideal to fill single or multiple grooves configurations in static face seal applications.

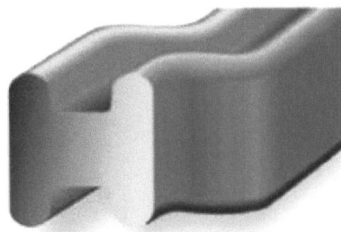

Fig. 4.18 - Quad®-O-Stat Brand Seals (www.mnrubber.com)

Figure 4.19 shows the construction of another *Custom-Ring,* example 3. Such a ring is designed specifically for static face sealing applications. Each of the six contact points serves as an individual seal with the corner lobes functioning as seal backups to the central lobes. If one lobe fails, the remaining lobes provide zero leakage sealing. They can be installed in standard O-Ring grooves.

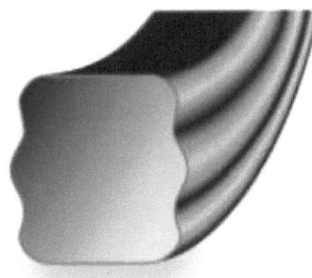

Fig. 4.19 - Quad® P.E. Plus Brand Seals (www.mnrubber.com)

4.2.8- Metallic Sealing Rings

Cross Sections: As shown in Fig. 4.20, *Metallic Sealing Rings* are available from various manufacturers in a variety of cross sections. Like the rubber-based seals, they could be none spring energized, or spring energized.

Material: Materials include Steel, Stainless Steel, Cast Iron, Ductile Iron and Bronze. Piston ring coatings of Manganese Phosphate, Tin Nickel, PTFE/Nickel, Chrome, and Silver Lead Indium are available to maximize performance for lubricating requirements, corrosion resistance, wear, and friction reduction.

Fluid Compatibility: They are compatible with petroleum base and synthetic fluids and phosphate esters among others.

Major Advantages:

- Reliable under severe operating conditions for extended periods of time.
- High pressure sealing up-to 1700 bar psi (25,000) without the risk of blow-by.
- High service temperature up to 1800 °F (982 °C).

Selection: The figure shows the selection criteria based on the operating conditions.

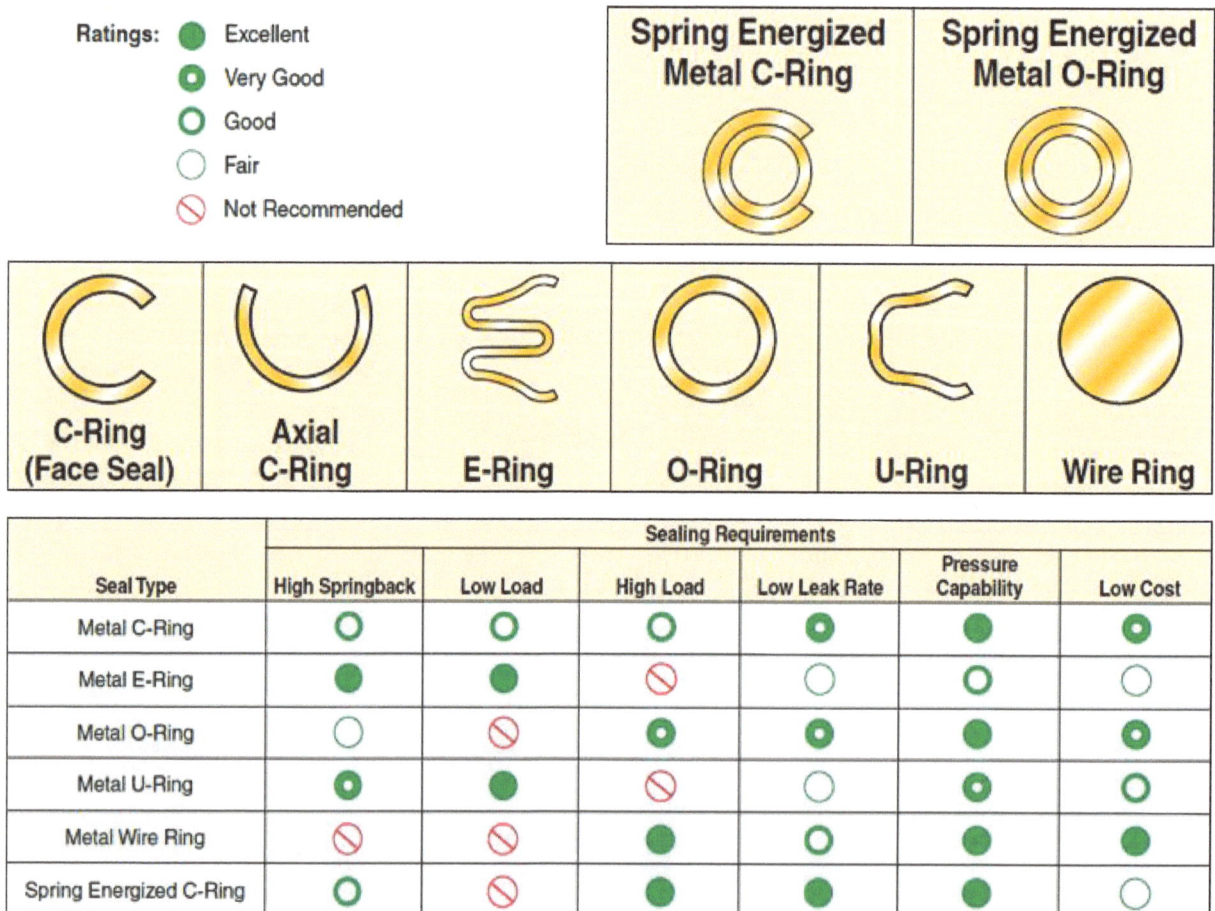

Seal Type	High Springback	Low Load	High Load	Low Leak Rate	Pressure Capability	Low Cost
Metal C-Ring	Good	Good	Good	Very Good	Excellent	Very Good
Metal E-Ring	Excellent	Excellent	Not Recommended	Good	Good	Fair
Metal O-Ring	Good	Not Recommended	Very Good	Very Good	Excellent	Very Good
Metal U-Ring	Very Good	Excellent	Not Recommended	Good	Very Good	Good
Metal Wire Ring	Not Recommended	Not Recommended	Excellent	Good	Excellent	Excellent
Spring Energized C-Ring	Good	Not Recommended	Excellent	Excellent	Excellent	Fair

Fig. 4.20 - Metallic Sealing Rings (Courtesy of Parker)

4.3- Cup Seals

Figure 4.21 shows the basic *Cup Seal* shape. Such a seal performs the sealing function in one direction where the pressure increases the sealing force. This cup seal requires a backing plate to retain it in position and used for pressure up to 70 bar (1000 psi).

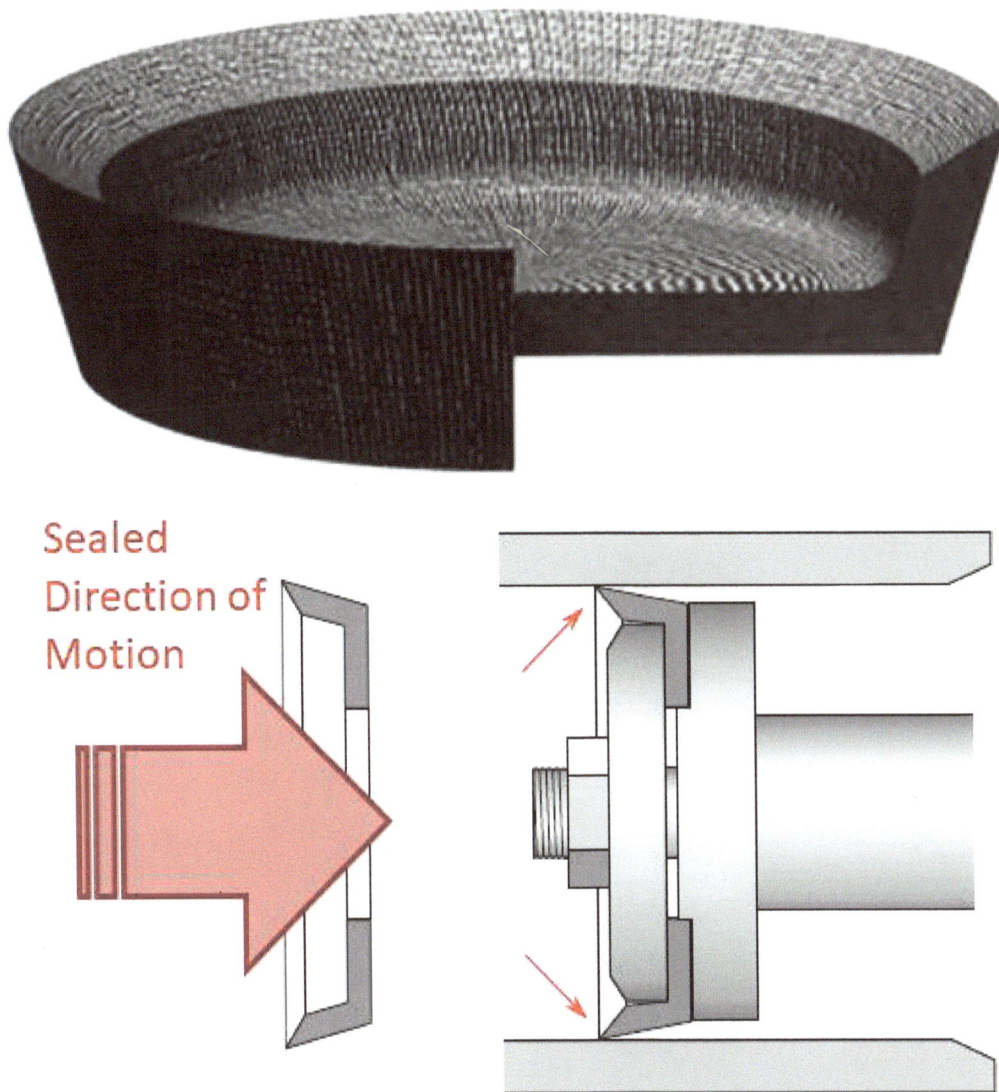

Fig. 4.21 - Cup Seals (Curtesy of MFP Seals)

4.4- U- Cup Seals

Figure 4.22 shows an asymmetric *U-Cup Seal*. U-Cups are very popular in industrial cylinders and can be used as piston or rod seals. When used with back-up rings, system pressure should not exceed 173 bar (2,500 psi).

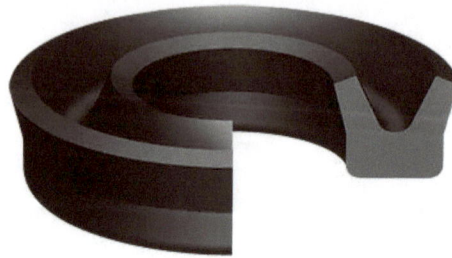

Fig. 4.22 - U-Cup Seal (Courtesy of MFP Seals)

Figure 4.23 shows the sealing mechanism of a U-Cup seal. As shown in the figure, the U-Cup Seal performs both static and dynamic sealing. It consists of a static sealing lip, and a dynamic sealing lip. Like Cup seals, it seals in one direction and relaxes in the other.

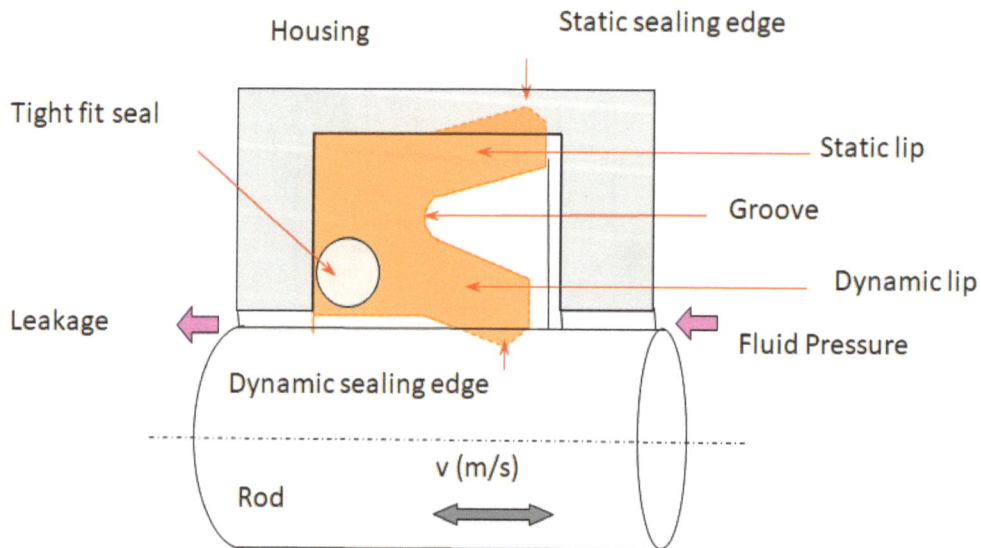

Fig. 4.23 - U-Cup Sealing Mechanism

Figure 4.24 shows a symmetric O-Ring loaded squeeze U-Cup seal with and without anti-extrusion Back-up Ring. Back-up rings are available and required for use with pressure up to 400 bar (5,800 psi).

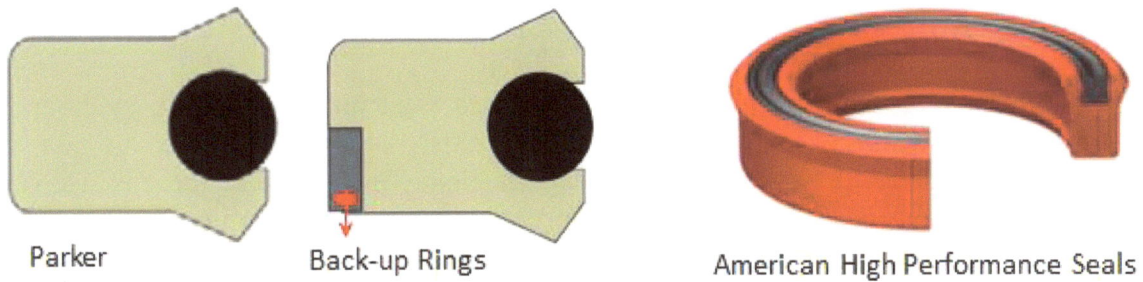

Parker   Back-up Rings   American High Performance Seals

Fig. 4.24 - Symmetric U-Cup Seals

4.5- T- Shaped Seals

Figure 4.25 shows various types of *T-Shaped* seals. They are used for sealing both cylinder rods and pistons for pressure up to 700 bar (10152 psi). Each of the shown below sealing packages consists of a T-Shaped rubber sealing ring and the two anti-extrusion Back-up Rings.

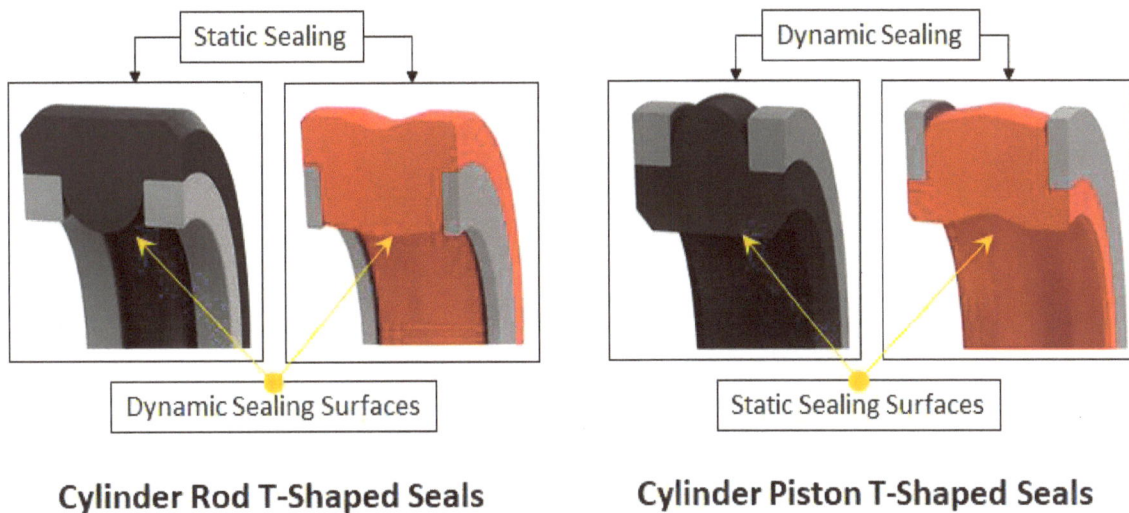

Cylinder Rod T-Shaped Seals   **Cylinder Piston T-Shaped Seals**

Fig. 4.25 -T-Shaped Seals (Courtesy of American High-Performance Seals)

Figure 4.26 shows a T-Shaped seal produced by other manufacturer. The use of the supporting Back-up Rings on both sides of the seal eliminates rolling or spiraling in long stroke cylinders or dry rod conditions.

Fig. 4.26 - T-Shaped Sealing Package (Courtesy of MFP Seals)

4.6- V-Packings

The *V-Packing* is a multi-part sealing set. They are used for sealing both cylinder rods and pistons. The V-Packing, as shown in Fig. 4.27, consists of a one (top) female adaptor and a one(bottom) male adaptor, and V-Rings between them. The female adaptor is also referred to as *Backup Ring* or *Base Ring*. It is manufactured from an elastomer with good extrusion resistance. The male adaptor is also referred to as *Compression Ring* or *Energizing Ring*. It ensures the uniform loading pressure distribution on the other rings. The number of the V-Rings depends on the operating pressure. The material of the V-Rings depends on the application. As shown in Fig. 4.28, V-Rings of different material can be used in the same pack to get the best performance. This type of seals is usually used for high pressure applications.

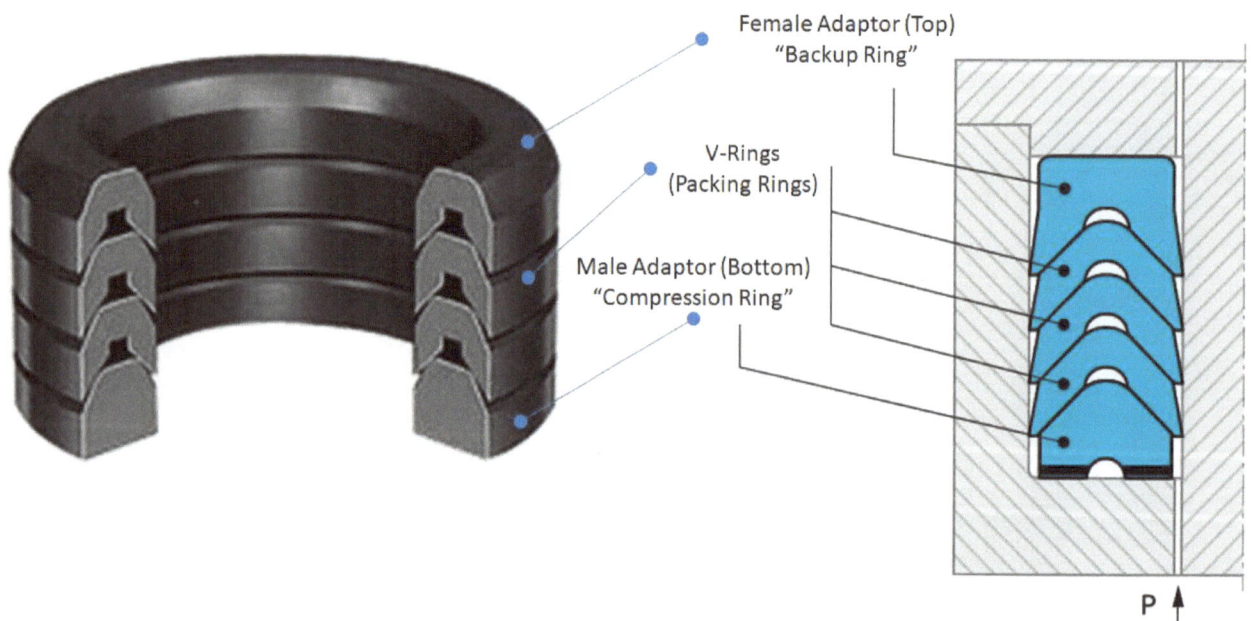

Fig. 4.27 - V-Packings (Courtesy of Trelleborg)

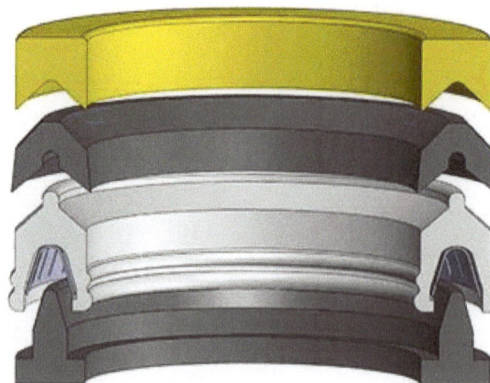

Fig. 4.28 - V-Packings of Different Materials (www.hydrapakseals.com)

4.7- Spring-Energized Seals

Figure 4.29 shows various configurations of *Spring-Energized* seals. They are used for static and dynamic sealing. Spring-Energized seals are more dynamically stable and typically used in areas where elastomeric seals cannot meet the frictional, temperature, pressure, or chemical-resistance requirements of the application.

Common uses for Spring-Energized Seals are:
- Explosive decompression resistant applications.
- Applications with extreme operating temperature and/or pressure.
- Application with high surface speeds.
- Non-lubricated surfaces.

Fig. 4.29 - Spring-Energized Seals (Courtesy of MFP Seals)

Figure 4.30 shows a typical example for Spring-Energized Plastic U-Cup seals with the following advantages:

- Suitable for reciprocating and rotary applications.
- Low coefficient of friction.
- Stick-slip free operation.
- High abrasion resistance.
- Dimensionally stable.
- Resistant to most fluids, chemicals, and gases.
- Withstands rapid changes in temperature.
- Excellent resistance to aging.
- Interchangeable with O-Ring and Back-up Ring combinations.

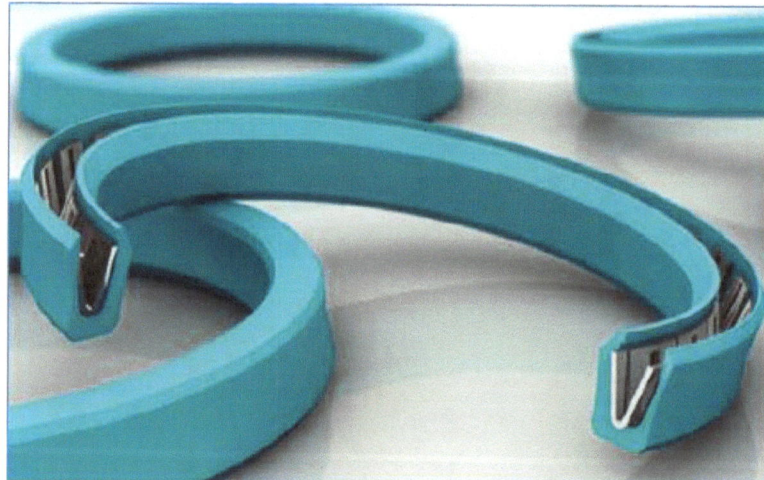

OPERATING CONDITIONS

Pressure:	Maximum dynamic load: 20 MPa. Maximum static load: 40 MPa (207 MPa with back-up ring)
Speed:	Reciprocating up to 15 m/s. Rotating up to 1.27 m/s
Operating temperature:	-70 °C to +300 °C. Special Turcon and Zurcon® materials as well as alternative spring materials are available for applications outside this temperature range.
Media compatibility:	Virtually all fluids, chemicals and gases

Fig. 4.30 - Spring-Energized U-Cup Seals (Courtesy of Trelleborg)

Figure 4.31 shows the sealing mechanism for the seals shown in the previous figure. The spring supplies the load required for sealing at low pressures. Fluid pressure energizes the sealing lips, so total sealing force rises with increasing operating pressure.

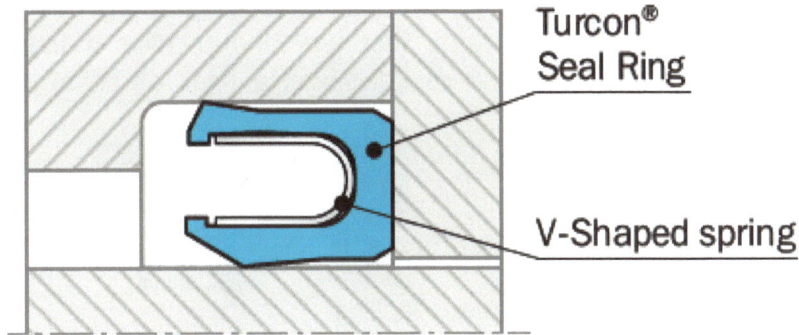

Spring Force without system pressure

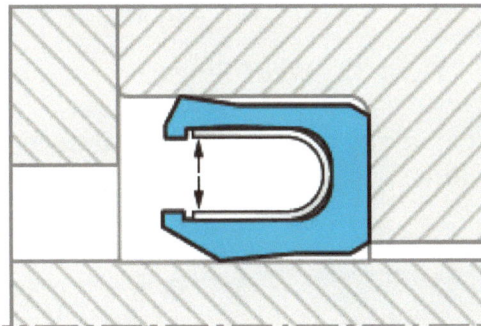

Sealing Force after applying system pressure

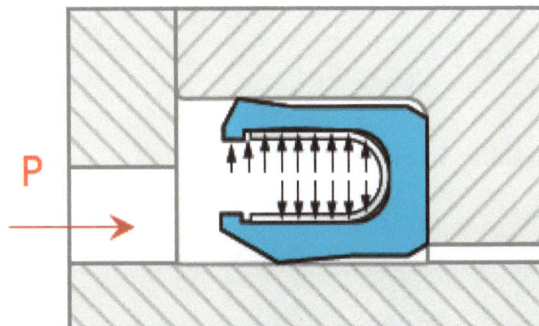

Fig. 4.31 - Sealing Mechanism of a Typical Spring-Energized Seal (Courtesy of Trelleborg)

4.8- Wear-Rings

Hydraulic seals are not designed to provide bearing surfaces or to carry lateral loads. As shown in Fig. 4.32, the function of *Wear-Rings* is to:
- Provide bearing surfaces to absorb side loads of cylinder rods and/or pistons.
- Prevent metal-to-metal contact that would otherwise damage and score the sliding surfaces and eventually cause seal damage, leakage, and component failure.

Wear-Rings are also referred to as *Guide-Rings* or *Wear-Bands.*

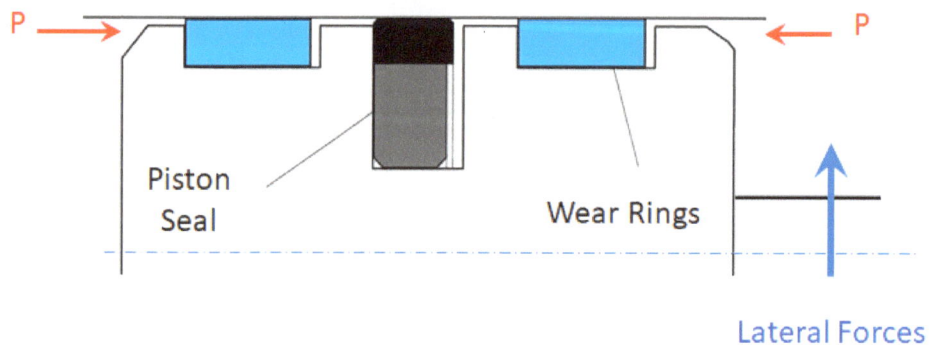

Fig. 4.32 - Wear-Rings

Figure 4.33 shows some typical Wear-Rings that are made of different material such as Nylon, Teflon and other plastics. Like O-Rings, as shown in Fig. 4.34, Wear Rings are available as pieces or as strips.

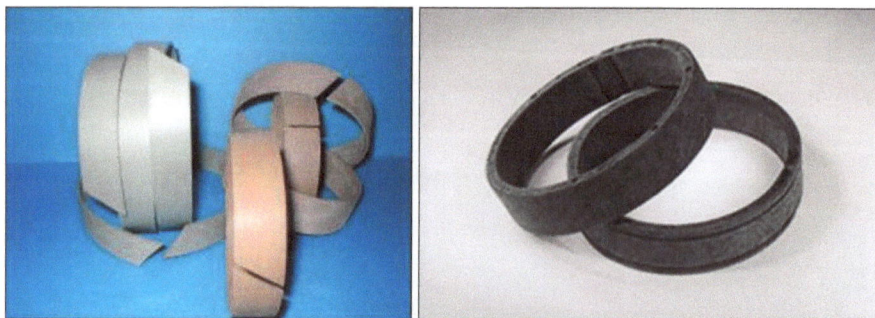

Fig. 4.33 - Examples of Wear Rings

Fig. 4.34 - Wear-Strip (Courtesy of MFP Seals)

Wear Rings are available, as shown in Fig. 4.35, with various cut designs. Figure 4.36 shows a Wear-Ring design with a special texture on the sliding surface. It is named *TEARDROP* profile. Such a profile comprises small lubricant pockets on the surface which improve the initial lubrication and promotes the formation of a lubricant film. They also help to protect the seal system through their ability to embed any foreign particles.

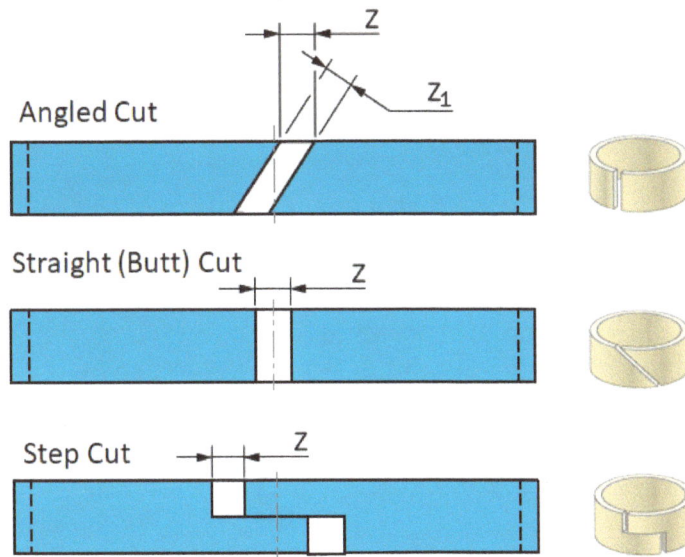

Fig. 4.35 - Wear-Rings Cut Design (Courtesy of Trelleborg)

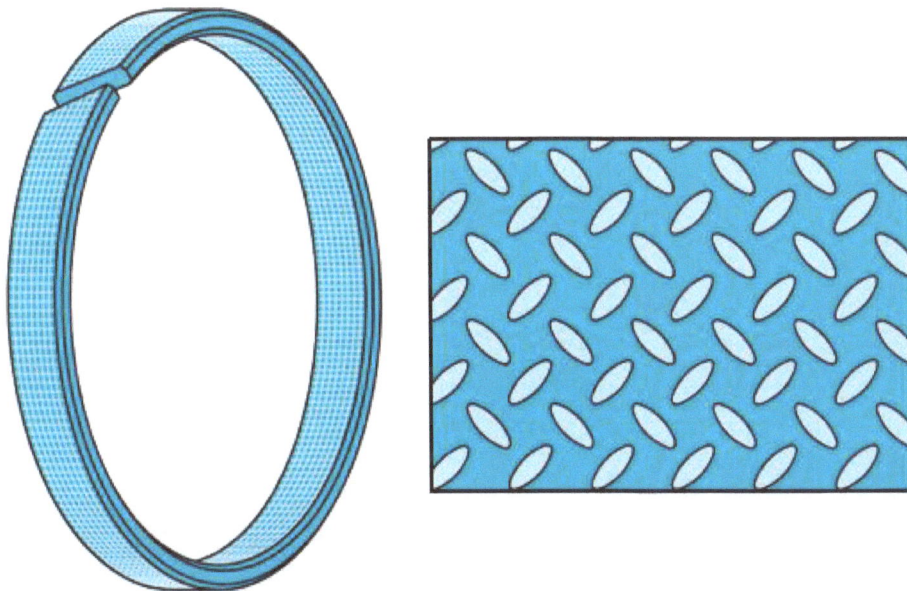

Fig. 4.36 - Wear-Rings with TEARDRP Profile on the Sliding Surface (Courtesy of Trelleborg)

Equation 4.1 shows how to calculate the minimum bearing length.

$$P_R = \frac{F \times f}{\text{Projected Area}} = \frac{F \times f}{D \times T} \rightarrow T = \frac{F \times f}{D \times P_R}$$

4.1

Where (as shown in Fig. 4.37):

- P_R = Design Pressure for a specific Wear-Ring.
- F = maximum estimated lateral force based on cylinder design and application.
- f = Safety factor (default = 2).
- D = Diameter of cylinder rod or piston.
- T = minimum bearing length.

Example: A wear ring is required for a cylinder rod of 60 mm diameter and 40,000 N maximum estimated radial force. A specific Wear-Ring material is selected that has 100 N/mm² design pressure. Then the minimum bearing length is:

$$T \text{ (mm)} = \frac{40.000 \times 2}{60 \times 100} = 13.3$$

Assuming the manufacturer design tables shows Wear-Rings with a maximum length of 10 mm, installation of two rings is recommended to provide a longer guide length.

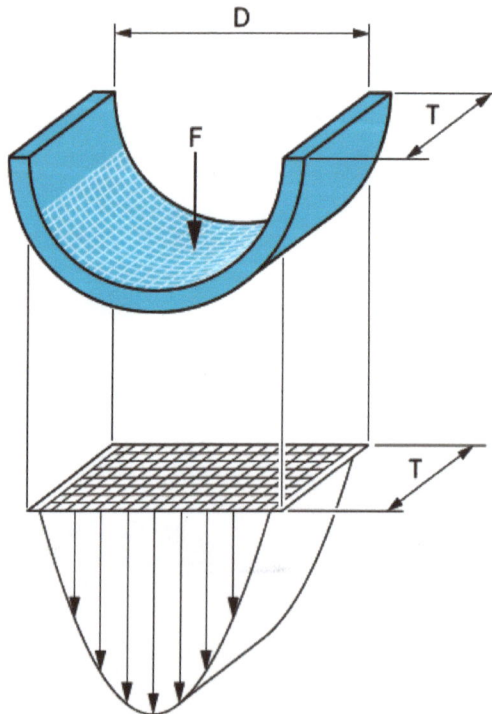

Fig. 4.37 - Calculation of the Bearing Length (Courtesy of Trelleborg)

Figure 4.38 shows typical examples of Wear-Rings with various cross sections that are characterized by:

- High bearing load capabilities.
- High operating temperatures and pressures.
- Cost effective.
- Easy installation and replacement.
- Wear-resistant and long service life.
- Low friction and self-lubrication.
- Wiping/cleaning effect.
- Ability to embed foreign particles possible.
- Damping of mechanical vibrations.

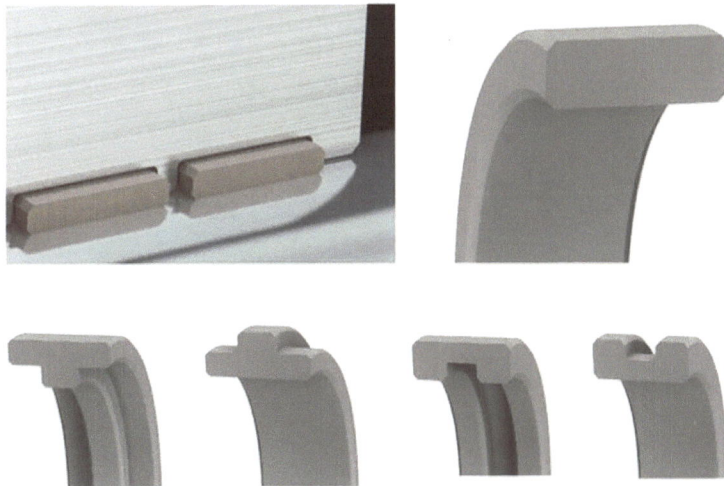

Fig. 4.38 - Wear-Rings of Various Cross Sections (Courtesy of American High-Performance Seals)

Figure 4.39 shows a typical example of Wear-Rings assembled on a cylinder piston (left) and in a cylinder rod gland (right).

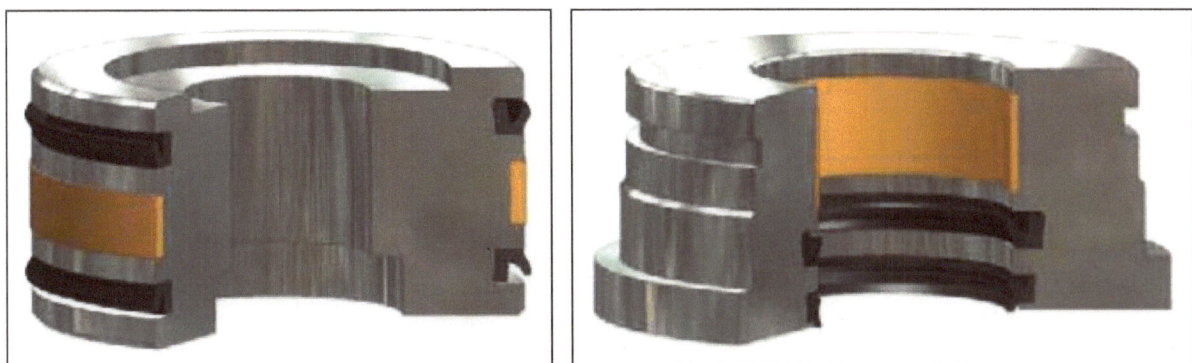

Fig. 4.39 - Typical Wear-Rings for a Cylinder Piston and Rod Gland (Courtesy of Parker)

4.9- Backup Rings

The main purpose of standard *Backup Rings* is to prevent seal extrusion. Backup rings are produced a variety of materials such as leather, rubber, elastomers, and Teflon. Figure 4.40 shows various types of the standard Backup Rings, each of which is produced with a wide range of sizes. Figure 4.41 shows examples of sealing rings supported by Backup Rings. Figure 4.42 shows a typical example of Back-up Rings assembled in a cylinder piston (left) and in a cylinder rod gland (right).

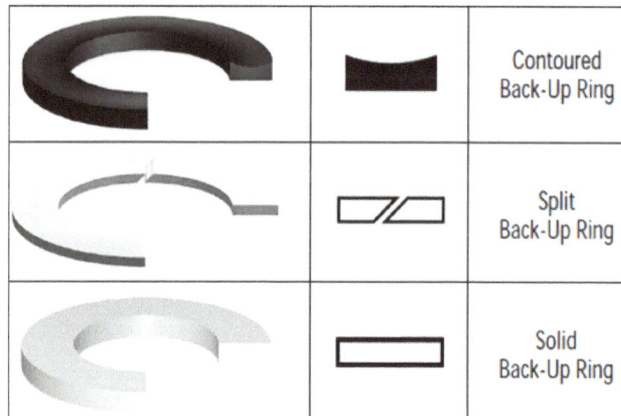

Fig. 4.40 – Cross Sections Standard Backup Rings (www.mfpseals.com)

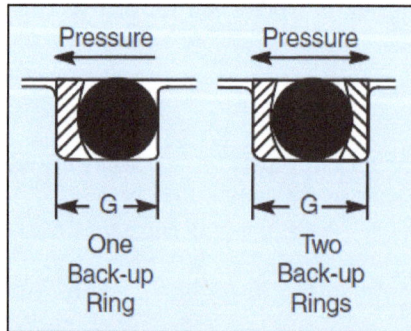

Fig. 4.41 - Examples of Sealing Rings Supported by Backup Rings (Courtesy of Parker)

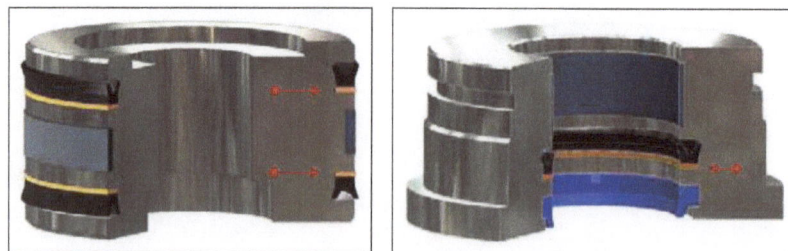

Fig. 4.42 - Typical Back-up Rings for a Cylinder Piston and Rod Gland (Courtesy of Parker)

4.10- Rod Wipers

Contaminated hydraulic fluid is a frequent cause of total system failures. As a result of increasingly sensitive elements within the hydraulic system, the Rod-Wipers function becomes ever more important.

Rod *Wipers* (also known as *Scrapers*, *Excluders* or *Dust Seals*) are mainly used as:
- The first line of defense against contamination that is settled on the cylinder rod.
- The last line of defense against external leaking.
- Host sufficient lubrication film for continuous self-lubrication of the cylinder rod.

As shown in Fig. 4.43, a traditional Rod Wiper is referred to as a single-acting Wiper because it prevents contamination from getting into the cylinder during rod retraction. It performs both static and dynamic sealing. They are usually made from a mix of brass and some other material to compromise the required hardness and the flexibility.

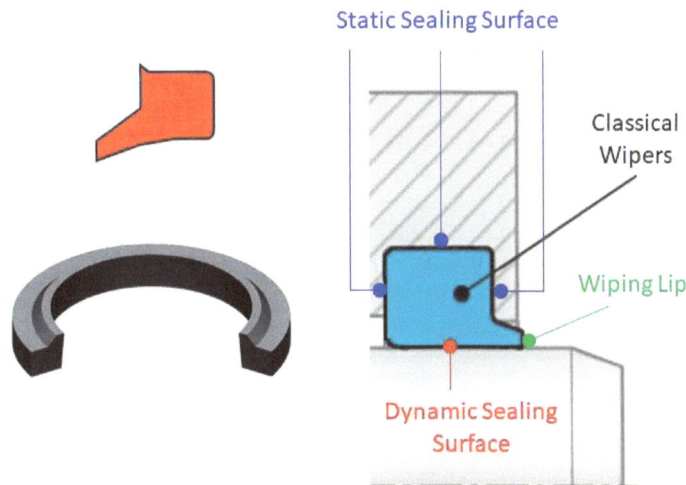

Fig. 4.43 - Traditional Single-Acting Rod Wiper Seal

Example 1: Figure 4.44 shows a typical example of a basic single-acting single-lip wiper seal.

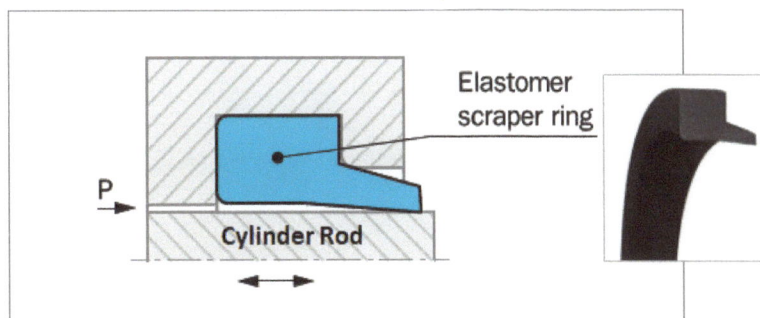

Fig. 4.44 - Typical Basic Single-Acting Rod Wiper Seal (Courtesy of Trelleborg)

Example 2: As shown in Fig. 4.45, Modern Rod Wiper Seals may contain more than one scraper lip. The figure shows a Single-Acting Rod Wiper Seal that contains *Redundant* sealing lips, a thin metallic lip and an elastomeric lip. The two scraper lips are engaged in tandem behind each other in a compact metal housing. The metallic lip is self-adjusting, lasts longer, and effectively removes contaminants from the rod surface. Rubber lip is the last line of defense against contaminants and provide better sealing.

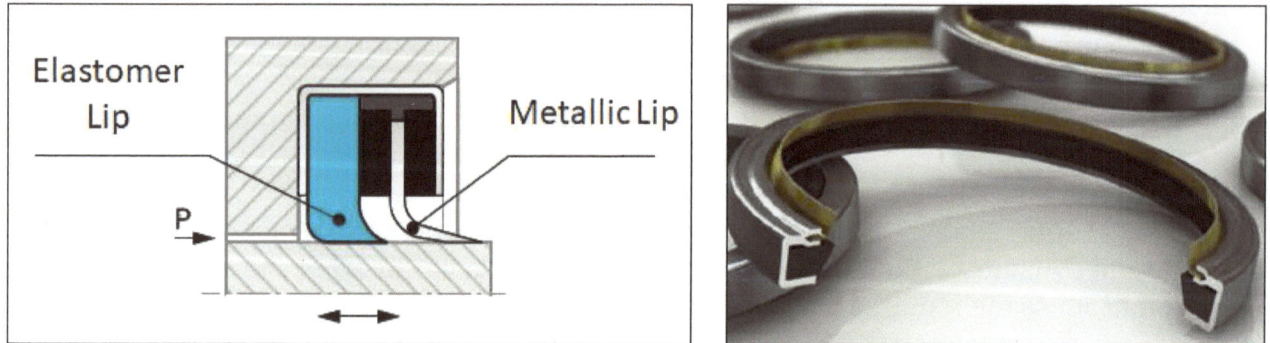

Fig. 4.45 - Single-Acting Wipers with Redundant Sealing Lips (Courtesy of Trelleborg)

Example 3: Figure 4.46 shows a typical example of a *Double-Acting Wiper Seal*. The scraper lip is designed in a particular way that it reliably scrapes off the dirt but leaves a residual oil film on the rod, which is required for correct operation. The radial squeeze is sufficient to remove particles, dust, and water. The scraping lip facing inwards is designed in a way that it assumes a sealing function even under low pressure. The static seal is achieved by a tight radial fit between the scraper body and the groove.

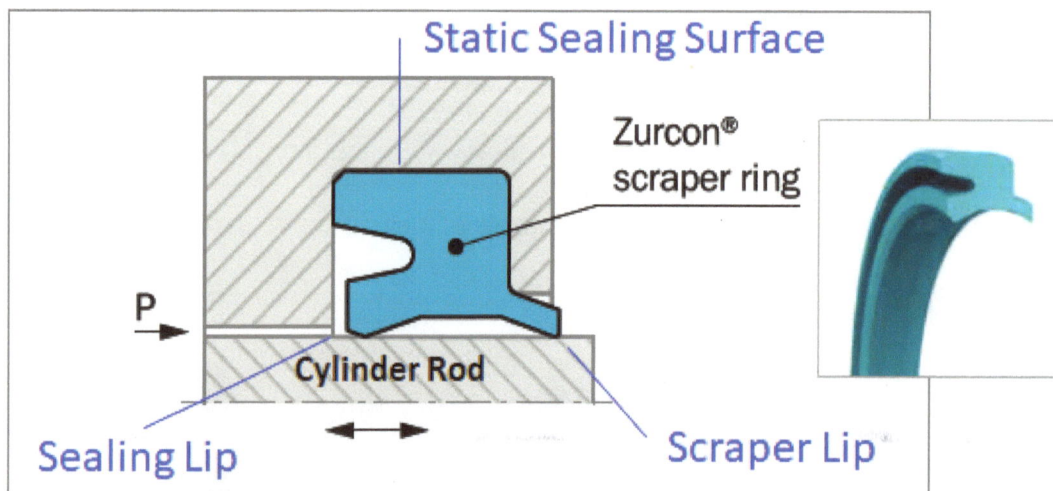

Fig. 4.46 - Double-Acting with Redundant Sealing Lips (Courtesy of Trelleborg)

Example 4: Figure 4.47 shows a high-performance rod wiper referred to as "*Umbrella Wiper Technology*". Such wipers are designed for harsh environment where the wiper is subjected to numerous contaminants. What makes this design unique is the protective guard that entirely cover the retaining groove. Typical applications are agriculture, off-highway, and forestry equipment.

Advantages of this type of wipers are as follows:

- Sheets water and debris away from rod and housing.
- Prevents traditional ingression of moisture and mud.
- Eliminate corrosion in wiper and gland housing.
- Unlike rod wipers with metallic lips normally specified for harsh environment, it can be installed without special tools.
- Made from material with good resistance to UV and chemical.
- Offers long life and sealability.

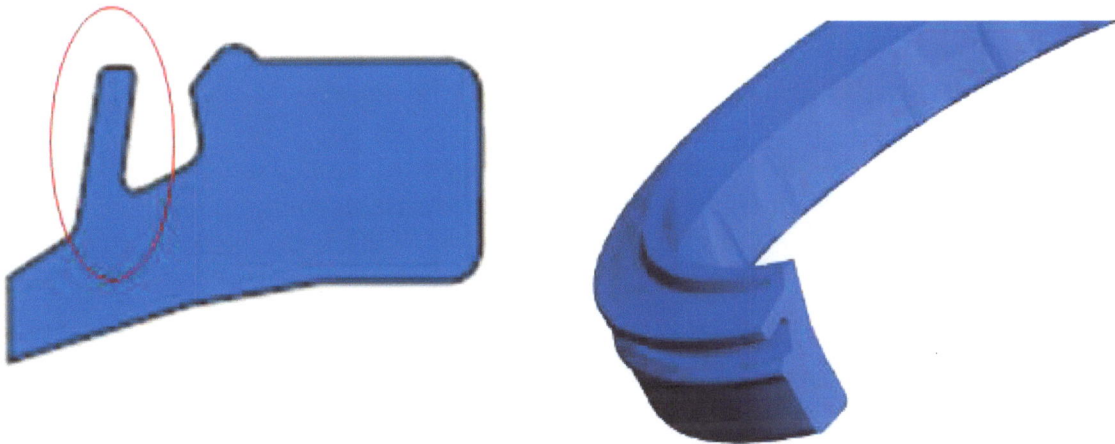

Fig. 4.47 - Umbrella Wiper Technology (Courtesy of Hallite Seals)

4.11- Materials for Hydraulic Sealing Elements

Seal materials play a major role in the performance and service life of hydraulic seals, and consequently the reliability of hydraulic components. Generally, hydraulic seals work for a variety of applications and working conditions, such as a wide range of temperature and pressure, contact with various hydraulic fluids, and exposure to the outside environment and side loads. A wide variety of hydraulic seal materials have been developed and used by seal manufacturers.

Like hydraulic components, seal design and performance prediction are based on modeling and simulation. Seal material development is a continuous process that depends on experimentation and field testing. The seal could be produced from one or combination of the following materials to optimize seal performance:

- Fabric.
- Rubber (natural and synthetic).
- Leather.
- Metal.
- Elastomeric Compounds (a mixture of base polymer and other chemicals that form a finished rubber material).
- Engineered Plastics.

A hydraulic seal can also be composed of a base compound and coated by industrial coating. As shown in Fig. 4.48, the most common coating is Polytetrafluoroethylene (PTFE). PTFE coating offer low frictional motion. Consequently, it eliminates sticking, eases seal installation, and minimize power losses.

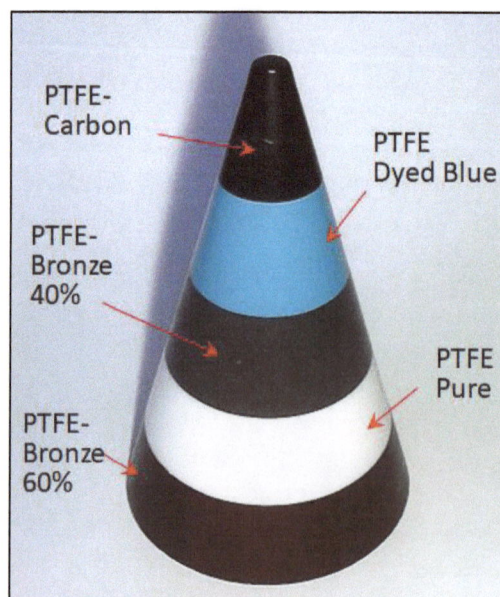

Fig. 4.48 - PTFE Seal Material

Every manufacturer has their own codes for seal materials. However, Table 4.1, shows the standard abbreviations for synthetic rubbers used in hydraulic seals manufacturing.

ELASTOMER RUBBER COMPOUNDSAND REFERENCES					
General Description	**Chemical Description**	**Abbreviation (ASTM 1418)**	**ISO/DIN 1629**	**Other Trade Names & Abbreviations**	**ASTM D2000 Designation**
Nitrile	Acrylonitrile-butadiene rubber	NBR	NBR	Buna-N	BF, BG, BK, CH
Hydrogenated Nitrile	Hydrogenated Acrylonitrile-butadiene rubber	HNBR	(HNBR)	HNBR, HSN	DH
Ethylene-Propylene	Ethylene propylene diene rubber	EPDM	EPDM	EP, EPT, EPR	BA, CA, DA
Fluorocarbon	Fluorocarbon Rubber	FKM	FPM	Viton ®, Fluorel ®	HK
Chloroprene	Chloroprene rubber	CR	CR	Neoprene	BC, BE
Silicone	Silicone rubber	VMQ	VMQ	PVMQ FC, FE, GE	FC, FE, GE
Fluor-silicone	Fluor-silicone rubber	FVMQ	FVMQ	FVMQ	FK
Polyacrylate	Polyacrylate rubber	ACM	ACM	ACM	EH
Ethylene Acrylic	Ethylene Acrylic rubber	AEM	AEM	Vamac ®	EE, EF, EG, EA
Styrene-butadiene	Styrene-butadiene rubber	SBR	SBR	SBR	AA, BA
Polyurethane	Polyester urethane / Polyether urethane	AU / EU	AU / EU	AU / EU	BG
Natural rubber Natural rubber	Natural rubber Natural rubber	NR	NR	NR	AA

Vamac ® and Viton ® are registered trademarks of E. I. du Pont de Nemours and Company or affiliates.
Fluorel ® is a registered trademark of Dyneon LLC

Table 4.1 - Standard Abbreviations for Synthetic Rubber (news.ewmfg.com)

4.12- Properties and Test Methods for Hydraulic Sealing Elements

Nowadays machineries are getting faster, and the operating conditions are becoming more severe. This increases the demand for seal material development and testing. The operation of hydraulic seals combines several disciplines such as physics, mechanics, thermodynamics, fluid dynamics, tribology, etc. There are standard test procedures for conducting most of the tests on elastomers. This section overviews the main properties, shown in Table 4.2, of hydraulic seals and the corresponding test methods. It is to be noted that some of these properties affect the dynamic sealing functions only, some affect the static sealing functions only, and others affect both functions. Additionally, some of the test methods are standardized, and others are based on manufacturer R&D activities but not standardized.

If uniform and repeatable results are to be obtained, it is important to follow the test procedures carefully along with the following considerations:

Test Specimens
Test methods include descriptions of standard specimens for each test. Often, two or more specimens are required due to specimen variability. Results from different specimens are rarely 100% agree.

Test Variables
Test results are only comparable if the following test variables are identical:
- Temperature under which the test was performed.
- Load or pressure used in the test.
- Fluid medium in which the seal is in contact with during the test.
- Duration and rate of applying the test procedure.
- Environmental conditions, such as humidity in the air.

4.12.1- Resilience

Definition: *Resilience* is essentially the ability of a compound to return quickly to its original shape after a temporary deflection.

Units: It is dimensionless property

Effect: Reasonable resilience is vital for moving seals. Resilience is primarily an inherent property of the elastomer. It can be improved somewhat by compounding. More important, it can be degraded or even destroyed by poor compounding techniques.

Testing: It is very difficult to create a standard laboratory test which properly relates this property to seal performance. Therefore, compounding experience and functional testing under actual service conditions are considered to insure adequate resilience. Tested specimens are inspected visually after test.

Hydraulic Sealing Elements and Test Methods
Change in Seal Shape
1- Resilience
Test Method: Non-Standard Test
Change in Seal Length
2- Modulus of Elasticity
3- Elongation
4- Yield Tensile Strength
5- Ultimate and Fracture Tensile Strengths
Test Method: Standard Test Method (ASTM D412 / DIN 53504)
Change in the Seal Cut
6- Tear Strength:
Test Method: Standard Test Method (ISO 34-1 / DIN 53507)
Change in Seal Thickness
7- Compression Set
Test Method: Standard Test Method (ASTM D395 / DIN ISO 815)
Change in Seal Volume
8- Swelling
Test Method: Standard Test Method (DIN ISO 1817)
9- Shrinkage
Test Method: Standard Test Method (TR-10 Low Temperature ASTM D1329-16)
Change in Seal Surface
10- Surface Hardness
Test Method: Standard Test Method (ASTM D 2240 / ISO 868 / DIN 53505)
Change in Seal Chemical Structure
11- Compatibility with Hydraulic Fluid
Test Method: Standard Test Methods (ASTM D6546-15 OR ISO 6072)
Change in Seal Performance
12-Exrusion Resistance
Test Method: Standard Test Method (ASTM C1183 / C1183M)
13-Explosive Decompression Resistance
Test Method: Standard Test Method (NACE TMO192-98)
14- Seal Friction
Test Method: Non-Standard Test
15- Wiper Performance
Test Method: Non-Standard Test

Table 4.2 - Hydraulic Sealing Elements Properties and Test Methods

4.12.2- Modulus of Elasticity

Definition: As shown in Fig. 4.49, *Modulus of Elasticity* of a hydraulic seal is the ratio between the stress (force per cross section area) acting on the seal body and the corresponding percentage longitudinal change of the seal. Also referred to as Young's modulus.

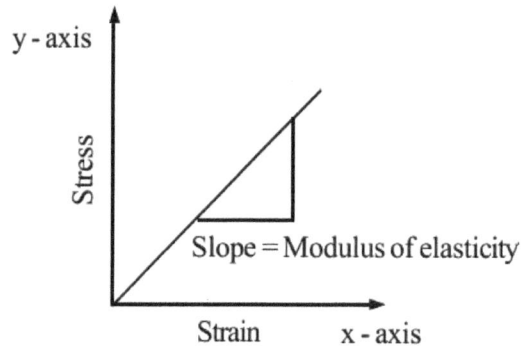

Fig. 4.49 – Modulus of Elasticity

Units: It is expressed in MPa (metric) or psi (English) per predetermined percentage elongation.

Effect and Importance:
- The higher the modulus of a compound, the more ability to recover after releasing the stress, and the better its resistance to extrusion.
- Modulus of elasticity is directly proportional to the seal hardness.
- Polyurethane and filled PTFE compounds generally have very high tensile strength, providing the associated excellent tear and abrasion resistance.

4.12.3- Elongation

Definitions: *Elongation* is defined as the percentage increase in length with respect to the original length.

Units: It is expressed as a % of the original length. For example, if a 1-inch sample was stretched to two inches, it would be 100% elongation.

Effect and Importance:
- This property primarily determines the stretch which can be tolerated during installation of a sealing element. However, easy to stretch also means easy to extrude.
- The change in the length of a compound is a clear sign of material degradation.

4.12.4- Yield Tensile Strength

Definitions: *Yield Tensile Strength* is the stress within which the seal maintain elastic performance.

Units: It is expressed in MPa (metric) or psi (English).

Effect and Importance:
This property determines the maximum strength beyond which the seal is plastically deformed. That helps to know the maximum loads such seal can withstand.

4.12.5- Ultimate and Fracture Tensile Strength

Definitions:
- *Ultimate Tensile Strength* is the stress required to reach ultimate elongation, and seal nicking starts.
- *Fracture Tensile Strength* is the stress at which the seal breaks.

Units: It is expressed in MPa (metric) or psi (English).

Effect and Importance:
- All properties (Elongation, Tensile Strength, Ultimate Strength, and Fracture Strength) $\rightarrow$
  - Determine the application load conditions for sealing elements.
  - Used throughout the industry as a quality assurance measures on production batches of elastomer materials.

Testing: (*Standard Test Method ASTM D412 / DIN 53504*)

<u>Test Purpose:</u> Determination of Tensile Strength values.

<u>Test Specimen:</u> The test specimens used for this purpose are usually tensile bars or standardized rings with rectangular specific cross-sections. Inconsistent values were found as a result of using specimens with various cross sections.

<u>Test Sequence:</u> As shown in Fig. 4.50, the test specimen is stretched at a constant speed to break. Tensile Strength curve (force versus change in the length) is developed.

In pulling specimens to find tensile strength, elongation, and modulus values, ASTM D412 requires a uniform rate of pull of 508 mm (20 inches) per minute. In one test, tensile strength was found to decrease 5% when the speed was reduced to 50.8 mm (2 inches) per minute, and it decreased 30% when the speed was further reduced to 5.08 mm (0.2 inches) per minute. Elongation and modulus values decreased also, but by smaller amounts.

High humidity in the air will reduce the tensile strength of some compounds.

<u>Test Evaluation:</u> As shown in Fig. 4.51, the following values can be calculated from the experiment:
- Yield Tensile Strength.
- Ultimate Tensile Strength.
- Fracture Tensile Strength.
- Elongation at break.

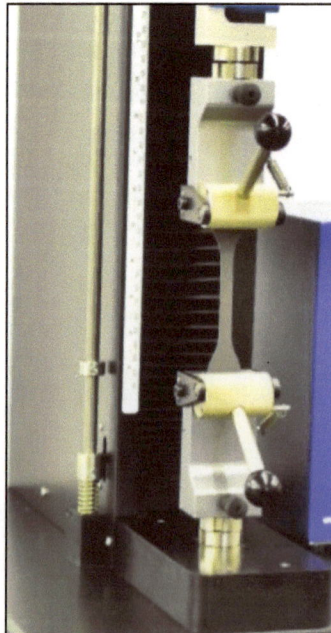

Fig. 4.50 - Elongation and Tensile Strength Test Machine

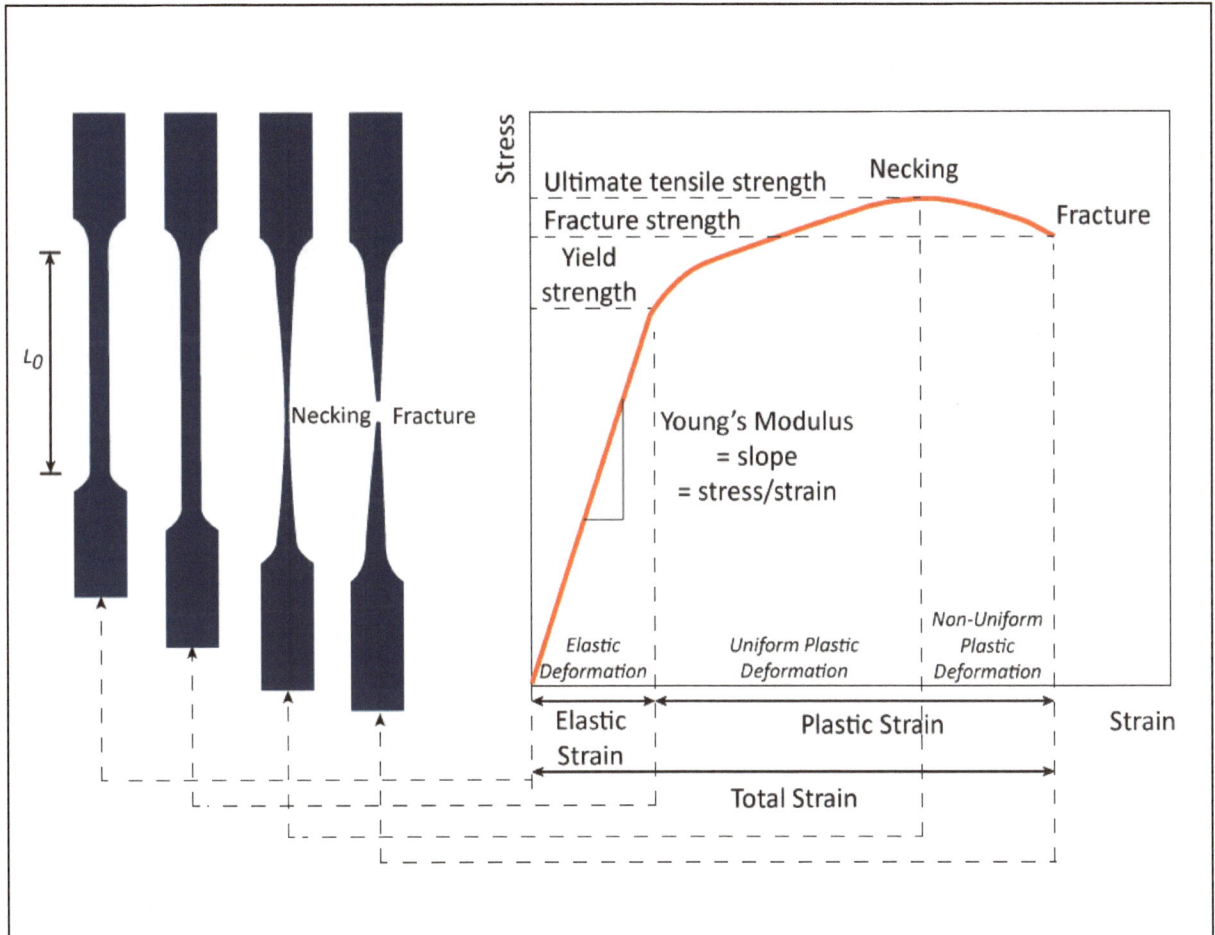

Fig. 4.51 - Results from Elongation and Tensile Strength Test

4.12.6 -Tear Strength

Definitions: *Tear Strength* is the ratio of the force achieved at the moment of rupture and the initial cross-section thickness of the specimen.

Units: Tear Strengths is expressed in pounds/inch (lb/in) or Kilo Newton/meter (kN/m).

Effect and Importance:
Tear Strength is a measure of its ability to resist tear during assembly and operation. Seals with poor tear resistance (less than 100 lbs./in.) (17.5 kN/m):
- Have the danger of tearing during assembly if it must pass over ports and sharp edges.
- Will fail quickly under further flexing or stress once a crack is started.

Testing (*Standard Test Method ISO 34-1 / DIN 53507*):
Test Purpose: Determination of the *Tear Propagation Resistance* (the sensitivity of elastomers in the event of cutting and tearing damage).

Test Specimen: As shown in Fig. 4.52, the specimens used for this test are usually tensile bars or standardized rings with rectangular cross-sections.

Test Sequence: A longitudinal cut is made in the material to be tested, the two half-strips are clamped in a pulling machine and pulled apart.

Test Evaluation: The force required to propagate the cut is measured in relation to the sample thickness.

Fig. 4.52 - Tear Resistance Test

4.12.7- Compression Set

Definition: The *Compression Set* (CS) refers to the ability of a sealing material to restore thickness and consequently the sealing force after a certain time in contact with a fluid medium under certain temperature.

Units: It is expressed in % in comparison with a targeted value of deformation.

Effect and Importance:
The lower the compression set value, the better restoring, sealing, and seal lifetime. In general, Compression Set is a result of one or more of the following conditions:
- Selection of seal material with inherently poor compression set properties.
- Improper gland design.
- Excessive temperature causing the seal to harden and lose its elastic properties.
- Volume swell of the seal due to system fluid.
- Excessive seal squeeze due to over tightening of adjustable glands.
- Incomplete curing (vulcanization) of seal material during production.
- Fluid incompatibility with the seal material.

Note: Magnitude of CS is inversely proportional to the ability or recovery. Hence, Poor CS means higher magnitude. Figure 4.53 shows a flattened O-Ring due to poor compression set. The seal shown in Fig. 4.54 exhibits nearly 100% compression set, i.e. no recovery (thickness restoration) after compression.

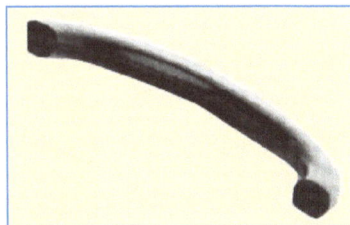

Fig. 4.53 - Flattened O-Ring due to Poor Compression Set (Courtesy of Parker)

Fig. 4.54 - Seal Exhibiting nearly 100% Compression Set (Courtesy of Parker)

Testing (*Standard Test Method ASTM D395 / DIN ISO 815*):
Test Purpose: Determination of the compression set.

Test Specimen: a specimen with a specific form.

Test Sequence (As shown in Fig. 4.55):
- Compression test must not be done earlier than 16 hours after elastomer manufacturing.
- Tests should only be carried out with samples which have not been previously stressed.
- Original thickness (h_0) of the test specimen is measured.
- The sample is compressed to the thickness (h_1) (default: $h_1 = 0.75\ h_0$).
- The compressed specimen is stored in an apparatus in a medium (default: air) and under a specified temperature for a predefined test time (default: 24 or 72 hours).
- Test temperature should be mentioned in the test report.
- 30 minutes after removal from storage and removing the compression, recovered thickness of the specimen (h_s) is measured.

Test Evaluation:
Compression set values are only comparable if the following test parameters are identical:
- Form of the test specimen.
- Deformation (default: 25%).
- Duration of the deformation.
- Temperature and medium during the deformation.
- Equation 4.2 shows how to calculate the compression set value.
- ASTM Compression Set D395 Test Method B, states, "The percentage of compression employed shall be approximately 25%." Significantly higher compression set values were found after compressing less than 25%, while results after 30 or 40% compression were sometimes smaller.

$$CS\ (\%) = \frac{(h_0 - h_S)}{(h_0 - h_1)} \times 100 \qquad\qquad 4.2$$

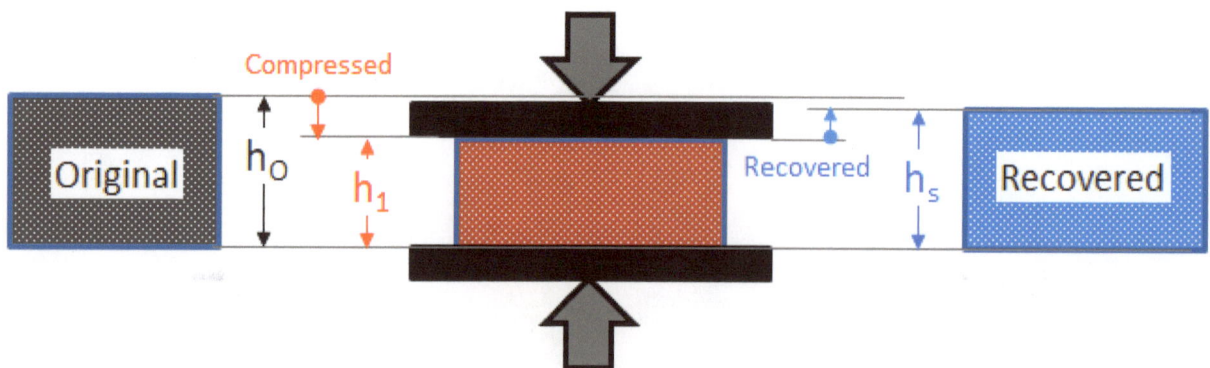

Fig. 4.55 - Compression Set Test

4.12.8- Swelling

Definition: Elastomers have a higher coefficient of thermal expansion than steel. This means the seal will expand more when hot. *Swelling* of a hydraulic seal is the volumetric increase of the seal after it has been in contact with a fluid medium.

Units: It is expressed as a % of the original volume.

Effect and Importance: Seal Swelling is associated with changes in the seal physical properties such as:
- Possibility of seal extrusion under high pressure.
- Reduced hardness, elasticity, and tensile strength.
- Marked softening of the elastomer.

The magnitude of the volume change depends on five factors:
1. Fluid in contact
2. Composition of the elastomeric compound.
3. Working conditions (temperature, pressure, humidity, and time).
4. Geometric form (thickness) of the seal.
5. Stress condition of the seal. (Volume change of stretched parts is greater than compressed parts).

Testing (*Standard Test Method DIN ISO 1817*):
Test Purpose: Determination of seal volume increase after a defined storage period in contact with specific fluid under certain temperature.

Test Sequence:
1. Measure the original volume of the seal.
2. The seal is immersed in a test fluid and stored according to the standard or to customer specifications.
3. At the end of the storage period (and after cooling down), the volume of the seal is measured again.
4. The result is expressed as a percentage of the initial volume.

Test Evaluation: As a rule-of-thumb (unless otherwise stated).
- For static seals, up to 50% volume swell can usually be tolerated.
- For dynamic applications, (10-20) % swell is a reasonable range.
- Seals with smaller cross-sections have been found to swell more than larger ones.

4.12.9- Shrinkage

Definition: Elastomers have a higher coefficient of thermal expansion than steel. This means the seal will shrink more when cold than steel does. *Shrinkage* of a hydraulic seal is the volumetric decrease of the seal after it has been in contact with a fluid medium.

Units: It is expressed as a % of the original volume.

Effect and Importance: Seal Shrinkage is associated with changes in the seal physical properties such as:
- As in seal swelling, reduced hardness, elasticity, and tensile strength.
- As in seal swelling, marked softening of the elastomer.
- Reduced retaining force between the seal and the static housing faces.
- Shrinkage may result in possible leak.

Testing (Low-Temperature *Standard Test Method ASTM D1329-16 / TR-10*):
Test Purpose: Determination of elastomer retraction and viscoelastic properties at low temperature.

Test Specimen:
- Figure 4.56 shows several specimens of various lengths mounted and ready to be elongated. Standard specimens have lengths of 1", 1.5", or 2".

Test Sequence:
- A small dog-boned specimen is held in an elongated condition.
- The test fixture is used to apply 100% elongation.
- The specimen is subjected to a low temperature for certain time.
- The specimen is then allowed to retract freely while raising the temperature at a uniform rate of 1°C (1.8 °F) per minute.
- Measure the temperature at 10% and 70% retraction (shrinkage).
- Comparison is based on the % shrinkage at a specific temperature or the temperature at specific % shrinkage.

Test Evaluation: Elastomers are compared based on temperatures corresponding to 10% and 75% shrinkage because this is the temperature window within which the elastomers start and end the retraction. Maximum tolerated shrinkage for both static and dynamic seals is (3-4) %.

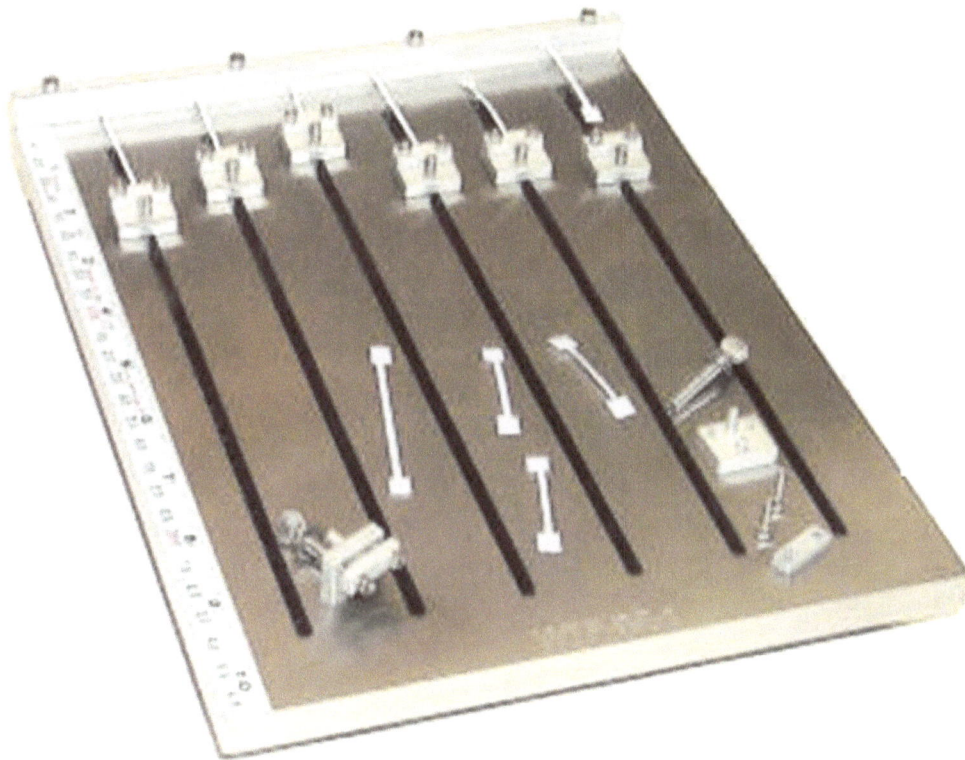

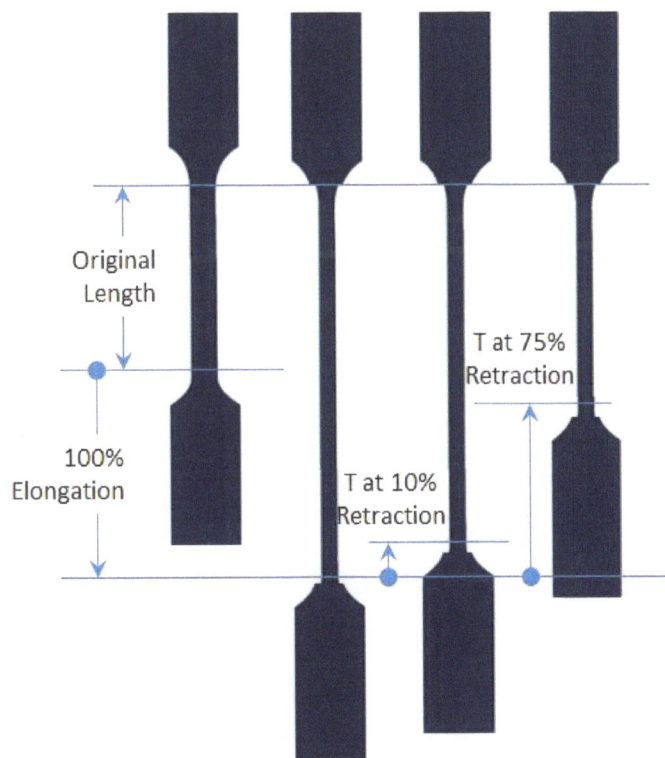

Fig. 4.56 - Elastomers Retraction Test (www.wyomingtestfixtures.com)

4.12.10- Surface Hardness

Definition: One of the most important parameters in rubber technology is the *Hardness.* Hardness is the resistance of a body against penetration of a harder body of a standard shape at a defined load.

Units: The hardness scale has a range of 0 (softest) to 100 (hardest). Common hydraulic seals have hardness that fall in two scales, *Shore A* (for soft-to-medium compounds) and Shore D for (for medium-to-hard compounds).

Effect and Importance:
This property is important for dynamic seals where harder elastomers have better ability to resist abrasion, wear, and surface scraping. For static seals, higher hardness helps avoid seal extrusion. It is also to be noted that:
- Seal hardness is directly proportional to the Modulus of Elasticity.
- Seal leakage is directly proportional to seal hardness.
- Seal friction is inversely proportional to seal hardness.

Testing:
- Measured by devices called *Durometer.*
- For Shore A" and "Shore D": **Standard Test Methods ASTM D 2240 / ISO 868 / DIN 53505.**
- For "Durometer IRHD (International Rubber Hardness Degrees)": **Standard Test Methods ASTM 1414 and 1415 / ISO 48.**

Test Purpose: determination of seal hardness

Test Specimen (Rubber Desk):
- Diameter min. 30 mm (1.181 inch).
- Specimens of softer elastomers should be thicker than harder Elastomers. Thickness min. 6 mm (0.240 inch).
- Upper and lower surfaces smooth and flat.
- The samples must not have been previously stressed.

Hardness Testers:
As shown in Fig. 4.57, the test can be done by manual or automated *Hardness Tester.*

The Shore type hardness testers are spring loaded indentation devices, in which values are obtained as a function of the viscoelastic property of the material. A calibrated spring force is applied on a standard pin (Indentor) against the specimen. Each 0.001 inch (0.0254 mm) of deflection of the pin is shown as 1-degree Shore (A). Therefore, harder material results in more material resistance to penetration, more spring compression and pin deflection, which means higher shore number.

As shown in Fig. 4.58, the shape of the pins depends on the estimated range of hardness. For Shore A, pin with conical head. For Shore D, pin with spike head.

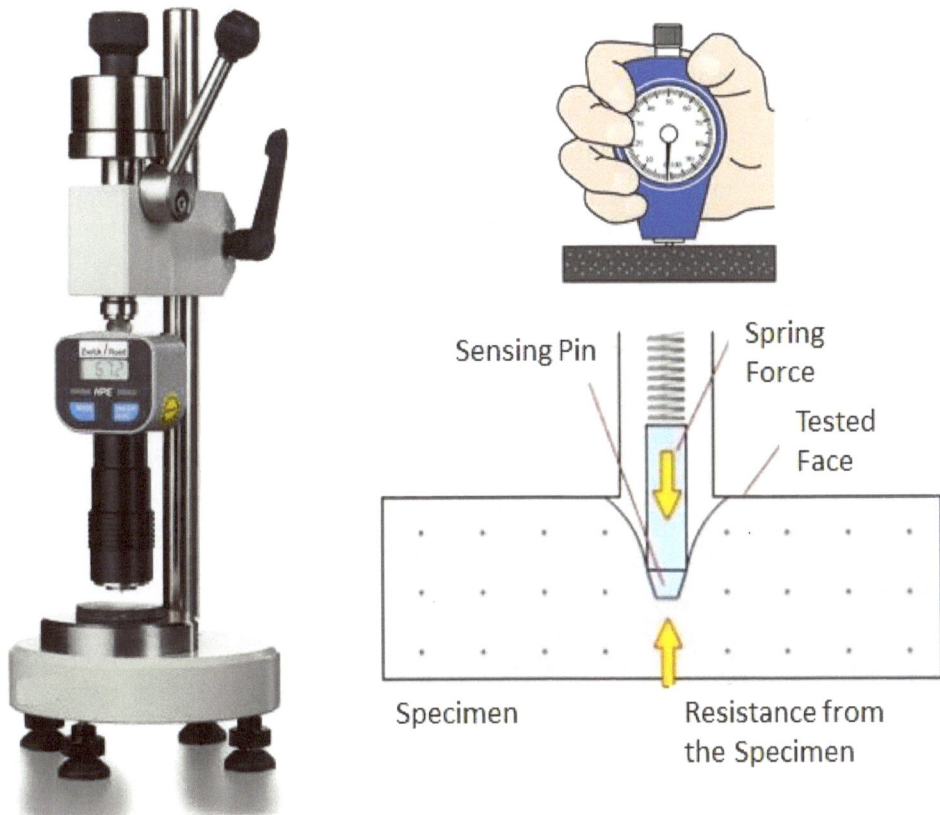

Fig. 4.57 - Hardness Test Devices for Elastomers

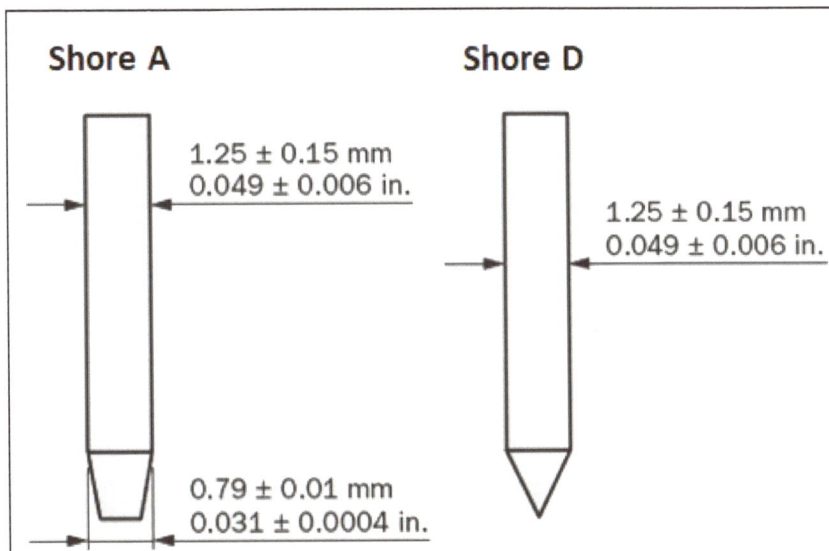

Fig. 4.58 - Pins for Hardness Tests (Courtesy of Trelleborg)

Test Sequence:
- The test must not be carried out earlier than 16 hours after elastomer manufacturing.
- The test should be carried out at temperatures of 23 ±2 °C (73.4 ±2 °F).
- The test temperature should be mentioned in the test report.
- In gauging the hardness of an O-Ring, which has no flat surface, accuracy of the test may vary depending on the measuring spot. As shown in Fig. 4.59, the actual crown of the O-Ring, the point that gives the most reliable reading.
- The test is carried out by applying a specific load on the pin at a certain speed. Inconsistent results were found as a result of performing the test at different speeds and different loads.
- The value is read after a holding time of three seconds.

Test Evaluation (as shown in Table 4.3):
- Shore A values of 60-75 are recommended for O-rings.
- Shore A values of 75-90 are recommended for seals of high resistant to abrasion.
- Harder compounds (above Shore A 90), Shore D measurements must be made.
- The scales overlapped in between the medium and hard rubbers. For example, Shore A 90 approximately equal Shore D 40.

Fig. 4.59 - Hardness Test for O-Rings

Shore A														
5	10	20	30	40	50	60	70	80	90					
Shore D														
									40	50	60	70	80	85
Very Soft			Soft			Medium				Hard				

Table 4.3 - Hardness Scales for Hydraulic Seals

4.12.11- Compatibility with Hydraulic Fluids

Definition: *Compatibility* with hydraulic fluids means the ability of a seal to resist chemical interaction with the working hydraulic fluids.

Effect and Importance:
This property is very important because any change in the seals chemical structure significantly affects the seal shape and physical properties. Seal deterioration due to chemical reaction with the hydraulic fluid causes clogging of control orifices, leakage, and the relevant consequences.

Table 4.4 shows the state of compatibility of common seals with common hydraulic fluids.

Seal materials	Fluid Types					
	Petroleum oil	Water-in-Oil Emulsion	Water Glycol	Phosphate Ester*	Chlorinated hydrocarbon	Synthetic with petroleum fractions
Buna-N (Acrylonitrile)	Excellent	Excellent	Very Good	Poor	Poor	Poor
Neoprene (Chloroprene)	Good	Good	Good	Poor	Poor	Poor
Butyl	Poor	Poor	Good	Fair to good	Poor	Poor
Silicone	Fair	Fair	Fair to poor	Fair to good	Poor to fair	Fair
Ethelene-Propylene	Poor	Poor	Good to excellent	Excellent	Fair	Poor
Viton® (Fluorocarbon)	Excellent	Excellent	Excellent	Good to Excellent	Good to Excellent	Good to Excellent
Metals	Conventional	Conventional	**	Conventional	Conventional	Conventional
Pipe Sealants	Conventional, Loctite® or Teflon® tape	Conventional, Loctite® or Teflon® tape	Loctite® or Teflon® tape	Loctite® or Teflon® tape	Loctite® or Teflon® tape	Loctite® or Teflon® tape

- *Many types and blends of fluids are sold under the designation "phosphate ester." Check with fluid supplier to verify exact compatibility.
- **Avoid zinc, cadmium, or galvanized materials.
- Viton® and Teflon® are trademarks of E.I DuPont DeNemours & Co., Inc.
- Loctite® is a trademark of the Loctite Corp.

Table 4.4 - Compatibility of Common Hydraulic Fluids with Common Seal Materials (www.schoolcraftpublishing.com)

Testing (*Standard Test Methods ASTM D6546-15* and *ISO 6072*):

Test Purpose: This test is used for determining compatibility of elastomeric seals for industrial hydraulic fluid applications.

Test Procedure: The test procedure, as shown in Fog. 4.60, includes exposing an O-ring test specimen to industrial hydraulic fluids under definite conditions of temperature and time.

Test Evaluation: The resulting deterioration of the O-ring material is determined by comparing the changes in work function, hardness, physical properties, compression set, and seal volume after immersion in the test fluid to the pre-immersion values.

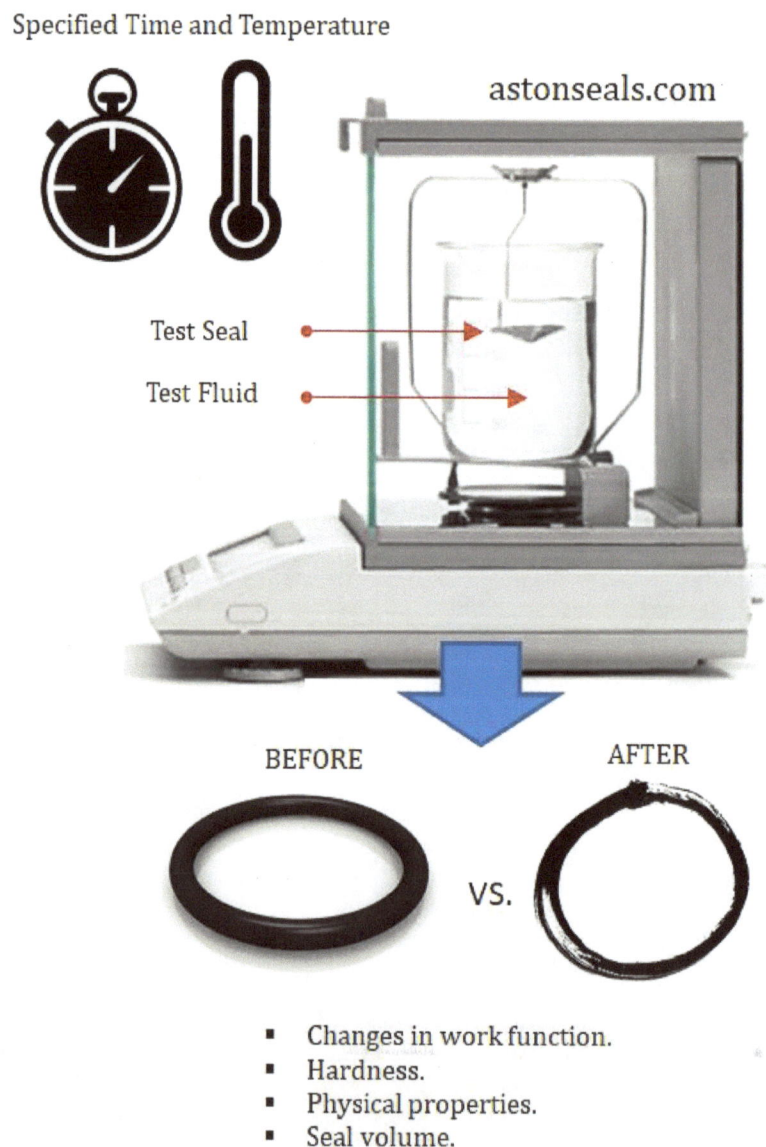

- Changes in work function.
- Hardness.
- Physical properties.
- Seal volume.

Fig. 4.60 - Fluid Compatibility Standard Test Method

4.12.12- Extrusion Resistance

Definition: *Extrusion Resistance* is the seal ability to resist extrusion through the gap between sealed surfaces as a result of increased working pressure or seal material softness.

Effect and Importance:
Seal extrusion causes seal failure, excessive leakage, and oil contamination.

Testing: (*Standard Test Method ASTM C1183*):
Test Purpose: This test method covers the procedures for determining the extrusion rate of elastomeric sealants. There is no known ISO equivalent to this test method.

Test Procedure: This test method measures the volume of sealant extruded over a given period of time at a given pressure (kPa or psi).

4.12.13- Explosive Decompression Resistance

Definition: *Explosive Decompression Resistance* is the seal ability to resist damage due to releasing of absorbed when the pressure is suddenly reduced.

Effect and Importance:
'Explosive decompression' is a commonly used term for the damage caused when a rubber seal, containing absorbed gas, is subjected to rapid decompression from a high pressure. The failure of a seal due to explosive decompression damage can lead to obvious safety and lost production consequences.

Testing: (*Standard Test Method NACE TMO192-98*):
Various oil companies and national standards organizations have developed test protocols aimed at defining elastomeric seal materials that can avoid damage during decompression in high pressure gas duty. This test is used for evaluating elastomeric materials in Carbon Dioxide decompression environments. Finite element analysis potentially provides a tool for reducing the quantity of costly and time-consuming performance testing which currently has to be carried out to prove seal integrity.

4.12.14- Hydraulic Seal Friction

4.12.14.1- Hydraulic Seal Friction Conditions

As shown in Fig. 4.61, based on the lubrication condition between the seal and the sealed surface, hydraulic seal friction modes are as follows:

- Boundary seal friction if no lubricant is found between the seal and the sealed surface.
- Mixed friction if little lubricant is found between the seal and the sealed surface.
- Viscous friction if the lubricating film can completely separate the seal from the sealed surface.

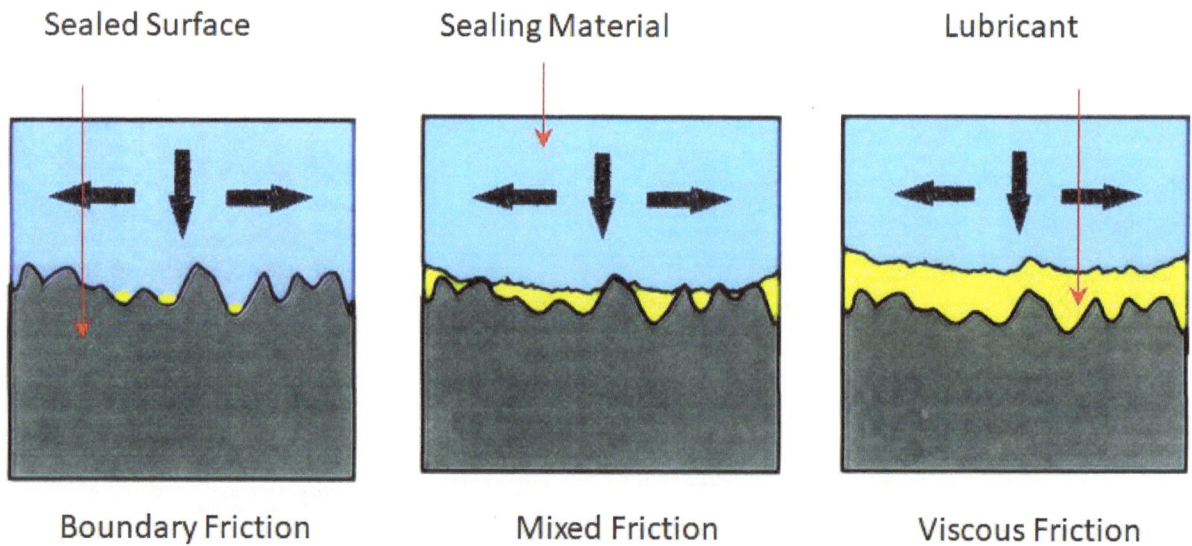

Fig. 4.61 - Hydraulic Seal Friction Conditions

4.12.14.2- Friction in Translational Seals

Examples of translational seals are cylinder rod and piston seals. This friction is usually mistakenly ignored in calculating a cylinder force or pressure. For a cylinder speed ranging from 5-15 cm/s (10 – 30 fpm), overcoming cylinder seal friction wastes approximately 5% of the cylinder input power. Seal leakage in hydraulic cylinders is approximately one tenth of the losses due to friction.

Figure 4.62 shows results of experimental work to investigate extending rod seal friction versus velocity at 75 bar (1088 psi) constant pressure. The figure shows that various seals have the same trend. Seal friction drops once the rod starts extending, then gradually increases with increasing velocity.

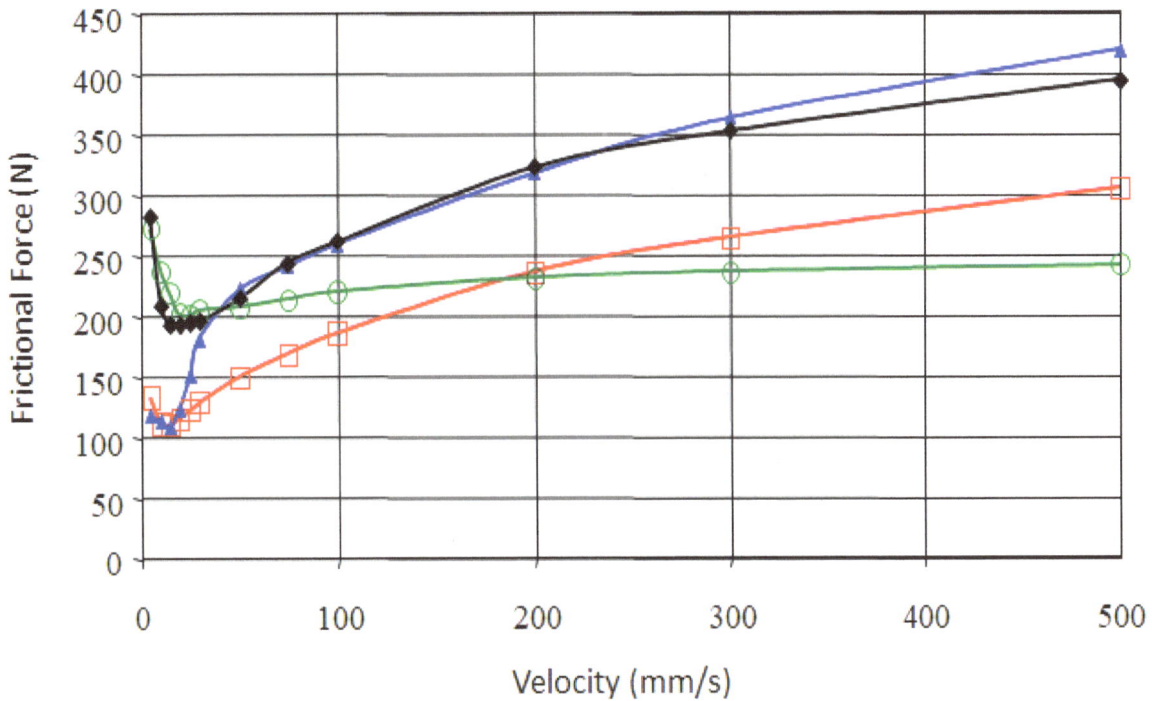

Fig. 4.62 - Rod Seal Friction at Constant Pressure

4.12.14.3- Friction in Rotational Seals

Examples of rotational seals are in hydraulic pumps and motors. Figure 4.63 and Equation 4.3 show how to calculate the surface (*Tangential*) speed on the sealed surface of a rotational shaft.

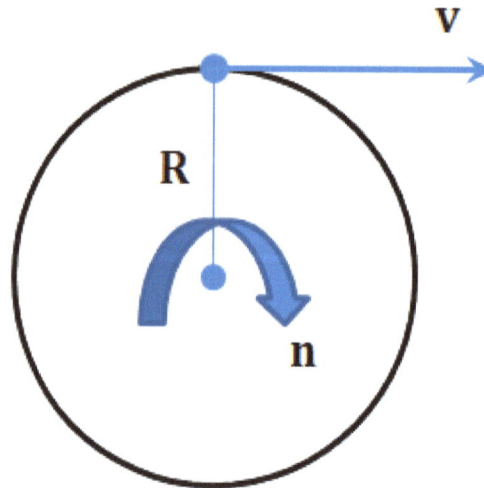

Fig. 4.63- Surface (Tangential) Speed of a Rotational Shaft

$$v = \omega \times R = 2\pi n \times R$$ 4.3

Where:

v = Tangential speed in (m/min) or (FPM) depends on the units of **R**.
R = Radius of the shaft in (m) or (foot).
n = Shaft rotational speed (RPM).

Figure 4.64 shows a Nomograph that was developed to provide a quick and approximate solution for the previous equation. For example:
- If shaft rotational speed = 4000 rpm, and shaft Diameter = 1 (in) ≈ 25 (mm), then:
- Surface speed (FPM) = 2 x 3.14 x 4000 x 0.5 / 12 = 1047
- Surface speed (m/min) = 2 x 3.14 x 4000 x 0.025 / 2 = 314

As shown in the previous example, surface speed is 314 m/min = 523 cm/s. this speed is almost 100 times greater than the recommended linear speed of a cylinder rod. Therefore, seal friction in rotational shafts is much higher than in translational shafts.

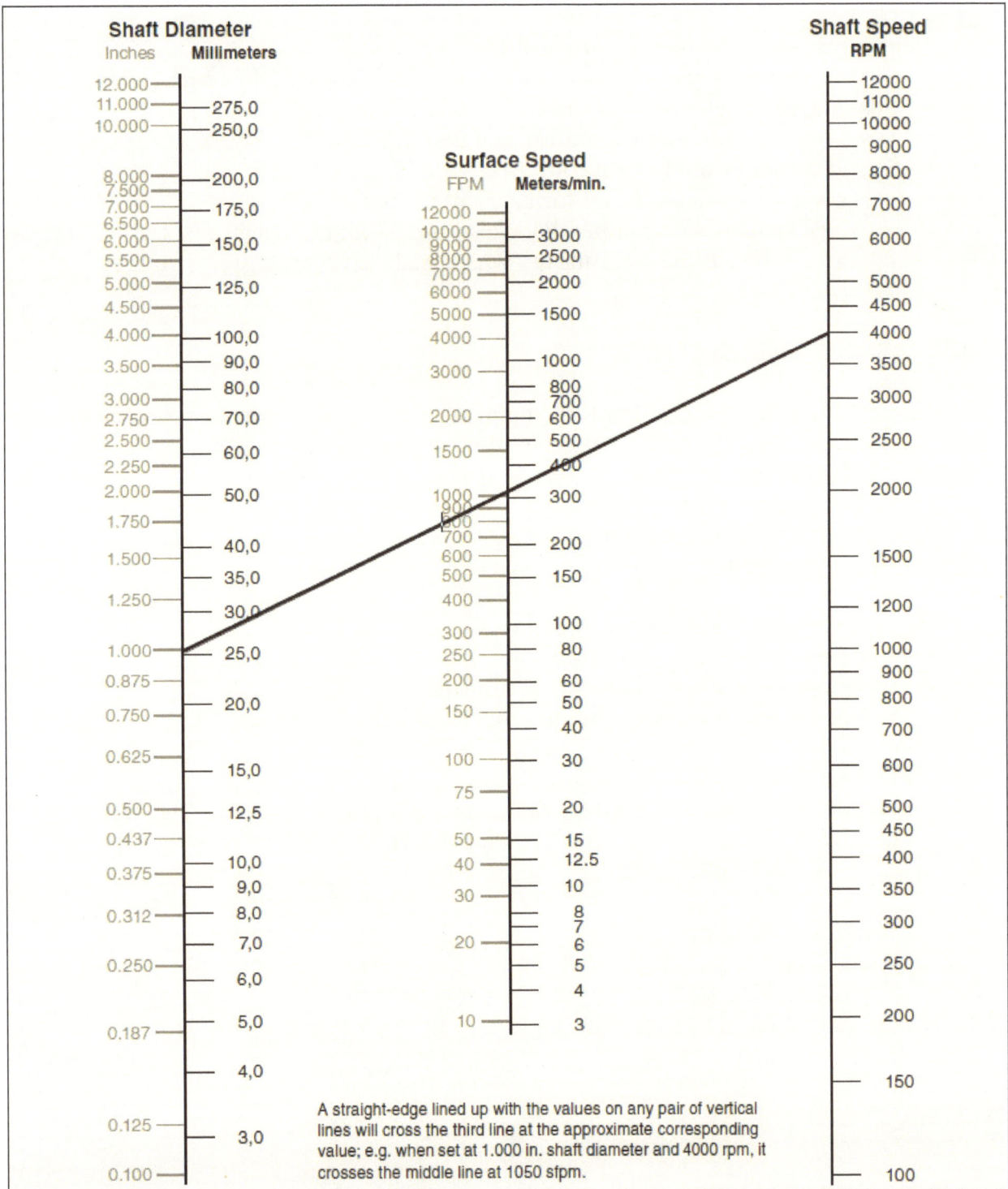

Shaft Diameter
Inches   Millimeters

Surface Speed
FPM   Meters/min.

Shaft Speed
RPM

A straight-edge lined up with the values on any pair of vertical lines will cross the third line at the approximate corresponding value; e.g. when set at 1.000 in. shaft diameter and 4000 rpm, it crosses the middle line at 1050 sfpm.

Fig. 4.64 - Nomograph for Calculating Rotational Shafts Surface Speed (Courtesy of Parker)

4.12.14.4- Factors Affecting Seal Friction

Friction of dynamic seals depends primarily on:
- **Seal Geometry:** shape, dimensions, exposed area, and dynamic surface roughness.
- **Seal Material:** type and properties.
- **Working Conditions:** temperature and pressure.
- **Speed (translational or rotational):**
  - **Low Speed:** affects the performance of the seal.
  - **High Speed:** causes a breakdown in the oil film between the seal and the sealing surface. Hence, the seal runs dry which leads to premature seal failure.

4.12.14.5- Controlling Seal Friction

❖ Basically, seal friction can't be 100% eliminated. However, lowering seal friction results in:
- Reduced energy loss and heat generation.
- Reduced wear.
- Reduced rate of chemical attack.
- Increased extrusion resistance.
- Increased seal life.

❖ Seal friction can be reduced by a combination of the following:
- Seal Design:
  - Waved seal surfaces to retain lubricating oil.
  - Seal design and placement should consider better heat dissipation.

- Seal Material:
  - Seals that work at low speed are coated by a dry lubricant such as Teflon layer.
  - Seals with high hardness have less fiction. However, seal hardness should be compromised with seal leakage.

- Working Conditions:
  - Select proper seals based on field working conditions.
  - Control working conditions within recommended limits.
  - Use compatible fluids with anti-friction additive packages.

4.12.15- Wiper Performance Test

Effect and Importance:
Dirt ingression is a major cause of hydraulic system inefficiency, degradation and failure. Although superior filtration systems exist and are used to limit exposure to contaminants, there are several locations in a typical hydraulic system that remain vulnerable. Breather ports, external couplers and power shaft systems that drive hydraulic pumps, external motors and cylinders are included among areas that may be compromised.

Testing:
Test Purpose: A new, innovative non-standard test method developed by **Hallite Seals Americas**, Inc. in cooperation with **Milwaukee School of Engineering** (MSOE) to assess the amount of dirt entering a simulated hydraulic system through the rod wiper located on the hydraulic cylinder.

Test Conditions:
MSOE's Fluid Power Institute™ built a test rig with the following parameters:
- Test Duration: 24.000 cycles, 24384 meters (80,000 feet) linear travel.
- Cycle Rate: 0.25 Hz
- Total Stroke Length: 101.6 (40 inches).
- Test Pressure: Atmospheric
- Test Temperature: 66 °C (150 °F).
- Test Oil: MIL-PRF-46170
- Test Contaminant: ISO 12103-1-A4 Course Test Dust

Test Procedure (as shown in Fig. 4.65):
1. The test rod wiper is installed in the rod end of the cylinder housing along with a TPE-faced, two-piece rod seal (Hallite Type 16 profile) to simulate typical boundary lubrication that is found on the rod in standard cylinder application.
2. The hydraulic fluid is heated to test temperature by heaters located in the reservoir.
3. The hydraulic fluid is pumped beneath the rod wiper.
4. Air supplied to the dust chamber. Dust moves at a high velocity inside the chamber, as air is supplied.
5. The rod is cycled with the specified frequency for a specified duration.
6. Oil drains back to reservoir.
7. Dirt content measured in the oil reservoir via the particle counter.

Test Evaluation:
MSOE tested the Hallite 520 and 820 wipers against two competitors. Based on these results, the Hallite 820 provided the most protection allowing the least amount of contamination into the test fixture. Note, the Hallite 820 utilizes a secondary protective structure outboard of the primary wiper lip (referred to Hallite's Umbrella Wiper Technology) which accounts for the better ingression protection. The Hallite 520 also provided a high level of protection against contamination as compared to similar competitor wipers.

0.25 Hz

Dust Chamber
(ISO 12103-1-A4
Course Test Dust)

4

Air Supply

1

Test Wiper

Rod Seal to
insure
proper
boundary
lubrication

3

5

6

Oil Reservoir

MIL-PRF-46170

Electronic
Particle
Counter

7

2

Electric Heaters
66 °C (150 °F)

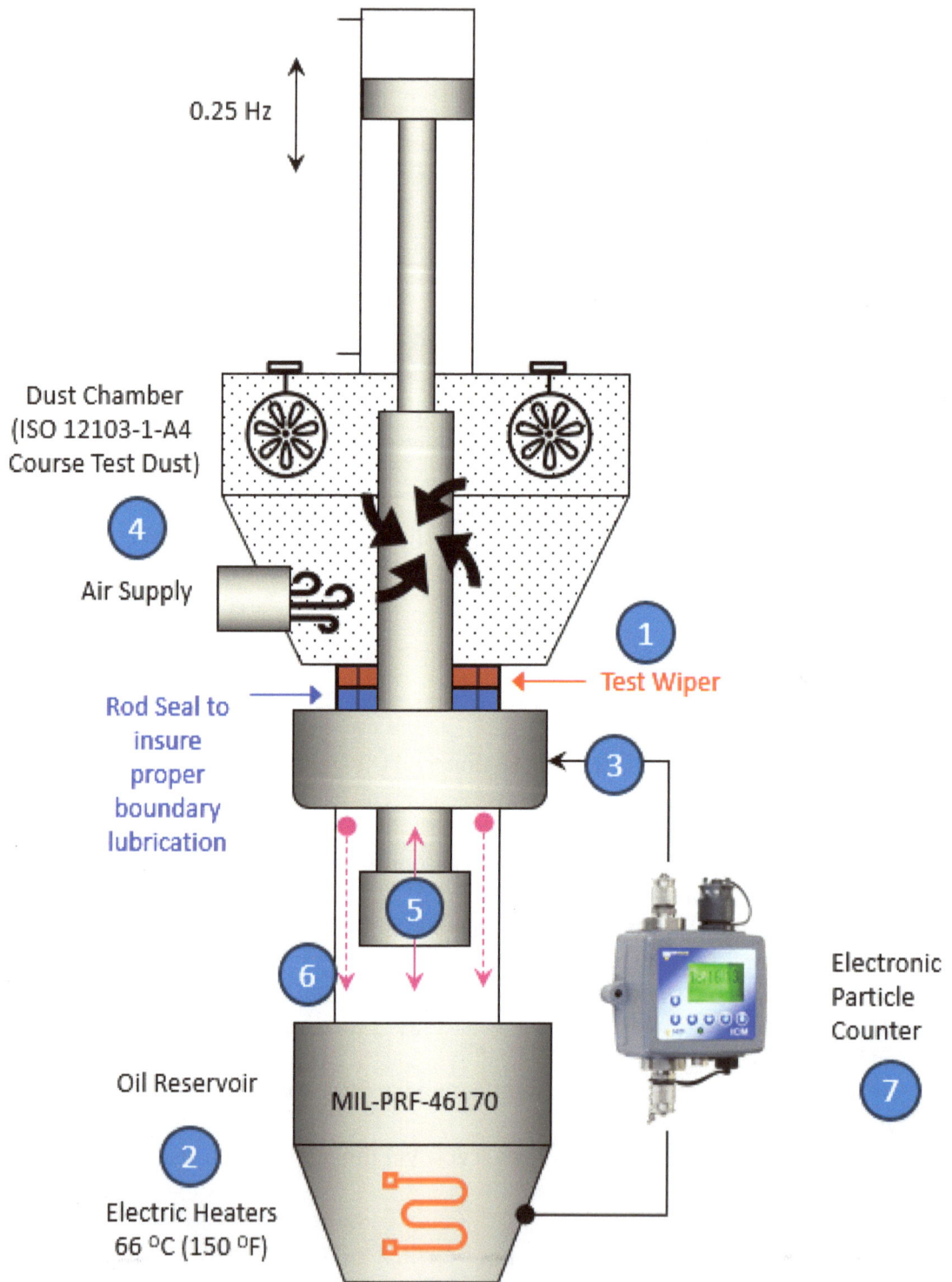

Fig. 4.65 - Cylinder Rod Wiper Performance (Courtesy of Hallite Seals)

4.13- Best Practices for Hydraulic Seals Selection

Hydraulic seals selection is a major step in designing a hydraulic system. Ideal hydraulic seal is the one that:

- Performs efficiently with low friction.
- Works under extended operating conditions (pressure and temperature).
- Has high tensile strength and resists twisting and spiral failures.
- Has high tear and abrasion resistance.
- Compatible with various types of hydraulic fluids.
- Resists chemicals and acids.
- Requires minimum space and is easy to install.
- Cost effective.

Unfortunately, there is no one seal that satisfies all the sealing requirements. Therefore, seal properties should be compromised based on the application. The following best practices list provides guidelines for selecting hydraulic seals.

1. Selection of Seal Type Based on Application.
2. Selection of Seal Dimensions.
3. Selection of Seal Lip Geometry.
4. Selection of Seal Crossectional Shape.
5. Selection of Seal Material Based on Working Temperature.
6. Selection of Seal Material Based on Working Pressure.
7. Selection of Seal Material Based on Working Fluid.
8. Selection of Seal Material Based on Hardness.
9. Selection of Seal Material Based on General Properties.

The following subtitles provide interpretation for the items contained in the best practice list.

4.13.1- Selection of Seal Type

Figure 4.66 shows a guide for selecting the general type of hydraulic seal. As shown in the figure, the general type of the seal depends on the application and the sealed components.

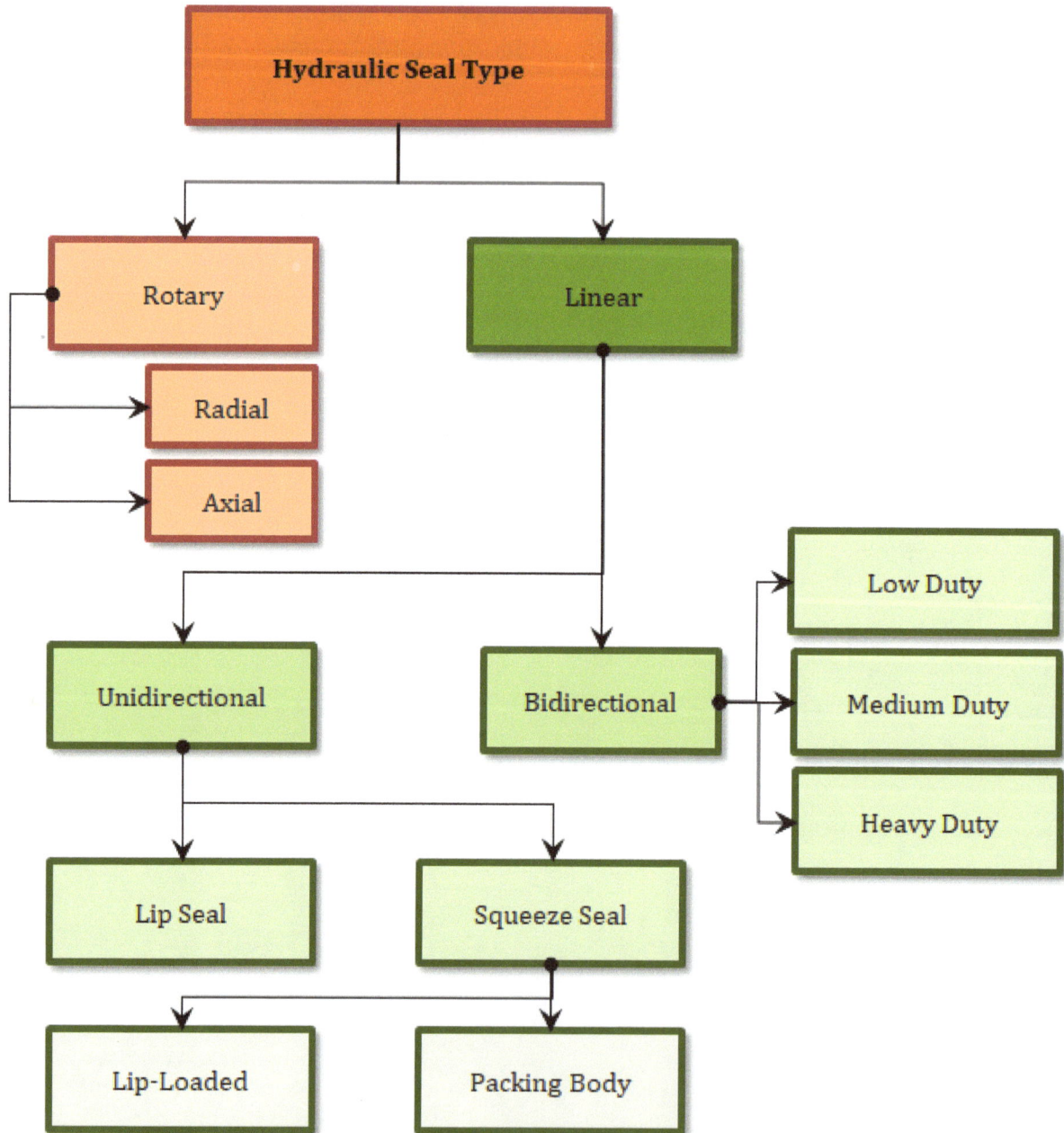

Fig. 4.66 - Hydraulic Seal Type Selection Based on Application

4.13.2- Selection of Seal Dimensions

O-Ring Cross Section versus its Stability: In designing an O-ring seal, there are usually several standard cross section diameters available. For any given piston or rod diameter, O-rings with smaller cross section diameters are inherently less stable than larger cross sections, tending to twist in the groove when reciprocating motion occurs. This leads to early O-ring spiral failure and leakage.

4.13.3- Selection of Seal Lip Geometry

Lip geometry will determine several functions of the seal. Force concentration on the shaft, sealing, lubricating film, hydroplaning, and contamination exclusion are all factors dependent on lip shape. Table 4.5 shows four different lip shapes and provides helpful insights for choosing an appropriate lip geometry as follows:

- **Example 1:** Rounded Cut is recommended for film stability, and self-lubrication.
- **Example 2:** *Straight Cut* is recommended for best contamination exclusion, Rod Wipers.
- **Example 3:** *Beveled Cut* is recommended for best film breaking for rod sealing.
- **Example 4:** *Square Cut* is recommended for piston guiding and sealing.

Contact Shape	Rounded	Straight Cut	Beveled	Square
Seal Lip Shape / Shape of Contact Force/ Stress Profile				
Film Breaking Ability	Low	High	Very High	Medium
Contamin-ation Exclusion	Low	Very High	Low	High
Tendency to Hydroplane	High	Very Low	Low	Medium
Typical Uses	Pneumatic U-cups	Wipers and Piston Seals	Rod Seals	Piston Seals

Table 4.5 - Effect of Lip Geometry on Seal Function (Courtesy of Parker)

4.13.4- Selection of Seal Crossectional Shape

The crossectional shape of a seal dramatically affects the sealability, especially at low pressure. Figure 4.67 shows that, for dynamic seals, sealability is traded off with low friction performance at low pressure. In other words, the seal cross section that offers better sealability at low pressure experiences increased friction.

With this in mind, seals are often categorized as either "*Lip Seals*" or "*Squeeze Seals*" and many other seals fall somewhere in between. Lip seals are characterized by low friction and low wear; however, they also have poor low pressure sealability. Squeeze seals are characterized by just the opposite: high friction and high wear, but better low pressure sealability.

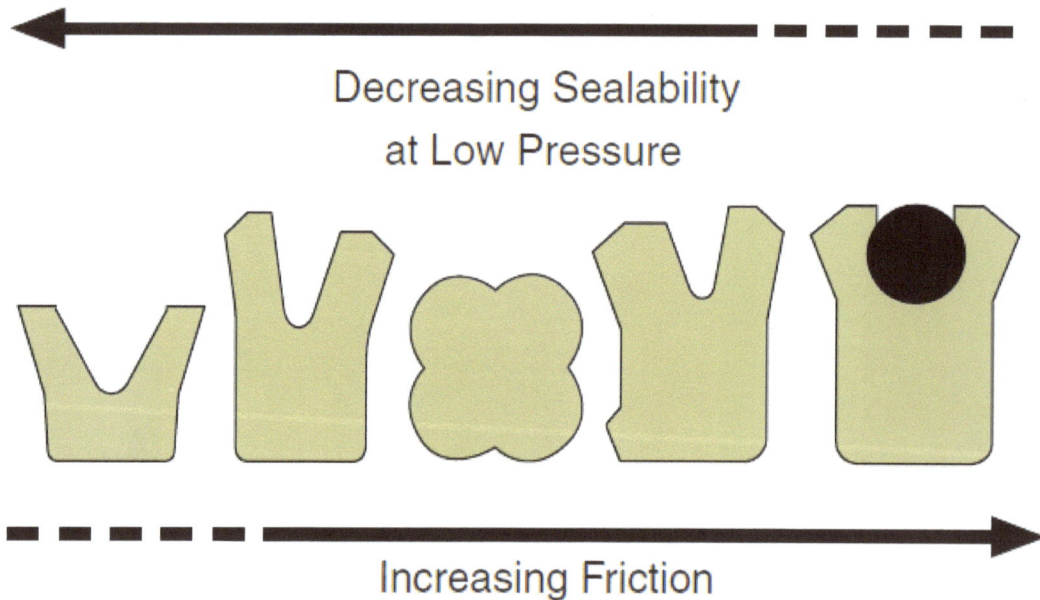

Fig. 4.67 - Lip vs. Squeeze Sealing (Courtesy of Parker)

4.13.5- Selection of Seal Material Based on Working Pressure

Range of working pressure for a hydraulic seal must be checked during selecting a seal. Working pressure affects the lifetime of the seal. For example, the seal that is designed to work at 150 bar (2175 psi) can last for 2 years. If it works at 200 bar (2900 psi) it will last for 2 months. If it works at 350 bar (5076) it will last for 2 days.

4.13.6-Selection of Seal Material Based on Working Temperature

Performance of a hydraulic seal is highly affected by working temperature. Synthetic rubber can be formulated for continuous use at high or low temperatures, or for occasional short exposure to wide variations in temperature.

Figure 4.68 shows the recommended working temperature range for common elastomeric materials. As shown in the figure, the most used seal material (NBR) can work at temperature range from -50 to 100 °C (-58 to 212 °F). Silicon rubber (Viton Seals) can work at the most extended temperature ranging from -50 to 200 °C (-58 to 392 °F).

Temperature Effects on Sealing Elements

High Temperatures:
At a high working temperature, a combination of the following effects could happen:
- Some of the oil at the interface surface evaporates resulting in dry-running condition.
- Lack of lubrication will cause greatly accelerated seal wear and leakage rate.
- High temperature softens the seal body resulting in seal extrusion.
- Prolonged exposure to excessive heat causes permanent surface hardening and leakage.
- Thermal expansion of synthetic rubber makes it difficult to design the grooves for the seals.

Low Temperatures:
At low temperature environment, a combination of the following effects could happen:
- When cooled, elastomer compounds lose their elasticity and may be damaged due to insufficient elastic memory to overcome the seal shrinkage.
- At very low temperatures, seals are hardened, have glasslike brittleness, and a dynamic seal may shatter if mechanically struck or hit with pressure spike.

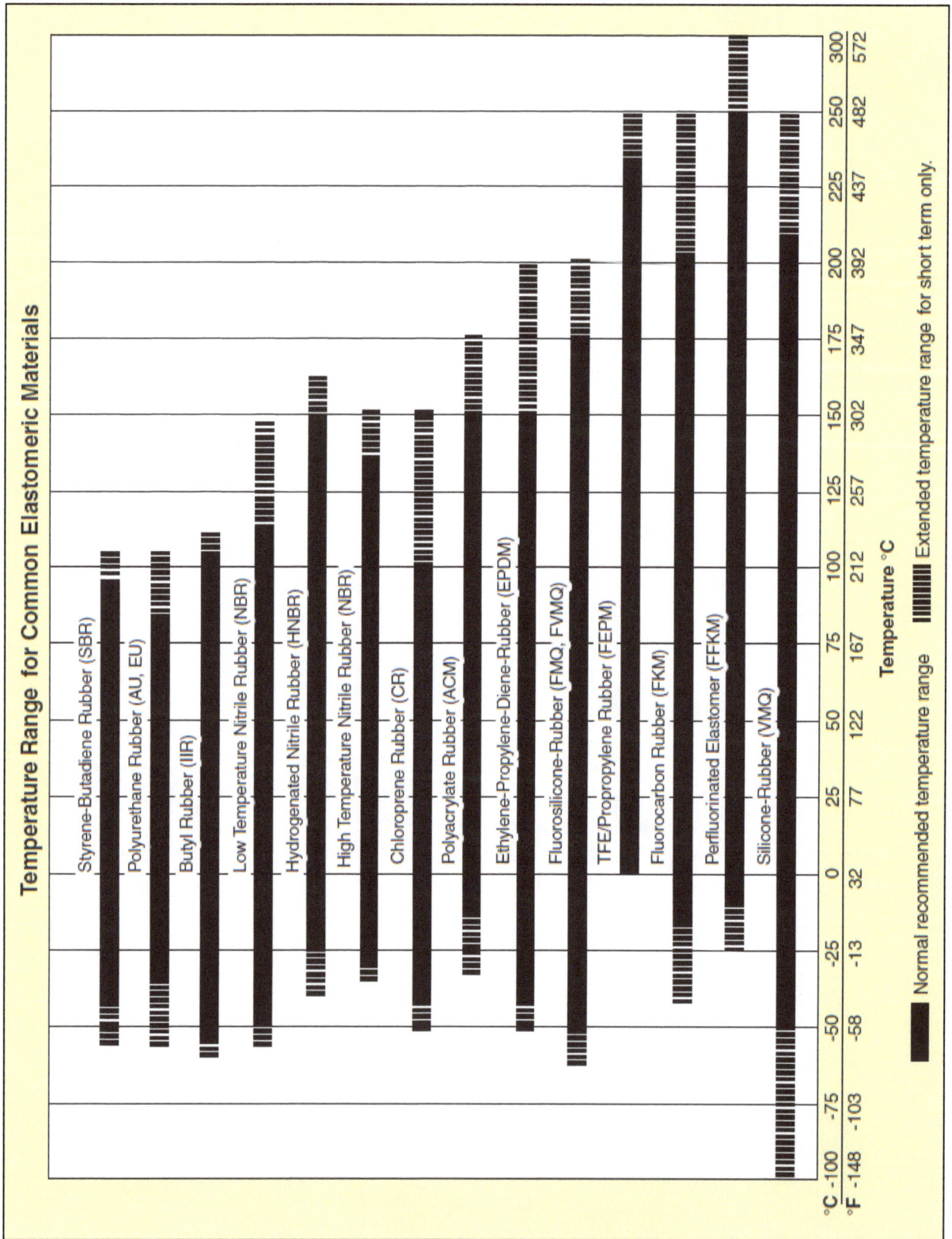

**Fig 4.68 - Recommended Working Temperature Range for Common Elastomeric Materials
(Courtesy of Parker)**

4.13.7- Selection of Seal Material Based on Working Fluid

Chemical compatibility is an important consideration when choosing the hydraulic fluid for the system. Hydraulic sealing elements must not react chemically with the working fluid.

Seal manufacturers should provide recommendations, as shown in Table 4.6, for the best seal material for fire-resistant fluids.

Water also can adversely affect the seals, so special seals must be applied where water-based fire-resistant fluids are used. Table 4.7 shows fluid compatibility for common sealing materials.

Properties of the Four Groups of Non-Flammable Fluids			
Properties	**HFA/HFB**	**HFC**	**HFD**
kinematic viscosity (mm²/s) to 50°C (122°F)	0.3 to 2	20 to 70	12 to 50
viscosity/temperature relationship	good	very good	bad
density at 15°C (59°F)	ca. 0.99	1.04 to 1.09	1.15 to 1.45
temperature range	3°C to 55°C (37°F to 131°F)	-25°C to 60°C (-13°F to 140°F)	-20°C to 150°C (-4°F to 302°F)
water content (weight %)	80 to 98	35 to 55	none
stability	emulsion poor solution very good	very good	very good
life of bearings	5 to 10%	6 to 15%	50 to 100%
heat transfer	excellent	good	poor
lubrication	acceptable	good	excellent
corrosion resistance	poor to acceptable	good	excellent
combustion temperature	not possible	after vaporizing of water under 1000°C (1832°F)	ca. 600°C (1112°F)
environmental risk	emulsion: used oil synth.: dilution	special waste	special waste
regular inspection	pH-level concentration water hardness micro-organisms	viscosity water content pH-level	viscosity neutral pH spec. gravity
seal material	**NBR, FKM**	**NBR**	**FKM, EPDM**[1]
(1) only for pure (mineral oil free) phosphate-ester (HFD-R)			

Table 4.6 - Recommended Seal Materials for Fire-Resistant Fluids (Courtesy of Parker)

Seal Material	Compatible Fluids	Temperature Range
1. Metallic piston rings	Petroleum base and synthetic fluids, phosphate esters - for high pressure and severe conditions	Low to 500°F (260°C)
2. Leather	Petroleum base and some synthetics, phosphate esters - for medium to high pressure	-65°F to 225°F -54°C to 107°C
3. Neoprene rubber	General purpose industrial use, Freon™ 12; weather and salt water resistant	-65°F to 300°F -54°C to 149°C
4. Nitrile rubber (Buna N™)	Petroleum base fluids and mineral oils - used for some rotating seals, extrusion resistant	-65°F to 225°F -54°C to 107°C
5. Silicone rubber	Water and petroleum base fluids, phosphate esters; low tensile strength and tear resistance recommended for static seals only	-80°F to 450°F -62°C to 232°C
6. Fluoro-Elastomers (Viton™ and Fluorel™)	Petroleum base, synthetic, diester, silicate ester, and halogenated hydrocarbon fluids - for high temperature fluid applications	-20°F to 400°F -29°C to 204°C
7. Polyurethane	Petroleum base fluids - high resistance to ozone, sunlight and weathering; low water resistance	-65°F to 200°F -40°C to 93°C

**Table 4.7 - Fluid Compatibility for Common Sealing Materials
(Hydraulic Specialist Study Manual, IFPS)**

4.13.8- Selection of Seal Material Based on Hardness

As it has been stated previously, while seal friction is inversely proportional to the seal hardness, leakage is directly proportional to seal hardness. Therefore, seal hardness should be compromised based on the seal application.

Example 1: Seal hardness should be high for Guide-Rings because their main objective is not to seal, but to guide a piston or a rod at low friction.

Example 2: Seal harness for piston seals should be lower for better sealability.

Figure 4.69 shows that elastomers have the same trend that the seal hardness is inversely proportional, at different rates, with the working temperature. Various Nitrile compounds were used in this analysis.

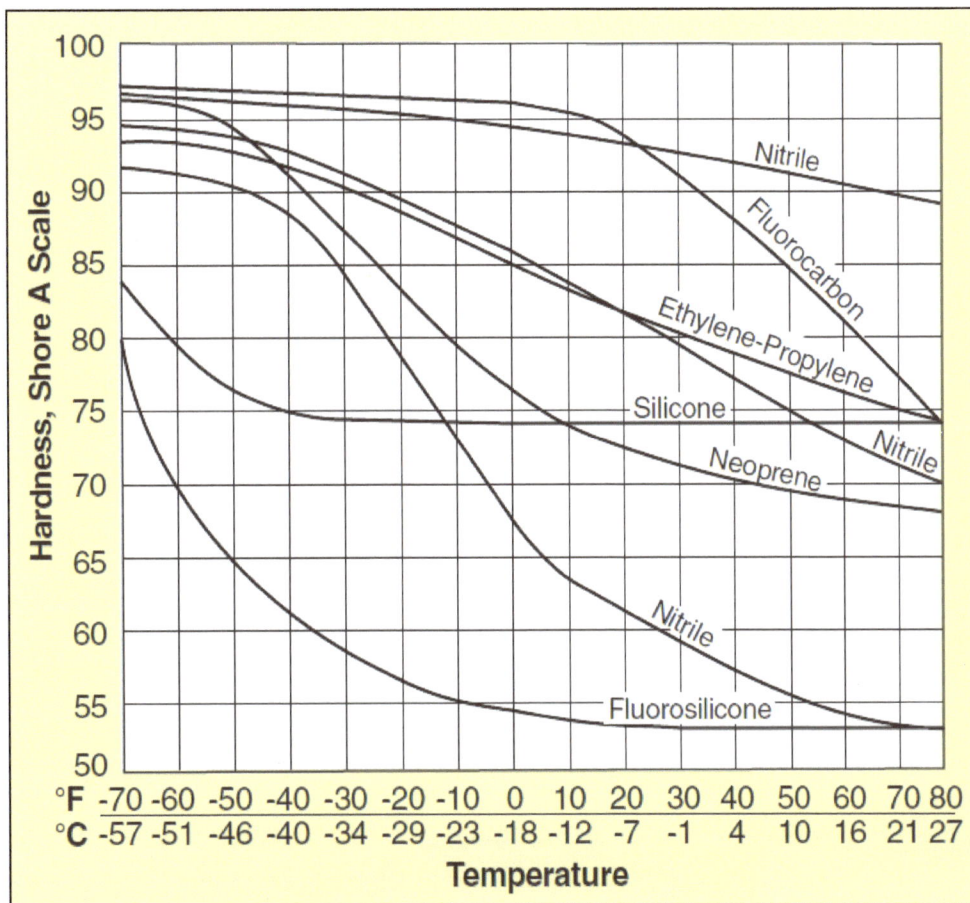

Fig 4.69 - Effect of Low Temperature on Elastomer Hardness (Courtesy of Parker)

4.13.9- Selection of Seal Material Based on General Properties

The general properties of a hydraulic seal are identified based on the application to achieve a reasonable service life and optimum performance. Table 4.8 shows the general properties of commonly known elastomers.

Comparison of Properties of Commonly Used Elastomers
(P = Poor – F = Fair – G = Good – E = Excellent)

Elastomer Type (Polymer)	Parker Compound Prefix Letter	Abrasion Resistance	Acid Resistance	Chemical Resistance	Cold Resistance	Dynamic Properties	Electrical Properties	Flame Resistance	Heat Resistance	Impermeability	Oil Resistance	Ozone Resistance	Set Resistance	Tear Resistance	Tensile Strength	Water/Steam Resistance	Weather Resistance
AFLAS (TFE/Prop)	V	GE	E	E	P	G	E	E	E	G	E	E	PF	PF	FG	GE	E
Butadiene		E	FG	FG	G	F	G	P	F	F	P	P	G	GE	E	FG	F
Butyl	B	FG	G	E	G	F	G	P	G	E	P	GE	FG	G	G	G	GE
Chlorinated Polyethylene		G	F	FG	PF	G	G	GE	G	G	FG	E	F	FG	G	F	E
Chlorosulfonated Polyethylene		G	G	E	FG	F	F	G	G	G	F	E	F	G	F	F	E
Epichlorohydrin	Y	G	FG	G	GE	G	F	FG	FG	GE	E	E	PF	G	G	F	E
Ethylene Acrylic	A	F	F	FG	G	F	F	P	E	E	F	E	G	F	G	PF	E
Ethylene Propylene	E	GE	G	E	GE	GE	G	P	G	G	P	E	GE	GE	GE	E	E
Fluorocarbon	V	G	E	E	PF	GE	F	E	E	G	E	E	E	F	GE	F	E
Fluorosilicone	L	P	FG	E	GE	P	E	G	E	P	G	E	G	P	F	F	E
Isoprene		E	FG	FG	G	F	G	P	F	F	P	P	G	GE	E	FG	F
Natural Rubber		E	FG	FG	G	E	G	P	F	F	P	P	G	GE	E	FG	F
Neoprene	C	G	FG	FG	FG	F	F	G	G	G	FG	GE	F	FG	G	F	E
HNBR	N, K	G	E	FG	G	GE	F	P	E	G	E	G	GE	FG	E	E	G
Nitrile or Buna N	N	G	F	FG	G	GE	F	P	G	G	E	P	GE	FG	GE	FG	F
Perfluorinated Fluoroelastomer	V, F	P	E	E	PF	F	E	E	E	G	E	E	G	PF	FG	GE	E
Polyacrylate	A	G	P	P	P	F	F	P	E	E	E	E	F	FG	F	P	E
Polysulfide		P	P	G	G	F	F	P	P	E	E	E	P	P	F	F	E
Polyurethane	P	E	P	FG	G	E	FG	P	F	G	G	E	F	GE	E	P	E
SBR or Buna S		G	F	FG	G	G	G	P	FG	F	P	P	G	FG	GE	FG	F
Silicone	S	P	FG	GE	E	P	E	F	E	P	FG	E	GE	P	P	F	E

Table 4.8 - General Properties of Commonly Known Elastomers (Courtesy of Parker)

4.14- Sealing Solutions for Hydraulic Cylinders

4.14.1- Considerations for Hydraulic Cylinders Reliable Sealing

Hydraulic cylinders are required to operate leak-freely in a variety of applications and environmental conditions, including exposure to abrasives, debris and both high and low temperatures.

Therefore, hydraulic seal manufacturers aggressively compete against each other to develop a comprehensive range of fluid power seals made from high quality materials with optimized designs to meet application requirements.

For a cylinder to perform reliably under leak-free conditions, the following set of features should be considered during putting together a sealing package for a cylinder:
- Seals must be able to function at specified pressure and temperature.
- Seals must be able to withstand expected pressure spikes.
- Seals must be able to carry expected lateral loads.
- Seals must be compatible with type of hydraulic fluid used.
- Seals must offer conditions of low friction.
- Seals must be designed for easy installation and port passing.

To help cylinder seals to perform reliably:
- Cylinder must be mounted coaxially with the load to minimize the lateral force.
- Cylinder mounting points should be attached to non-vibrating frames.
- Cylinder should operate in a clean environment.
- Cylinder should operate within the recommended working temperature and pressure.

There is no one sealing solution that is good for all cylinders. Design of a cylinder sealing package is a case-by-case and highly depends on the application and working conditions.

Table 4.9 shows some working conditions and typical applications for hydraulic cylinders that should be taken into considerations when designing a sealing solution.

CYLINDER SPECIFICATION		LIGHT-DUTY		MEDIUM-DUTY		HEAVY-DUTY	
PRESSURE	Max	350 bar	5000 psi	500 bar	7500 psi	700 bar	10000 psi
	Normal Working	160 bar	2300 psi	250 bar	3625 psi	400 bar	5800 psi
		No pressure peaks		Intermittent pressure peaks		Regular pressure peaks	
Design		Lower operating stresses. Rigid well- aligned mounting, minimal side loading.		Steady operating stresses with intermittent high stress, some side loading.		Highly stressed for the majority of its working life. Side loading common.	
Condition of Fluid		Good system filtration. No cylinder contamination likely.		Good system filtration, but some cylinder contamination likely.		Contamination unavoidable from internal and external sources.	
Working Environment		Clean and inside a building. Operating temperature variations limited.		Mixture of indoors and outdoors but some protection from the weather.		Outdoors all the time or dirty indoor area. Wide variations in temperature, both ambient and working. Difficult service conditions.	
Usage		Irregular with short section of stroke at working pressures. Regular usage but at low pressure.		Regular usage with most of the stroke at working pressure.		Large amount of usage at high pressure with peaks throughout the stroke.	
Typical Applications		Machine tools Lifting equipment Mechanical handling Injection moulding machines Control and robot equipment Agricultural machinery Packaging equipment Aircraft equipment Light duty tippers		Heavy duty lifting equipment Agricultural equipment Light duty off-road vehicles Cranes and lifting platforms Heavy duty machine tools Injection moulding machines Some auxiliary mining machinery Aircraft equipment Presses Heavy duty tippers (telescopic) Heavy duty mechanical handling		Foundry and metal fabrication plant Mining machinery Roof supports Heavy duty earthmoving machinery Heavy duty off-road vehicles Heavy duty presses	

**Table 4.9 – Working Conditions and Typical applications for Hydraulic Cylinders
(Courtesy of Hallite Seals)**

Figure 4.70 shows the main components of a typical sealing solution for a light-duty hydraulic cylinder. As shown in the figure, piston seal package contains a piston dynamic seal and a wear or a guide ring. Piston static seal is used if the piston head is assembled with the piston rod.

Like pistons, cylinder rod seal package contains a dynamic seal and a Guide Ring. Unlike pistons, cylinder rod has additional sealing elements such as Rod Wiper and a Buffer Seal. A static seal is used at the rod side to seal the cylinder head against the barrel to prevent external leakage.

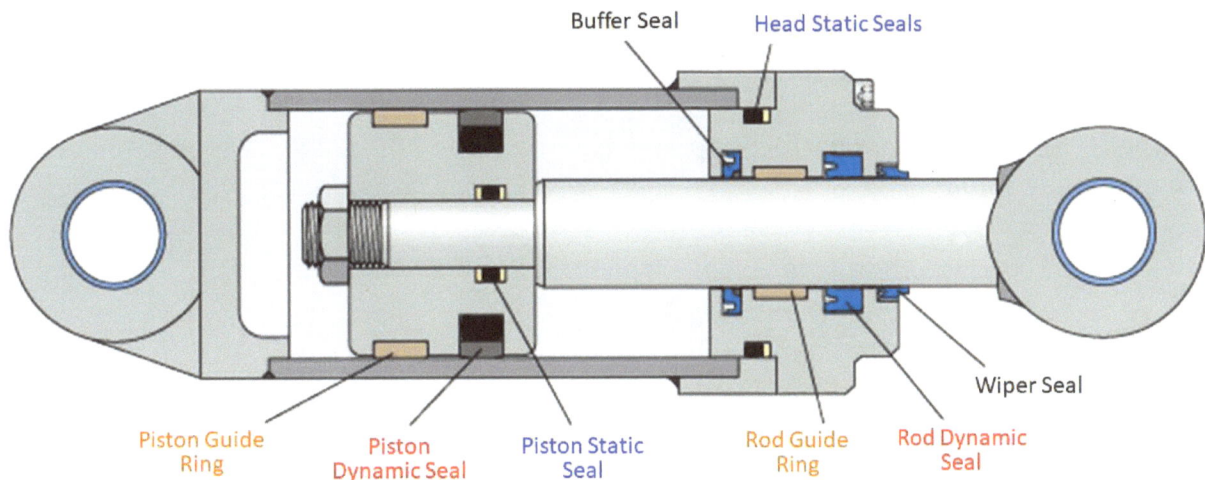

Fig. 4.70 - Basic Components of Cylinder Sealing Solutions

The following sections overview the common piston and rod sealing solutions. Presented examples cover cylinders for light-duty applications [pressure up to 350 bar (5000 psi)], medium-duty applications [pressure up to 500 bar (7500 psi)], high-duty applications (pressure up to 700 bar (10000 psi)]. Piston and rod seals are generally classified as *Unidirectional* (*Single-Acting*) and *Bidirectional* (*Double-Acting*) seals.

4.14.2- Sealing Solutions for Cylinder Rods

4.14.2.1- Unidirectional Rod Sealing Solutions

Rod seals must exhibit no dynamic leakage to the atmosphere side under all operating conditions and must be completely leak-free when the cylinder rod is at a standstill. The following rod sealing solutions are good to seal cylinder rods against external leakage to the atmospheric side.

Example 1: U-Cup Seals for Unidirectional Rod Sealing

Figure 4.71 shows various types of U-Cup seals for cylinder rod sealing as follows:

Type 1: The U-Cup has two sealing lips in the dynamic sealing zone. The two sealing lips provide improved sealing at low system pressures. Furthermore, the second sealing lip prevents entry of dirt from the atmospheric side.

Type 2: The O-Ring loaded U-Cup seal has stronger radial contact forces with increased system pressure. Hence, it has excellent sealing behavior with and without pressure activation. The short sealing lip reduces friction compared to common U-Cups.

Type 3: The spring energized U-Cup seal has the same advantage as type 2 plus it is more dynamically stable.

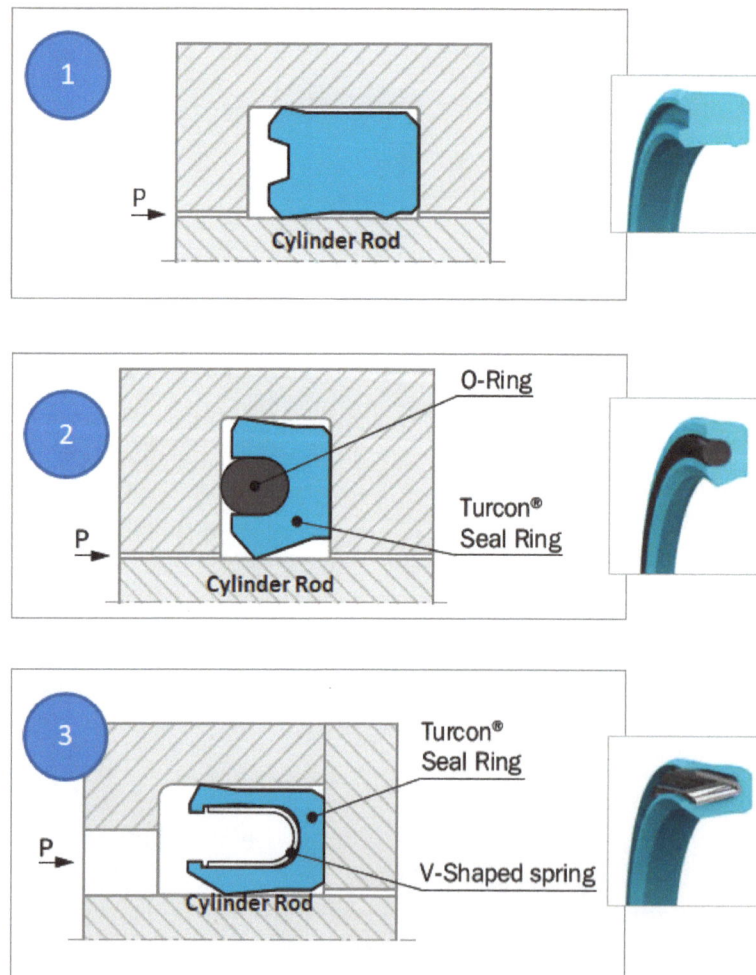

Fig. 4.71 - U-Cup Seals for Unidirectional Rod Sealing (Courtesy of Trelleborg)

Example 2: V-Packing Seals for Unidirectional Rod Sealing

This rod sealing package, as shown in Fig. 4.72, utilizes a set of V-Packings as the dynamic seal. This solution is recommended for applications with extremely high-pressure and possible pressure spikes.

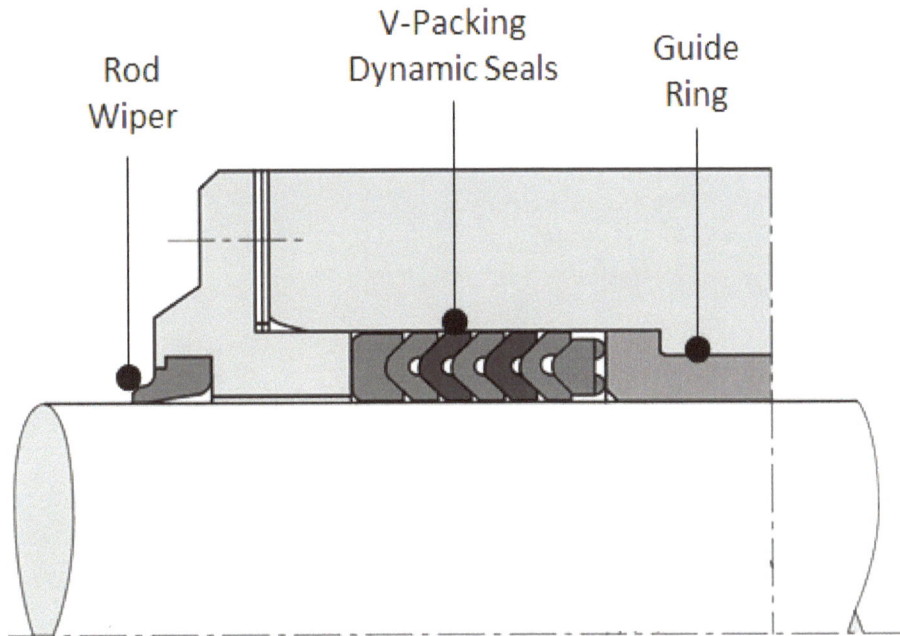

Fig. 4.72 - V-Packing Seals for Unidirectional Rod Sealing

Example 3: Low Friction Rod Seals for Better Positioning Accuracy

Figure 4.73 shows a single-acting rod seal where high demands are made on cylinder positional accuracy and stick-slip-free movement, e.g. closed-loop position-controlled servo cylinders.

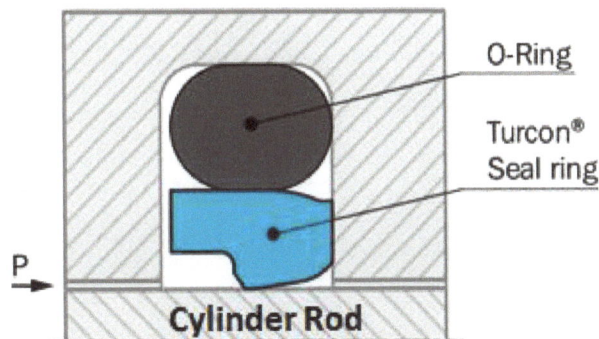

Fig. 4.73 - Unidirectional Rod Sealing Solution for Better Positioning Accuracy (Courtesy of Trelleborg)

Example 4: Cylinder Rod Unidirectional Sealing Solution for Light-Duty Applications
This rod sealing package, as shown in Fig. 4.74, contains a single U-Cup Seal to prevent external leakage. The U-Cup seal can be *O-Ring-loaded* or *Spring-Energized.* The sealing package also contains a Rod-Wiper to remove dirt from the rod surface during retraction.

The figure shows a Guide-Ring to carry lateral forces. As shown in the figure, the Guide-Ring position changes depending on the lateral force. The Guide-Ring should be closer to the wiper as the lateral force increases. As shown in Fig. 4.75, to improve the operational safety under high lateral loads, up to three Wear-Rings can be installed.

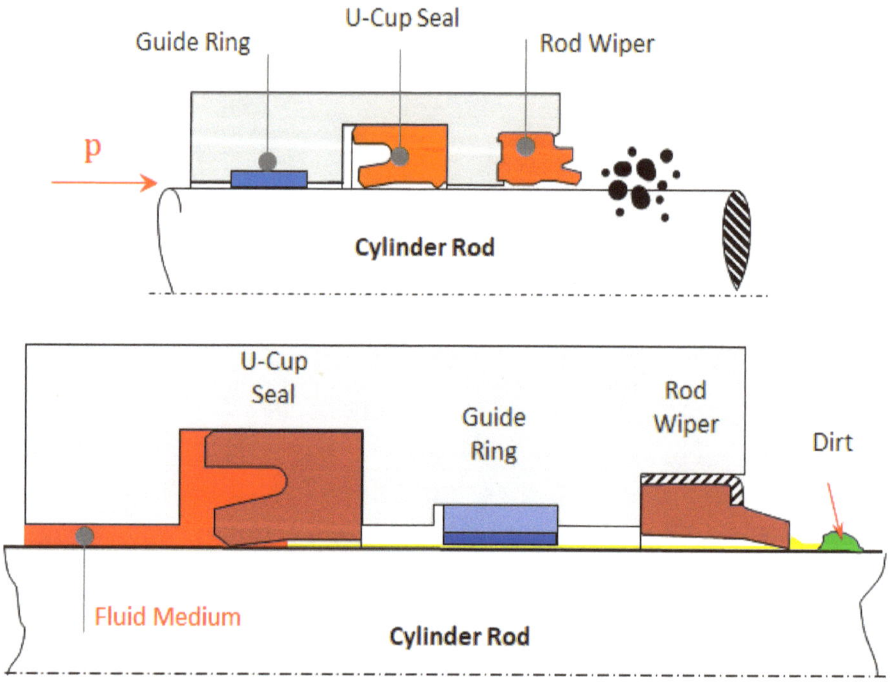

Fig. 4.74- Cylinder Rod Unidirectional Sealing Solution for Light-Duty Applications (Courtesy of Trelleborg)

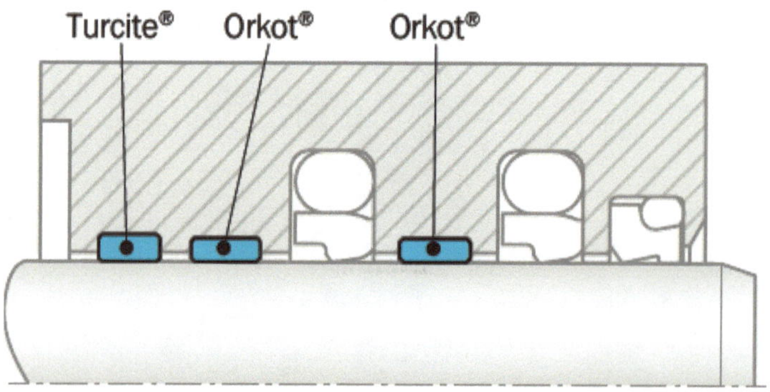

Fig. 4.75- Wear-Rings for Cylinder Designs

Example 5: Cylinder Rod Unidirectional Sealing Solution for Medium-Duty Applications
This rod sealing package, as shown in Fig. 4.76, contains two dynamic seals, a primary (high-pressure) seal and a secondary U-Cup (low-pressure) seal. When two dynamic seals are used, they are referred to as a *Redundant Sealing System*. Obviously, the package contains a Rod Wiper and a Guide-Ring. Figure 4.77 shows an example for that same rod sealing solution but from a different manufacturer.

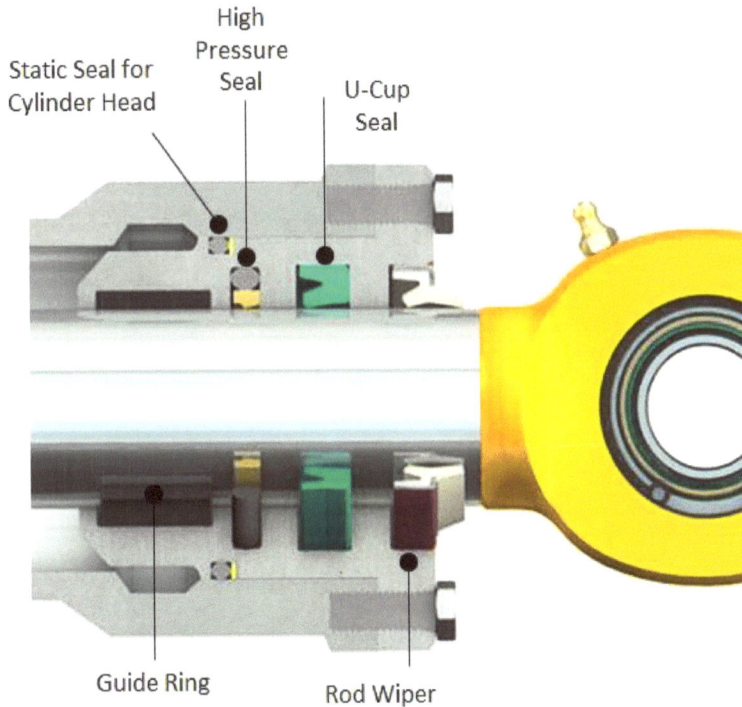

**Fig. 4.76 - Cylinder Rod Unidirectional Sealing Solution for Medium-Duty Applications
(www.skf.com)**

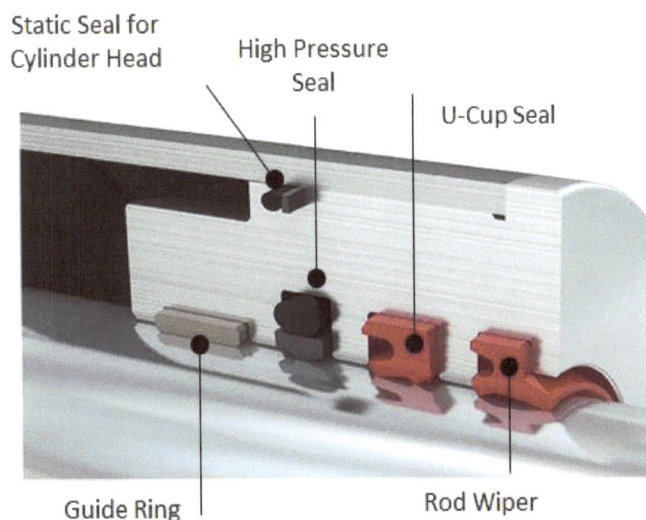

**Fig. 4.77 - Cylinder Rod Unidirectional Sealing Solution for Medium-Duty Applications
(Courtesy of American High-Performance Seals)**

Example 6: Redundant Unidirectional Rod Sealing Solution for Heavy-Duty Applications
This rod sealing package, as shown in Fig. 4.78, contains a *Redundant* sealing system. It is also referred to as *"Double Sealing"* or *"Tandem Sealing"* solution. As shown in the figure, two sealing stages, each is constructed from an O-Ring for static sealing and a custom-designed seal for dynamic sealing.

Redundant sealing systems are used where:
- Environmentally harmful fluids are used.
- A single seal can't withstand the application conditions over the demanded service life.
- A machine starts at low temperature (*Cold Starts*). During cylinder extension, due to the very high viscosity, the oil is heated because the friction at the primary seal and is then reliably wiped off - at a now lower viscosity - by the secondary seal.

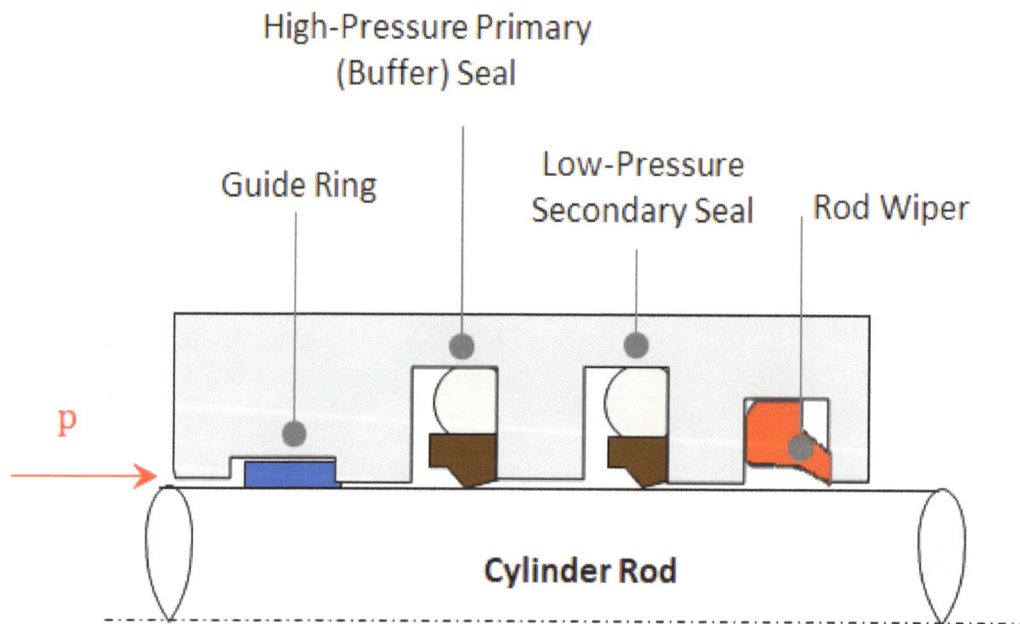

Fig. 4.78 - Redundant Unidirectional Rod Sealing Solution for Heavy-Duty Applications

Figure 4.79 shows a typical industry example for a redundant sealing solution. During cylinder retraction, the oil is stored in the reservoir between the primary and the secondary seals. As the rod continue to retract, the oil is pumped back hydrodynamically through the primary seal clearance against the system pressure. This design is known as the *Back-Pumping Effect*.

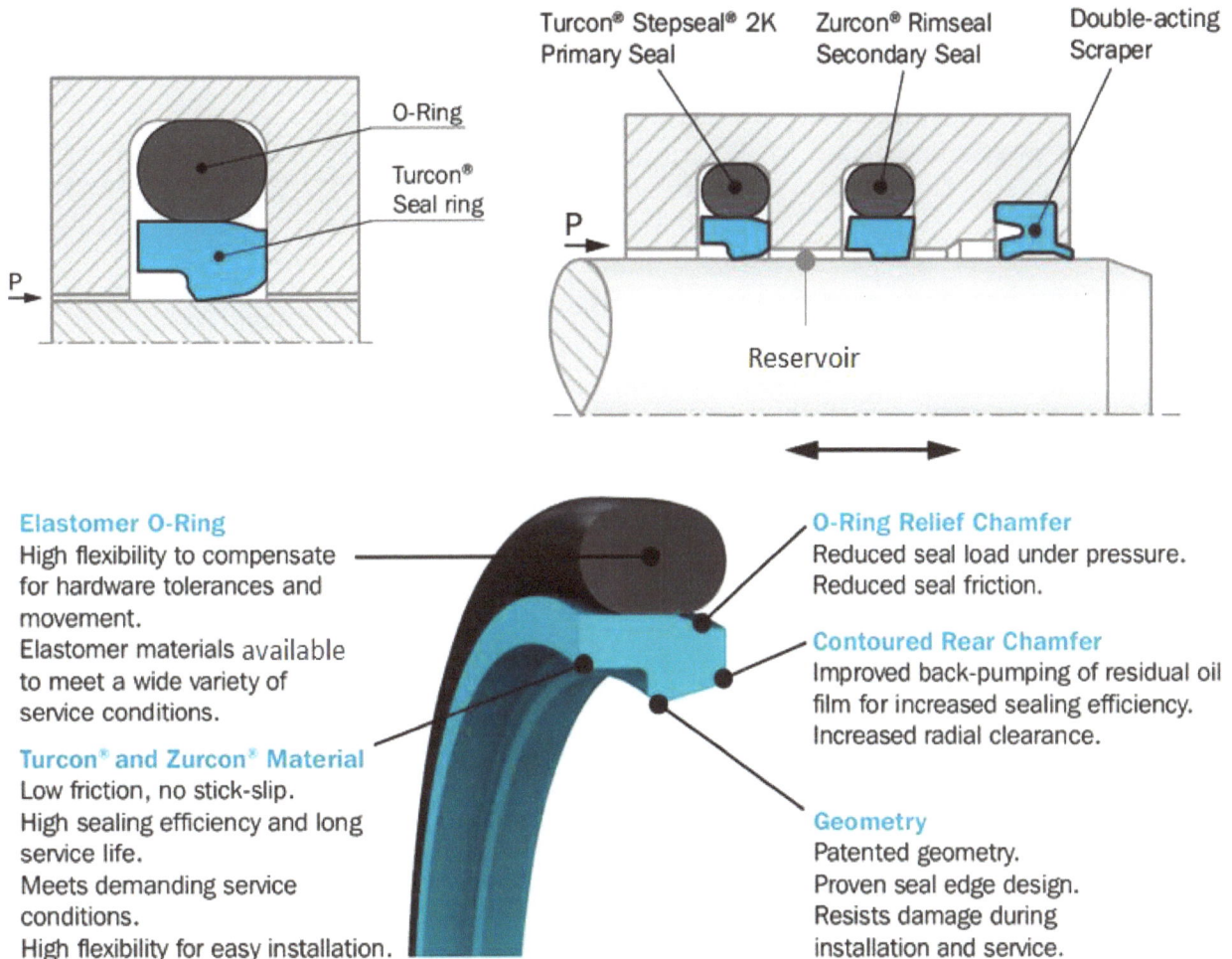

Elastomer O-Ring
High flexibility to compensate for hardware tolerances and movement.
Elastomer materials available to meet a wide variety of service conditions.

Turcon® and Zurcon® Material
Low friction, no stick-slip.
High sealing efficiency and long service life.
Meets demanding service conditions.
High flexibility for easy installation.

O-Ring Relief Chamfer
Reduced seal load under pressure.
Reduced seal friction.

Contoured Rear Chamfer
Improved back-pumping of residual oil film for increased sealing efficiency.
Increased radial clearance.

Geometry
Patented geometry.
Proven seal edge design.
Resists damage during installation and service.

Fig. 4.79 - Typical Redundant Unidirectional Rod Sealing Solution for Heavy-Duty Applications (Courtesy of Trelleborg)

Example 7: Redundant Unidirectional Sealing Solution for Slow and Long Stroke Cylinder Rods

In long-stroke cylinders and equipment operating with slow speed during retraction, it has been found that hydrodynamic back-pumping may become insufficient to prevent build-up of pressure in the seal system behind the primary seal. Pressure build-up in the seal system leads to leakage, increased friction and wear, and may ultimately require frequent replacement of the seals. The usual solution in such equipment has been to provide space for a buffer volume behind the primary seal or to install a drain line. As shown in Fig. 4.80, first invented by *Trelleborg Sealing Solutions*, the built-in check valve function eliminates pressure build-up and so prevent pressure built up in the reservoir volume. Hence, improve sealing performance with outstanding sliding and wear resistance properties.

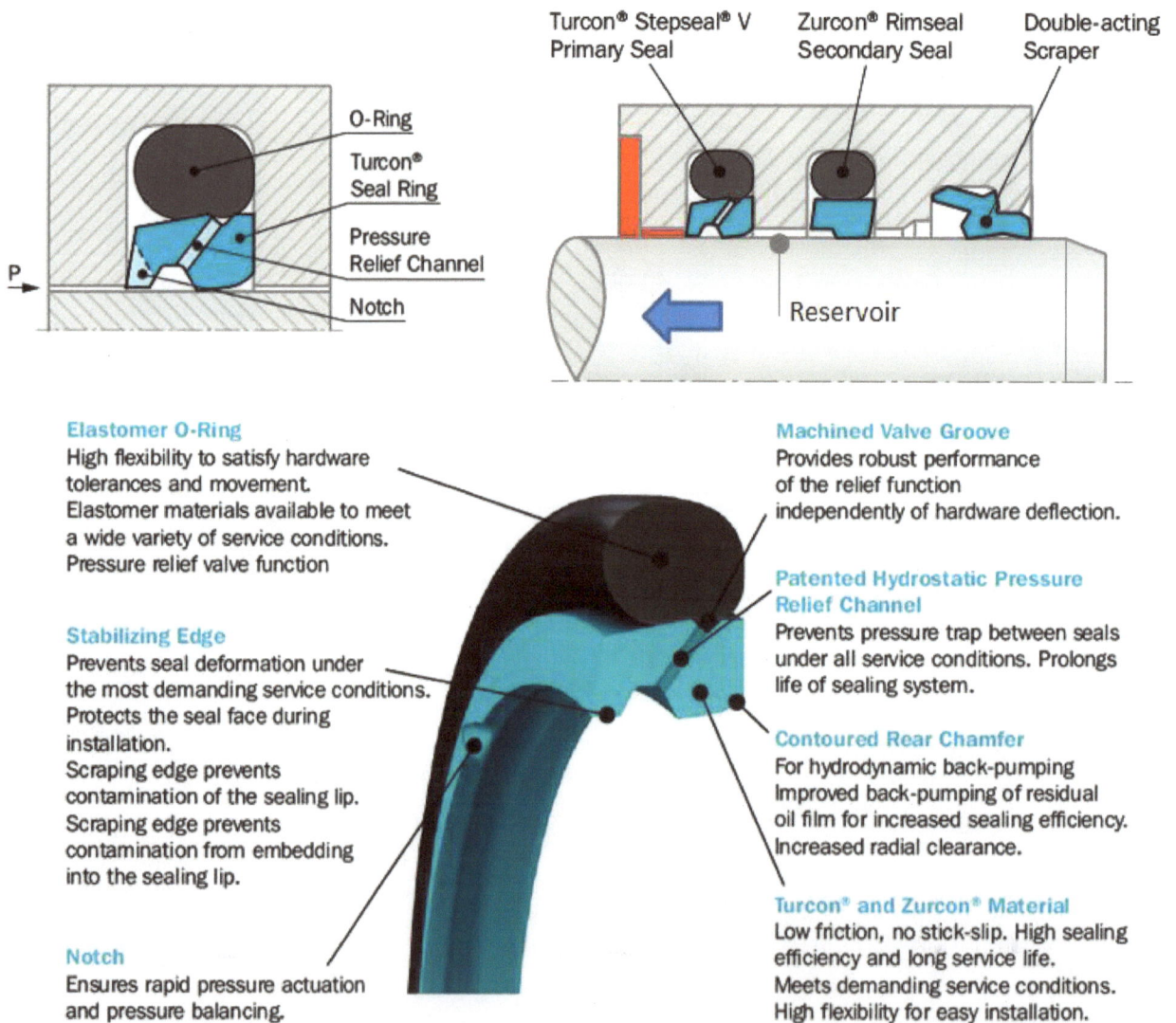

Elastomer O-Ring
High flexibility to satisfy hardware tolerances and movement. Elastomer materials available to meet a wide variety of service conditions. Pressure relief valve function

Stabilizing Edge
Prevents seal deformation under the most demanding service conditions. Protects the seal face during installation. Scraping edge prevents contamination of the sealing lip. Scraping edge prevents contamination from embedding into the sealing lip.

Notch
Ensures rapid pressure actuation and pressure balancing.

Machined Valve Groove
Provides robust performance of the relief function independently of hardware deflection.

Patented Hydrostatic Pressure Relief Channel
Prevents pressure trap between seals under all service conditions. Prolongs life of sealing system.

Contoured Rear Chamfer
For hydrodynamic back-pumping Improved back-pumping of residual oil film for increased sealing efficiency. Increased radial clearance.

Turcon® and Zurcon® Material
Low friction, no stick-slip. High sealing efficiency and long service life. Meets demanding service conditions. High flexibility for easy installation.

Fig. 4.80 - Redundant Unidirectional Sealing Solution for Slow and Long Stroke Cylinder Rods (Courtesy of Trelleborg)

Example 8: Redundant Unidirectional Rod Sealing Solution for Heavy-Duty and Large Lateral Force Applications

Figure 4.81 shows a couple of typical industry examples for a *Redundant* (*Tandem*) sealing system developed for to meet heavy duty demands. For large lateral forces, two Guide-Rings are used.

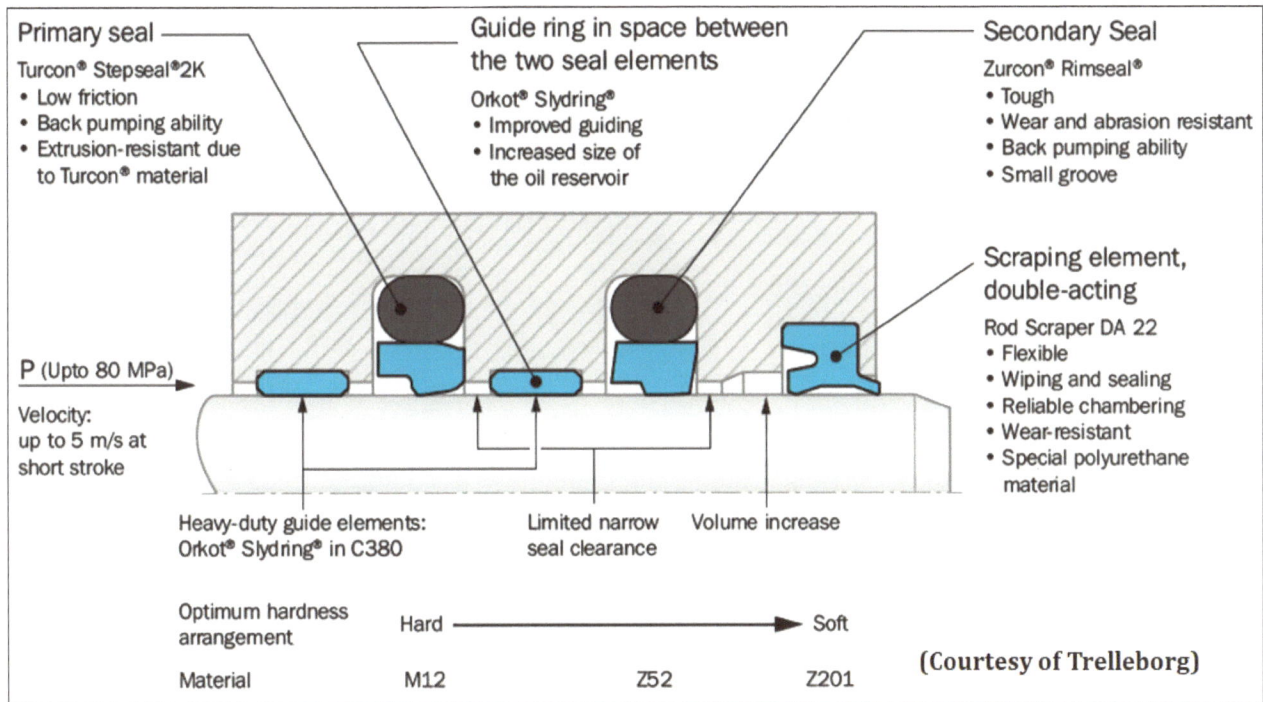

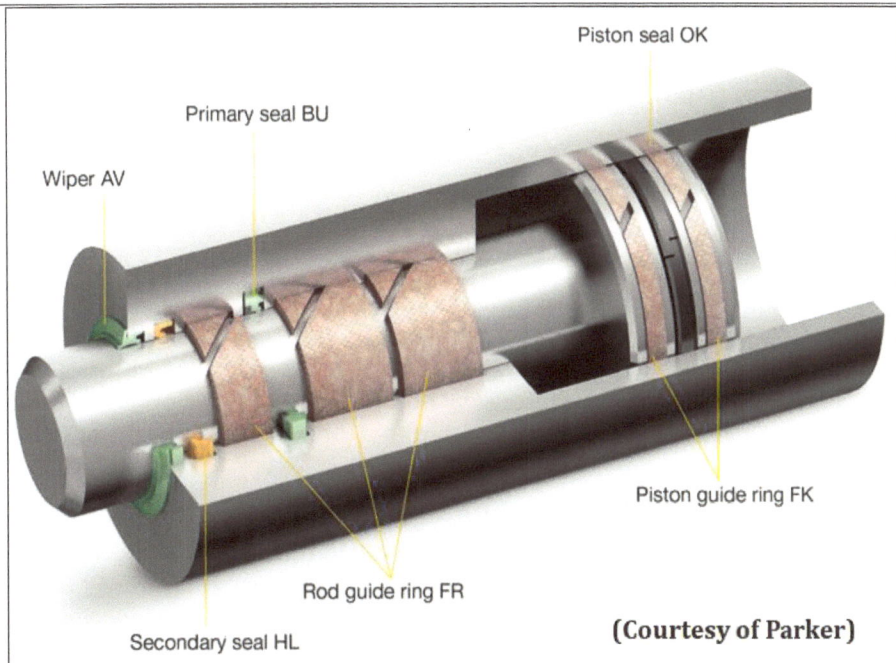

Fig. 4.81 - Redundant Unidirectional Rod Sealing Solution for Heavy-Duty
and Large Lateral Force Applications

4.14.2.2- Bidirectional Rod Sealing Solutions

Example 1: Cylinder Rod Basic Bidirectional Sealing Solution for Light-Duty Applications
In a tandem cylinder shown in Fig. 4.82, the double-acting seal is a combination of a custom sealing ring and an energizing O-Ring. Together, with the squeeze of the O-Ring, ensures a good sealing effect even at low pressure. At higher system pressures, the O-Ring is energized by the fluid, pushing the sealing ring face with increased force.

As shown in the figure, the trapezoidal profile cross section of the custom sealing ring allows the lubricating hydrodynamic fluid film to be built under the seal with stick-slip-free linear motion.

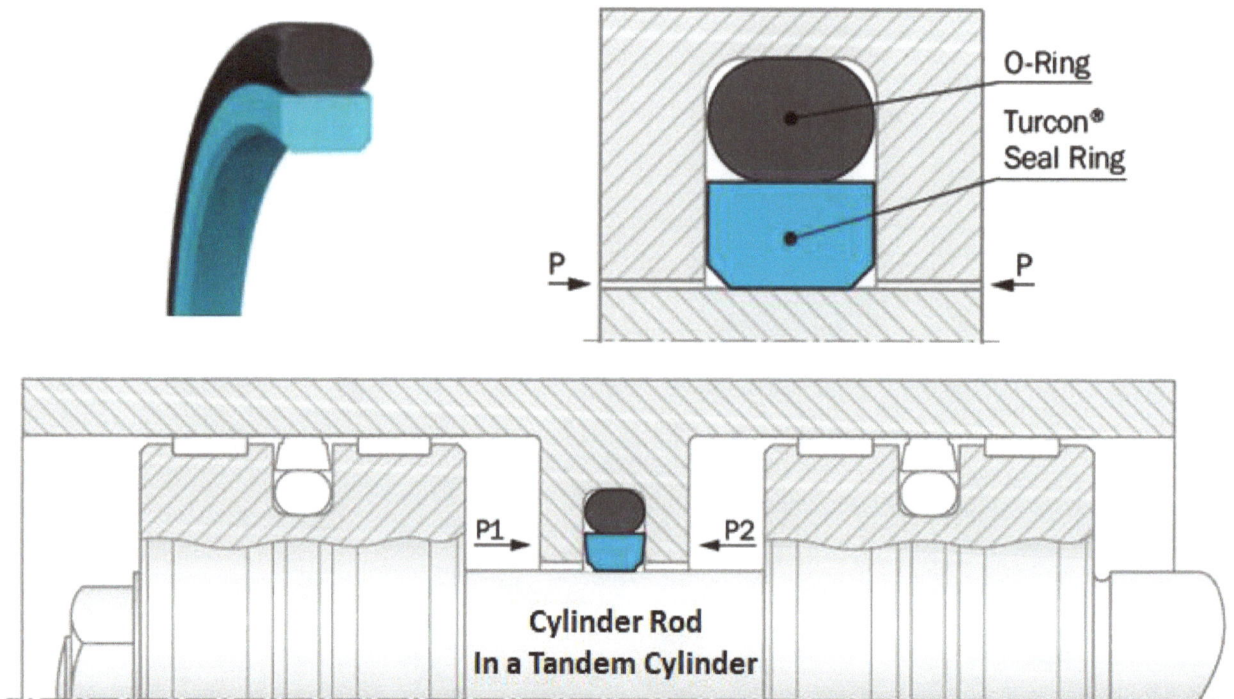

**Fig. 4.82 - Cylinder Rod Basic Bidirectional Sealing Solution for Light-Duty Applications
(Courtesy of Trelleborg)**

Example 2: Bidirectional Rod Sealing Enhanced Solution for Medium-Duty Applications
Figure 4.83 shows various enhancements for the basic bidirectional rod seal design.

Solution 1 for Blow-By-Effect: shows one radial notch is made on each side of the seal ring. The notches override the *Blow-By-Effect*. The notches are made to assure that a rapid energizing of the seal takes place at sudden changes in pressure and direction of motion.

Solution 2 for short-stroke and high-frequency: Reciprocating movement, with an increasing frequency above 5 Hz, the formation of lubrication under the contact face of the seal ring is reduced. High-frequency is most often occurring in connection with short-strokes. These two types of movements together accelerate the wear on hardware and seal. In this solution a symmetric seal is used with angled contact faces to ensure oil film is not scraped away from the surface. Oil is transported into the groove in the middle of the contact area forming an oil reservoir for lubrication. Wear particles are also likely to be captured in this groove, thus preventing them from embedding in the surface where the highest contact force occurs.

Solution 3 for better leakage control: The seal is designed with two seal edges and two lips. It acts as the primary seal for pressures from both sides, prevents build-up of hydrodynamic pressure over the seal profile, and prevents the risk of blow-by effect. The central sealing face increases the sealing effect.

Solution 4 for better leakage control and low friction: The sealing ring and the *Bean Seal* together create the dynamic sealing function while the O-Ring performs the static sealing function. The shown design incorporates a limited footprint Bean Seal in the dynamic sealing face. This optimizes leakage control while minimizing friction.

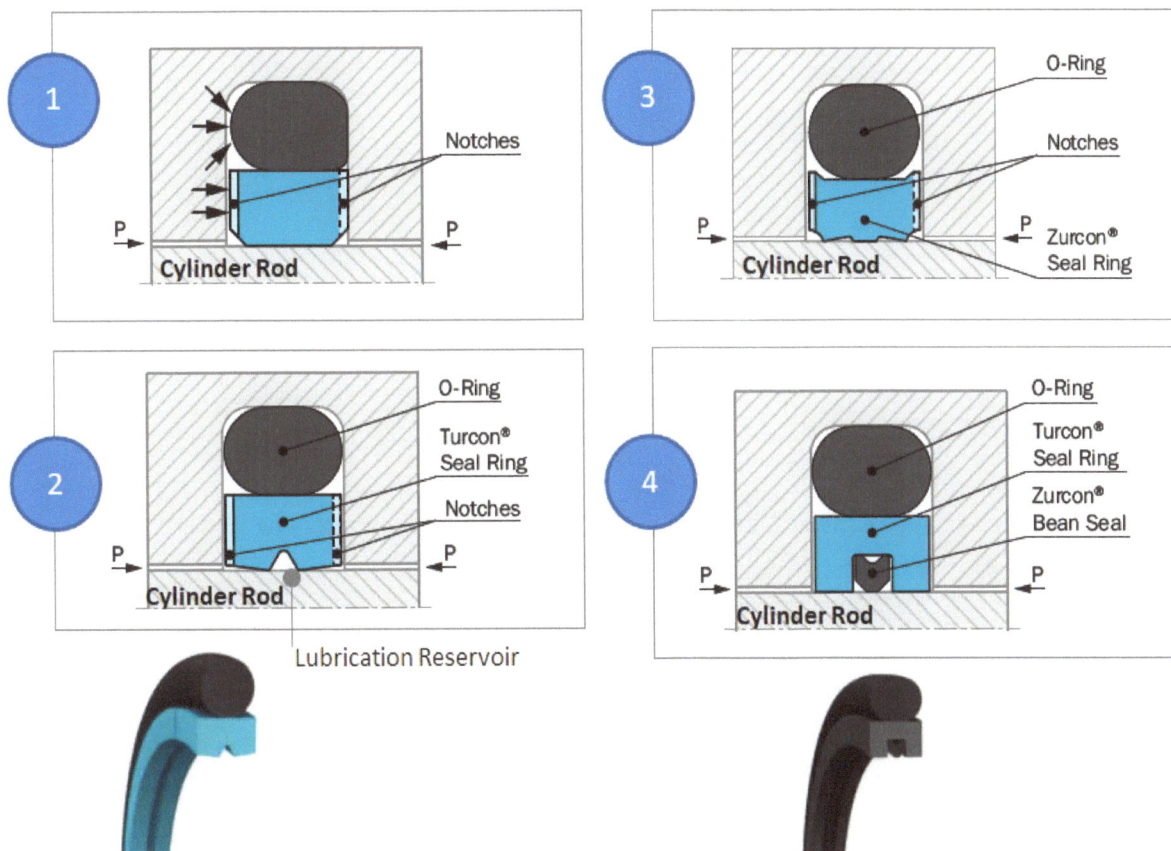

**Fig. 4.83 - Enhanced Bidirectional Rod Seal Designs for Medium-Duty Applications
(Courtesy of Trelleborg)**

Example 3: U-Cup Seals for Bidirectional Rod Sealing

Figure 4.84 shows two single-acting U-cup seals which are placed back-to-back to form bidirectional rod sealing solution. As shown in the figure, U-Cup seals could be O-Ring loaded or spring-loaded.

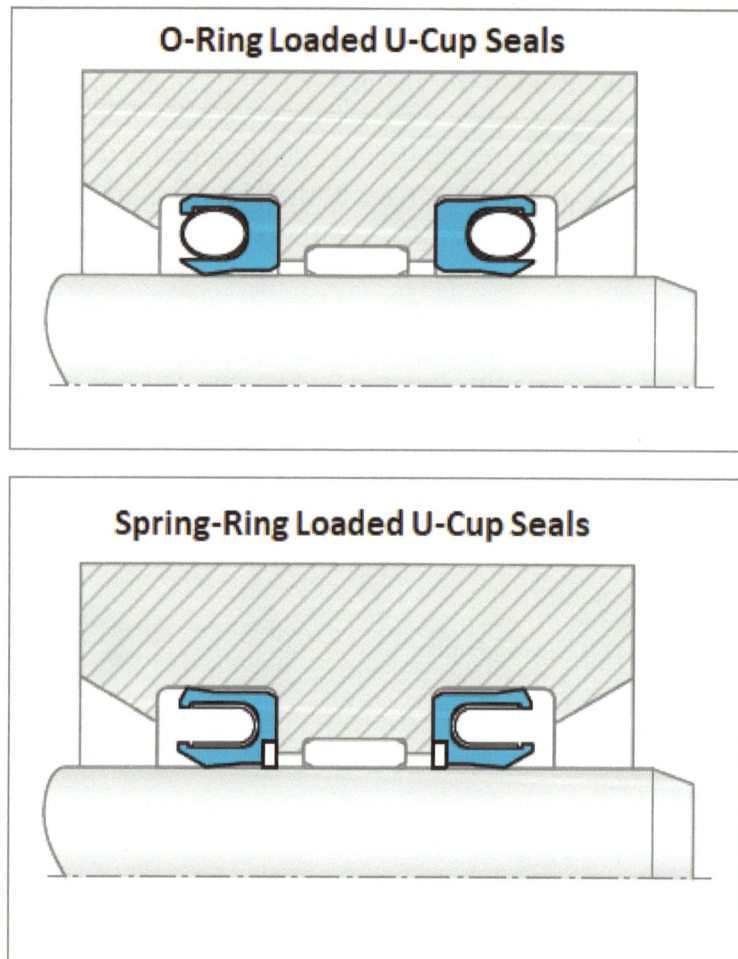

**Fig. 4.84 - Cylinder Rod Bidirectional Sealing Solutions using U-Cup Seals
(Courtesy of Trelleborg)**

4.14.3- Sealing Solutions for Cylinder Pistons

4.14.3.1- Unidirectional Piston Sealing Solutions

The following piston sealing solutions are good for cylinders that develop power during extension stroke and retracts load-free at low pressure.

Example 1: U-Cup Seals for Unidirectional Piston Sealing

Figure 4.85 shows various types of U-Cup seals for cylinder pistons seal as follows:

Type 1: The U-Cup shown in the figure is provided with a robust dynamic sealing lip and a wide contact area of the static lip, which guaranties an effective position in the groove. The profile is suitable for pressures up to 400 bar (5801 psi) provided that the extrusion gap is adapted to the pressure level.

Type 2: The O-Ring loaded U-Cup seal provides static sealing by the O-Ring. The O-Ring is protected from damage under high pressure and pressure cycles by the contoured O-Ring contact zone which supports the O-Ring. The seal shown in the figure is designed with hydrodynamic back-pumping effect which allows the seal to relieve pressure trapped between seals in tandem configurations.

Type 3: The spring energized U-Cup seal is a more dynamically stable.

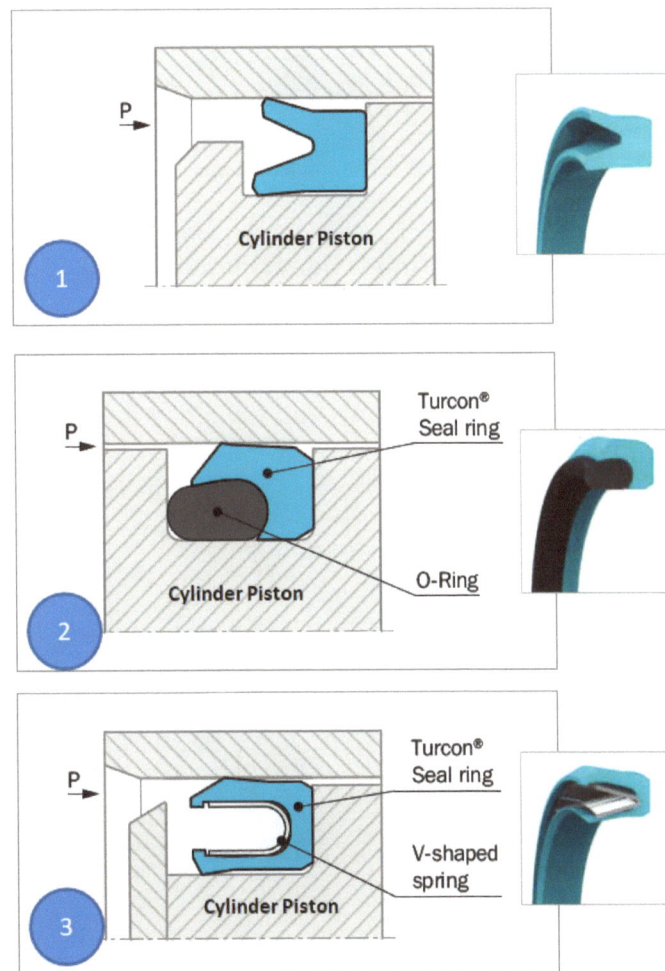

Fig. 4.85 - U-Cup Seals for Unidirectional Piston Sealing (Courtesy of Trelleborg)

Example 2: V-Packing Seals for Unidirectional Piston Sealing
This piston seal, as shown in Fig. 4.86, utilizes a set of V-Packings as the dynamic seal. This solution is assumed for applications with high extremely high-pressure and possible pressure spikes.

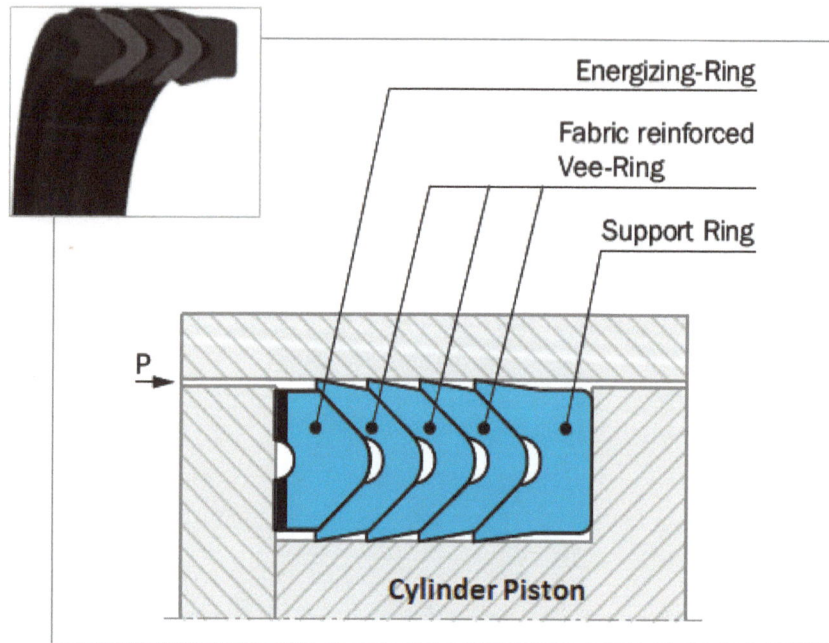

Fig. 4.86 - V-Packing Seals for Unidirectional Piston Sealing (Courtesy of Trelleborg)

Example 3: Piston Seals for Better Positioning Accuracy
Figure 4.87 shows a single-acting piston seal design where high demands are made on positional accuracy and stick-slip-free movement.

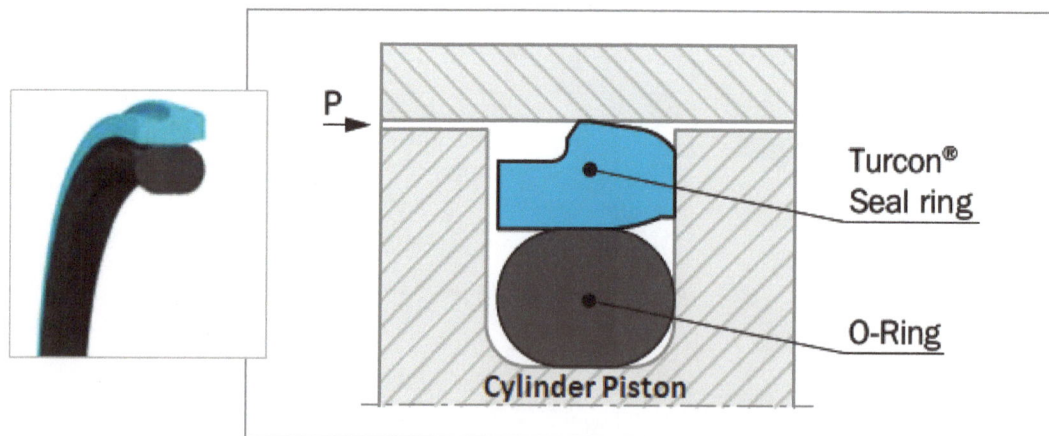

**Fig. 4.87 - Unidirectional Piston Sealing Solution for Better Positioning Accuracy
(Courtesy of Trelleborg)**

Example 4: Cylinder Piston Unidirectional Sealing Solution for Light-Duty Applications
This piston seal package, as shown in Fig. 4.88, contains a single U-Cup Seal to prevent external leakage. The U-Cup seal can be *O-Ring-loaded* or *Spring-Energized.* A piston guide ring offers bearing surface for lateral load.

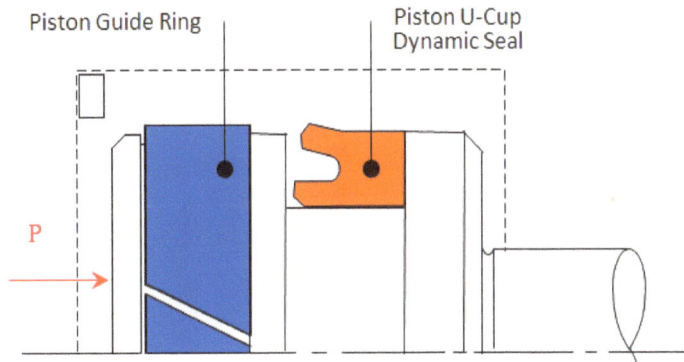

Fig. 4.88 - Cylinder Piston Unidirectional Sealing Solution for Light-Duty Applications

Example 5: Cylinder Piston Unidirectional Sealing Solution for Medium-Duty Applications
As shown in Fig. 4.89, for medium-high pressure applications, unidirectional piston U-Cup seals are O-Ring loaded or Spring-Energized.

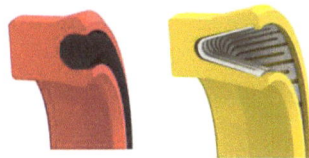

**Fig. 4.89 - Cylinder Piston Unidirectional Sealing Solution for Medium-Duty Applications
(Courtesy of American High-Performance Seals)**

Example 6: Cylinder Piston Unidirectional Sealing Solution for Heavy-Duty Applications
As shown in Fig. 4.90, for high pressure applications, unidirectional piston U-Cup seals are supported by Back-up Rings to prevent seal extrusion.

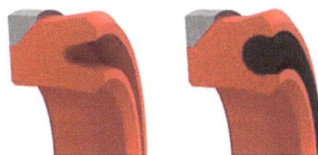

**Fig. 4.90 - Cylinder Piston Unidirectional Sealing Solution for Heavy-Duty Applications
(Courtesy of American High-Performance Seals)**

Example 7: Redundant Unidirectional Sealing Solution for Slow and Long Stroke Cylinder Pistons

In long-stroke cylinders and equipment operating with low speed during retraction, it has been found that hydrodynamic back-pumping may become insufficient to prevent build-up of pressure in the seal system behind the primary seal. Pressure build-up in the seal system leads to leakage, increased friction and wear, and may ultimately require replacement of the seals. The usual solution in such equipment has been to provide space for a buffer volume behind the primary seal or to install a drain line. As shown in Fig. 4.91, first invented by *Trelleborg Sealing Solutions*, the built-in check valve function eliminates pressure build-up in the reservoir volume. Hence, improving seal performance with outstanding sliding and wear resistance properties.

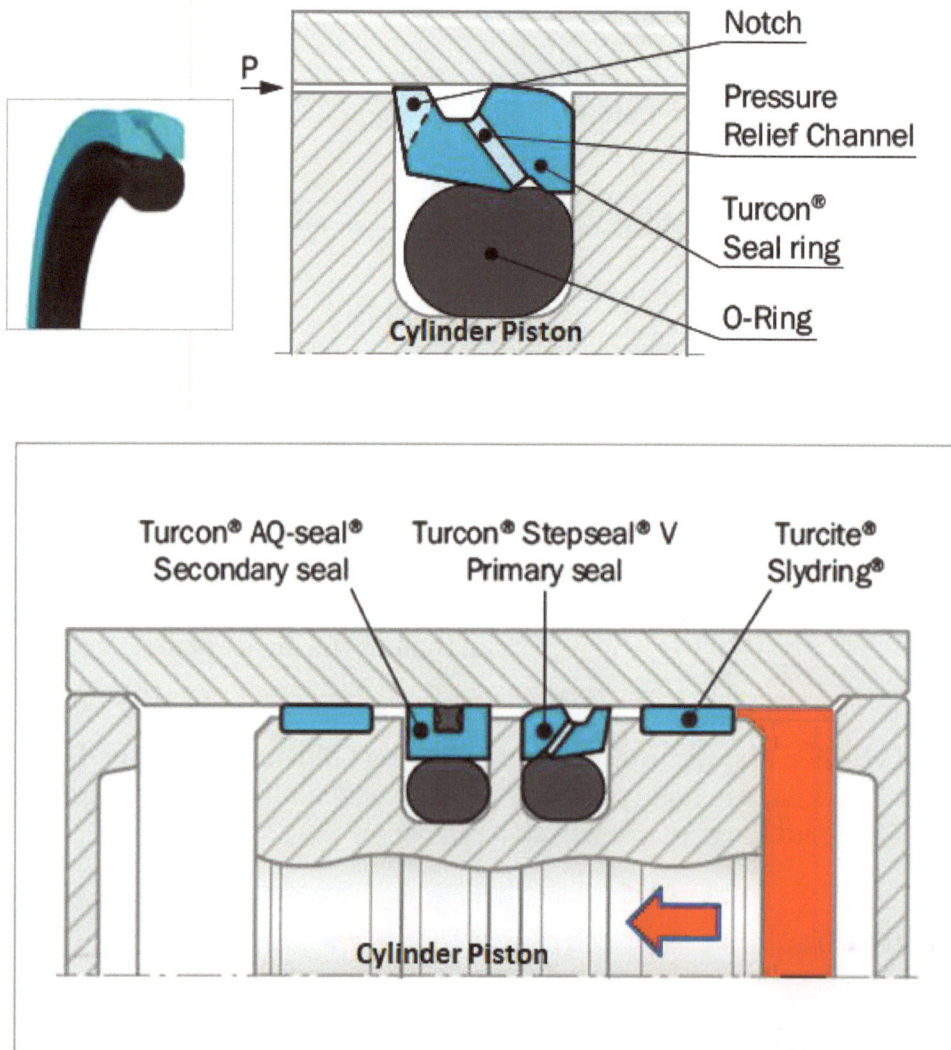

Fig. 4.91 - Redundant Unidirectional Sealing Solution for Slow and Long Stroke Cylinder Pistons (Courtesy of Trelleborg)

4.14.3.2- Bidirectional Piston Sealing Solutions

The following sealing solutions are good for cylinders that develops power during both extension and retraction directions.

Example 1: Basic Bidirectional Piston Sealing Solution for Light-Duty Applications
As shown in Fig. 4.92, the double-acting seal is a combination of a custom seal ring and an energizing O-Ring. Together, with the squeeze of the O-Ring, ensures a good sealing effect even at low pressure.

At higher system pressures, the O-Ring is energized by the fluid, pushing the seal ring face with increased sealing force. As shown in the figure, the trapezoidal profile cross section of the seal ring allows the lubricating hydrodynamic fluid film to be built above the seal with stick-slip-free linear motion. Two Guide-Rings are used to absorb lateral forces. Figure 4.93 shows a typical industry example for a piston with such seal package.

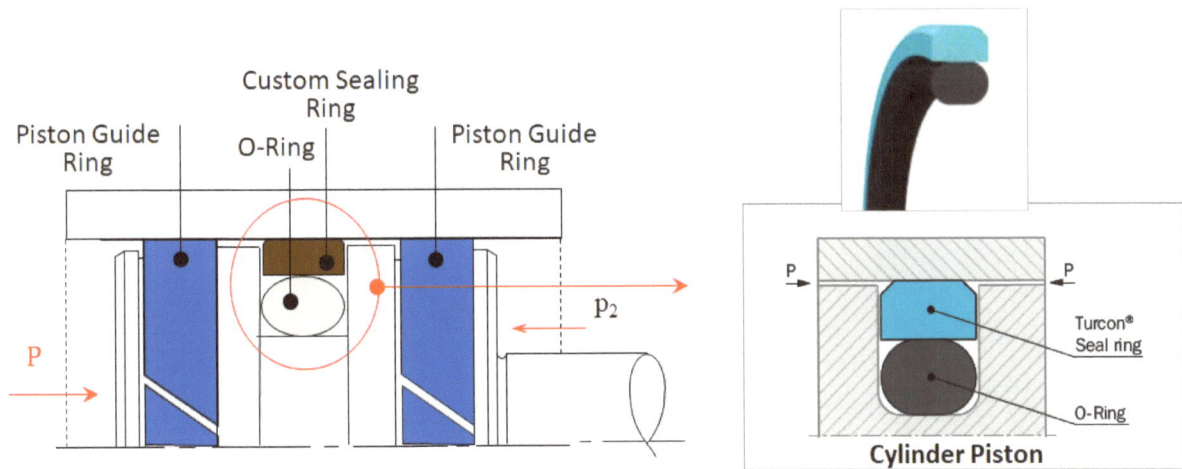

Fig. 4.92 - Basic Bidirectional Piston Sealing Solution (Courtesy of Trelleborg)

Fig. 4.93 - Basic Bidirectional Piston Sealing Solution for Light-Duty Applications

Example 2: Bidirectional Piston Sealing Enhanced Solution for Medium-Duty Applications

Figure 4.94 shows a typical example from industry for a bidirectional piston sealing solution with an enhanced sealing ring.

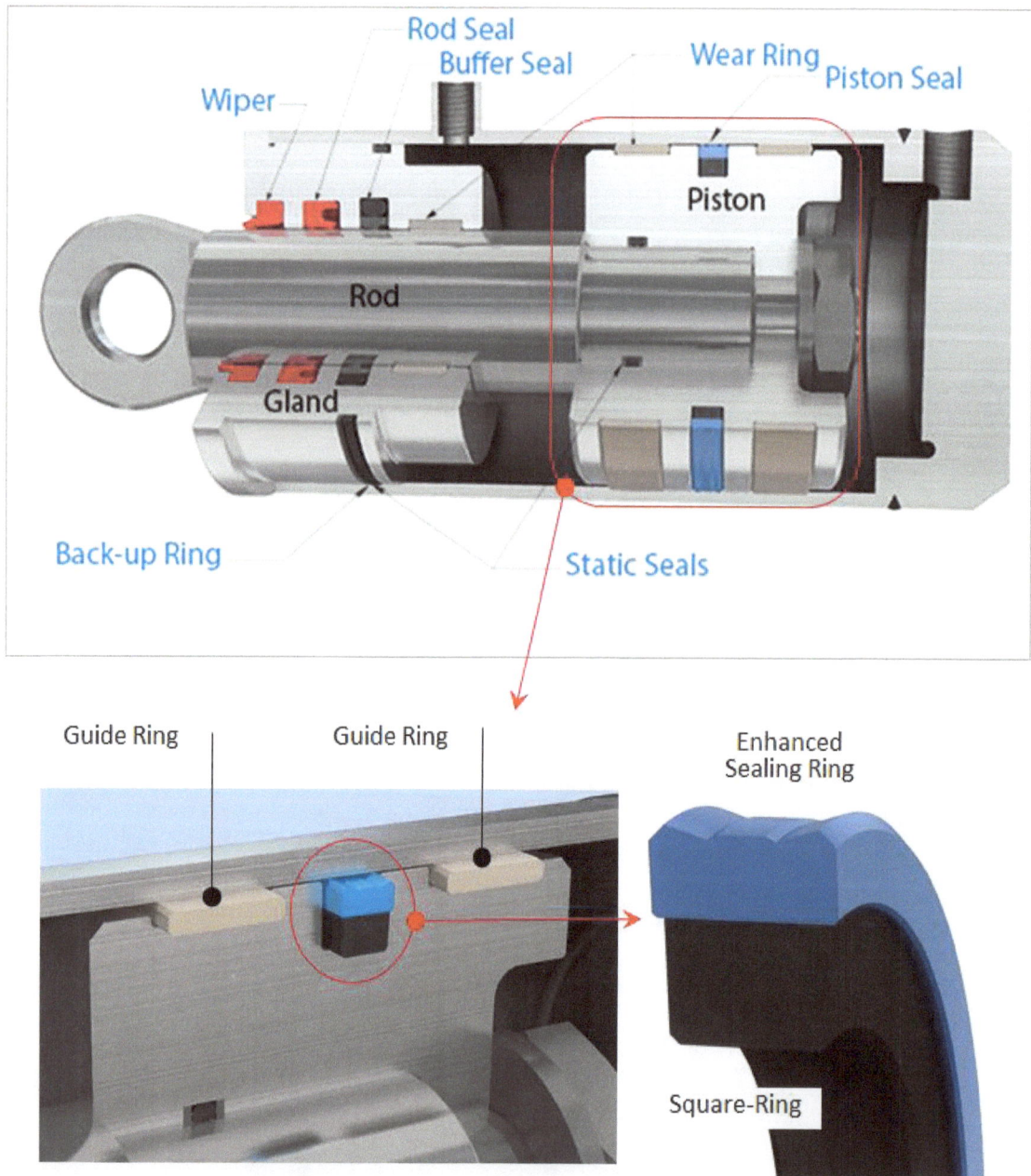

Fig. 4.94 - Enhanced Bidirectional Piston Sealing Solution for Medium-Duty Applications (Courtesy of American High-Performance Seals)

Figure 4.95 shows various enhancements for the basic bidirectional piton seal design.

Solution 1 for Blow-By-Effect: shows one radial notch is made on each side of the sealing ring. The notches override the *Blow-By-Effect*. The notches are made to assure that a rapid energizing of the seal takes place at sudden changes in pressure and direction of motion.

Solution 2 for short-stroke and high-frequency: Reciprocating movement, with an increasing frequency above 5 Hz, the formation of lubrication under the contact face of the seal ring is reduced. High-frequency is most often occurring in connection with short-strokes. These two types of movements together accelerate the wear on hardware and seal. In this solution a symmetric seal is used with angled contact faces to ensure oil film is not scraped away from the surface. Oil is transported into the groove in the middle of the contact area forming an oil reservoir for lubrication. Wear particles are also likely to be captured in this groove, thus preventing them from embedding in the surface where the highest contact force occurs.

Solution 3 for better leakage control: The seal is designed with two seal edges. It acts as the primary seal for pressures from both sides, prevents build-up of hydrodynamic pressure over the seal profile, and prevents the risk of blow-by effect. The central sealing face increases the sealing effect.

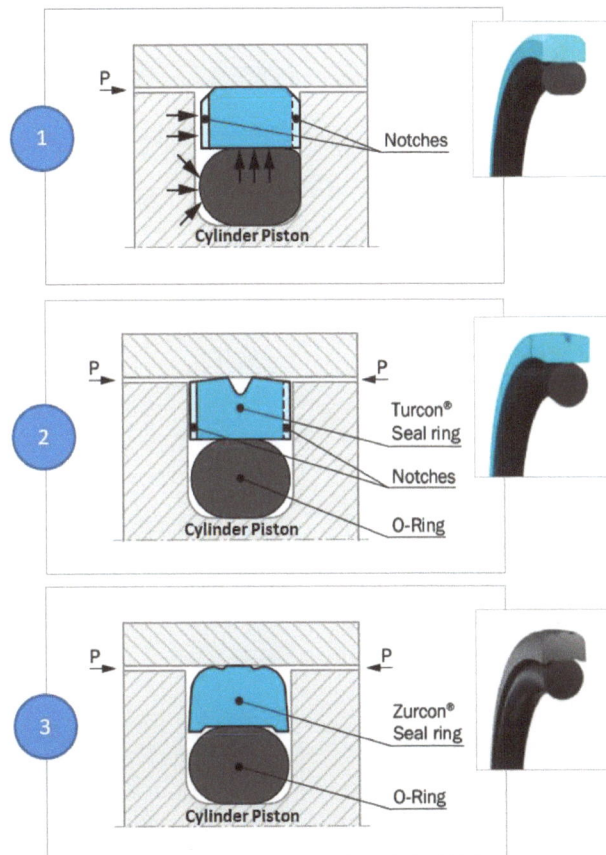

Figure 4.95 - Enhanced Bidirectional Piston Seal Designs for Medium-Duty Applications (Courtesy of Trelleborg)

Example 3: Bidirectional T-Shaped Piston Sealing for Heavy-Duty Applications
Figure 4.96 shows a T-Shaped seal with two standard Backup-Rings. Its compact design provides improved stability and extrusion resistance in dynamic fluid sealing applications.

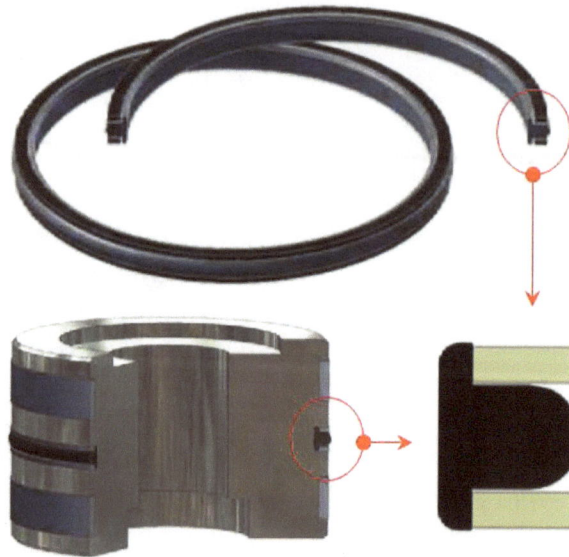

Fig. 4.96 - Typical T-Shaped Seals for Bidirectional Piston Sealing (Courtesy of Parker)

This enhanced piston seal package, as shown in Fig. 4.97, contains a heavy-duty squeeze-type T-Shaped piston seal and two L-Shaped Guide-Rings. This solution is good to seal across the piston during extension and retraction. The figure shows that the dynamic seal surface is designed with as wavy surface for better self-lubrication and reduced friction.

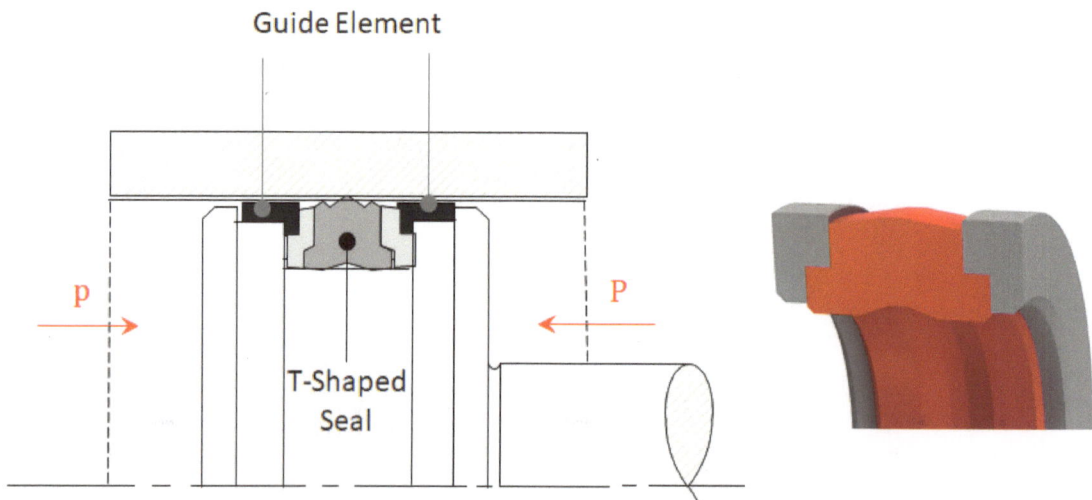

**Fig. 4.97 - T-Shaped Seals for Bidirectional Piston Sealing
(Courtesy of American High-Performance Seals)**

Example 4: U-Cup Seals for Bidirectional Piston Sealing
Figure 4.98 shows two single-acting U-cup seals are placed back-to-back to form bidirectional piston sealing solution. This piston sealing solution is good for cylinders that are pushing and pulling loads. The figure shows also a U-Cup seal for the cushioning head, a Guide-Ring, and an assembled magnet for cylinder position sensing.

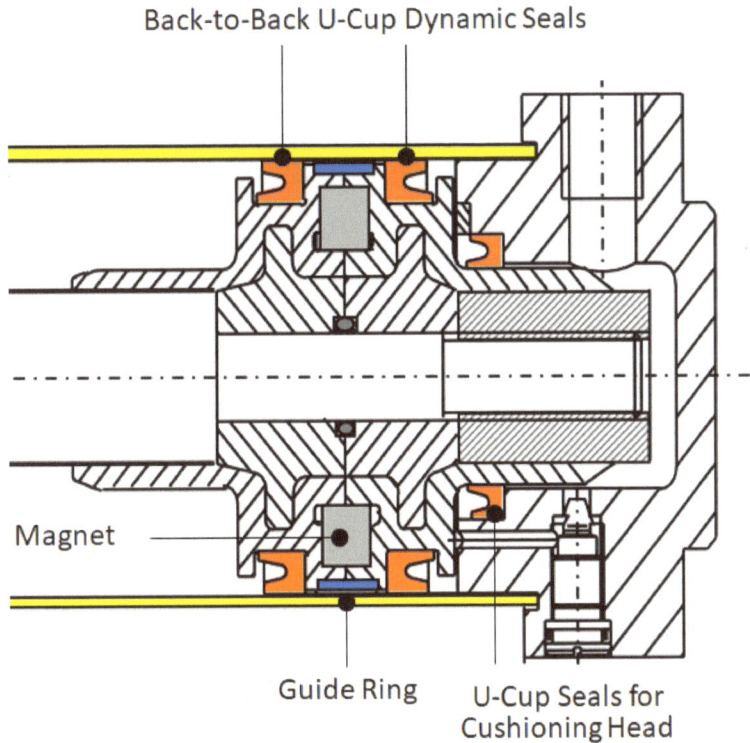

Fig. 4.98 - U-Cup Seals for Bidirectional Piston Sealing

Figure 4.99 shows a typical non-symmetrical hydraulic cylinder piston seal. Two seals can be placed on a piston, back-to-back, in separate glands offering bidirectional fluid sealing.

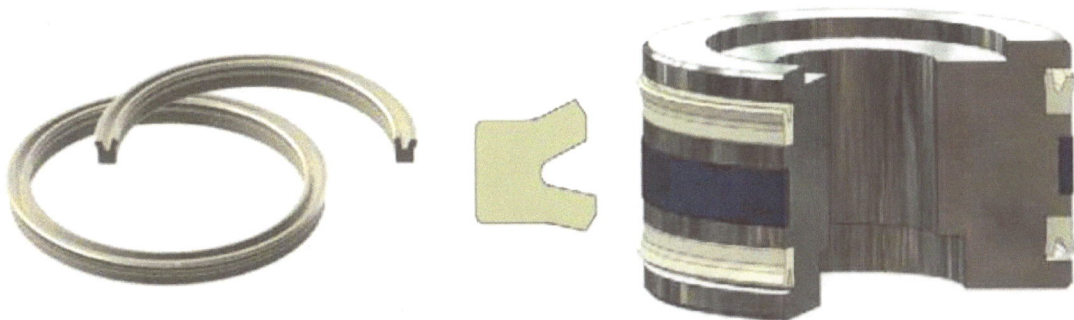

Fig. 4.99 - Typical U-Cup Seals for Bidirectional Piston Sealing (Courtesy of Parker)

Figure 4.100 shows a typical industry example for spring-energized U-Cup seals for bidirectional piston sealing.

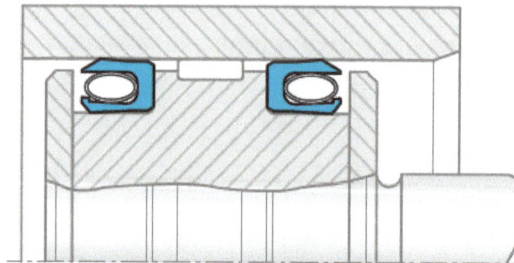

Fig. 4.100 - Typical U-Cup Seals for Bidirectional Piston Sealing (Courtesy of Trelleborg)

4.14.4- Piston and Rod Design for Proper Sealing

There are some design issues that must be considered for proper sealing functions, otherwise the seal may be damaged during assembly and operation. Theses design issues are:
- Design of *Seal Groove.*
- Design of cylinder piton and rod *Lead-in Chamfers.*
- Design of *Extrusion Gap.*
- Design of Mating *Surface Finish.*

4.14.4.1- Design of Seal Groove and Lead-In Chamfers

When designing a seal groove, sharp edges must be eliminated by proper rounding. Figure 4.101 shows the effect of improper lead-in chamfers on an O-Ring. During installation, the improper lead-in chamfer on cylinder piton and rod cause tearing of the O-Ring. Seals manufacturers also provide instructions for lead-in chamfers based on seal size.

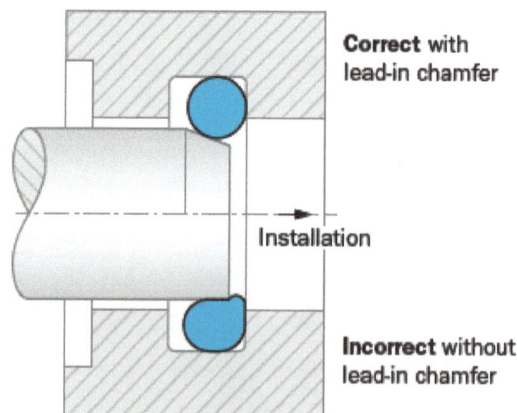

Correct with
lead-in chamfer

Installation

Incorrect without
lead-in chamfer

**Fig. 4.101 - Improper Lead-in Chamfer Cause Tearing of an O-Ring During Installation
(Courtesy of Trelleborg)**

Therefore, seal manufacturers must provide design instructions based on the cylinder size and working conditions. Figure 4.102 provide an industry example for a double-acting piston seal.

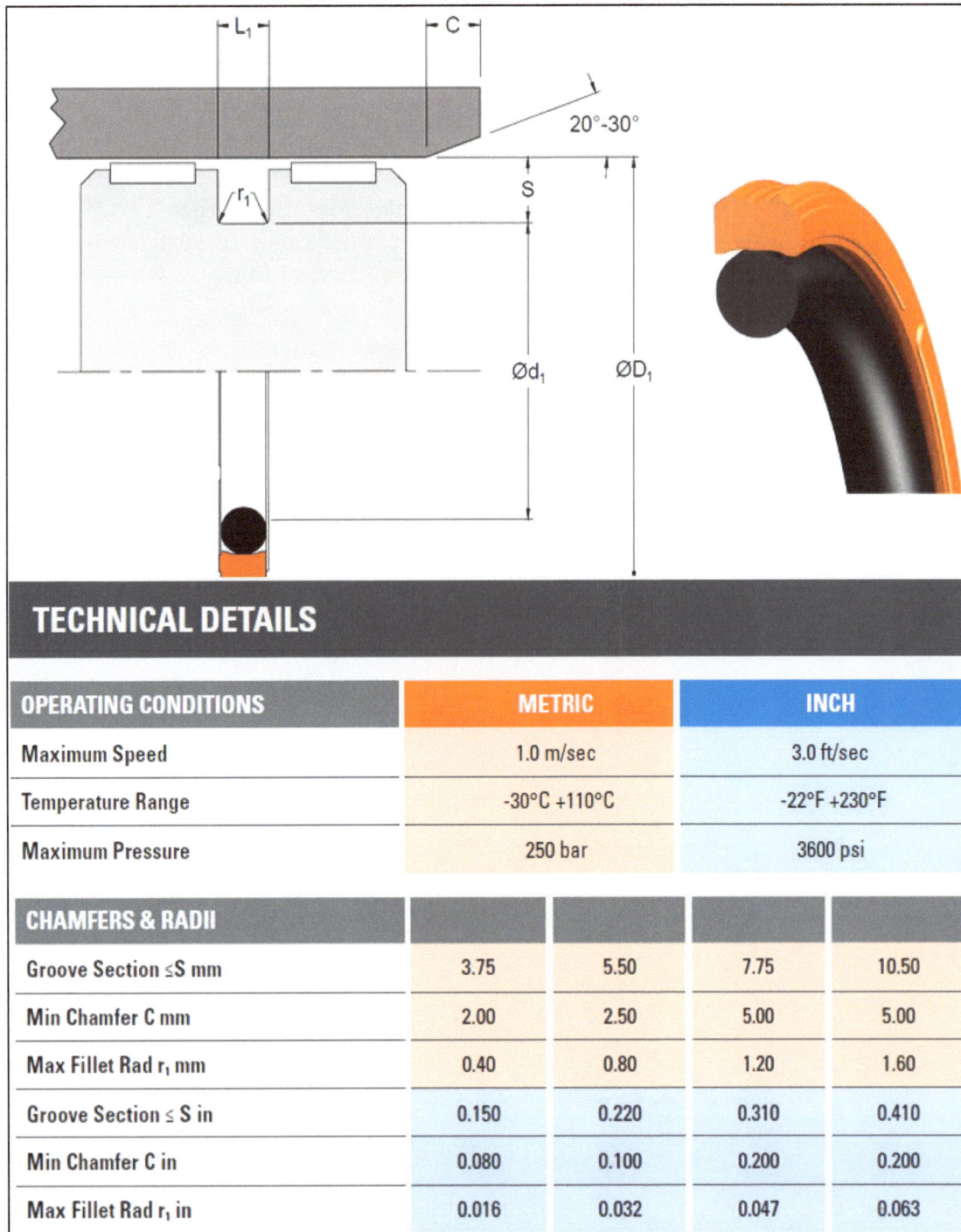

TECHNICAL DETAILS

OPERATING CONDITIONS	METRIC		INCH	
Maximum Speed	1.0 m/sec		3.0 ft/sec	
Temperature Range	-30°C +110°C		-22°F +230°F	
Maximum Pressure	250 bar		3600 psi	
CHAMFERS & RADII				
Groove Section ≤S mm	3.75	5.50	7.75	10.50
Min Chamfer C mm	2.00	2.50	5.00	5.00
Max Fillet Rad r_1 mm	0.40	0.80	1.20	1.60
Groove Section ≤ S in	0.150	0.220	0.310	0.410
Min Chamfer C in	0.080	0.100	0.200	0.200
Max Fillet Rad r_1 in	0.016	0.032	0.047	0.063

Fig. 4.102 - Data Sheet for a Double Acting Piston Seal 764 with Pre-Loaded O-Ring (Courtesy of Hallite Seals)

4.14.4.2- Extrusion Gap Design

Seal manufacturers provide guidelines for designing the extrusion gap.

<u>Example 1: Use of graphs provided by the manufacturer:</u>
Figure 4.103 shows that the *Extrusion Limits* are defined based on the seal hardness and working pressure. Softer seals extrude at low pressure. Therefore, at same extrusion gap (diametrical clearance), the required seal hardness is proportional to working pressure. Softer seals require less extrusion gap. Therefore, for same working pressure, the allowable extrusion gap is proportional to the seal hardness. The figure also shows that, at the same working pressure, the required extrusion gab is inversely proportional to seal hardness. The figure states to reduce the clearance shown by 60% when Silicon or Fluro silicone are used because the show less resistance to extrusion.

<u>Case Study:</u> For fluid pressure of 55.2 bar (800 psi) and a seal hardness 70 Shore A, maximum allowable gap extrusion is 0.3 mm (0.01 in)

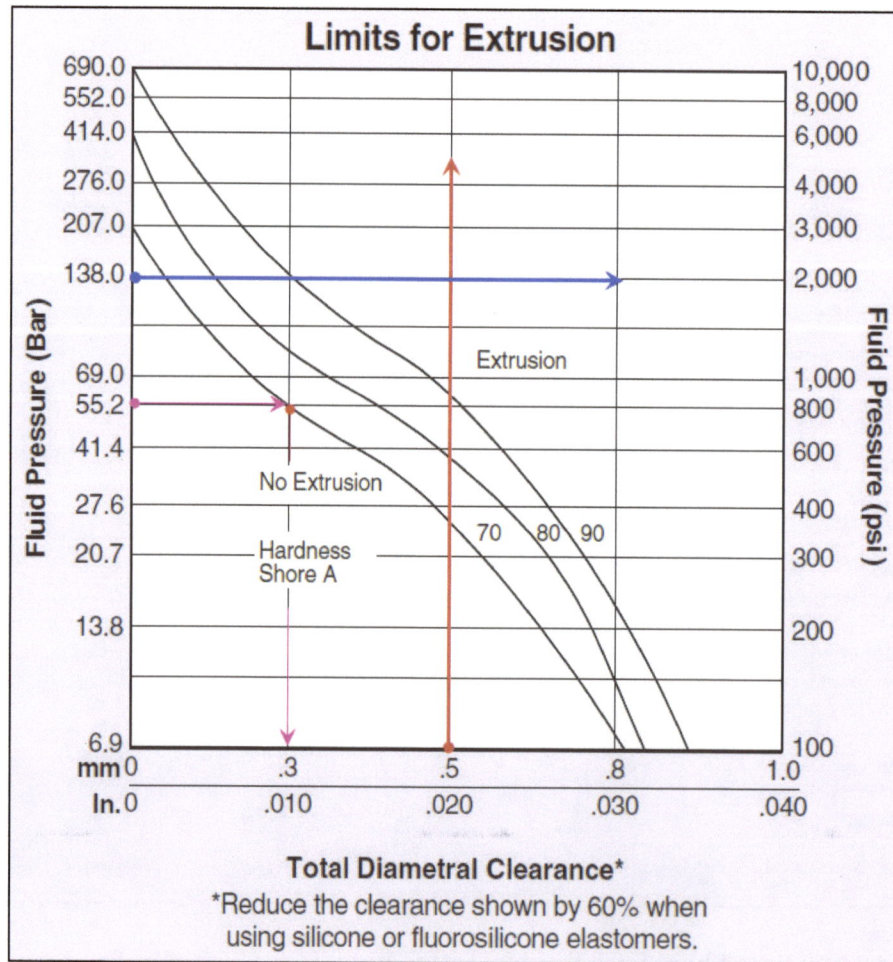

Fig. 4.103 - Limits of Extrusion (Courtesy of Parker)

<u>Example 2- Use of dimensional parameters and a nomograph provided by the manufacturer:</u>

Figure 4.104 shows the following dimensional parameters:

e = Maximum sealing and anti-extrusion gap.
D = Piton diameter.
d = Rod diameter.
S = Cross section.
P = Working pressure
T = Working Temperature.

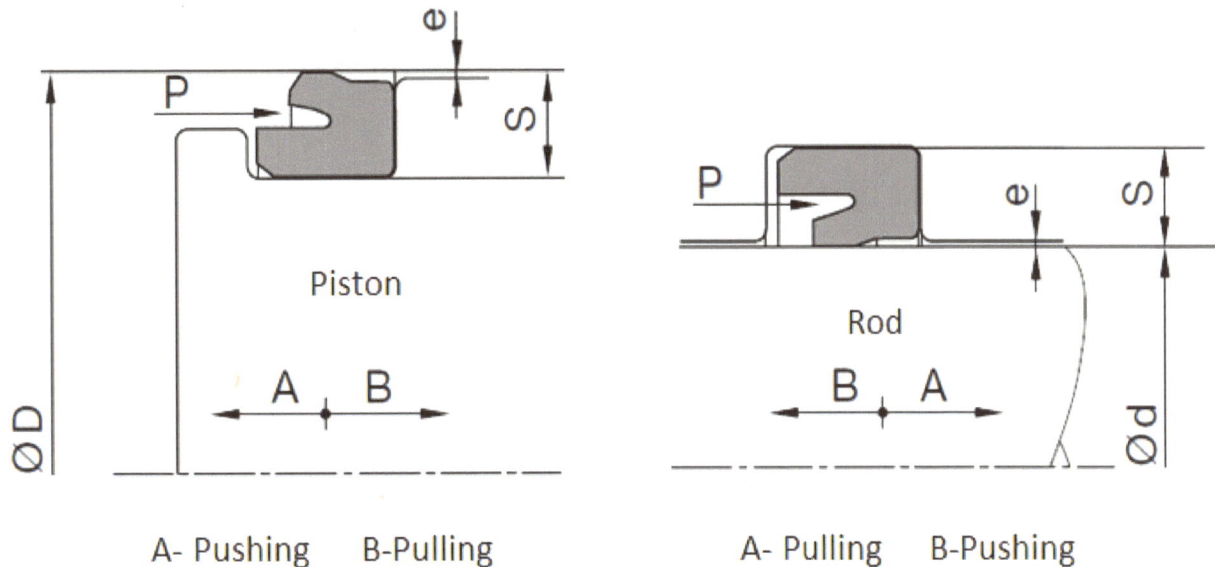

Piston

A- Pushing   B-Pulling

Rod

A- Pulling   B-Pushing

Fig. 4.104 - Anti-Extrusion Gap (Courtesy of Parker)

The nomograph, presented in Fig. 4.105, is developed for finding the maximum allowable sealing gap. The chart is based on the worst-case scenario such as the pushing case and the softest material. This chart is applicable for seals with 70-85 Shore A hardness.

Method of using the chart is as follows:
1- Draw the line connecting **d/D** to **S** and extend it until it interests with the line ξ**1**.
2- Draw the line connecting **P** to **T** and extend it until it interests with the line ξ**2**.
3- Connect the two intersections and read the allowable sealing gap.

<u>Case Study:</u> for d/D = 100 mm, S = 6 mm, P = 100 bar, and T = 80 OC, the sealing gap e = 0.18 mm

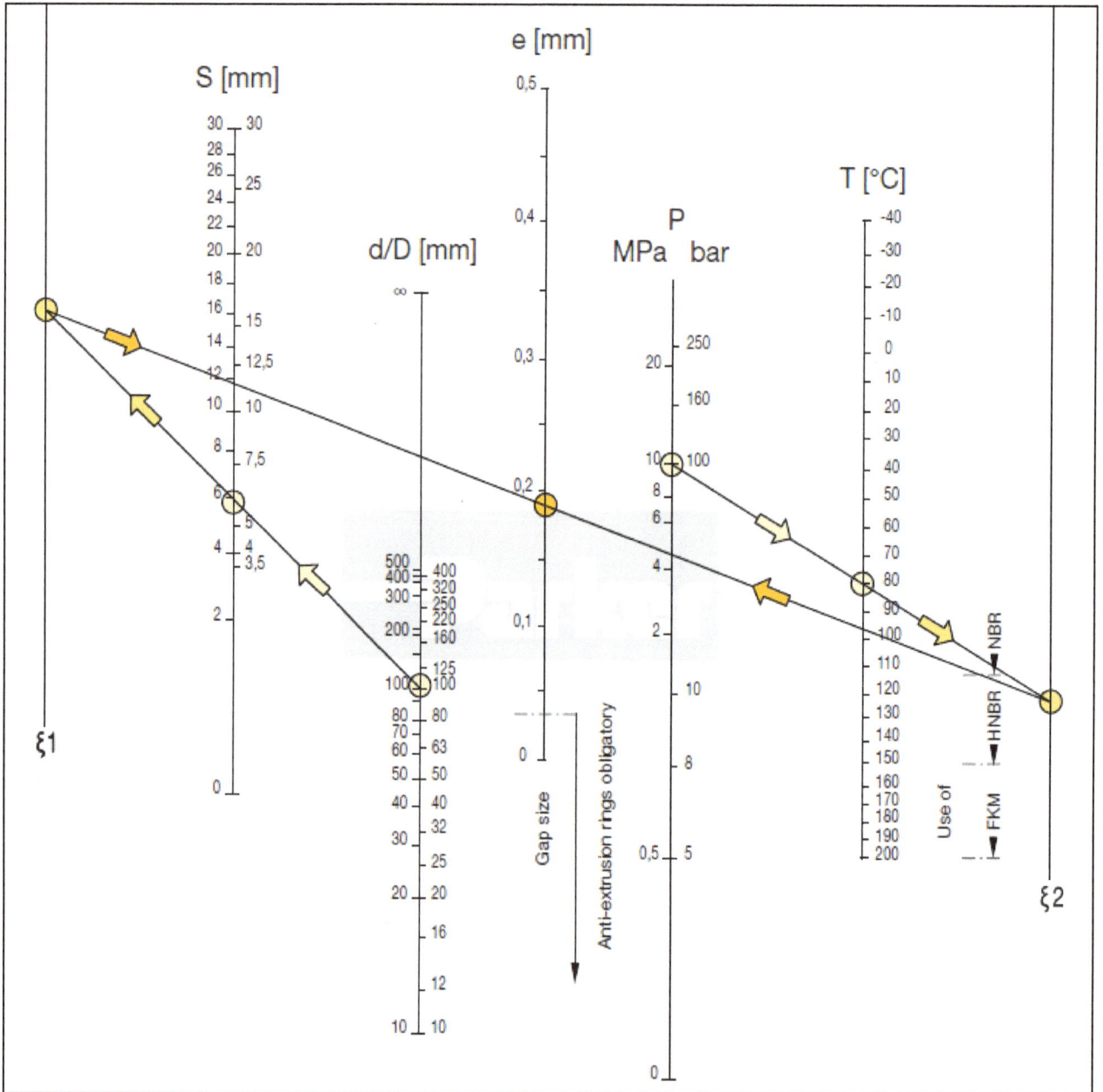

Fig. 4.105 - Anti-Extrusion Gap Nomograph (Courtesy of Parker)

<u>Example 3- Use of tabulated results provided by the manufacturer:</u>

Table 4.10 shows an industry example.

MAXIMUM EXTRUSION GAP			
Pressure bar	100	160	250
Maximum Gap mm	0.60	0.50	0.40
Pressure psi	1500	2400	3750
Maximum Gap in	0.024	0.020	0.016

**Table 4.10 - Data Sheet for a Double Acting Piston Seal 764 with Pre-Loaded O-Ring
(Courtesy of Hallite Seals)**

4.14.4.3- Design of Mating Surface Finish

Figure 4.106 shows the dynamic and static sealing surfaces in a hydraulic cylinder. Proper finish of mating surfaces in contact with the seals is critical in assuring maximum seal performance and service life within a given application. If the dynamic sealing surface is too smooth, it will improperly retain lubrication film. Poor seal lubrication causes excessive seal wear due to frictional heat. If the dynamic sealing surface is too coarse, premature seal failure may occur due to the roughness of the surface, hence causing small cuts or scores in the sealing lip. The static sealing surface finish must not be ignored in the control of leakage. Generally, static sealing surfaces should be free from chatter marks.

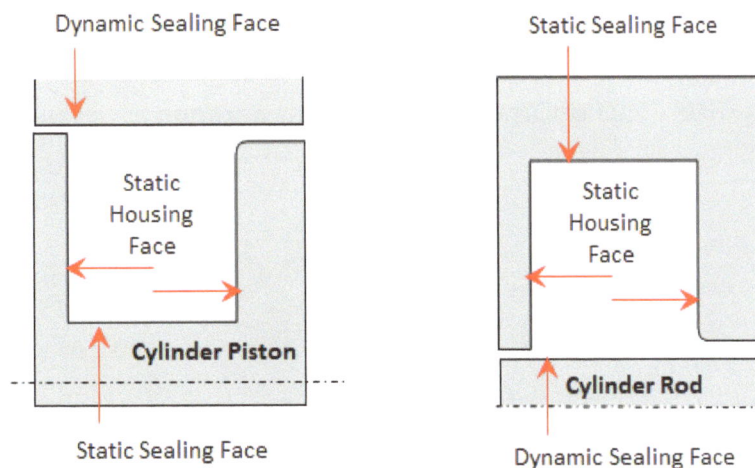

Fig. 4.106 - Dynamic and Static Sealing Surfaces in a Hydraulic Cylinder

Many parameters can be used to define surface finishes, which are explained in **ISO 4287** and **ISO 4288**. The following are the most commonly used surface finish measurements in the fluid power industry.

Figure 4.107 shows the surface finish measurement **Ra** is the arithmetical mean deviation of an absolute ordinate over the evaluation length. Figure 4.108 shows the surface finish measurement Rt is the Sum of height of the largest profile peak height **Rp** and the largest profile valley **Rv** over the evaluation length. Figure 4.109 shows the surface finish measurement **Rz(n)** is the Sum of height of the largest profile peak height **Rp** and the largest profile valley **Rv** within a sampling length. the surface finish measurement **Rz** is then the average of **Rz(n)** over the evaluated length.

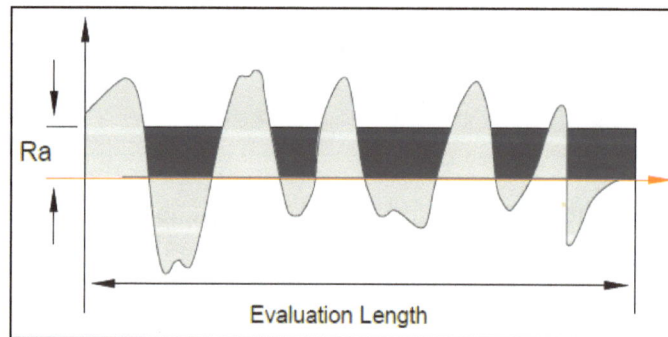

Fig. 4.107 - Surface Finish Measurement Ra (Courtesy of Hallite Seals)

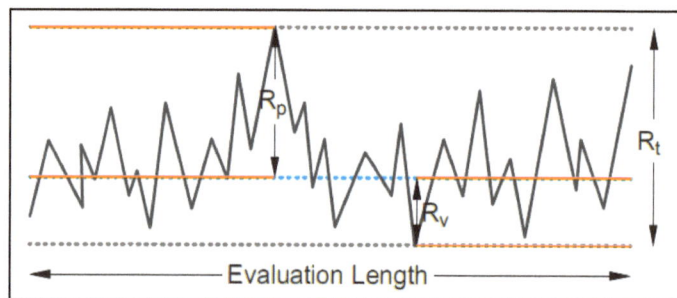

Fig. 4.108 - Surface Finish Measurement Rt (Courtesy of Hallite Seals)

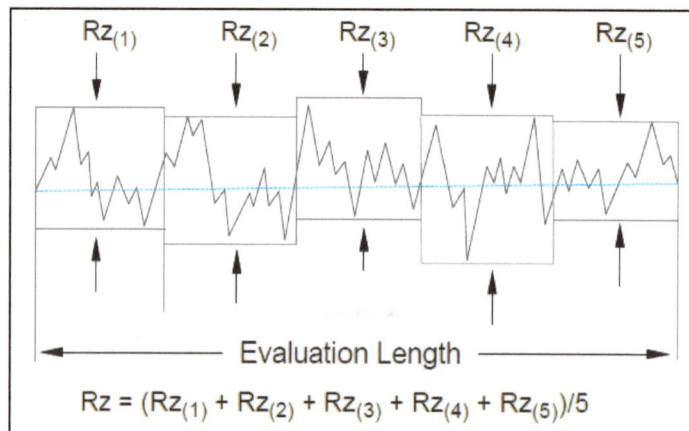

Fig. 4.109 - Surface Finish Measurement Rz (Courtesy of Hallite Seals)

Table 4.11 shows an industry example.

SURFACE ROUGHNESS	µmRa	µmRz	µmRt	µinRa	µinRz	µinRt
Dynamic Sealing Face ØD₁	0.1 - 0.4	1.6 max	4 max	4 - 16	63 max	157 max
Static Sealing Face Ød₁	1.6 max	6.3 max	10 max	63 max	250 max	394 max
Static Housing Faces L₁	3.2 max	10 max	16 max	125 max	394 max	630 max

Table 4.11 - Data Sheet for a Double Acting Piston Seal 764 with Pre-Loaded O-Ring
(Courtesy of Hallite Seals)

4.15- Sealing Solutions for Rotational Shafts

Rotational Dynamic Seals are used to prevent external leakage along rotating shafts in hydraulic pumps, motors, and rotary actuators. They are designed for axial, radial, or combined sealing directions. Like hydraulic seals for linear shafts, as shown in Fig. 4.110, materials and designs of a rotational shaft seals vary depending on working conditions and the applications. The following subtitles presents most commonly used hydraulic sealing solutions for rotational shafts.

Fig. 4.110 - Various Materials and Designs for Rotational Shaft Seals
(Courtesy of American High-Performance Seals)

4.15.1- Rotational Radial Seals

A *Rotational Radial Seal* acts radially against the rotating shaft to prevent axial leakage. Figure 4.111 shows light duty rubber lip radial seal. Figure 4.112 shows the construction details of heavy duty rotational radial seal shown in the previous figure. As shown in the figure, such a seal holds the rotating shaft radially preventing axial leakage.

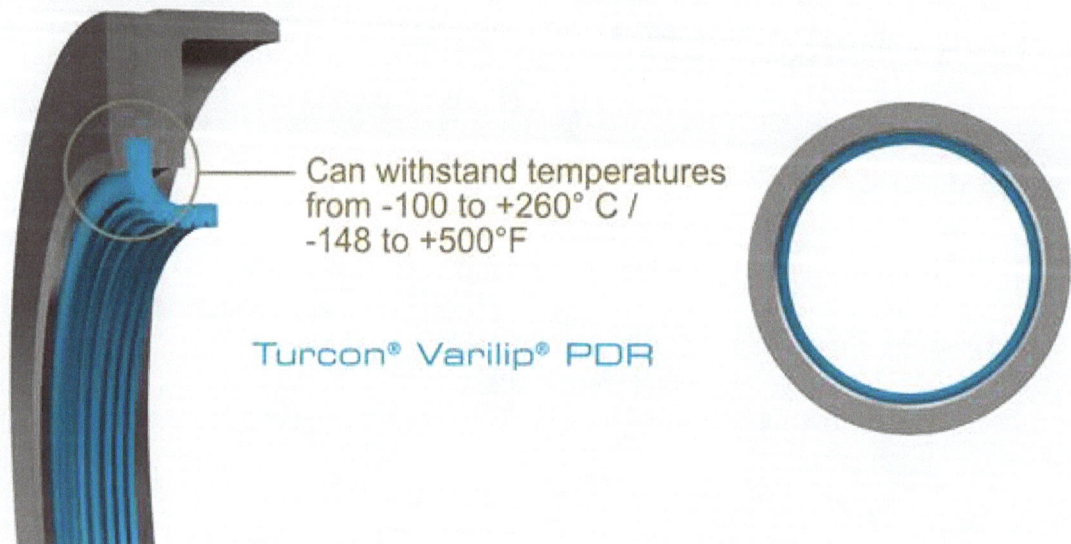

Can withstand temperatures from -100 to +260° C / -148 to +500°F

Turcon® Varilip® PDR

Fig. 4.111 – Light Duty Rubber Lip Rotational Radial Seal

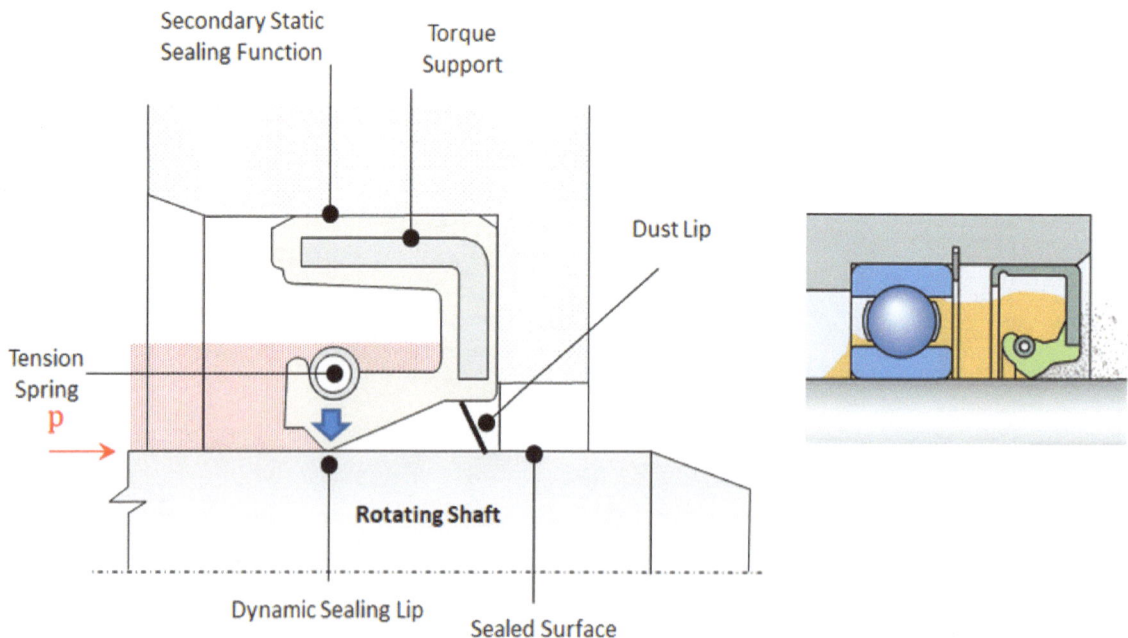

Secondary Static Sealing Function

Torque Support

Dust Lip

Tension Spring

p

Rotating Shaft

Dynamic Sealing Lip

Sealed Surface

Fig. 4.112 – Heavy Duty Rotational Radial Seals

Figure 4.113 shows typical rotational radial seals. They are available for wide range of pressure anywhere from 1 to 700 bar (15 to 10152 psi). These seals are comprised of standard and special seal designs of rubber, with or without metal supporting inserts, with or without a spring, and non-metal rotary shaft lip seals. The dynamic sealing lips are made of rubber or special elastomers to meet specific fluid, temperature, and other demanding operating requirements. Typical applications: oil and grease retention for power transmissions, motors, pumps, gearboxes, fans, and machine tools.

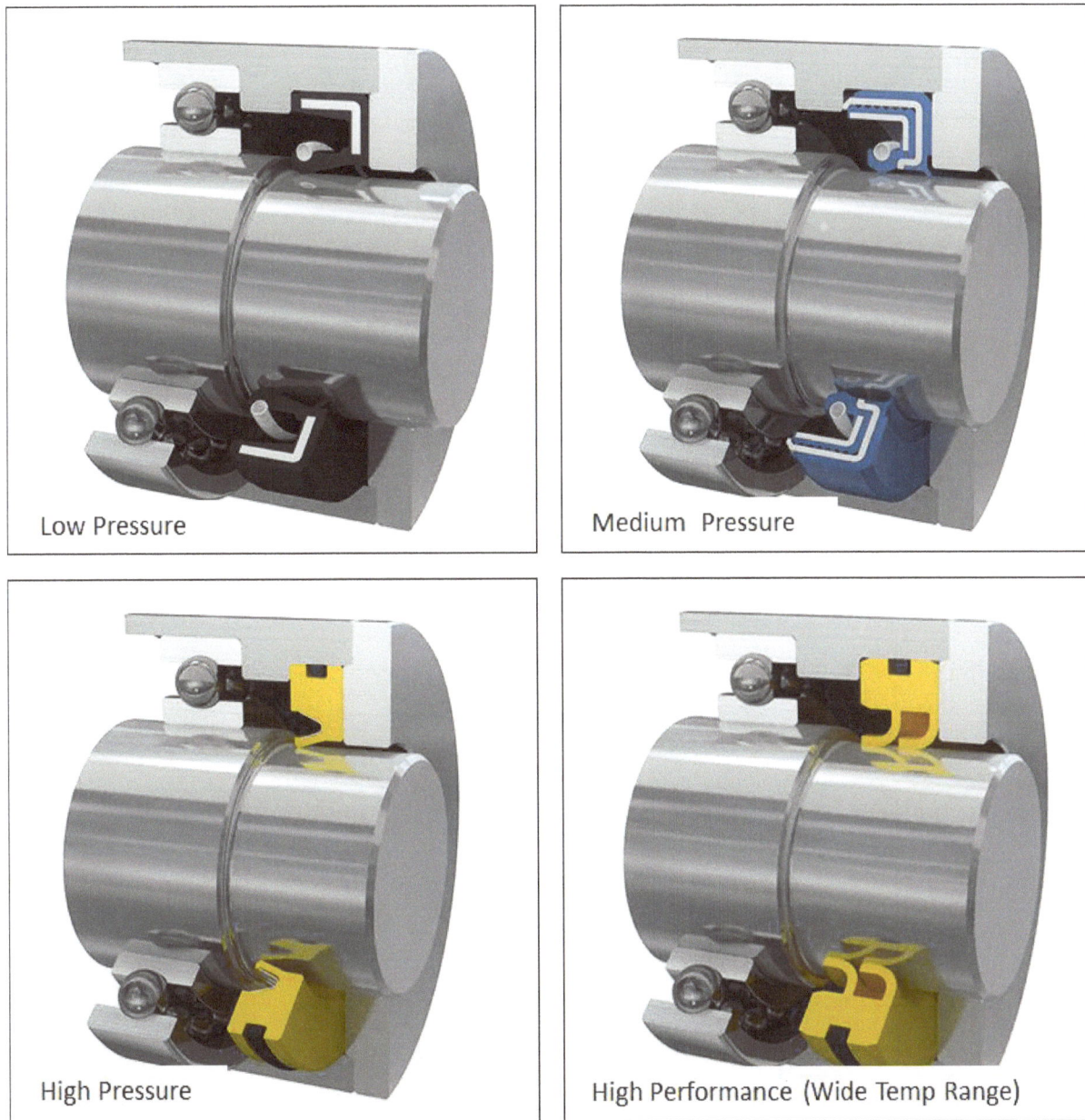

Low Pressure

Medium Pressure

High Pressure

High Performance (Wide Temp Range)

**Fig. 4.113 - Constructional Details of the Most Common Rotational Radial Seals
(Courtesy of American High-Performance Seals)**

Figure 4.114 shows standard configurations of rotational radial seals. Waved rubber om the static sealing surface improves the retaining force between the seal and the groove in the housing, hence preventing rotation of the seal in its groove.

With Rubber Case		With Metallic Case	
Flat Surface	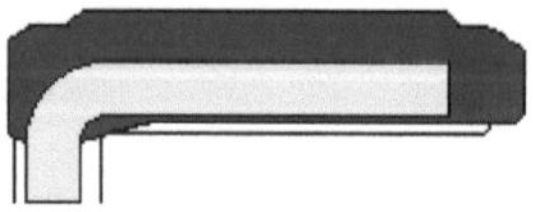	DIN Type B without Dust Lip without Support	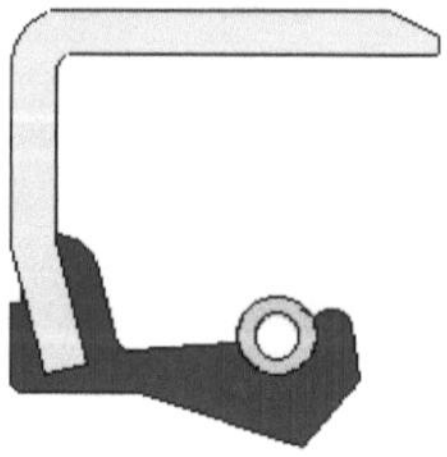
DIN Type A Waved Surface without Dust Lip	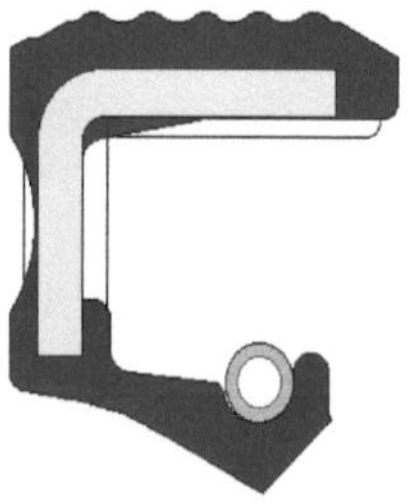	DIN Type BS with Dust Lip without Support	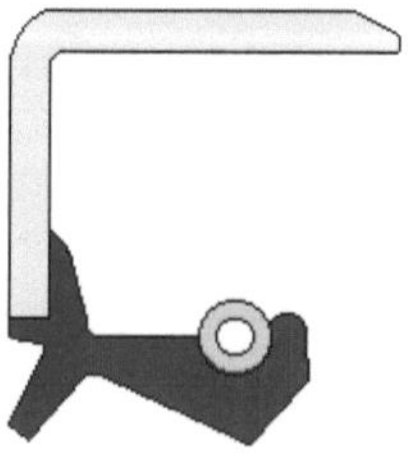
DIN Type AS Waved Surface with Dust Lip	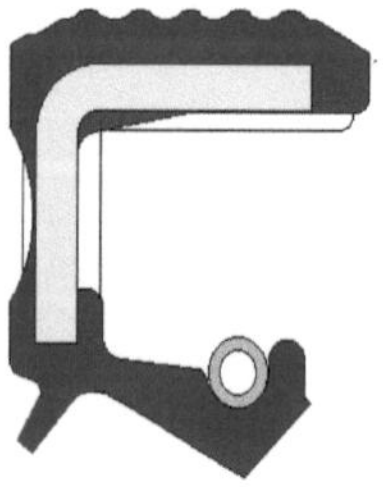	DIN Type C without Dust Lip with Support	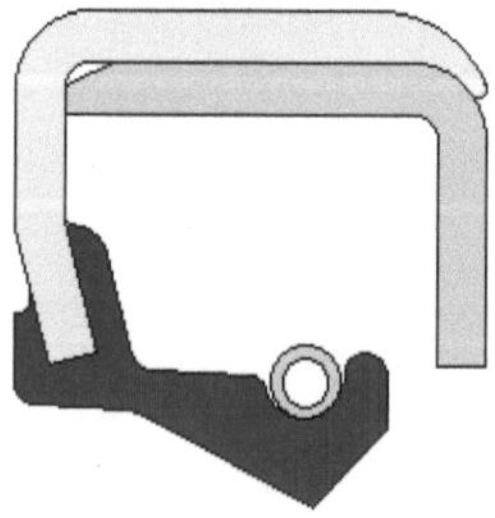
		DIN Type CS with Dust Lip with Support	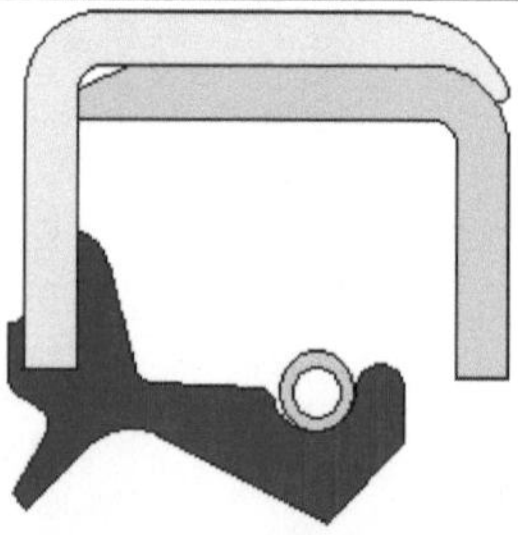

Fig. 4.114 - Configurations of Rotational Radial Seals to DIN 3760/ISO 6194 Standards

4.15.2- Rotational Axial Seals

A *Rotational Axial Seal* rotates with the shaft and seals axially against a stationary counter face that is perpendicular to the shaft. acts axially to prevent radial leakage. The following are common designs for rotational axial seals.

V-Ring Rotational Axial Seals: Figure 4.115, The *V-RING* axial seal is a unique all-rubber seal for rotary shafts. Such a seal is used for a broad range of applications. It can also be used as a secondary seal to protect a primary seal that do not perform well in hostile environments. This seal is normally stretched and mounted directly on the shaft, where it is held on the shaft by the inherent tension of the rubber body. The sealing lip is flexible and applies only a relatively light contact pressure against the counter-face and yet is still sufficient to maintain the sealing function.

As shown in Fig. 4.116, together with a *Clamping Band,* a V-Ring axial seal can be used for shaft diameters larger than 1500 mm.

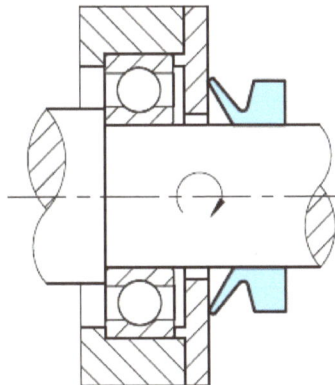

Fig. 4.115 - Classic Rotational Axial Seal (Courtesy of Trelleborg)

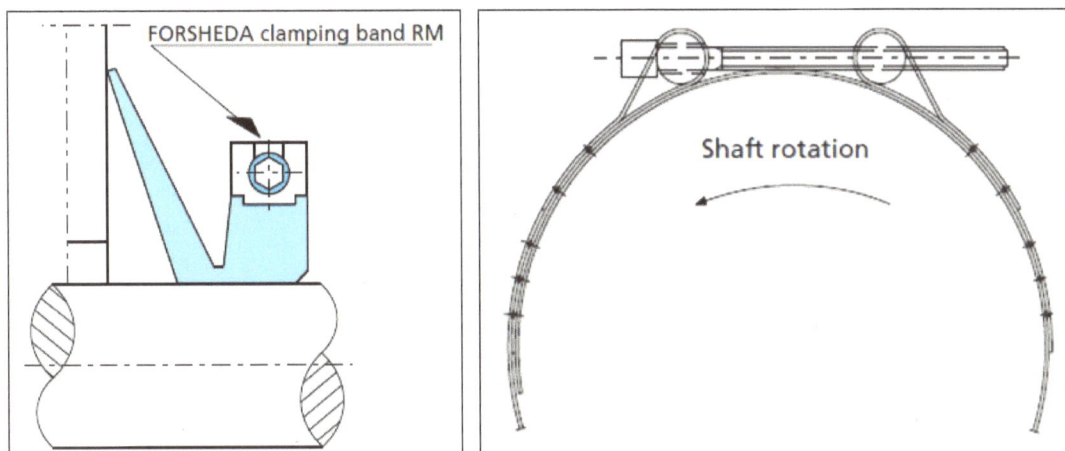

Fig. 4.116 - Classic Rotational Axial Seal (Courtesy of Trelleborg)

Power Losses in Rotational Shafts: A test was made for two types of V-Rings in contact with unhardened steel surface in dry-running condition. Figure 4.117 shows the power losses versus the peripheral (tangential) speed for various shaft diameters (d in mm). Once breakaway friction is overcome, the friction increases steadily until around the 12 m/s range, when it reduces quite quickly.

Due to the centrifugal force, the contact pressure of the lip decreases with increased speed. This means that frictional losses and heat are kept to a minimum, resulting in excellent wear characteristics and extended seal life. In the 15 - 20 m/s range the friction reduces to zero. The V-Ring then serves as a clearance seal and deflector.

Generally speaking, the power losses resulting from a V-Ring are always lower than a corresponding radial oil seal.

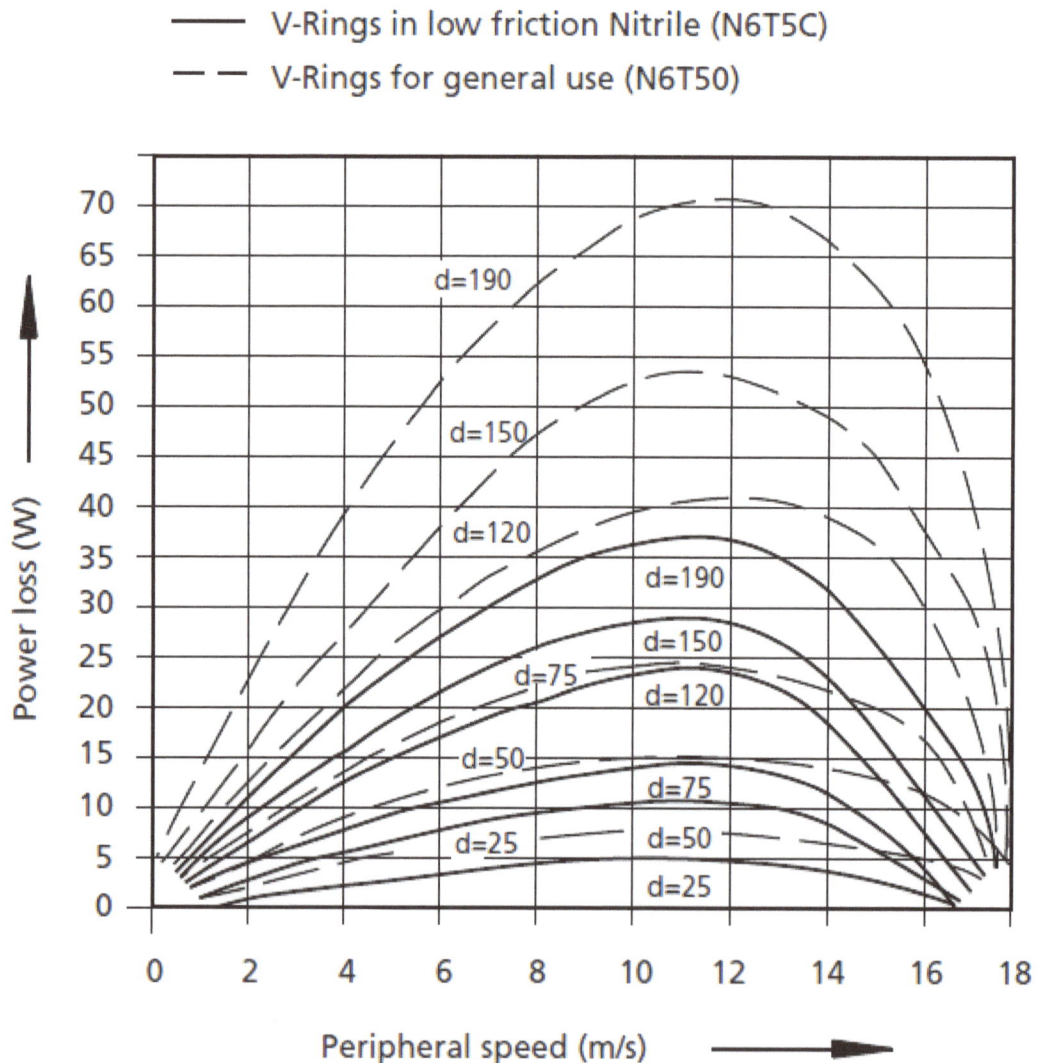

Fig. 4.117 - Power Losses versus Peripheral Speed (Courtesy of Trelleborg)

GAMMA Rotational Axial Seals: Figure 4.118 shows an enhanced rotational axial seal named *GAMMA Seal*. It consists of two parts, sealing element and metal case. The GAMMA seal is designed to be fixed on a rotational shaft at a predetermined distance from the sealing surface. The GAMMA seal is primarily intended for sealing against foreign matter, liquid splatter, and grease.

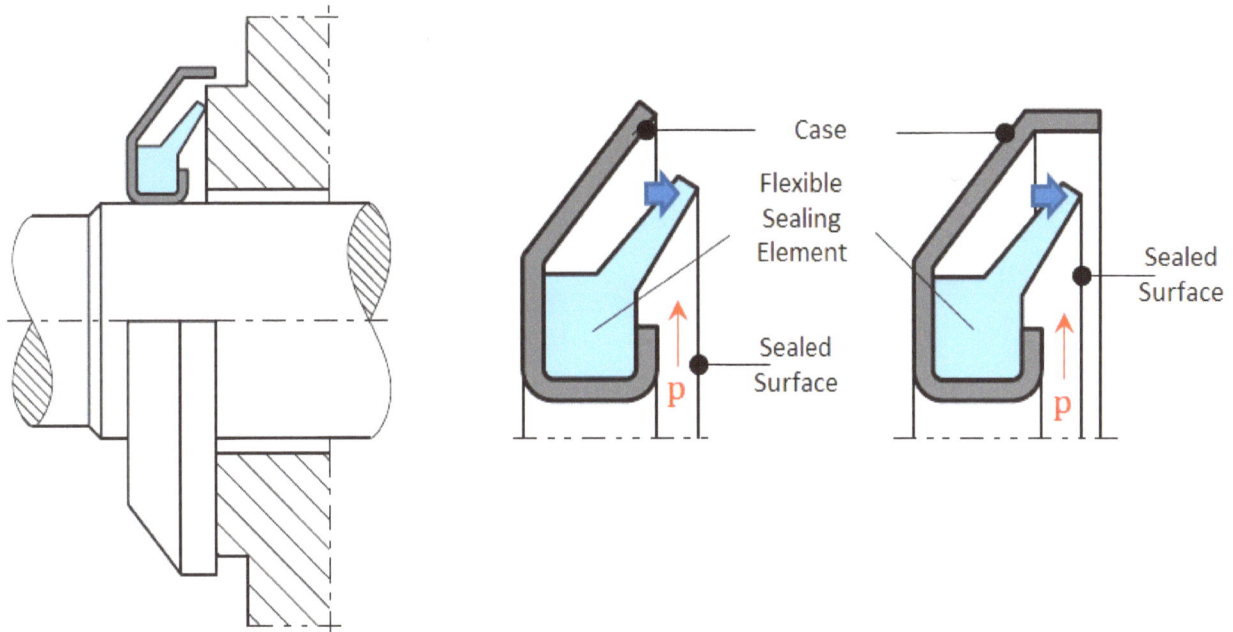

Fig. 4.118 - GAMMA Rotational Axial Seals (Courtesy of Trelleborg)

Enhanced Axial Shaft Seals: As shown in Fig. 4.119, axial shaft seals are used primarily as a protective seal for bearings. Their sizes are matched to those of roller bearings. If fluids are to be prevented from escaping, a design with an internal seal lip, is preferred. The design with external sealing lip is suitable for sealing grease and for protection against dirt entering from the outside.

Both types consist of an elastomer-elastic membrane with a metallic reinforcement ring. The membrane has an axial sealing lip. The elastomeric sealing lip is axially spring-loaded against the opposite mating face by a spider spring. The sealing lip is designed in a conical form to obtain a minimum contact area, thus considerably reducing friction, heat and wear.

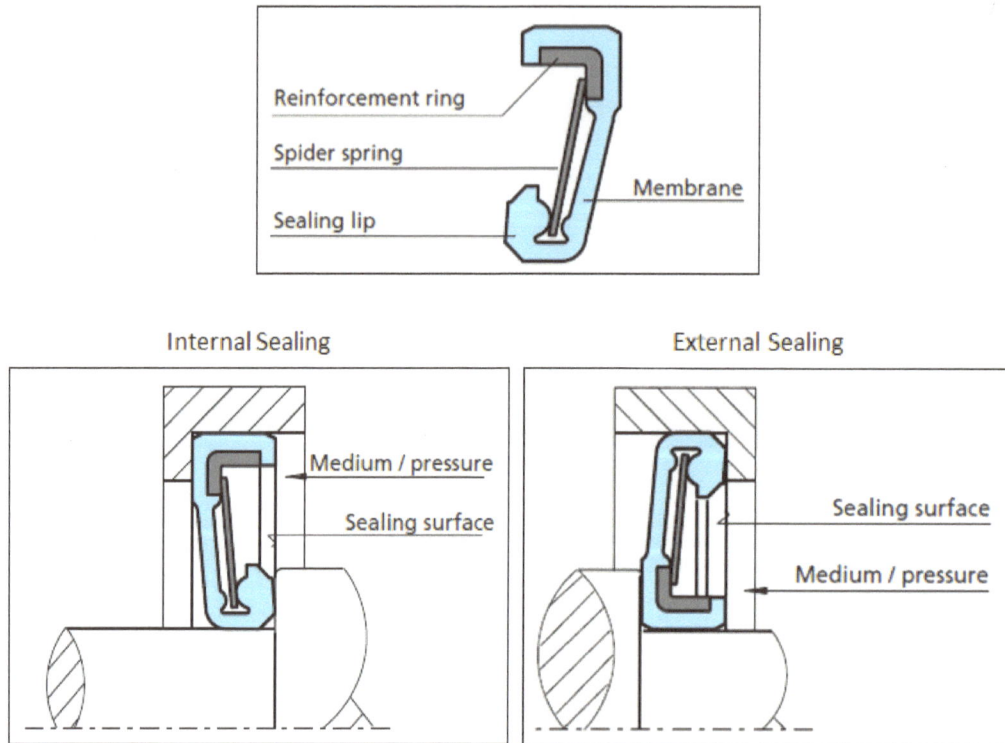

Fig. 4.119 - Enhance Rotational Axial Seals (Courtesy of Trelleborg)

4.15.3- Combined Axial/Radial Sealing Solutions for Rotational Shafts

As shown in Fig. 4.120, a flexible rubber V-Ring rotational axial seal is stretch-fit onto the shaft and rotate with the shaft against a counter face. It is used as a primary seal to protect the shaft against contaminants. The spring-energized rotational radial seal is a secondary seal to prevent external leakage.

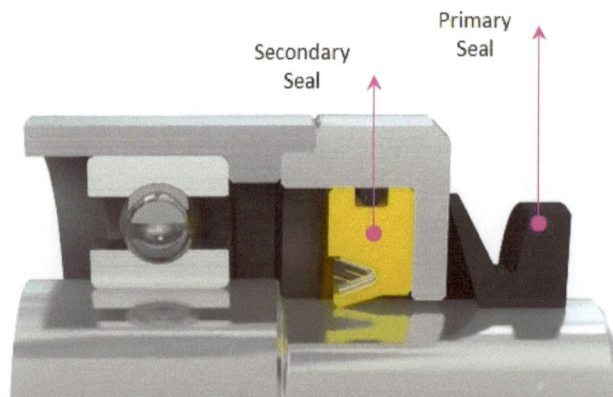

**Fig. 4.120 - Combined Axial/Radial Sealing Solutions for Rotational Shafts
(Courtesy of American High-Performance Seals)**

Chapter 5

Hydraulic Heat Exchangers

Objectives

This chapter overviews various types of heat exchangers including air-type, water-type, and plate-type. Construction, operation, features, applications, and sizing calculations are discussed.

The following topics are discussed in Chapter 11 in Volume 5 "Maintenance and Safety" of this series of textbooks:
- BP-Heat Exchangers-01-Selection and Replacement
- BP-Heat Exchangers-02-Maintenance Scheduling
- BP-Heat Exchangers-03-Installation and Maintenance
- BP-Heat Exchangers-04-Standard Tests and Calibration

The following topics are discussed in Chapter 11 in Volume 6 "Troubleshooting and Failure Analysis" of this series of textbooks:
- Heat Exchangers Inspection
- Heat Exchangers Troubleshooting
- Heat Exchangers Failure Analysis

Brief Contents

5.1- Contribution of Heat Exchangers:
5.2- Air-Type versus Water-Type Oil Coolers
5.3- Determination of Cooling Capacity for an Oil Cooler
5.4- Air-Type Oil Coolers
5.5- Shell-and-Tube Water-Type Oil Coolers
5.6- Plat-Type Oil Coolers
5.7- Cooling-Filtration Units
5.8- Oil Cooling Circuit Diagram
5.9- Oil Temperature Automatic Control Solutions
5.10- Electrical Oil Heaters

Chapter 5: Hydraulic Heat Exchangers

5.1- Contribution of Heat Exchangers

Contribution: *Heat Exchangers* are used in hydraulic systems either to remove heat or add heat to the hydraulic fluid so that the change in working temperature is maintained within the acceptable limit around the recommended ideal temperature, and consequently:

- Maintain fluid viscosity and other important properties within the recommended range.
- Prevent hydraulic fluid break-down and extend its service life.
- Maintain proper performance of hydraulic fluid additives.
- Improve machine startup in cold weather.
- Maintain sufficient lubrication for the moving parts inside hydraulic components.
- Reduce the rate of component wear and hence extend their lifetime.
- Maintain acceptable pressure drop through the components and transmission lines,
- Improve overall system energy efficiency.

Symbols: Heat exchangers shall be used when passive cooling cannot control the system fluid temperature within the permissible limits or if precise control of fluid temperature is required. Figure 5.1 shows symbols for hydraulic heat exchangers according to DIN ISO 1219 standard. As shown in the figure, a variety of heat exchanger designs are available Some of these designs can only cool or heat the fluid. Other designs can do both heating and cooling. Process of cooling can be water-based or air-based. As shown in the figure, an oil cooler can be equipped by an integrated temperature *Bypass Valve*. The bypass valve leaves the bypass channel open so that a part of the fluid bypasses the oil cooler. Only when the fluid temperature reaches the required value, the valve closes the bypass channel and the fluid is cooled down. So, operation at low temperature and cold starts are avoided.

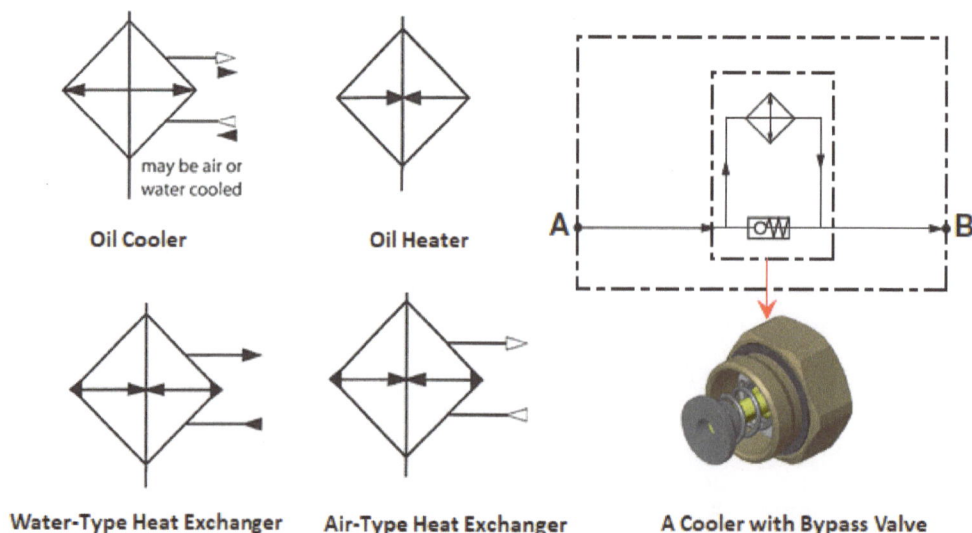

Oil Cooler   Oil Heater

Water-Type Heat Exchanger   Air-Type Heat Exchanger   A Cooler with Bypass Valve

Fig. 5.1- Symbols for Hydraulic Heat Exchangers

5.2- Air-Type versus Water-Type Oil Coolers

Heat Removal: Heat is removed from hydraulic systems <u>passively</u> by heat dissipation from the reservoir, transmission line, and other components, or <u>actively</u> by heat transfer using air-type or water-type oil coolers. Referring to Fig. 5.2, both *Air-Type* and *Water-Type* oil coolers are widely used in the industry. The following set of bullets summarizes the main features of both coolers:

Air-Type Oil Cooler   Water-Type Oil Cooler

Fig. 5.2- Air-Type versus Water-Type Oil Coolers

Application: Water-type oil coolers are commonly used in industrial applications where cooling water supply is available. Air-type oil coolers are commonly used for mobile applications where supplying cooling water is difficult.

Efficiency: Water-type oil coolers are more efficient than air-type coolers at higher ambient temperature.

Size: Water-type coolers have higher cooling capacity for same physical size or same cooling capacity for less size. So, it can be said that water-type coolers have a *Cooling Density* higher than that of air-type coolers.

Noise: Water-type oil coolers are quieter than air-type oil cooler. Air-type coolers typical noise level is 60-90 db because of the cooling fan.

Maintenance: Frequent cleaning is required for both coolers, especially when sea water is used (for water-type oil cooler) and when work environment is highly contaminated (for air-type coolers).

Fluid Compatibility: Regardless the type of oil cooler, if aggressive type of oil is used, consult the cooler manufacturer for oil compatibility. For water-type oil coolers, de-mineralized and untreated water may be used without concern. If sea water or chemically treated water is used, consult the manufacturer for water compatibility.

General Heat Transfer Terminologies: Heat transfer could occur by any of the following:
- **Conduction:** Transfer of heat by *Conduction* means transfer of heat between solid bodies.
- **Convection:** Transfer of heat by *Convection* means transfer of heat to or from fluids. A fluid could be liquid or gas.
- **Radiation:** Transfer of heat by *Radiation* means heat is transferred through vacuum in straight lines from hot spot to cold surrounding.

Heat Transfer in Water-Type Oil Coolers (Fig. 5.3): Heat transfer in water-type oil coolers is composed of several mechanisms as follows (assuming cooling water passes in the tubes and hot oil pass in the shell):
1. Convective heat transfer from the hot oil to the outer surface of the tube wall. Turbulent flow and high velocity improve heat transfer rate.
2. Conductive heat transfer through the tube wall. Most heat exchanger tubes are built from copper or aluminum alloys or similar materials that exhibit high thermal conductivity.
3. Convective heat transfer from the inner surface of the tube to the cooling water in the tube. Use of multi-pass flow patterns takes advantage of the fluid velocity and turbulence for increased *coefficient of heat transfer* values.

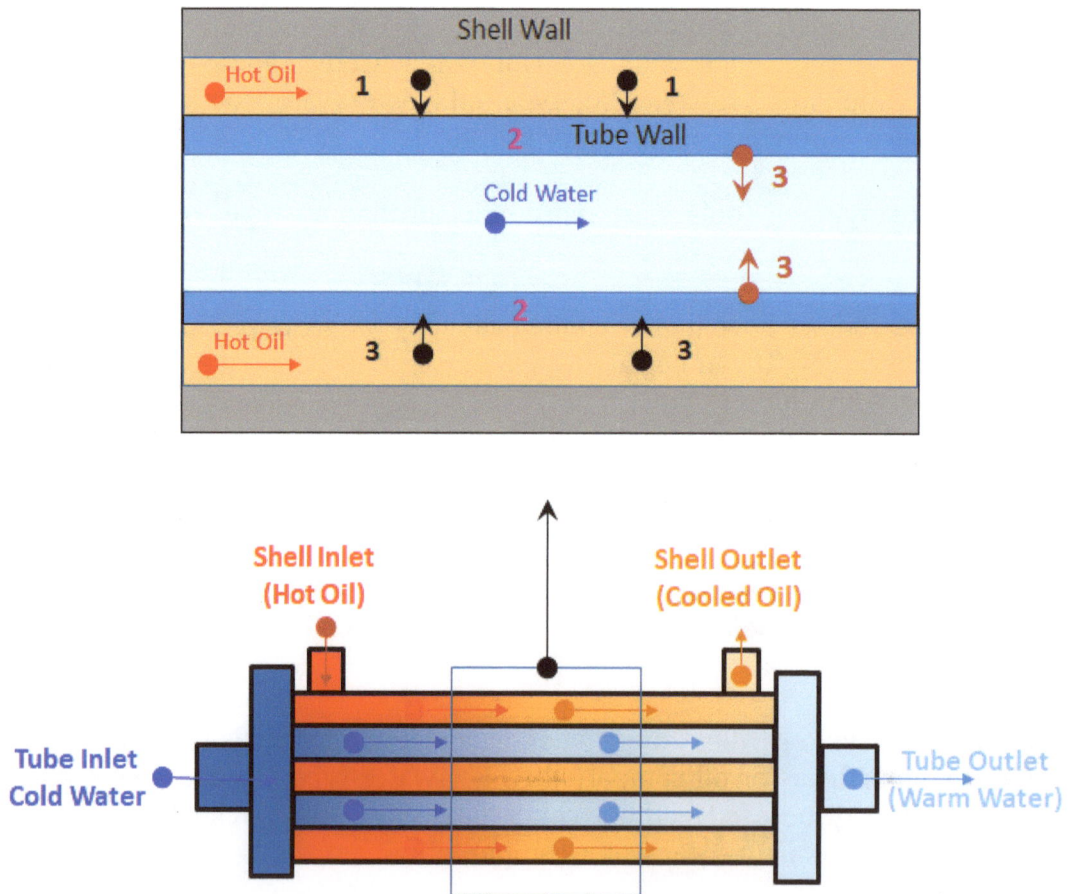

Fig. 5.3 - Heat Transfer Mechanisms in in Water-Type Oil Coolers

Heat Transfer in Air-Type Oil Coolers (Fig. 5.4): Heat transfer in air-type oil coolers is composed of several mechanisms as follows:

1. Convective heat transfer from the hot oil to the inner surface of the tube wall and the fins surrounding the tubes. Turbulent flow and high velocity improve heat transfer.
2. Conductive heat transfer through the tube wall.
3. Convective heat transfer from the outer surfaces of the tubes and fins to the cooling air that is forced by a fan.

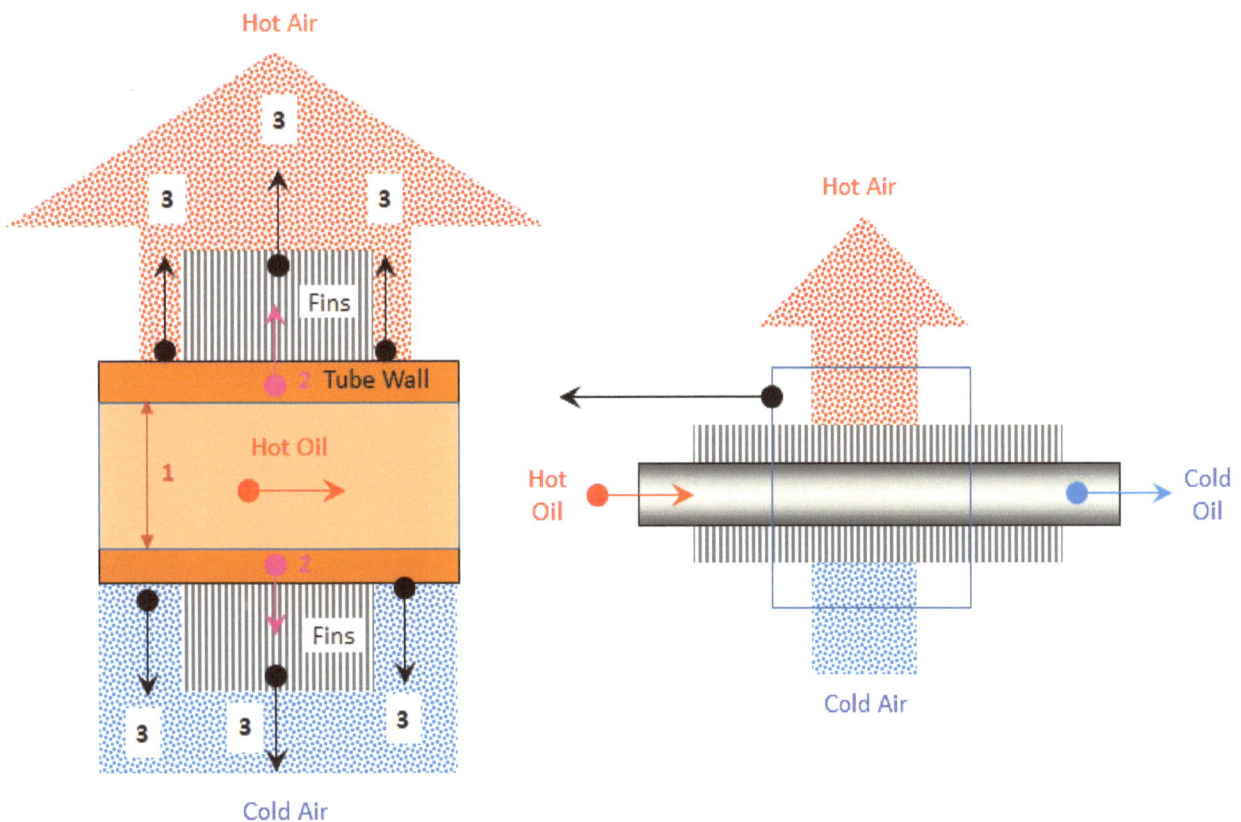

Fig. 5.4 - Heat Transfer Mechanisms in Air-Type Oil Coolers

5.3- Determination of Cooling Capacity for an Oil Cooler

All heat in a hydraulic circuit comes from wasted energy, which is the power put into the hydraulic system that does not convert into useful work. Determination of the *Cooling Capacity* required for a hydraulic system is the step number 1 in the process of sizing and selection of oil cooler. Such cooling capacity is based on the total power loss in the hydraulic system. It can be estimated approximately or calculated precisely as follows:

Approximate Estimation of Cooling Capacity: Hydraulic systems are approximately 75% efficient. Therefore, as a rule of thumb, cooling capacity of a hydraulic system can be estimated as one fourth of the max power required to drive the system in all phases of the machine operation.

Example: A hydraulic driven die casting molding machine has a 5-phases duty cycle. Maximum power required to drive this machine is in phase 4 was found to be 20 HP, then:

Cooling Capacity q = 20/4 = 5 HP = 5 HP x 2,545 = 12725 BTU/HR
Where a rate of 2,545 BTU/HR is equivalent to 1 horsepower.

Precise Calculation of Cooling Capacity: To do so, a detailed thermal analysis must be conducted, and the following *Thermal Balance* equation 5.1 must be solved

Cooling Capacity q = Total Heat Gained – Total Heat Dissipated   **5.1**

Sources of Gaining Heat: In a hydraulic system is discussed thoroughly in Chapter 3 of Volume 3 "Hydraulic Fluids and Contamination Control" of this series of textbooks. However, as a reminder, a hydraulic system gains heat due to internal power losses in hydraulic components and transmission lines. This analysis considers all the spots of heat generation in the system during the different phases of operation. It can also gain heat from external sources.

Sources of Heat Dissipation: A hydraulic system dissipates heat through the surfaces of transmission lines, walls of the reservoir, and surfaces of other components. Only heat dissipated through the reservoir walls can be considered for safe sizing of the cooler.

Example: A thermal analysis of a hydraulic system is conducted and resulted in:
- Fixed displacement pump inefficiency (10 HP) losses.
- Flow control valve with 15 gpm at a pressure drop of 750 psi (losses = 6.5HP).
- Directional control valve with 45 gpm and 100 psi pressure drop (losses 2.5HP).
- Hydraulic tubing with 30 gpm at 30 psi pressure drop (losses 1HP).
- Total Power Losses = 10 + 6.5 + 2.5 + 1 = 20 HP.
- Total Heat Gained = 20 HP x 2,545 = 50,900 BTU/HR.
- Total Heat Dissipated from Reservoir = 5 HP = 5 HP x 2,545 = 12725 BTU/HR.
- Eq. 5.1→ Cooling Capacity q = 50,900 – 12725 = 38,175 BTU/HR.

Oil Cooler Sizing: Eq. 5.2 shows how to find the proper size of an oil cooler for a given cooling capacity.

$$A\ (ft^2) = \frac{q\ (BTU/HR)}{\Delta T(\,^{\circ}F) \times U\left[\frac{BTU/HR}{^{\circ}F \times ft^2}\right]}$$

5.2

Where: $^{\circ}F$

- **A (ft²)** = Heat transfer surface area of the cooler.
- **q (BTU/HR)** = Cooling capacity required by the cooler.
- **ΔT (°F)** = Temperature difference of fluids at the entrance of the cooler.
- **U [(BTU/HR)/(°Fxft²)]** = *Coefficient of heat transfer* of the cooler.

Example: A hydraulic system requires an oil cooler of cooling capacity of 9,000 BTU/HR. The hot oil and cold water enter the cooler at temperature difference of 30 °F. A cooler designed with overall coefficient of heat transfer of 100 [(BTU/HR)/(°F x ft²)].

Eq. 5.2 → A (ft²) = 9000 / (30 x 100) = 3

However, for sizing an oil cooler of a particular design (brand), manufacturers provide selection charts that inherently include the overall heat transfer coefficient.

5.4- Air-Type Oil Coolers

5.4.1- Construction and Operation of Air-Type Oil Coolers

Construction: As shown in Fig. 5.5 the basic components forming an air-type oil cooler are a *Cooling Fan* and a *Radiator*. the radiator (as in vehicular equipment) consists of a bunch of tubes mad from high heat conductive metals with metal fins on its exterior.

Operation: The hot oil is circulated through the radiator that is subjected to a large air flow blown by the fan.

Heat Transfer: Heat is convected from the hot oil to inner surface of tubes wall, then conducted through the walls of the tubes and the fins that surrounds the tubes, and finally convected into a flow of air forced by a fan. Turbulent flow results in better heat transfer. Therefore, in some radiators, devices are placed inside the tubes to generate turbulence in the oil flow to improve the cooling efficiency. On the other hand, more pressure drop is expected in such radiators.

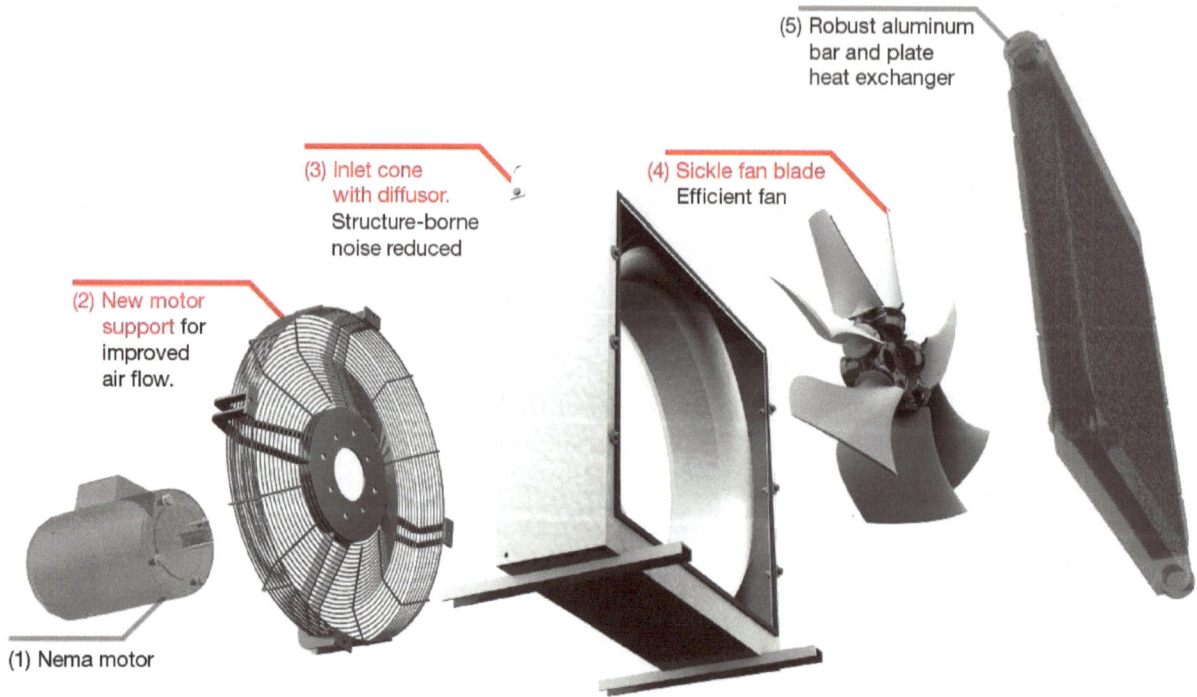

(5) Robust aluminum bar and plate heat exchanger

(3) Inlet cone with diffusor. Structure-borne noise reduced

(4) Sickle fan blade Efficient fan

(2) New motor support for improved air flow.

(1) Nema motor

Fig. 5.5- Construction of Air-Type Oil Coolers (Courtesy of Hydac)

Driving Motor: As shown in Fig. 5.6, the cooling fan can be driven by an AC or a DC electric motor. It can also be driven by a hydraulic motor.

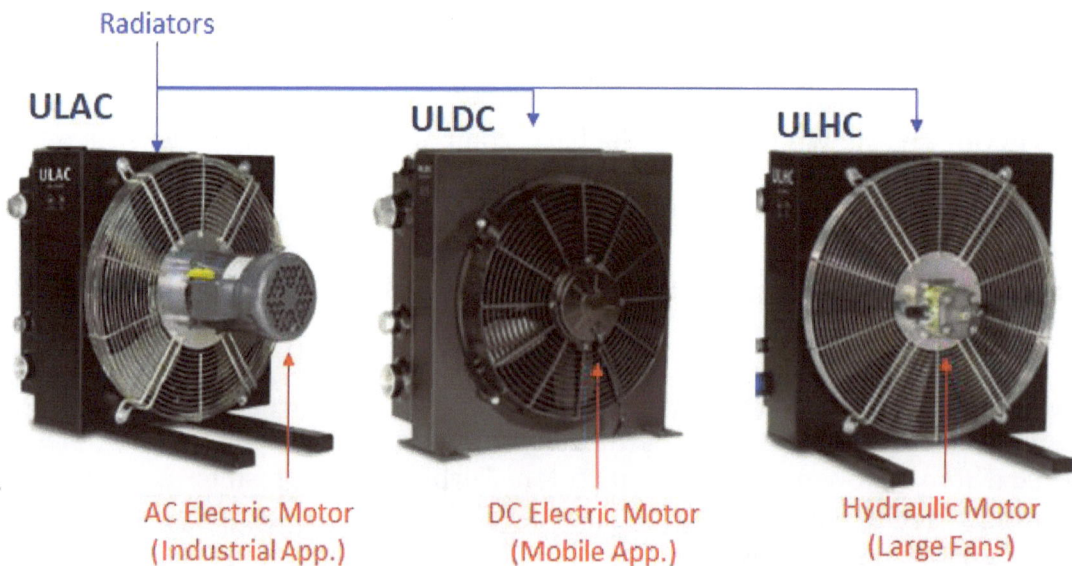

Radiators

ULAC   ULDC   ULHC

AC Electric Motor (Industrial App.)   DC Electric Motor (Mobile App.)   Hydraulic Motor (Large Fans)

Fig. 5.6- Driving Motors for Air-Type Oil Coolers (Courtesy of Hydac)

5.4.2- Sizing of Air-Type Oil Cooler

Air-type oil coolers are available in wide range of sizes to meet various cooling capacities. Selected size must meet the cooling capacity of the system. As shown in Fig. 5.7, cooling capacity of a specific size of air cooler depends on the oil flow through the cooler and the temperature difference between the ambient air and the oil. It is to be noted that the cooling performance curves shown in the figure are normalized per $1^{o}F$ temperature difference. As shown in Fig. 5.8, after selecting a proper size of a heat exchanger, the second important specification to determine is the pressure drop across the heat exchanger considering correction for oil viscosity.

Example:
- A hydraulic system requires cooling capacity q = 30,000 BTU/HR.
- Entering Temperature Difference (EDT) between oil and ambient air is 30 ^{o}F.
- Normalized cooling capacity q = 30,000 (BTU/HR) / 30 ^{o}F = 1000 (BTU/HR)/ ^{o}F.
- Oil Flow through the cooler is 40 gpm.
- **Size:** By locating the operating point on the performance curve, the best size is the cooler with part number "ULAC-016B".
- The cooler is a bit larger than needed, but it is ok since the curve reported a cooling tolerance in the order of 10%.
- **Pressure Drop:** Oil flow = 40 gpm → pressure drop = 18 psi.
- Pressure Drop Correction (based on oil Viscosity = 250 SSU) = 18 x 1.5 = 27 psi.

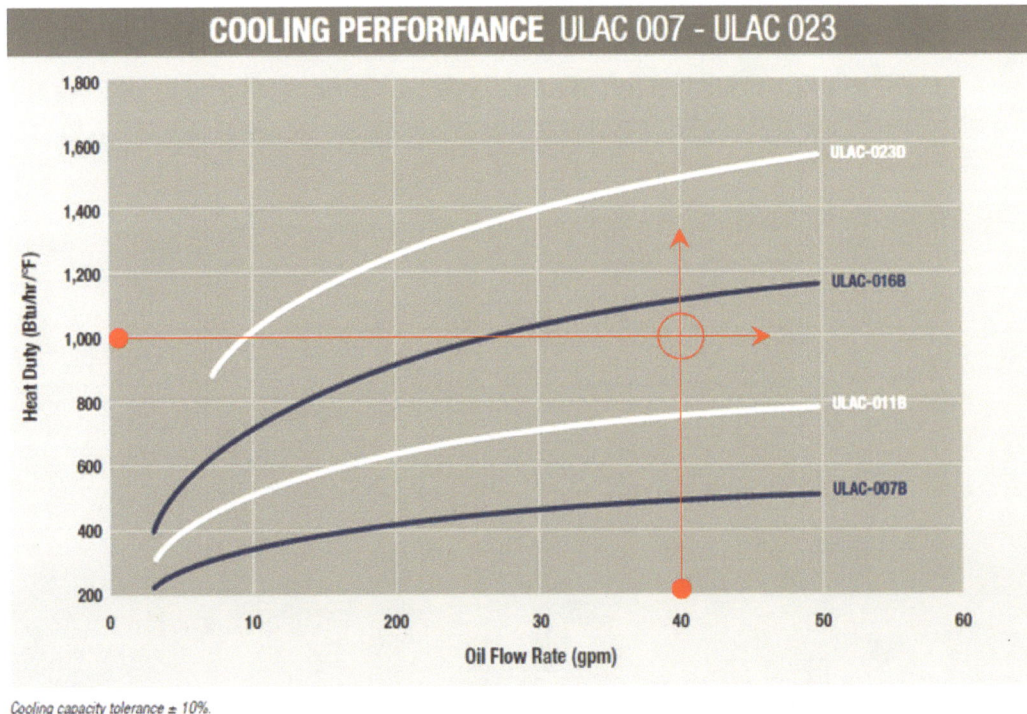

Cooling capacity tolerance ± 10%.

Fig. 5.7 - Cooling Performance of Air-Type Oil Coolers (Courtesy of Parker)

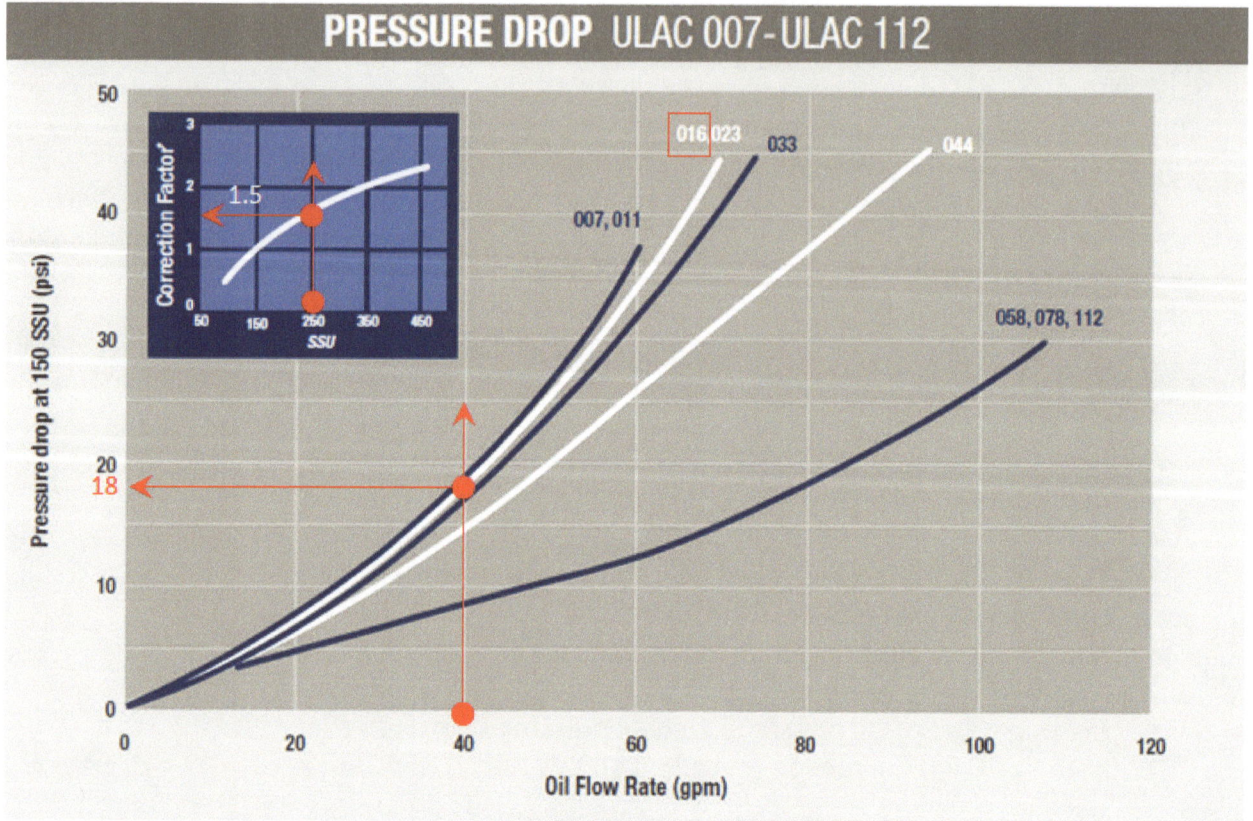

Fig. 5.8- Pressure Drop Across Air-Type Oil Coolers (Courtesy of Parker)

5.5- Shell-and-Tube Water-Type Oil Coolers

5.5.1- Construction and Operation of Shell-and-Tube Water-Type Oil Coolers

Construction (Multi-Pass & Counter-Flow): As shown in Fig. 5.9, *Shell-and-tube Water-Type Oil Coolers* are constructed of a bundle of tubes enclosed in a metal shell. Tubes are made of material that has high thermal conductivity.

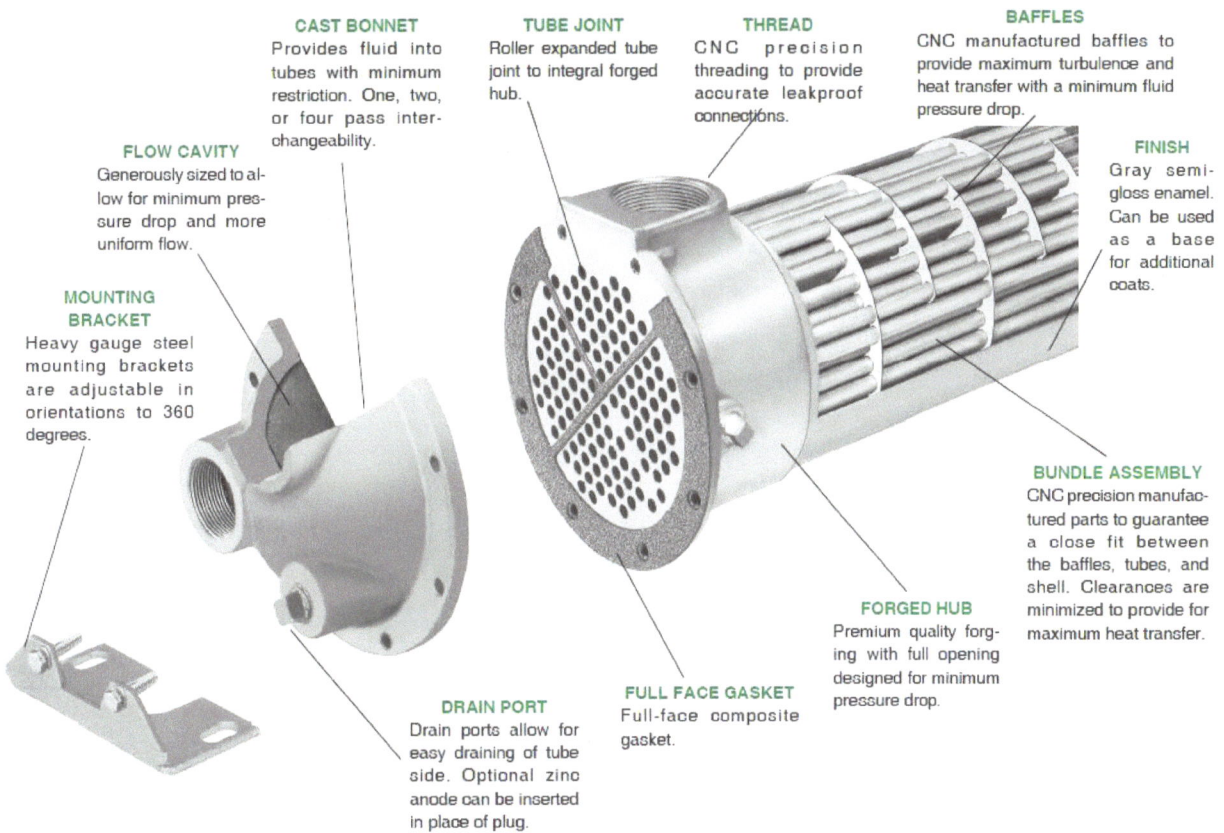

CAST BONNET
Provides fluid into tubes with minimum restriction. One, two, or four pass inter-changeability.

TUBE JOINT
Roller expanded tube joint to integral forged hub.

THREAD
CNC precision threading to provide accurate leakproof connections.

BAFFLES
CNC manufactured baffles to provide maximum turbulence and heat transfer with a minimum fluid pressure drop.

FLOW CAVITY
Generously sized to allow for minimum pressure drop and more uniform flow.

FINISH
Gray semi-gloss enamel. Can be used as a base for additional coats.

MOUNTING BRACKET
Heavy gauge steel mounting brackets are adjustable in orientations to 360 degrees.

BUNDLE ASSEMBLY
CNC precision manufactured parts to guarantee a close fit between the baffles, tubes, and shell. Clearances are minimized to provide for maximum heat transfer.

FORGED HUB
Premium quality forging with full opening designed for minimum pressure drop.

DRAIN PORT
Drain ports allow for easy draining of tube side. Optional zinc anode can be inserted in place of plug.

FULL FACE GASKET
Full-face composite gasket.

Fig. 5.9- Construction of Shell-and-Tube Oil Coolers (www.SouthwestThermal.com)

Baffles: As shown in the figure, baffle plates are used to generate turbulence in oil flow and elongate the contact time between the oil and tube metal aiming to improve the heat transfer.

Drain: There should be couple of drain plugs for draining both circuits of heat exchangers.

Leakage: shell and tube water cooled heat exchanger may leak water into the oil if water pressure is higher than oil pressure. It will leak oil into the water if oil pressure is higher.

Operation: As shown in Fig. 5.10, hot oil flows through the shell and cooling water flows through the tube because:

- Having hot oil in the shell promote heat dissipation from the shell to surroundings.
- If water causes rust, it will be kept inside the tube without contaminating the oil.
- Removing the tube bundle outside the shell and cleaning inside the tubes is easier.
- Shell side can tolerate a higher oil pressure that may result due to surges in tank line flow or back pressure in the return line (over 100 PSI).
- Shell is of less volume change due to hot oil.
- Avoid galvanic corrosion between the shell metal and the tube metal if non-desalinated cooling water is flowing in the shell.
- If cold water flows through the shell, condensation or humidity could occur on the outer surface of the shell that may cause water ingression into the system.

Heat Transfer: Heat is convected from the hot oil to the tube walls, then conducted through the tube wall, and finally convected to the cooling water in the tube.

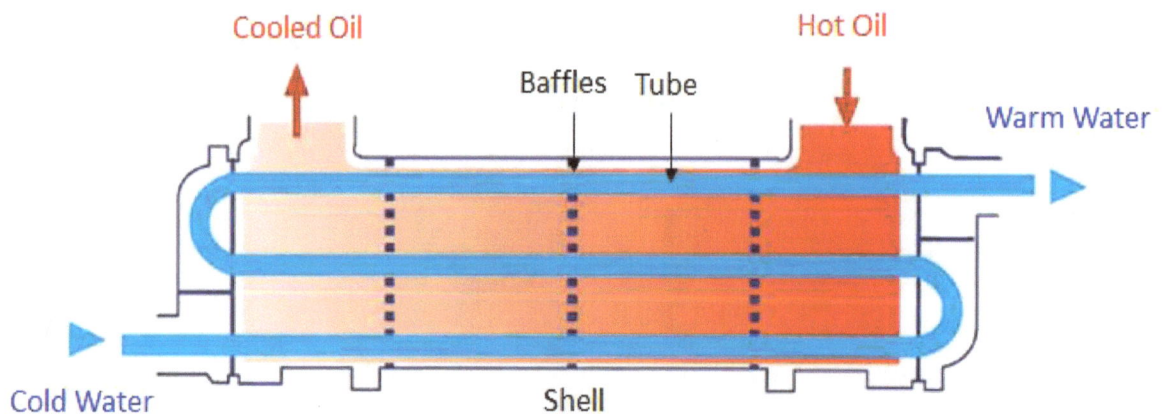

Fig. 5.10- Basic Construction of Shell-and-Tube Oil Coolers

Oil-Water Flow: The following rules of thumb shows better efficiency:

- Rule-of-thumb 1: Oil flows at a rate of 1 m/s (3 feet/s) or less.
- Rule-of-thumb 2: Water flows approximately at a rate equal one-half of the oil flow rate. This means the oil/water flow ratio is 2:1.

Galvanic Corrosion: In situation where sea water is used as cooling water (such as in marine & offshore applications) and it flows in the shell (not in the tube), a phenomenon called *Galvanic Corrosion* occur. When sea water flows between two different metals, conditions are created whereby they effectively become a battery. This means that an electrical current flow between the two metals. The result is that electrons, which make up the current, are supplied from one of the two metals, which in effect loses small particles of itself in the form of metal ions to the sea water. This creates something known as 'galvanic corrosion' which if ignored can damage the metals and ultimately cause them to fail. As shown in Fig. 5.11, *Zinc Anodes* work by introducing a third metal into the circuit that gives up or releases its electrons faster than the other metals in the heat exchanger. So, in effect, it sacrifices itself to protect the other metals in the heat exchanger. As shown in Fig. 5.12, Because zinc anodes wear out over time, they must be routinely inspected and replaced, if necessary, during heat exchanger maintenance.

Drain Plug   Zinc Anod   New
                                         Zinc Anod   Corroded
                                                            Zinc Anod

**Fig. 5.11- Zinc Anodes to Prevent Galvanic Corrosion
(https://ej-bowman.com/knowledge-centre/zinc-anodes)**

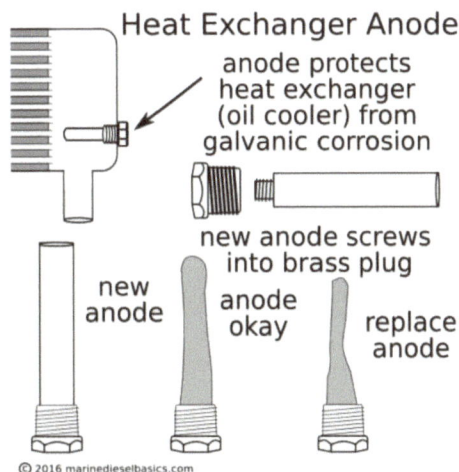

Heat Exchanger Anode

anode protects
heat exchanger
(oil cooler) from
galvanic corrosion

new anode screws
into brass plug

new
anode   anode
               okay   replace
                             anode

© 2016 marinedieselbasics.com

Fig. 5.12- Zinc Anodes Inspection (www.marinedieselbasics.com)

Number of Passes: Shell-and-tube oil coolers are configured in single-Pass, double-Pass, and Four-Pass designs. The term "Pass" here indicates the number of times the cooling water travels through the length of the cooler after it entered from the inlet port and before it is finally discharged from the outlet port.

- Single-Pass Shell and Tube Oil Coolers: As shown in Fig. 5.13, in a *Single-Pass* water-type heat exchangers, cooling water passes between the inlet and outlet ports without any extra looping inside the heat exchanger.

Fig. 5.13- Single-Pass Shell and Tube Oil Coolers

- Multi-Pass Shell and Tube Oil Coolers: As shown in Fig. 5.14, the tube bundle can form two or four passes. In such cases, full parallel or counter flow is not possible. Increasing the number of passes through an exchanger increases the amount of heat transfer.

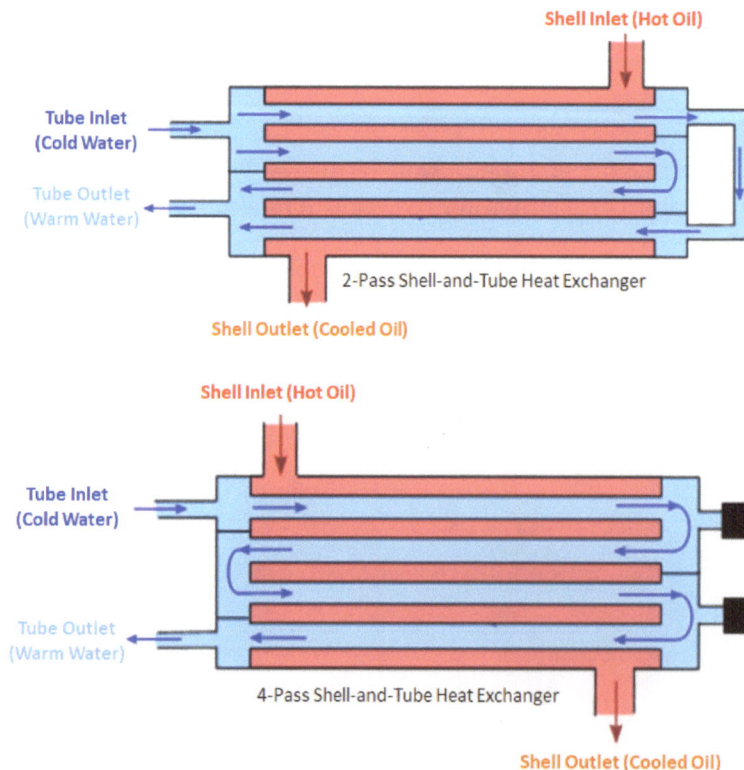

Fig. 5.14- Multi-Pass Shell and Tube Oil Coolers

As shown in Fig. 5.15, in relation to the number of passes, location of fluid ports is different.

FOUR
PASS

TWO
PASS

SINGLE
PASS

Fig. 5.15- Fluid Ports in Relation with the Number of Passes

Flow Direction: In relation to the direction of oil and water flow, it may be either parallel-flow or counter-flow as follows:

1. _Parallel-Flow_ Shell-and-Tube Oil Coolers: as shown in Fig. 5.16, hot oil and cooling water flow in the same direction. On one side of the cooler, the inlet cold water will meet the inlet hot oil. Both fluids will continue to flow beside each other along the whole length of the tube. Therefore, the temperature difference between the tube and shell will be highest at the inlet and is reduced along the tube towards the outlet. That helps cooling the oil <u>rapidly</u> but results in nonhomogeneous thermal stress on the cooler parts, which may reduce the cooler reliability.

2. _Counter-Flow_ Shell-and-Tube Oil Coolers: As shown in the figure, hot oil and cooling water flow oppositely. On one side of the cooler, the inlet cold water will meet the outlet cold oil. On the other side, the outlet hot water will meet the inlet hot oil. Therefore, the pipes on either side subject to almost constant temperature difference. That results in homogeneous thermal stress on the cooler parts, which improves the <u>reliability</u> and heat transfer <u>efficiency</u> of the cooler.

3. _Cross-Flow_ Shell-and-Tube Oil Coolers: As shown in Fig. 5.17, hot oil flows through the shell perpendicularly on the tube. Time for the hot oil to dissipate heat is less and so the heat transfer efficiency is less.

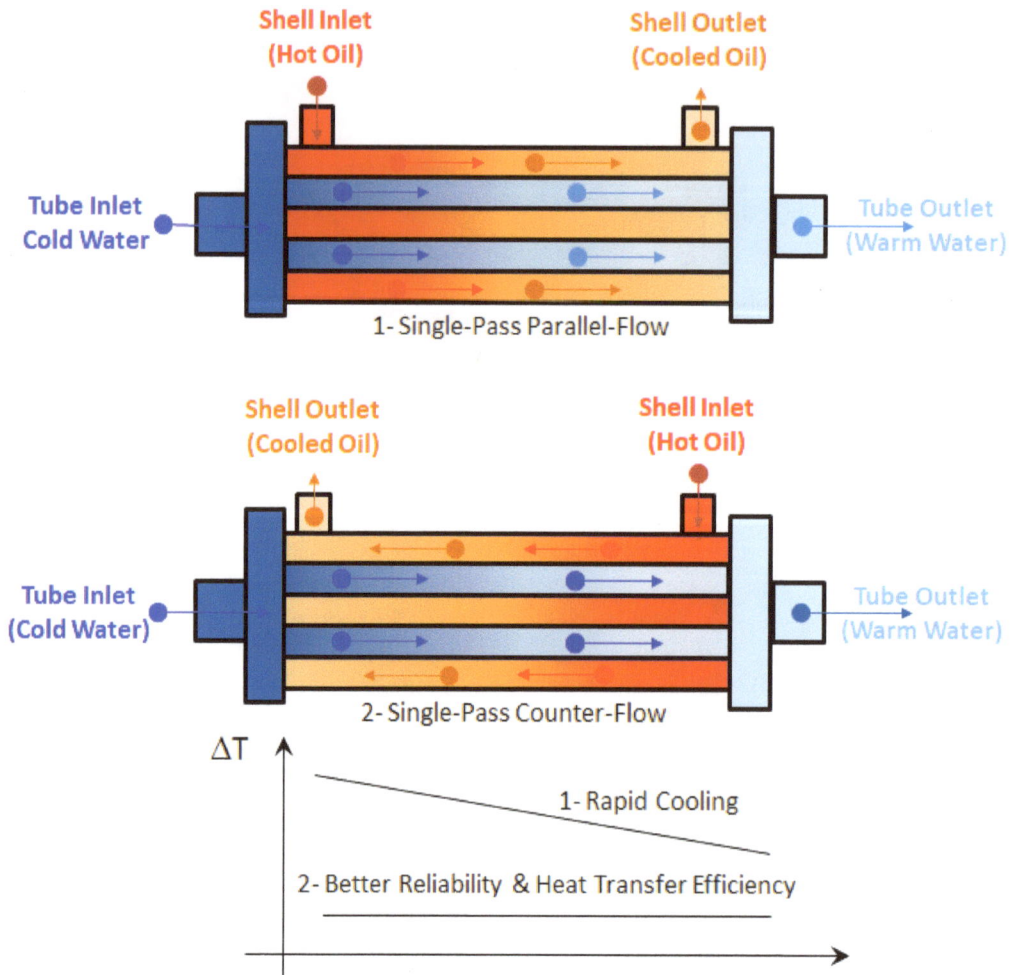

Fig. 5.16- Parallel-Flow versus Counter-Flow in Single-Pass Shell-and-Tube Oil Coolers

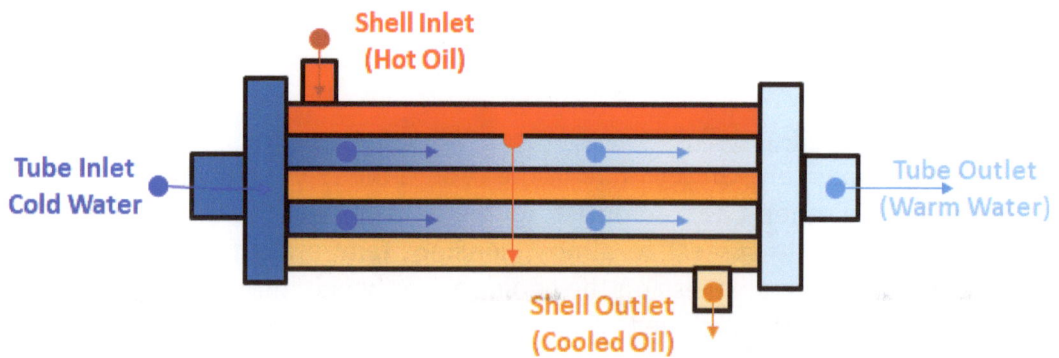

Fig. 5.17- Cross-Flow in Single-Pass Shell-and-Tube Oil Coolers

5.5.2- Oil/Water Safety Heat Exchanger

Oil/Water *Safety Heat Exchanger* have been developed to prevent the danger of mixing the cooling water to oil. As shown in Fig. 5.18, it consists of two coaxial tubes, the outer tube is filled with a sealing liquid surrounding the inner tube through which the cooling water is flowing. A pre-charged diaphragm accumulator compensates the thermal change in volume of the sealing liquid and keeps the sealing liquid pressure almost constant. Any leaks in the bank of tubes are indicated at once by the pressure switches. A special liquid enables high heat transfer.

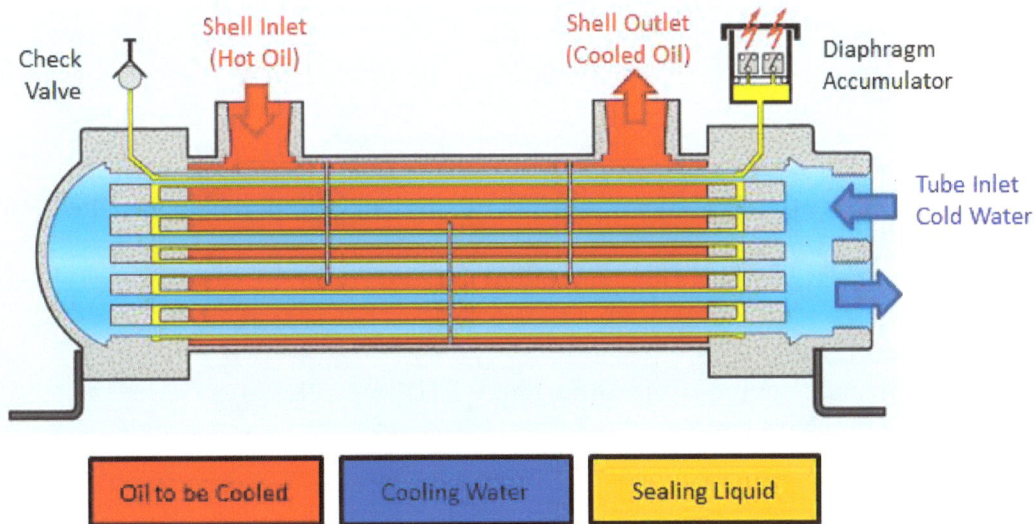

Fig. 5.18- Oil/Water Safety Heat Exchanger (universalhydraulik.com)

5.5.3- Sizing of Shell-and-Tube Water-Type Oil Coolers

As shown in Fig. 5.19, Shell and Tube heat exchangers are robust and reliable and come in a range of sizes; from large custom-built units to small off-the-shelf units.

Fig. 5.19- Custom and Off-the-Shelf Shell and Tube Heat Exchangers (www.fluiddynamics.com)

As shown in Fig. 5.20, performance curves for shell-and-tube heat exchangers are developed under the shown below specific conditions. If a typical application has different conditions, correction must be made based on correction factors provided by the manufacturer.
- Oil Viscosity = 100 SUS.
- Oil/Water Flow Ratio = 1:1 or 2:1 to match the curve results.
- Design Entrance Temperature Difference (DETD) = 40 °F.
- Oil Pressure Drop Coding is as: + = 5 psi, ☆= 10 psi, O = 20 psi, Δ= 50 psi.

Example:

Given Data:
- Required Cooling Capacity q_r = 19.8 hp as calculated based on the system energy analysis.
- Oil Viscosity is 200 SUS (for equivalent ISO VG, refer to conversion table provided in Volume 3).
- Oil Flow = 60 gpm.
- Actual Entrance Temperature Difference (ETD) = 30 °F
- Allowable Pressure Drop = 10 psi.

Step 1 (Cooling Capacity and Pressure Drop Correction Factors):

The figure shows that, for the given oil viscosity (200 SUS), the pressure drop correction factor from Column A (Kp = 1.3) and cooling capacity correction factor from Column B (Kv = 1.14).

Step 2 (Calculation of the Design Cooling Capacity q_d):

Equation 5.3 shows the design cooling capacity that considers corrections based on which the cooler size will be selected.

Design Cooling Capacity q_d = qr x (DETD/EDT) x Kv   **5.3**

$\rightarrow$ **q_d =** 19.8 x (40/30) x 1.14 = 30 hp

Step 3 (Selecting Proper Size Cooler based on Design Cooling Capacity q_d):

The figure shows that the proper size cooler for the calculated design cooling capacity (30hp) and given oil flow rate (60 gpm) is Model size 9 that has design pressure drop (legend +) of **DPoil** = 5 psi.

Step 4 (Check for the Actual Pressure Drop):

Equation 5.4 shows the how to calculate the actual pressure drop based on the pressure drop correction factor.

Actual Pressure Drop Δp = DPoil x Kp   **5.4**

$\rightarrow$ **Δp** = 5 x 1.3 = 6.5 psi

Since the actual pressure drop is less than the given allowable pressure drop (10 psi), then the selected size is ok. Otherwise, the next size (model 12) will be selected.

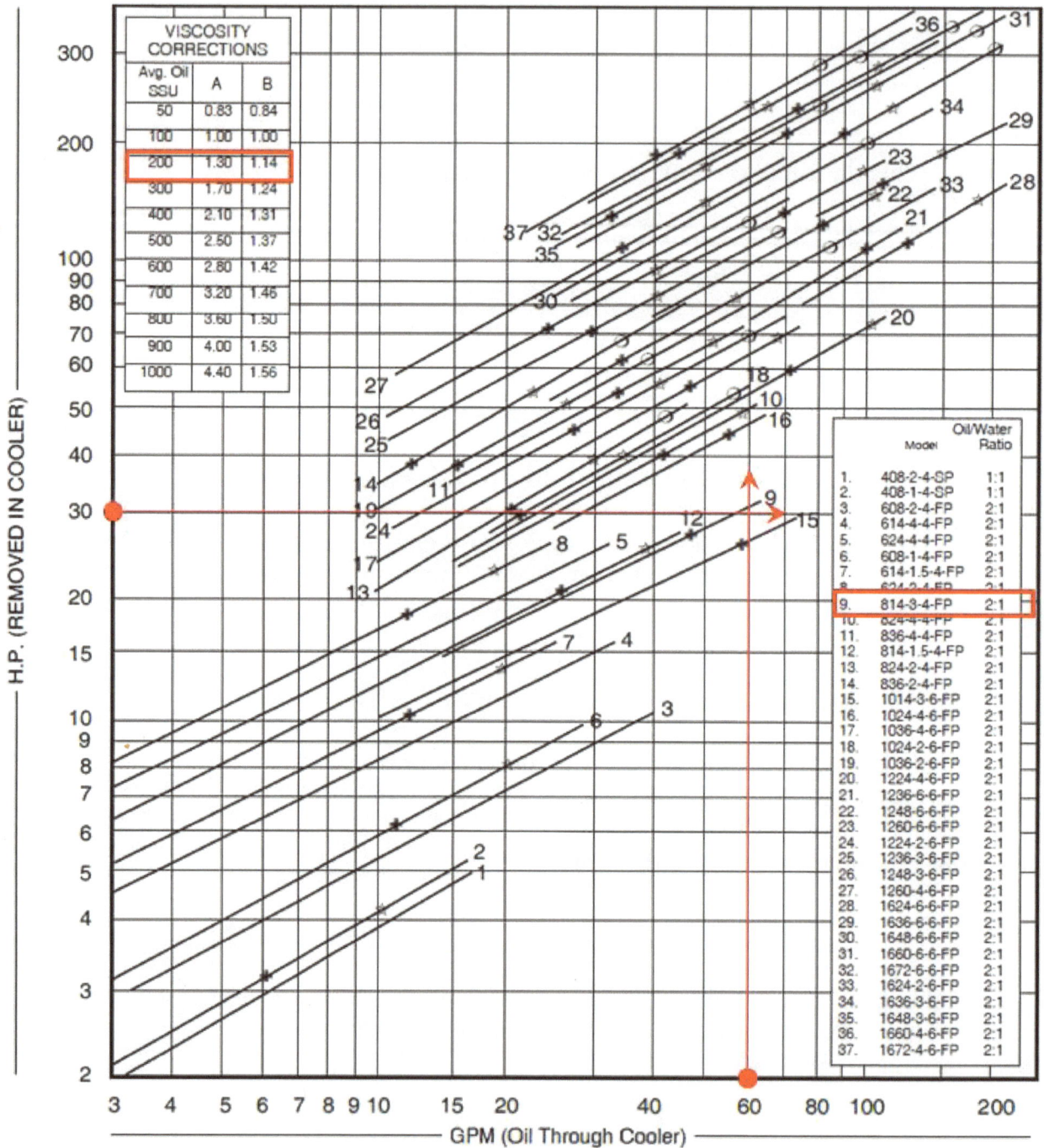

Fig. 5.20- Performance Curves of Shell-and-Tube (www.SouthwestThermal.com)

5.6- Plat-Type Oil Coolers

Plate type heat exchangers consist of a stack of stamped heat exchange plates which are either brazed together or bolted together in a frame with gaskets.

5.6.1- Brazed Plate-Type Oil Coolers

Construction: As shown in Fig. 5.21, a *Brazed Plate* oil cooler consists of a stack of stamped heat transfer stainless steel plates. The plates have smoothed edges and the end plate is provided with edge protection. The plates are brazed with spacers of copper or nickel. The whole assembly is brazed in a high-temperature oven.

Operation: Heat is transferred from the hot oil to the cold water through the heat transfer plates. The special stamp pattern of the plates induces turbulent flow which is necessary for optimum heat transfer and which in addition has a self-cleaning effect because the high level of wall friction reduces deposits on the surface. Figure 5.22 shows a cross-section in a plate-type oil cooler.

Direction of Flow: Concepts of parallel-flow and counter-flow are applicable in this type of oil coolers. It is possible to reverse the operation from parallel-flow to counter-flow without adversely affecting the function. However, changing the cold and hot side is not recommended. The plate heat exchangers are designed to have hot oil always flowing through the outer plate.

If cooling water flows through the outer plate, condensation could occur on the outside of the heat exchanger that increases possibility of water ingression into the system. Inlets and outlets are clearly identified by stamps or stickers affixed to the front of the unit. If no marks are shown, pour water in one port, the water outlet is where the water comes out of.

Features and Applications: Such type of coolers is particularly compact, lightweight, have smaller footprint, and efficient which makes it more suitable for mobile applications. On the other hand, it has limited fluid capacity, smaller heat transfer surface area, and therefore less cooling capacity.

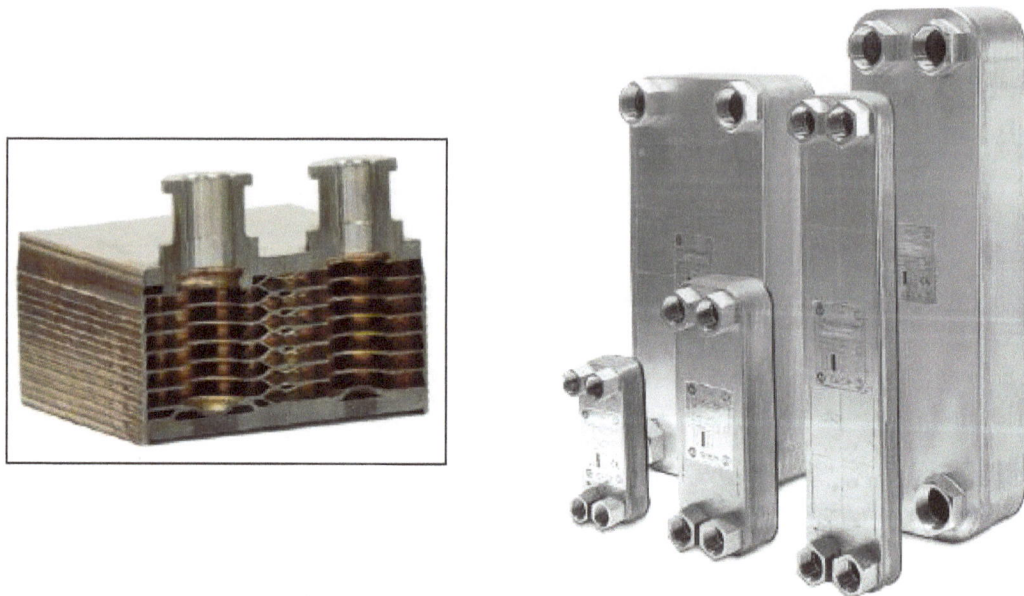

Fig. 5.21- Construction of Brazed Plate-Type Oil Cooler (Courtesy of Hydac)

Extremely Compact:
85-90% Reduction in volume and weight of a shell-and-tube heat exchanger of the same capacity.

LOW WATER CONSUMPTION. ECONOMICAL OPERATION COMPACT.

Corrugated:
Plates made of 316 stainless steel brazed with pure copper.

TURBULENT WATER FLOW PREVENTS CLOGGING AND REDUCES MAINTENANCE. SMALLER SIZE MAKES IT EASY TO INSTALL.

SAE O-Ring Connections:
Good for ease of assembly and leak proof operation.

Maximum Efficiency:
Maximum material efficiency. No "Dead Zone" because there is no need for gaskets. Up to 25% more capacity utilization.

Fig. 5.22- Cross-Section of a of Brazed Plate-Type Oil Coolers (Courtesy of Parker)

5.6.2- Gasketed Plate-Type Oil Coolers

Construction: As shown in Fig. 5.23, a *Gasketed Plate-Type* oil cooler consists of:

1- A stack of corrugated steel plates. Other materials can be used to make such a cooler suitable for different applications, e.g. for applications with seawater, titanium heat transfer plates are used.
2- Gaskets for consecutive plates "A" and "B".
3- Front fixed end plate (cover).
4- Rear moveable end plate (cover). It moves backward for disassembling process.
5- Guide and carrying bars.
6- Fluid connections.
7- Clamping bolts.

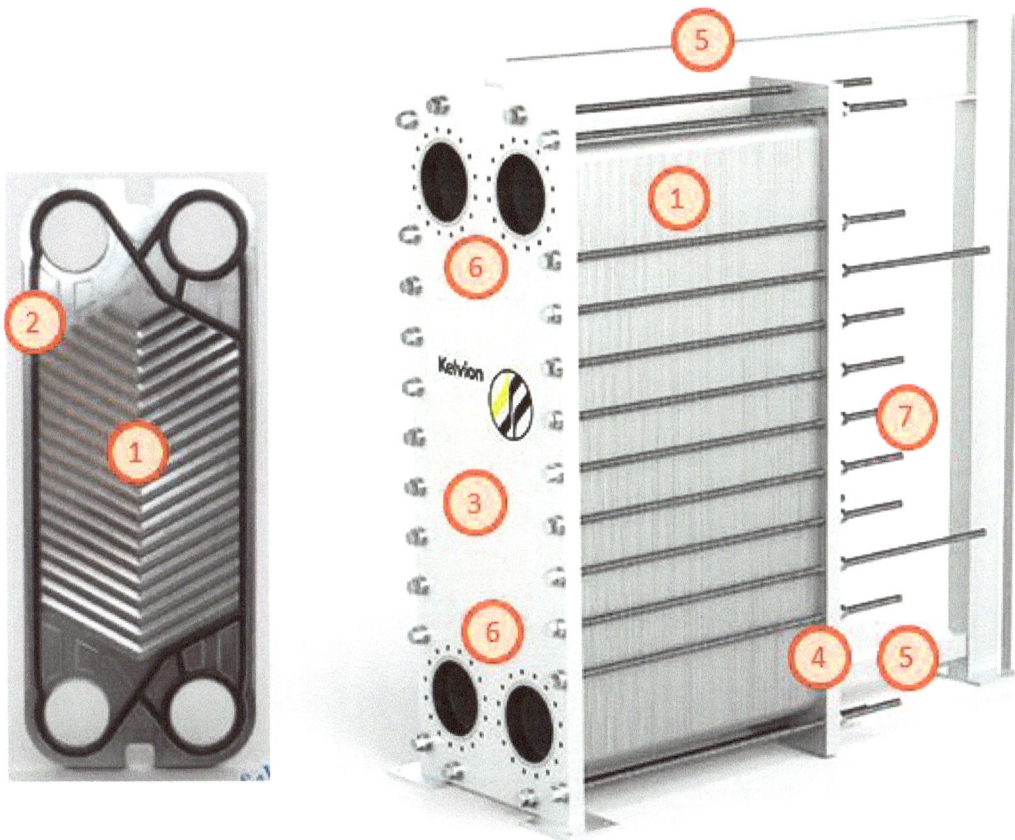

Fig. 5.23- Construction of Gasketed Plate-Type Heat Exchangers

Operation: As shown in Fig. 5.24, when the plates are assembled, they are perfectly aligned. When they are bolted, gaskets are squeezed sealing the clearances between plates and forming the flow channels. Gaskets are designed to force the hot oil to flow in front of plates marked "A" and cooling water are forced to flow in front of the plates marked "B". For correct operation of the coolers, plates must be installed in the correct order and so the gaskets. Plates are marked for better identification. Obviously, cooling capacity is proportional to the number and size of the plates. However, the number of plates must always be an odd number in order to have the hot oil adjacent to the end plates. That helps dissipating heat from end plates to the surrounding environment and avoids condensing humidity on the end plates if cooling water is flowing adjacently to the end plates.

Fig. 5.24- Operation of Gasketed Plate-Type Heat Exchangers (savree.com)

Plate Design: As shown in Fig.5.25, plates are corrugated in a pattern that generates turbulent flow so that the heat transfer efficiency is improved. Corrugation also strengthens the plate so than thinner plates can be used.

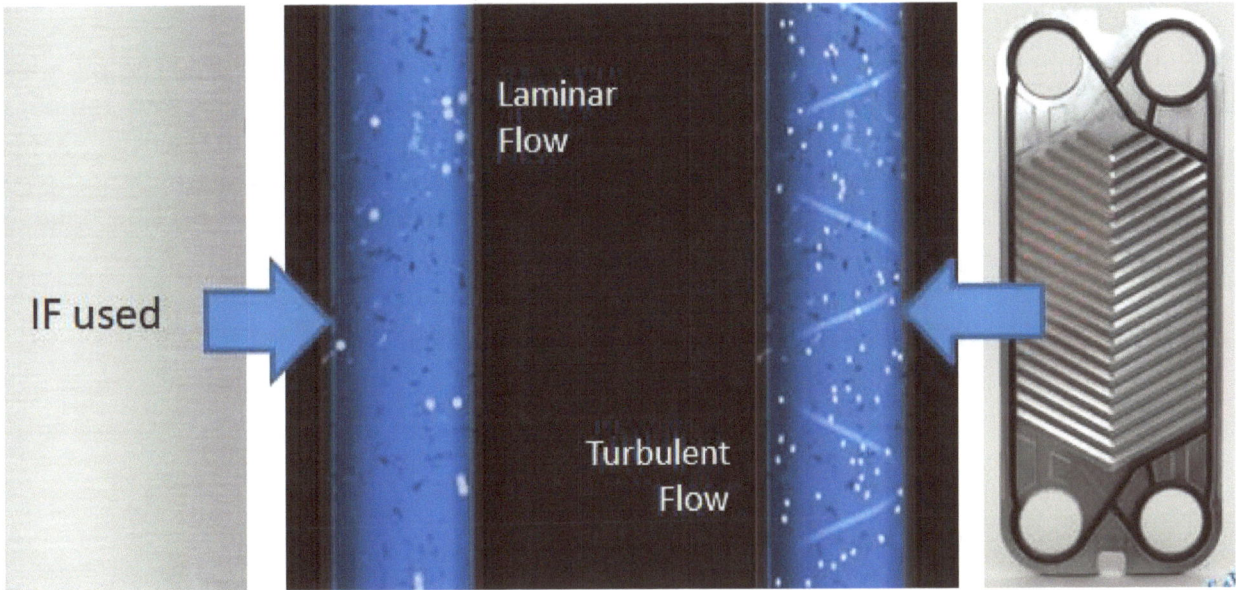

Fig. 5.25- Plate Design in Gasketed Plate-Type Heat Exchangers (savree.com)

Gasket Design: As shown in Fig.5.26, for correct operation of the coolers, correct gasket must be used for a plate. To guarantee the correct fluid flow direction and so the maximum heat transfer efficiency, different gasket templates are used for the hot oil plates "A", cooling water plates "B", and the end plates. Gaskets should be routinely inspected for leakage. Leaking gaskets reduces the heat transfer efficiency and, more important, allows mixing water with oil.

Fig. 5.26- Gasket Design in Gasketed Plate-Type Heat Exchangers (savree.com)

Flow Direction: As shown in Fig. 5.27, the concepts of parallel-flow and counter-flow are applicable in this type of oil coolers with the same considerations mentioned for shell-and-tube oil coolers. In parallel-flow, oil inlet will be beside the water inlet. In counter-flow, oil inlet should be located on the diagonal from the water inlet.

Better cooling efficiency is achieved by counter flowing through the plates. The oil inlet and water inlet being located on a diagonal. As show in the figure, in parallel-flow, temperature difference between the two fluids are high at the entrance and is reduced over the pass of the fluids. In counter-flow, temperature difference is almost constant along the path of the passes of the fluids. Hence, less thermal stresses on the plates and better reliability of the cooler.

Unlike brazed plat-type of coolers, it is not possible to change the direction of flow because of the more complex plate stamp pattern. Consult manufacturer to determine the connections before selecting the cooler.

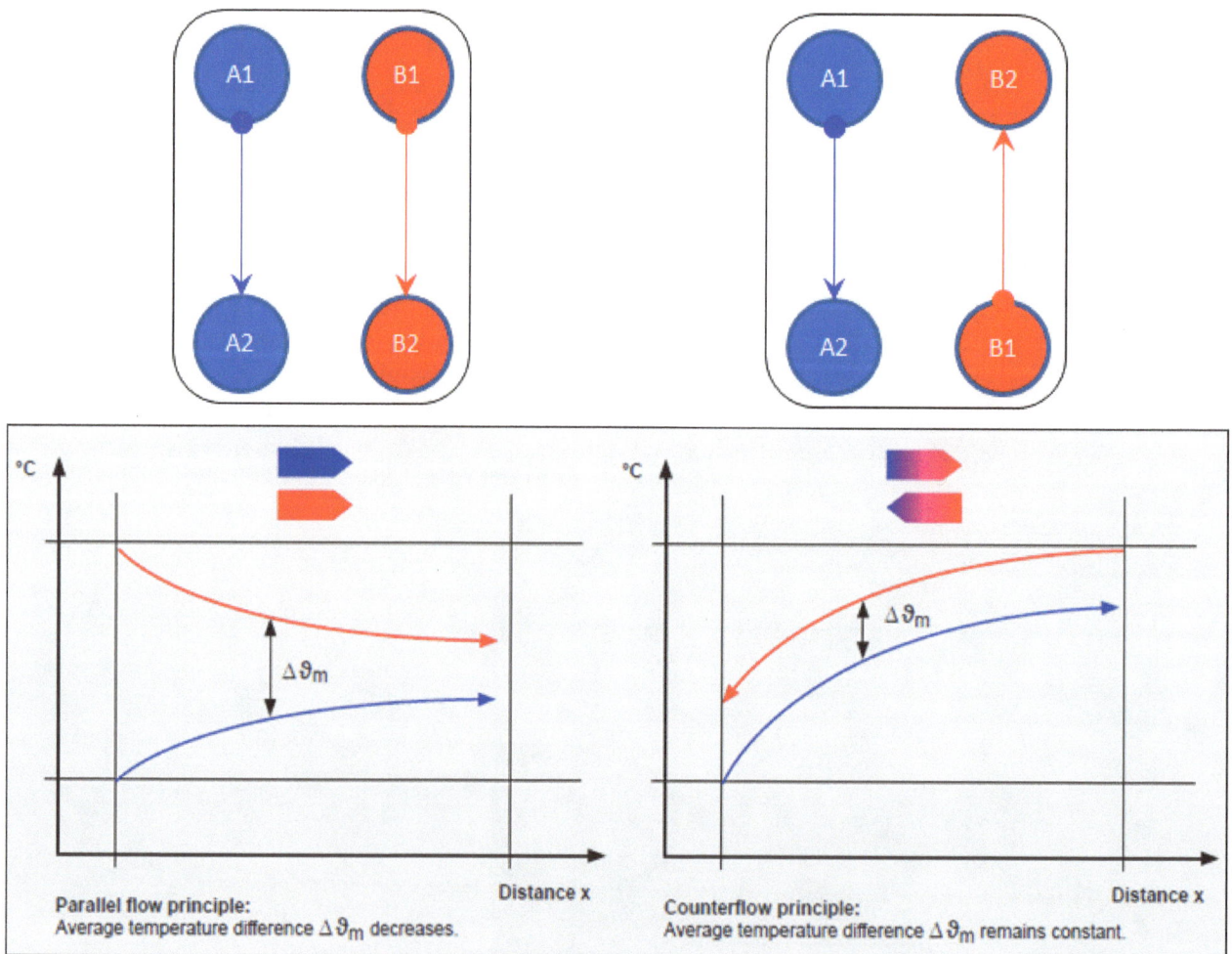

Fig. 5.27- Parallel-Flow versus Counter-Flow in Plate-Type Oil Coolers (Courtesy of Hydac)

Heat Transfer: As shown in Fig. 5.28, heat transfer in counter-flow plate-type oil coolers is composed of several mechanisms as follows:

1. <u>Convective</u> heat transfer from the hot oil to the adjacent plates.
2. <u>Conductive</u> heat transfer through the wall of the plates.
3. <u>Convective</u> heat transfer from the plates to the cooling water.

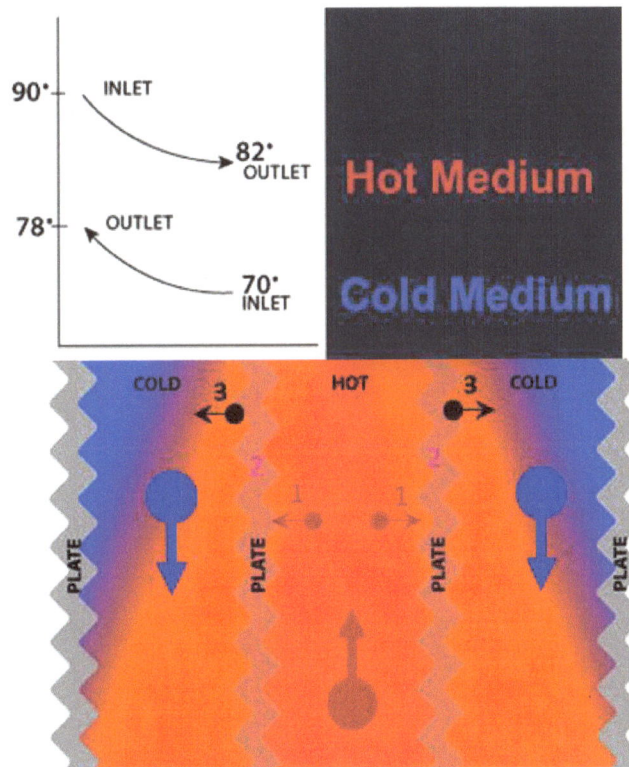

Fig. 5.28- Counter-Flow in Plate-Type Oil Coolers (savree.com)

Number of Passes: As shown in Fig. 5.29, the concepts of *single-pass* and *multi-pass* are applicable in this type of oil coolers.

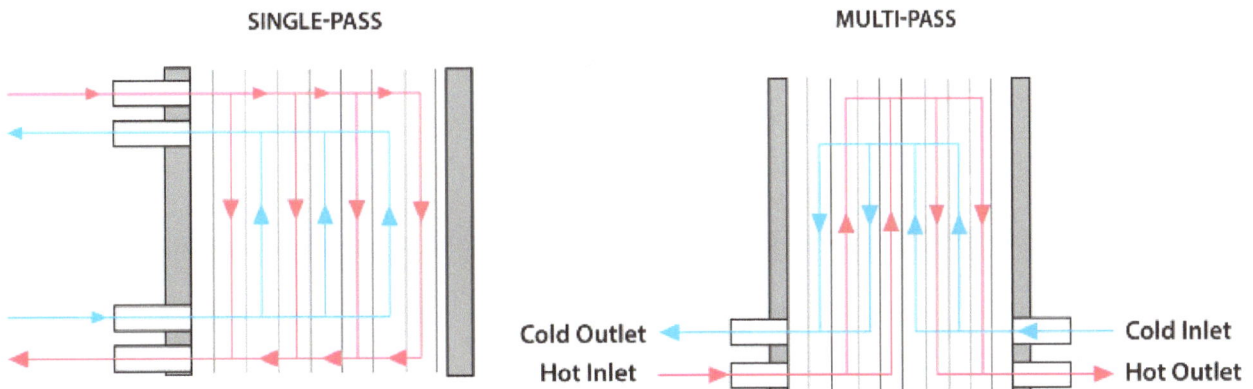

Fig. 5.29- Single-Pass and Multi-Pass in Plate-Type Oil Coolers (Courtesy of Hydac)

Advantages of Gasketed Plate Heat Exchangers:
- They are available in wide range of sizes anywhere from small to quite large sizes. However, they are more suitable for large flows.
- Have high cooling capacities.
- The stack of heat transfer plates and gaskets are clamped together with bolts in a frame. This means that the plate heat exchanger can also be disassembled for cleaning and maintenance.
- For applications with seawater, the heat transfer plates can be made of titanium.

Disadvantages of Gasketed Plate Heat Exchangers:
- More expensive.
- Leak is difficult to locate.
- Replacing gaskets is problematic.
- Plate clearances are easily blocked by contamination.
- Overtightening due to human error squeezes the gasket and reduces the gap between the plates. Hence, reduces the flow between the plates, reduces the heat transfer efficiency, and increase the pressure drop.

5.6.3- Sizing of Plate-Type Oil Coolers

Performance curves for plate-type heat exchangers are developed under the shown below specific conditions. If any of these conditions in different for an application, then correction must be made based on correction factors provided by the manufacturer.
- Oil Viscosity = ISO VG 32.
- Oil/Water Flow Ratio = 2:1.
- Oil Inlet Temperature = 140 $^\circ$F.
- Water Inlet Temperature = 80 $^\circ$F.

Example:

Given Data:
- Oil Viscosity is ISO VG 68.
- Oil Flow = 40 gpm.
- Available water flow 10 gpm.
- Required Cooling Capacity q_r = 40 hp as calculated based on the system energy analysis.
- Oil Inlet Temperature To = 140 °F.
- Water Inlet Temperature Tw = 100 °F.
- Allowable Pressure Drop = 30 psi.

Step 1 (Cooling Capacity and Pressure Drop Correction Factors):

Table 5.1 (provided by manufacturer) shows the cooling capacity correction factor (K_v = 1.2) and pressure drop correction factor (K_p = 1.7) based on actual oil viscosity as provided by the cooler manufacturer.

Viscocity Class	Cooling Capacity Factor, Kv	Pressure Drop Factor, Kp
ISO VG 22	0.95	0.9
ISO VG 32	1.0	1.0
ISO VG 46	1.05	1.3
ISO VG 68	1.2	1.7
ISO VG 100	1.35	2.2
ISO VG 150	1.6	3.0
ISO VG 220	1.9	4.3

Table 5.1- Cooling Capacity and Pressure Drop Correction Factors (Courtesy of Parker)

Step 2 (Entering Temperature Difference):

Actual Entrance Temperature Difference **(EDT) = To - Tw =** 140°F - 100°F = 40°F

Step 3 (Entrance Temperature Difference Correction Factors):

Table 5.2 (provided by manufacturer) shows the temperature difference correction factor (**Kt** = 1.43) based on **EDT**.

ETD	30	40	50	60	70
Kt	1.87	1.43	1.17	1.0	0.88

Table 5.2- Temperature Difference Correction Factors (Courtesy of Parker)

Step 4 (Oil/Water Flow Ratio Correction Factors):

Fig 5.30 shows the oil/water flow correction factor (**Kr** = 0.83 for oil/water flow = 40/10 = 4) as provided by the cooler manufacturer. Curve for series 61 is selected to check on smaller size cooler first.

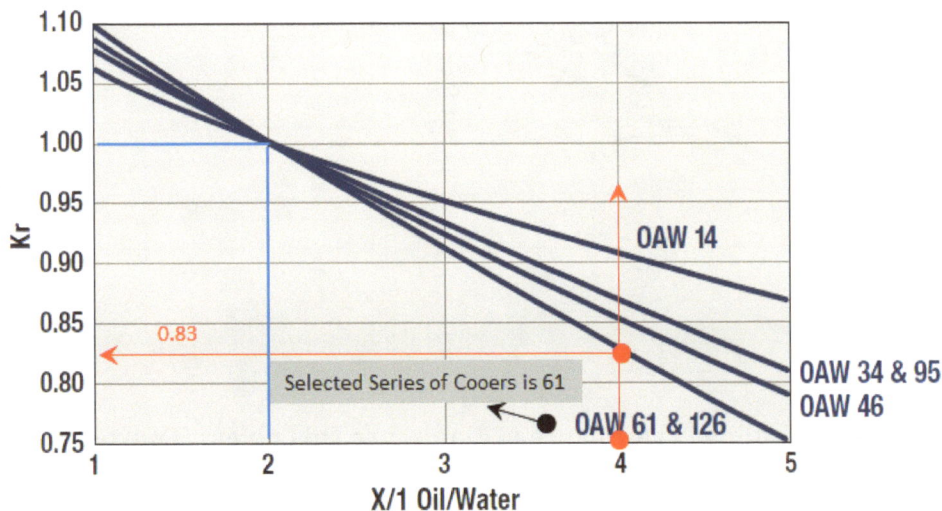

Fig. 5.30- Oil/Water Flow Correction Factors (Courtesy of Parker)

Step 5 (Calculation of the Design Cooling Capacity q_d):

Equation 5.5 shows the design cooling capacity that considers all the correction factors and based on which the cooler size will be selected.

Design Cooling Capacity q_d = (qr x Kv x Kt)/Kr 5.5

→ **q_d** = (40 x 1.2 x 1.43)/0.82 = 83 hp

Step 6 (Selecting Proper Size Cooler based on Design Cooling Capacity q_d):

Figure 5.31 shows performance curves for various sizes of plate-type heat exchangers. Based on the calculated design cooling capacity, the minimum size cooler for these conditions is OAW 61-40.

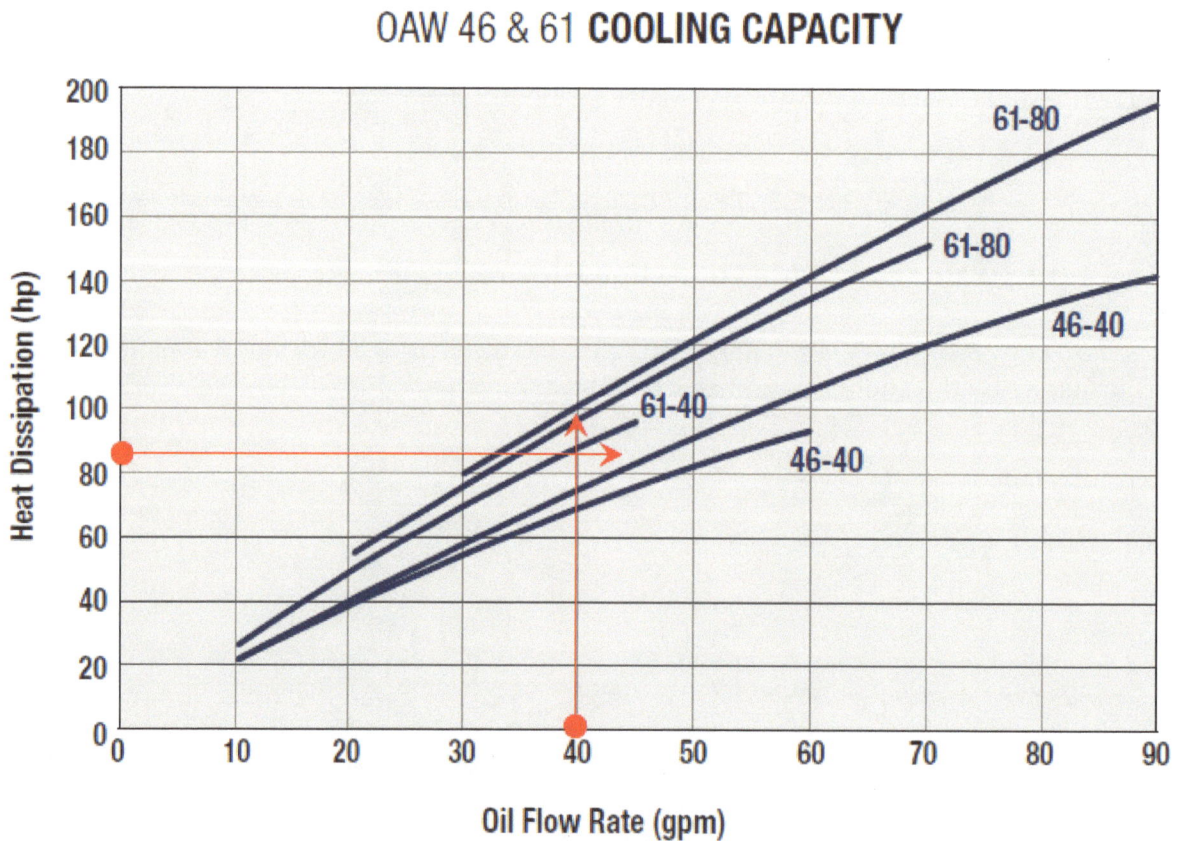

Fig. 5.31- Cooling Performance of Plate-Type Oil Coolers (Courtesy of Parker)

Step 7 (Check on Pressure Drop Δp based on Pressure Drop Correction Factor):

Figure 5.32 shows design pressure drop **DPoil** = 23 psi for the selected size of oil cooler and the assigned flow rate. However, pressure drop still needs to be corrected based on the fluid viscosity as follows:

Equation 5.6 shows the how to calculate the actual pressure drop based on the pressure drop correction factor

Actual Pressure Drop Δp = DPoil x Kp   **5.6**

$\rightarrow$ **Δp** = 23 x 1.7 = 39.1

Since this pressure drop exceeds the allowable pressure drop (30 psi), then the next larger size (61-50) cooler will be selected that has Design Pressure Drop (DPoil) = 12 psi.

$\rightarrow$ Actual Pressure Drop **Δp** = 12 x 1.7 = 20.4

This pressure drop is close to the allowable pressure drop.

OAW 46 & 61 **PRESSURE DROP**

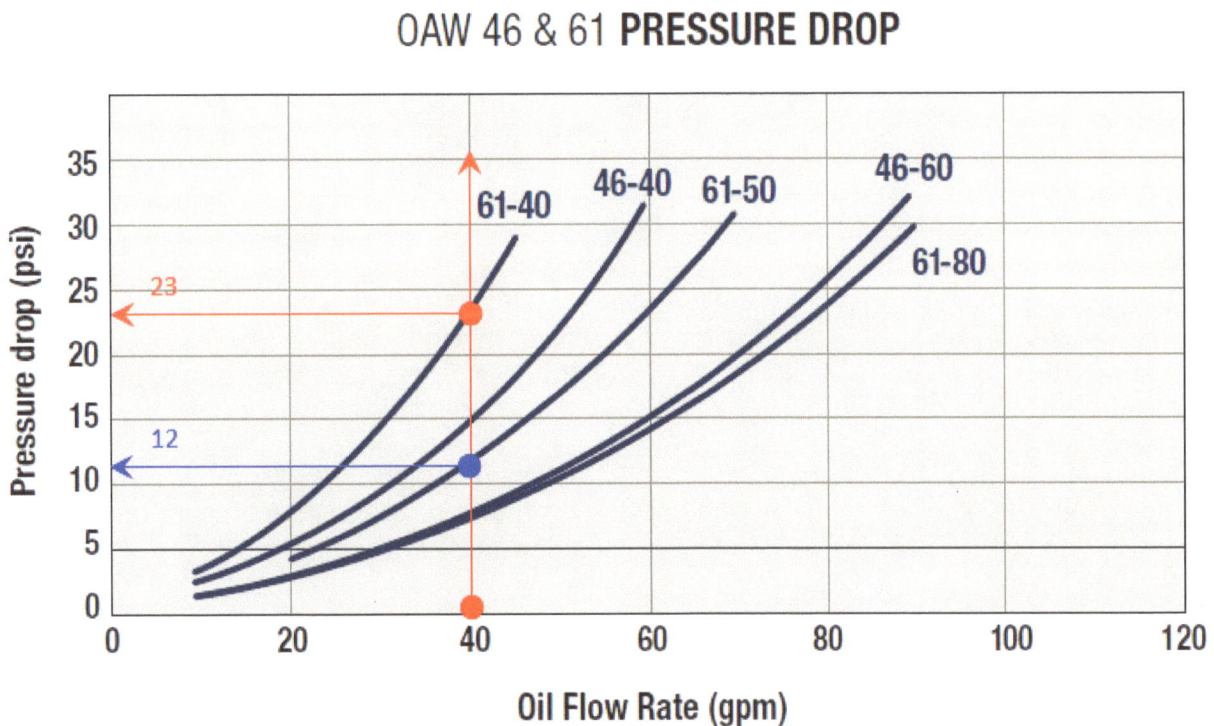

Fig. 5.32- Performance Curves of Plate-Type Oil Coolers (Courtesy of Parker)

5.7- Cooling-Filtration Units

Figure 5.33 shows a cooling-filtration unit for offline use where a plate-type cooler is integrated with an oil filter. Figure 5.34 shows a cooling-filtration unit for offline use where an air-type cooler is integrated with an oil filter.

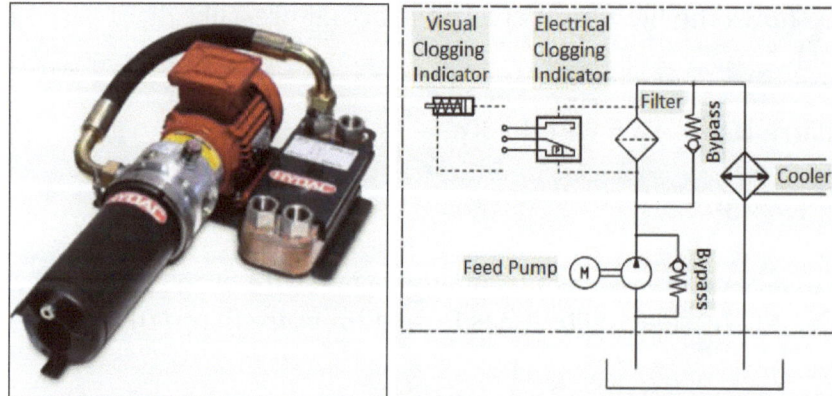

Fig. 5.33- Cooling-Filtration Unit (Courtesy of Hydac)

Fig. 5.34- Cooling-Filtration Unit (Courtesy of Donaldson)

5.8- Oil Cooling Circuit Diagrams

There is no single place for an oil cooler to be placed in a hydraulic system. It is a case-by-case type of solution. The following section provide examples of cooling circuit solutions. No matter the solution is, the following couple design considerations are common:

- In any case, if a filter and a cooler are place in series, the filter must be before the cooler. The reasons of that are hot oil is filtered easier than cold oil, only clean oil enters the cooler, and back pressure on cooler piping becomes less.

- The oil cooler is protected against back pressure by one of two devices placed in parallel to the oil cooler. Either a bypass valve or a low-pressure non-adjustable relief valve where both have cracking pressure of 25 psi.

5.8.1- Full Cooling on Return Line in Open Circuits

Figure 5.35 shows a cooling circuit solution where the fluid volume is circulated in full through the cooler that is placed on the return line of main system circuit. The cooler in this case must be sized based on the maximum flow that will pass through the return line. Therefore, a thorough flow distribution analysis is required. It is to be noted that, if a differential cylinder is used, the flow through the return line will be more than the pump flow.

This solution is recommended when:
- The main pump is small.
- The flow in the main circuit isn't surged by an accumulator or variable displacement pump.

The advantages of this solution are as follows:
- Using the main pump saves the cost of additional setup for cooling and filtration.
- Fluid is circulated faster and so heat removal is faster.

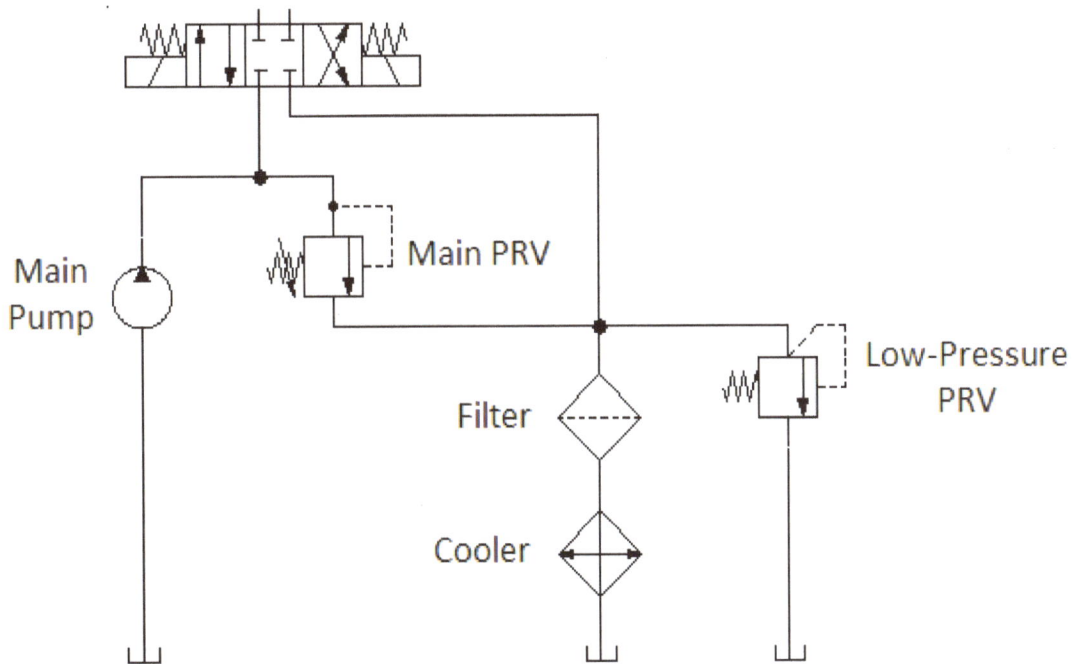

Fig. 5.35- Full Cooling on Return Line in Open Circuits

5.8.2- Full Offline Cooling

Figure 5.36 shows an *Offline Cooling* circuit solution where the full fluid volume is circulated through the cooler but using offline cooling circuit separated from the main system circuit.

In the offline cooling, as a rule of thumb, the cooling pump is sized to circulate the full volume of oil in the tank every 10-15 minutes. For example, if the oil volume in the tank is 150 liters, then the pump size should be 10-15 liter/minutes. since the working pressure in this circuit is low, just a gear pump or even a centrifugal pump are used.

This solution is recommended when:
- The main pump is quite large.
- The flow in the main circuit isn't stable or is surged by an accumulator or variable displacement pump.

The advantages of this solution are as follows:
- The cooler is sized to fit the heat load and not the full return flow of the main circuit.
- A stable cooling and filtration performance irrespective of variations in flow and duty cycle of the main hydraulic circuit.
- The cooler is completely isolated from surge pressures in the return line that can potentially damage the cooler.
- Maintenance can be performed without the need to shut down the main system.

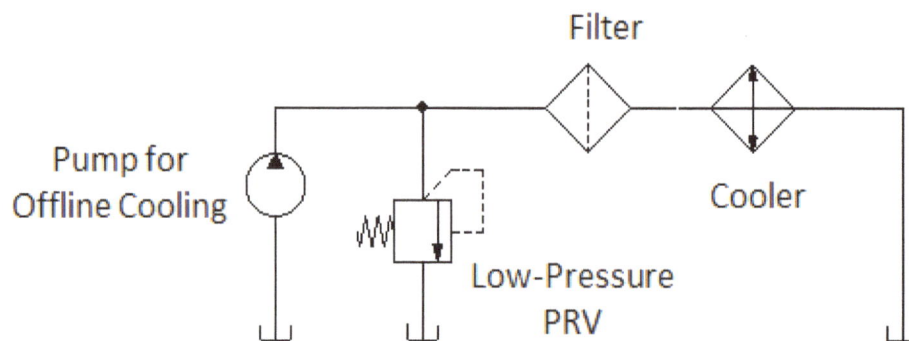

Fig. 5.36- Offline Full Cooling Circuit Diagram

5.8.3- Partial Cooling on Boosting Pump Line in Open Circuits

As shown in Fig. 5.37, a cooling circuit solution can be used where the full fluid volume is circulated through the oil cooler that is placed in series with the boosting pump. In this case, the low-pressure relief valve is set at (5 to 10 psi) cracking pressure to meet the maximum intake pressure at the main pump. This solution is recommended when the main pump is large and requires *boosting (supercharge)* the pressure of its intake line to avoid cavitation. The concern in this solution is introducing cold oil to the pump suction port isn't favorable for pump operation.

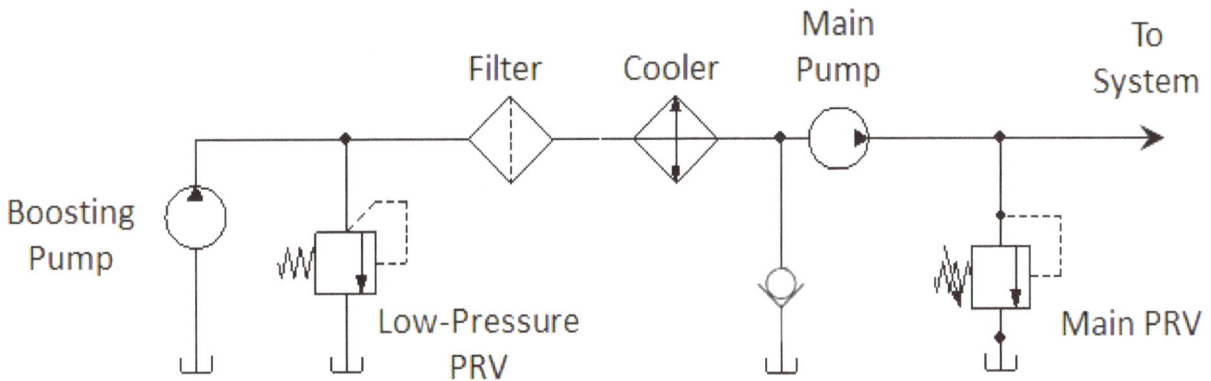

Fig. 5.37- Partial Cooling on Boosting Pump Line in Open Circuits

5.8.4- Partial Cooling on Main Relief Valve Return Line in Open Circuits

In the system shown in Fig. 5.38, the main relief valve on the main pump is frequently opened and more than quarter of the fluid volume passes through the relief. In such systems, the relief valve becomes the spot of the most heat generation in the system. In such a case, placing the oil cooler in the return line of the relief valve will be more adequate for better heat removal. However, in such a case, the cooling will be partial since the full volume of the oil will not pass through the cooler.

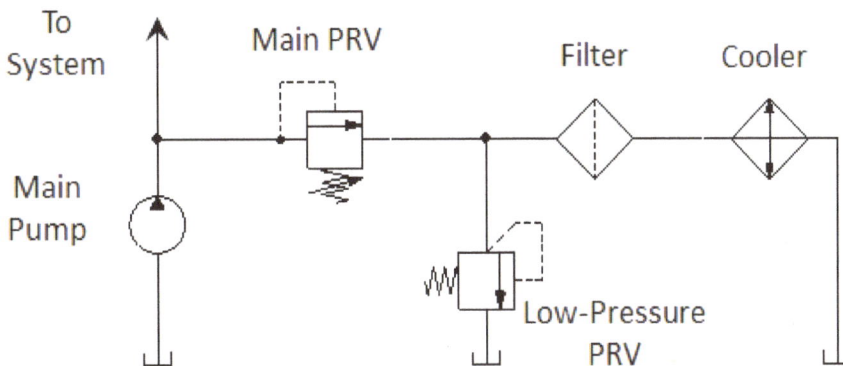

Fig. 5.38- Partial Cooling on Main Relief Valve Return Line in Open Circuits

5.8.5- Partial Cooling on Case Drain Line in Closed Circuits

As shown in Fig. 39, in a closed hydraulic circuit, the combined case drain from the pump and motor provides sufficient flow at low pressure to pass through an oil cooler. The oil cooler receives also the extra flow from the boosting pump through the charge pressure relief valve.

Caution: Cooler must be sized to prevent case drain back pressure from exceeding manufacturer maximum case pressure rating.

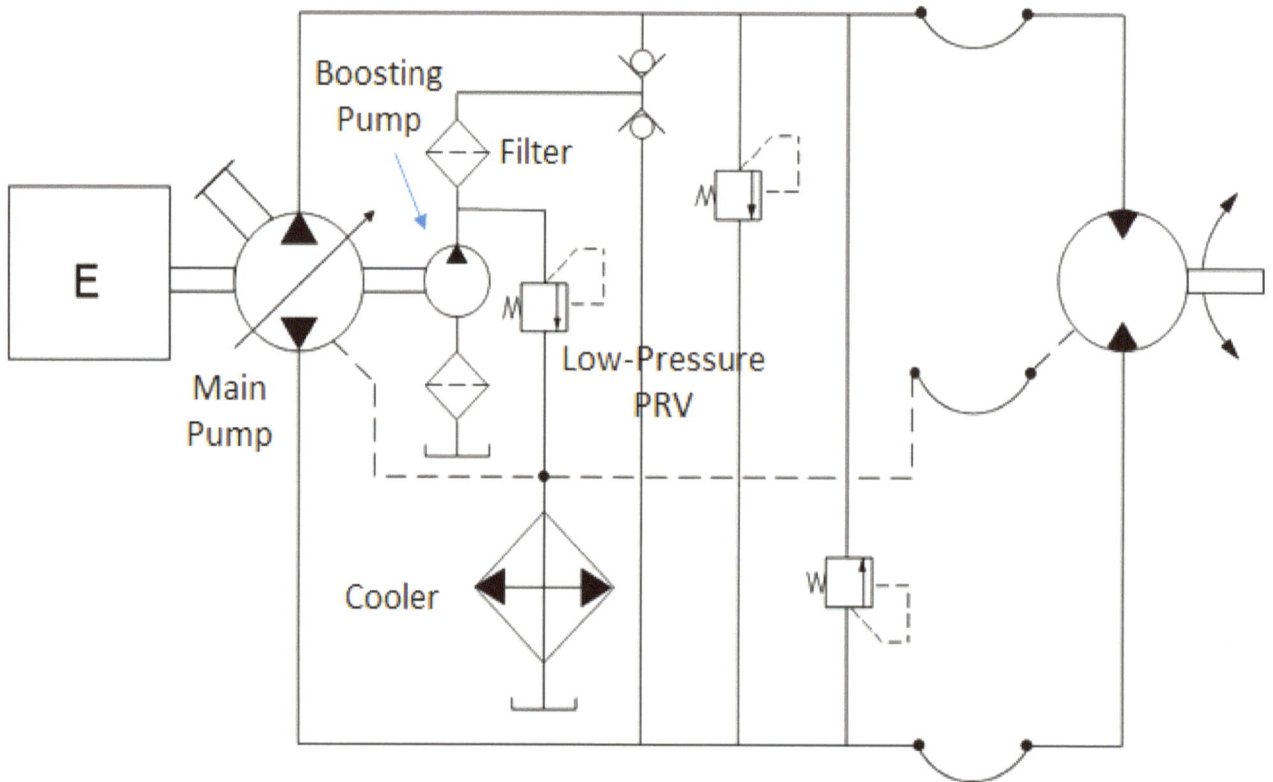

Fig. 5.39- Partial Cooling on Case Drain Line in Closed Circuits

5.9- Oil Temperature Automatic Control Solutions

Hydraulic fluid temperature can be automatically controlled using an On/Off or proportional control modes.

5.9.1- ON/Off Automatic Oil Temperature Control

In the system shown in Fig. 5.40, both heating and cooling was activated based on *On/Off Temperature Control* mode. The following bullets summarizes the construction of such a system:

- **Cooling Water Control Valve:** On/Off solenoid-operated control valve should be placed on the cooling water inlet line.

- **Cooling Water Shut-Off Valve:** A shut-off valve (not shown in the figure) should precede the cooling water valve for purposes of maintenance.

- **Temperature Sensing Elements:** Temperature sensing element for heating and cooling, should be placed in most representative point to the system temperature and good accessibility for service. The sensing element is placed inside a Sensing Bulb so that the sensing element can be removed for inspection without the need for draining the reservoir.

- **Cooling Water Temperature Switch:** A *Switch* is a devise that generates a binary signal. It must have an adjustable *Threshold* (*Hysteresis*). When the oil temperature rises above the maximum temperature, the temperature switch closes the electrical circuit to energize the solenoid valve so that the cooling water flows in the oil cooler. When the oil temperature falls below the minimum temperature, the temperature switch de-energizes the water valve stopping the flow of cooling water.

- **Heater Temperature Switch:** It should also have operating threshold. For example, it turns the heater on when the temperature gets lower than 20 °C and Turn it Off when it gets higher than 60 °C.

- **Performance:** As shown in Fig. 5.41, operating temperature fluctuates around a set point within a range proportional to the temperature switch threshold.

- **Air Cooling:** In air cooling systems, On/Off control concept of operation is applicable, except that the temperature switch operates a cooling fan rather than cooling water control valve. Alternatively, an On/Off control valve can be used on the oil line that flows through the radiator of air-type oil cooler.

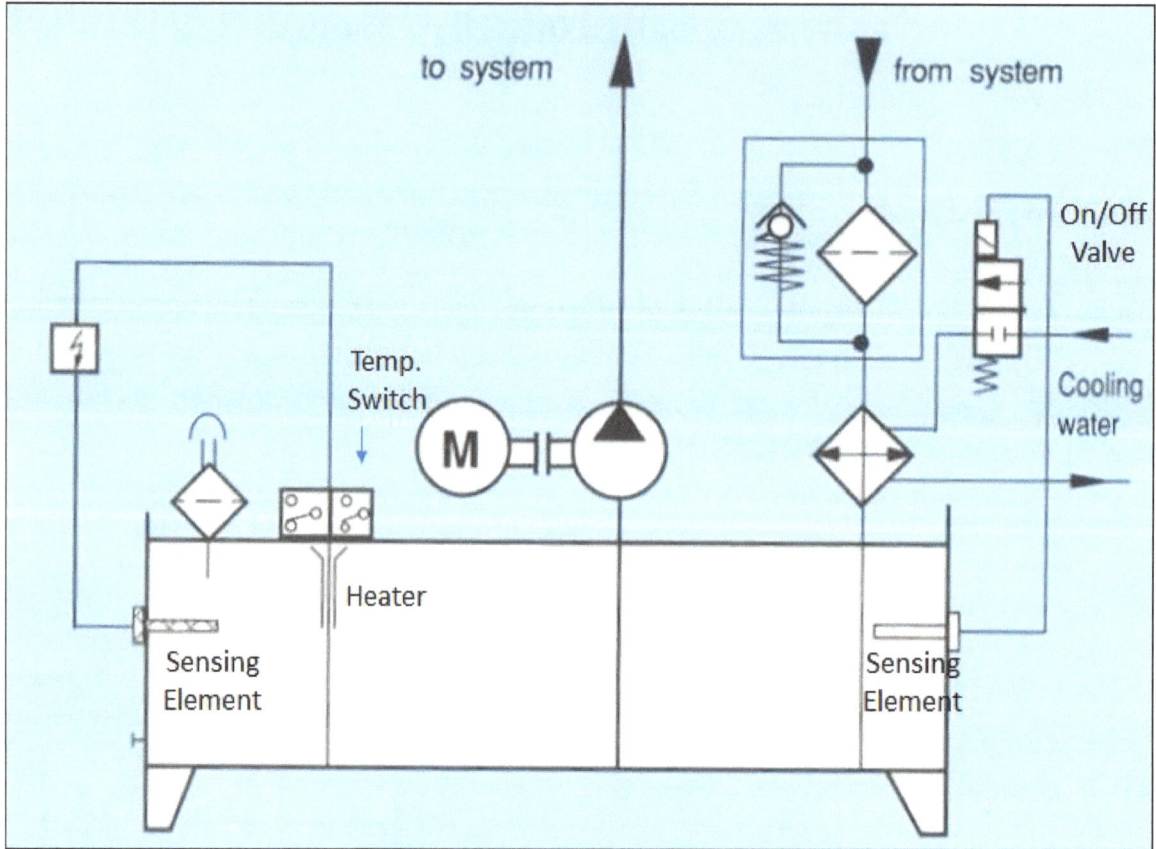

Fig. 5.40- ON/Off Automatic Temperature Control Systems (Courtesy of Bosch Rexroth)

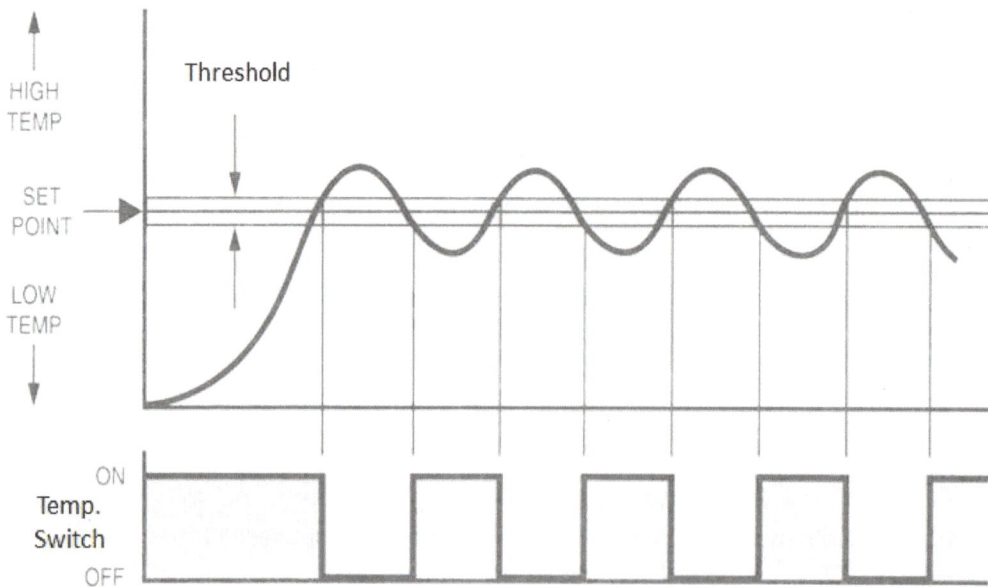

**Fig. 5.41- Performance of On/Off Automatic Temperature Control Systems
(Courtesy of Womack)**

5.9.2- Proportional Automatic Oil Temperature Control

The following bullets summarizes the difference between On/Off and *Proportional Temperature Control* systems:

- **Cooling Water Control Valve:** Instead of an on/off cooling water control valve, a proportional valve to control the flow rate of the cooling water automatically through a closed loop control system.

- **Temperature Sensors:** Instead of using temperature switches, temperature sensors are used that A *Sensor* is a device that generates analogue output signal proportional to temperature.

- **Performance:** As shown in Fig. 5.43, The *Closed Loop* control system maintains the oil temperature at the desired value with an acceptable steady state error margin. The performance of the temperature control system depends on tuning the PID controllers.

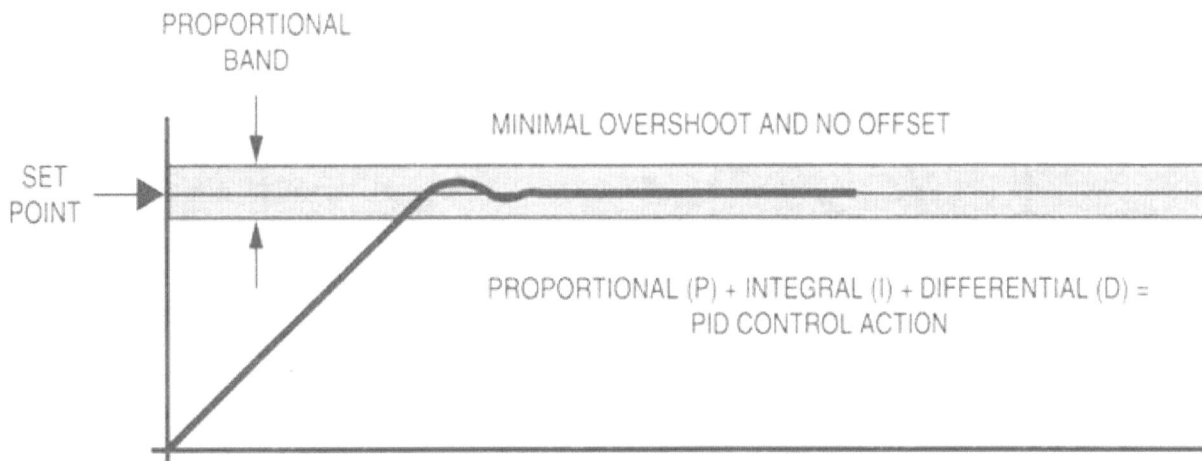

Fig. 5.42- Performance of Proportional Automatic Temperature Control Systems (Courtesy of Womack)

- **Air Cooling:** As shown in Fig. 5.42, the same of operation is applicable except that the control system controls the speed of the fan. In such a case, the fan should be driven by an electrical servo motor or a variable speed hydraulic motor. Alternatively, a hydraulic proportional valve is used to control the flow of oil to the radiator. At low temperature, most of the oil bypasses the radiator. At high temperatures, most of the flow is forced into the radiator.

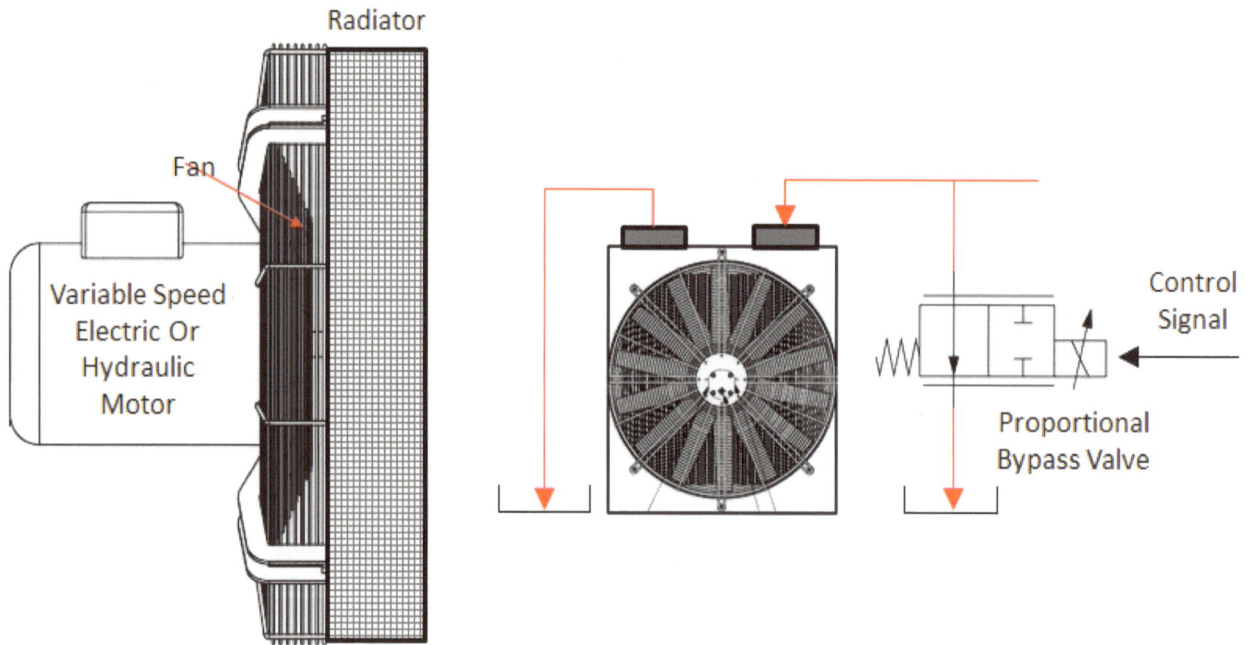

Fig. 5.43- Proportional Automatic Temperature Control Strategies for Air-Type Cooling Systems (Courtesy of Womack)

5.10- Electrical Oil Heaters

5.10.1- Construction and Operation of Electrical Oil Heaters

Heat Addition: Heat is added to hydraulic systems <u>passively</u> by running the pump over the relief valve for some time, or <u>actively</u> by powering a heater immersed in the oil.

Construction: Figure 5.44 shows a typical oil heater. The heating cartridge is placed inside a steel or stainless-steel enclosure "housing" so that the it can be removed for inspection without the need for draining the reservoir. Some oil heaters can be configured to have a built-in temperature switch or sensor for on/Off or proportional control; respectively.

Electrical Connections: Choice of AC or DC electrical power supply is connected via the use of terminals within the terminal box. Connecting terminals must be at a size suitable to the watt size of the heater and also should meet the client's requirements.

Operation: The heater is immersed in the reservoir, so that the whole oil volume in the reservoir is involved in the heating process. Overheating petroleum-based hydraulic fluid may result in ignition. It is suggested that the heater control system is interlocked with the pump. This means that heat can't turn on unless there is oil flow through the reservoir.

Heater Placement:
- It is preferred to distribute several small heaters in a reservoir rather than using just one large size heater. That helps for heating up the oil homogeneously and to avoid local overheating in one spot.
- The heater should be located where there is sufficient oil flow around the encloser to prevent overheating and breaking down the fluid close to the enclosure.

Heat Transfer in Electrical Oil Heaters: Heat transfer in these heaters are composed of the following mechanisms:
1. <u>Convection</u> from the heating element to the air surrounding the heating element filling the inside volume of the enclosure (bulb).
2. <u>Conduction</u> through the walls of the enclosure.
3. <u>Convection</u> from the walls of the enclosure to the cold oil.

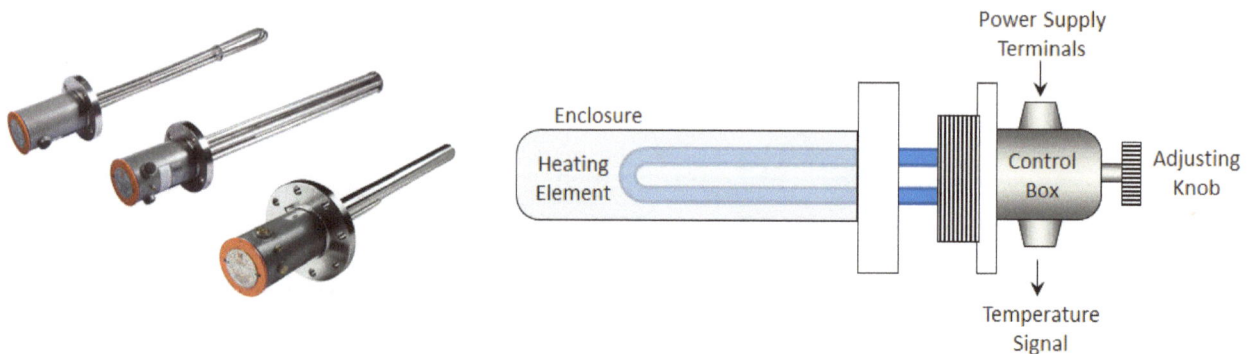

Fig. 5.44- Construction of Electrical Oil Heaters

5.10.2- Sizing of Electrical Oil Heaters

Heating Capacity: Equation 5.7 shows that the heating capacity is directly proportional to the size of the reservoir and inversely proportional to the allowable heating time. Design heating time should be kept within 1-3 hours to keep the heater size reasonable.

HeatingCapacity(kW) =

$$\frac{\text{Tank Capcity (Gallons)} \times \left[\text{Desired min Temp.}(^{0}F) - \text{Ambient Temp.}(^{0}F)\right]}{800 \times \text{Design Heating Time (Hours)}} \qquad 5.7$$

Heater Size: knowing the heating capacity, the heater surface area should be limited to

$$2 \text{ W/cm}^2 \ (10 \text{ W/in}^2)$$

Otherwise, the oil that is close to the body of the heater will be overheated and possible breakdown.

Example: A 100-gallon hydraulic reservoir is placed outdoor where the ambient temperature is around freezing temperature. The minimum temperature for such a machine is 60 ^{0}F. Allowable time for heating is one hour. Then,

- Eq. 5.3 → Heating Capacity (kW) = [100 (60-32)]/(800 x 1) = 3.5 x 1.341 = 4.7 kW.
- Heater Surface Area = 4700/2 = 2350 cm².

Chapter 6

Introduction to Hydraulic Filters

Objectives

This chapter presents an overview of hydraulic filters including the contribution of filters in hydraulic systems, ISO1219 symbols, construction, and operating principles. The chapter also presents various types of filters based on application in which the filter is used, type of connection to the circuit, body style of the filter, placement in the hydraulic circuit. The chapter also discusses the added accessories to the filter such as bypass valve and clogging indicators. Examples from industry are presented.

Brief Contents

6.1 - Contribution of Filters in hydraulic Systems

6.2 - Types of Filters Based on Application

6.3 - Types of Filters Based on Types of Contamination

6.4 - Interpretation of ISO 1219 Symbols for Hydraulic Filters

6.5 - Basic Construction and Operation of Hydraulic Filters

6.6 - Types of Filters Based on Hydraulic Connections

6.7 - Types of Filters Based on the Filter Body Style

6.8 - Filter Clogging Indicators

6.9 - Types of Filters Based on their Placement in the Circuit

Chapter 6 – Introduction to Hydraulic Filters

6.1 - Contribution of Filters in hydraulic Systems

Every hydraulic system has suspended particles in its fluid. Maintaining clean hydraulic fluid is 80% of the effort required to maintain a reliable hydraulic system.

Hydraulic filters are classified based on:
- Application.
- Place in a hydraulic circuit.
- Body style.
- Types of contamination.

Filters are also available in:
- Different sizes.
- Dirt holding capacity.
- Contamination removal efficiency.

A properly selected filter must perform the following tasks:
- Remove particulate contaminants from the hydraulic fluid.
- Maintain required cleanliness level as determined by the system manufacturer.
- Remove chemical contaminants and their products (varnish, sludge, etc.) from the fluid.
- Prevent aging of the hydraulic fluid due to chemical contaminants.
- Maintain the lubricity of the fluid.
- Extend the life of the hydraulic fluid.
- Remove water content in the fluid.
- Absorb moisture from breathing reservoirs.
- Permit preventive maintenance.
- Increase component life and system reliability.
- Increase intervals between scheduled maintenance.
- Prevent unexpected failures and consequent costs of unplanned shutdown.

6.2 - Types of Filters Based on Application

As shown in Fig. 6.1, hydraulic filters serving hydraulic systems in a wide range of both industrial and mobile applications. Wherever a hydraulic system is used, at least one filter is required.

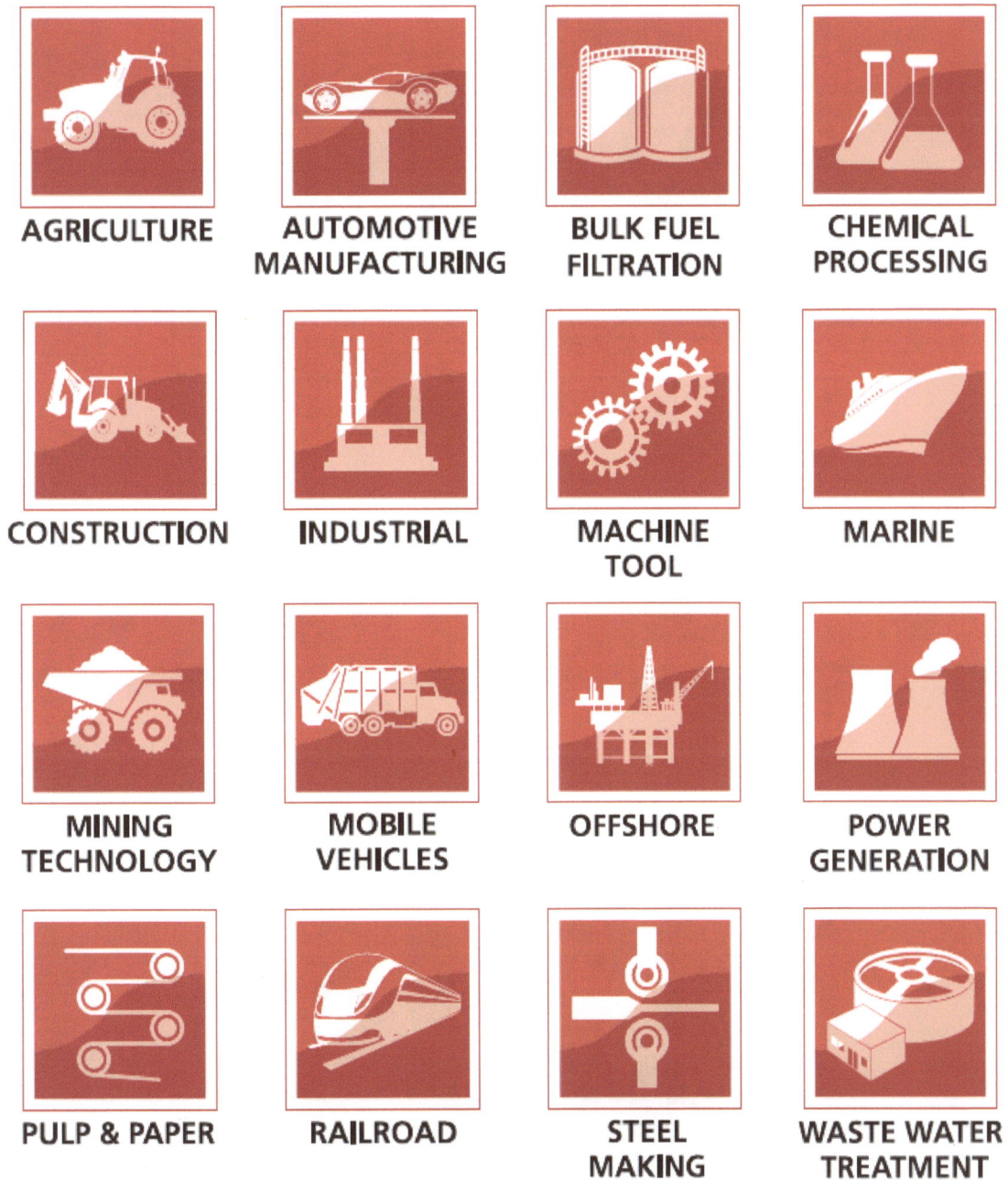

AGRICULTURE AUTOMOTIVE MANUFACTURING BULK FUEL FILTRATION CHEMICAL PROCESSING

CONSTRUCTION INDUSTRIAL MACHINE TOOL MARINE

MINING TECHNOLOGY MOBILE VEHICLES OFFSHORE POWER GENERATION

PULP & PAPER RAILROAD STEEL MAKING WASTE WATER TREATMENT

Fig. 6.1- Applications of Hydraulic Filters (Courtesy of Schroeder)

Figures 6.2 and 6.3 show hydraulic fluid filters among filtration solutions for a tractors and excavators; respectively.

Fig. 6.2- Filtration Solutions for Tractors (Courtesy of Donaldson)

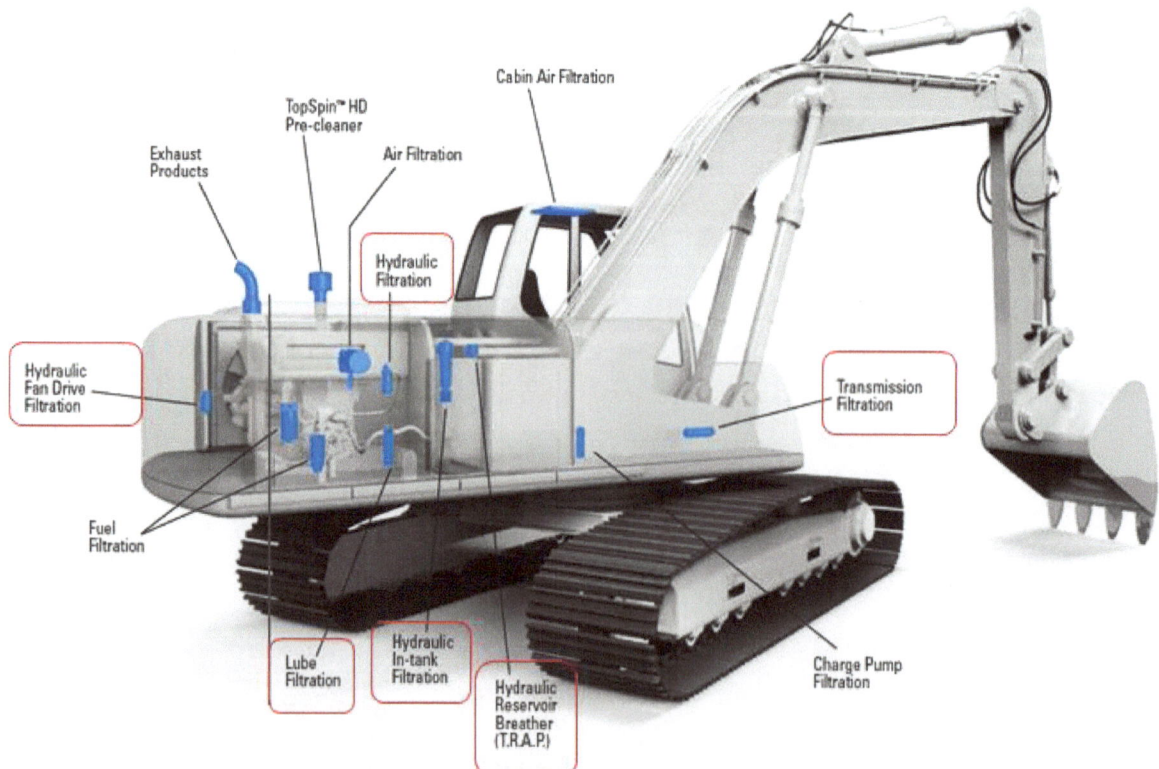

Fig. 6.3- Filtration Solutions for Excavators (Courtesy of Donaldson)

Figures 6.4 and 6.5 show hydraulic fluid filters among filtration solutions for a Dump Trucks and industrial hydraulic power units; respectively.

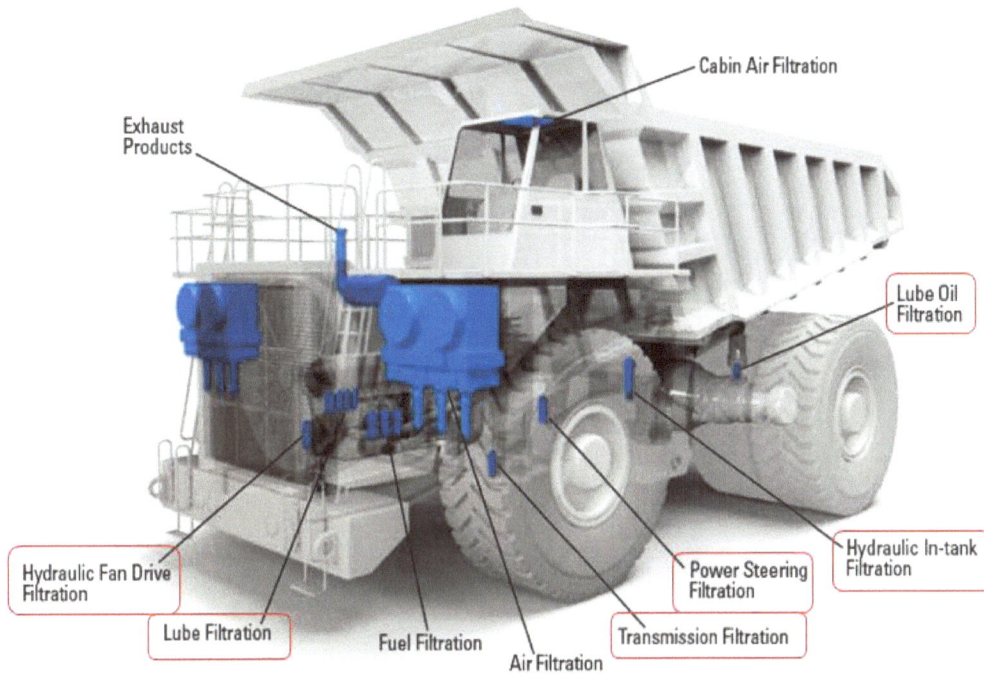

Fig. 6.4- Filtration Solutions for Dump Trucks (Courtesy of Donaldson)

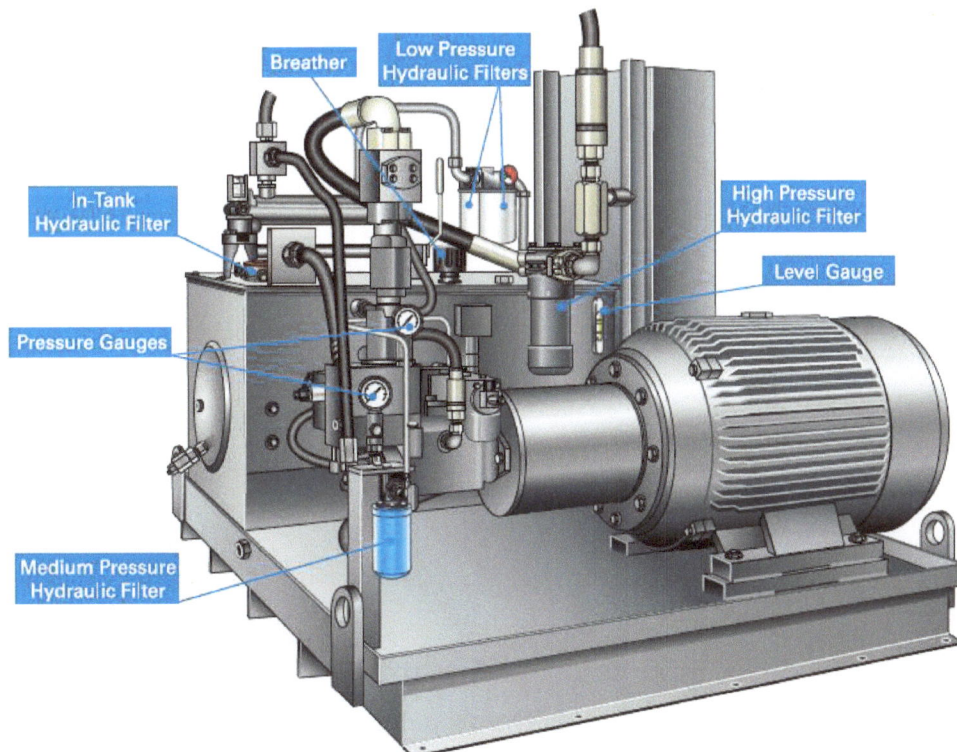

Fig. 6.5- Filtration Solutions for Industrial Hydraulic Power Units (Courtesy of Donaldson)

6.3 - Types of Filters Based on Types of Contamination

Hydraulic filters are classified based on the types of contamination as follows:

- Filters for *Fluidic Contaminants*. These filters are used to remove water content in the hydraulic fluids.

- Filters for *Chemical Contaminants*. These filters are used to remove products of hydraulic fluids breaking down, such as oxidations, varnish and sludges.

- Filters for *Particulate Contaminants*. These filters are used to remove particulate contaminates whether solid or elastic, abrasive or nonabrasive, and metallic or nonmetallic.

6.4 - Interpretation of ISO 1219 Symbols for Hydraulic Filters

Like all hydraulic components, hydraulic filters are presented in symbols that reflect the basic construction and function of the filter. Figure 6.6 shows the symbols for hydraulic filters as follows:

1. Non-Bypass-Filter.
2. Bypass-Filter is a hydraulic filter with built-in bypass valve.
3. Bypass-Filter with visual clogging indicator based on inlet pressure.
4. Bypass-Filter with visual clogging indicator based on differential pressure.
5. Sandwich-mounted non-bypass-Filter with visual clogging indicator based on differential pressure.
6. Bypass-Filter with Self-Cleaning feature. Self-Cleaning filter is a large filter that is equipped with a manual or an automatic self-cleaning mechanism. This mechanism wipes the outer surface of the filter element and the inner surface of the filter housing without stopping the machine. Such a filter contains surface type filter media.
7. Non-bypass-filter with electrical clogging indicator.
8. Bypass-to-Tank Filter. Used when element collapse Pressure < max working pressure. It protects sensitive components by directing dirty fluid to tank.

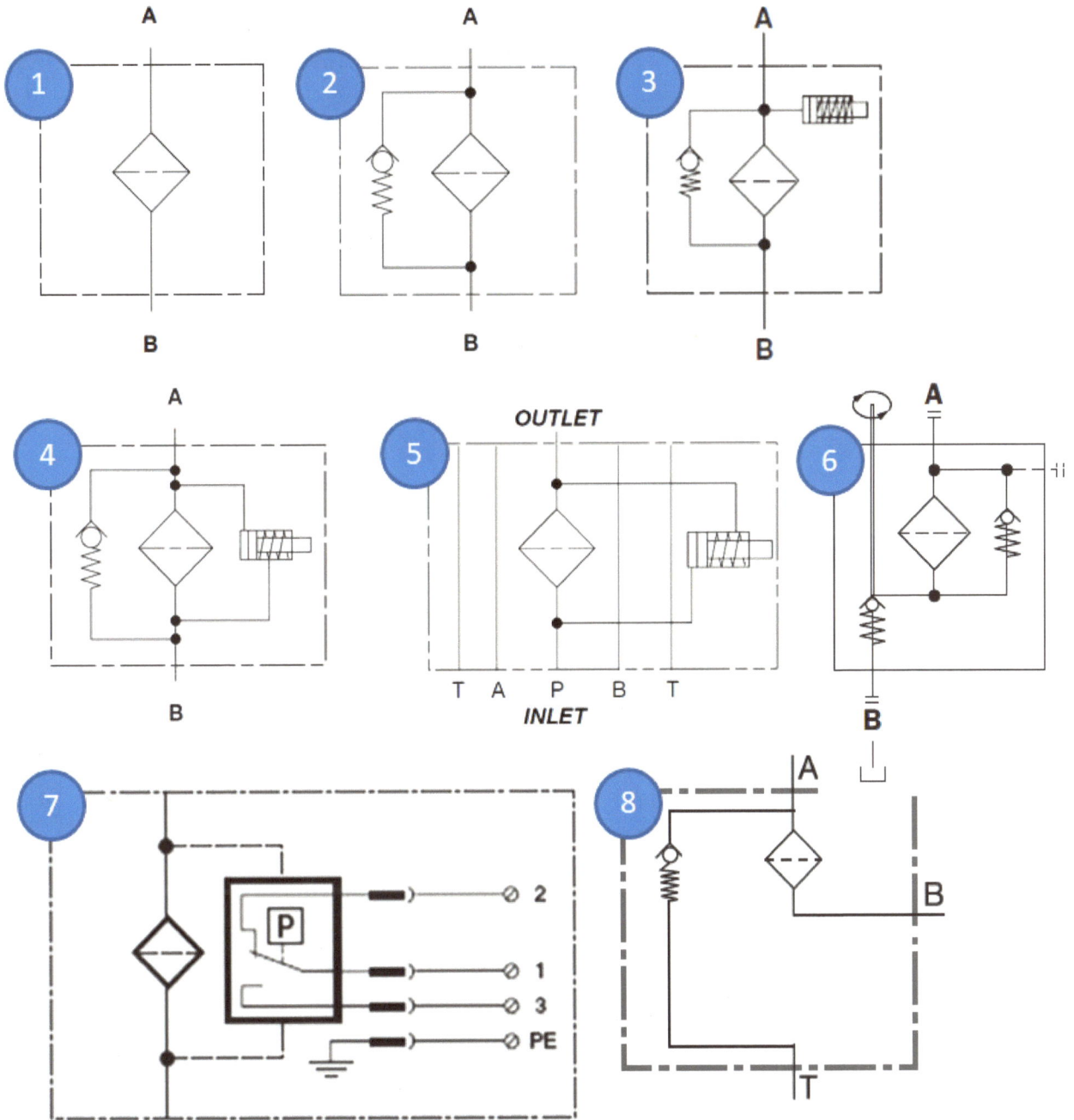

Fig. 6.6- Examples of Hydraulic Filters Symbols

6.5- Basic Construction and Operation of Hydraulic Filters

As shown in Fig. 6.7, basic construction of hydraulic filters contains the following elements:

Filter Housing: Is also referred to as "*Bowl*". It is the pressure vessel of the filter, secures the element and creates a seal between the element inlet and outlet areas. It also has a drain plug for draining before disassembling the fitter.

Filter Head: The head contains the port and possibly other options such as a bypass valve and clogging indicator. It also contains the filter mounting method. The primary concern in selecting the housing is the pressure rating. This should be determined before the housing style is selected.

Filter Element: Is also referred to as "*Filter Cartridge*". It consists of a central tube, media to remove specific type of contaminants, and end caps.

Clogging Indicator: Is to show the level of clogging of the filter media. It provides an alarm to replace the element before the bypass valve setting is reached. Replacing the element before the bypass opens prevents significant reduction in filtering efficiency.

Bypass Valve: A bypass valve is an optional feature in hydraulic filters. So, a filter may be specified as follows:

- Bypass-Filters: The *bypass-filter* allows the fluid to flow directly to the system when the filter media is clogged in order to limit the differential pressure across the media and prevents media collapse. Standard filter assemblies normally have a bypass valve cracking pressure between 1.5 – 7 bar (25 - 100 psi). When specifying a bypass-filter, it can generally be assumed that the manufacturer has designed the element to withstand the bypass valve differential pressure when the bypass valve opens. It is to be noted that, when the bypass valve opens, all contaminants will get into the system. Some systems can withstand short term operation with this contamination.
- Non-Bypass Filters: When sensitive components are used, such as servo and proportional valves, *Non-Bypass-Filter* prevents any unfiltered flow from going downstream. A low collapse pressure element should never be used in a non-bypass type housing. When contaminate buildup causing excessive differential pressure, the element may rupture or collapse. Hence, large quantities of contamination are immediately induced into the system, with severe problems. Therefore, when specifying a non-bypass filter design, make sure that the element has a differential pressure rating greater than maximum operating pressure on the filtration line.
- Bypass-to-Tank Filters: Alternative to non-bypass filters, and to avoid using high collapse pressure filter elements, a *Bypass-to-Tank* is another option. This allows the unfiltered bypass flow to return to tank through a third port, preventing unfiltered bypass flow from entering the system.

Figure 6.8 shows that hydraulic fluid introduced to the inlet port of the filter. The fluid is then forced into the cartridge from the outside surface to inside surface in order to utilize the larger area at the outside surface to retain maximum amount of dirt. Clean fluid is then pass through the central tube to the outlet port. In case of clogged filter media, bypass valve opens.

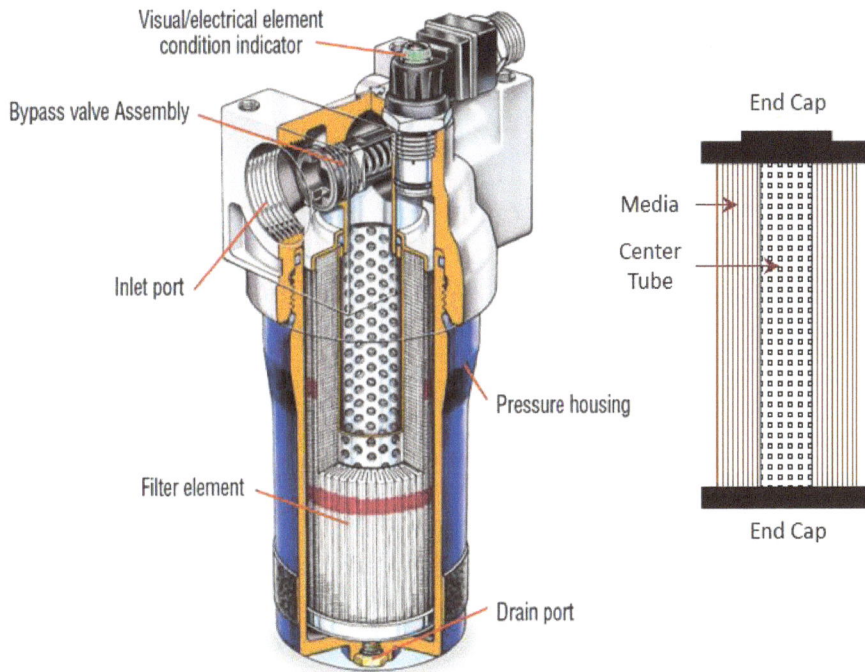

Fig. 6.7- Basic Construction of Hydraulic Filter (Courtesy of Parker)

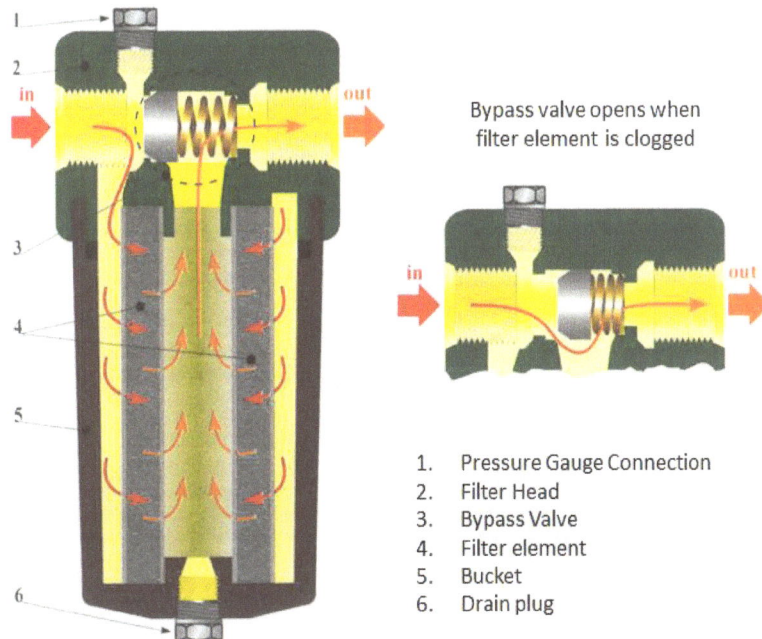

1. Pressure Gauge Connection
2. Filter Head
3. Bypass Valve
4. Filter element
5. Bucket
6. Drain plug

Fig. 6.8- Basic Construction of Hydraulic Filter (Courtesy of Assofluid)

6.6 – Types of Filters Based on Hydraulic Connections

As shown in Fig. 6.9, hydraulic filters can be connected to hydraulic system by one of the following methods:

Line-Mounted Filters: *Line Mounted* filters are assembled directly on a hydraulic line.

Flange-Mounted Filters: *Flange-Mounted* filters are assembled on one side of a manifold.

Sandwich-Mounted Filters: *Sandwich-Mounted* filters are assembled in between two standard manifolds or subplates.

Depending on the filter body style, a filter element in any filter can be replaced without disassembling the filter head from the system.

Line-Mounted   Flange-Mounted   Sandwich-Mounted

Fig. 6.9- Types of Filters Based on Hydraulic Connections

6.7 – Types of Filters Based on the Filter Body Style

As shown in Fig. 6.10, hydraulic filters are configured in various body styles in order to ease the assembly and to make it accessible during maintenance. The following sections shows examples of hydraulic filters body styles. Both pressure and return filters are available in a duplex version.

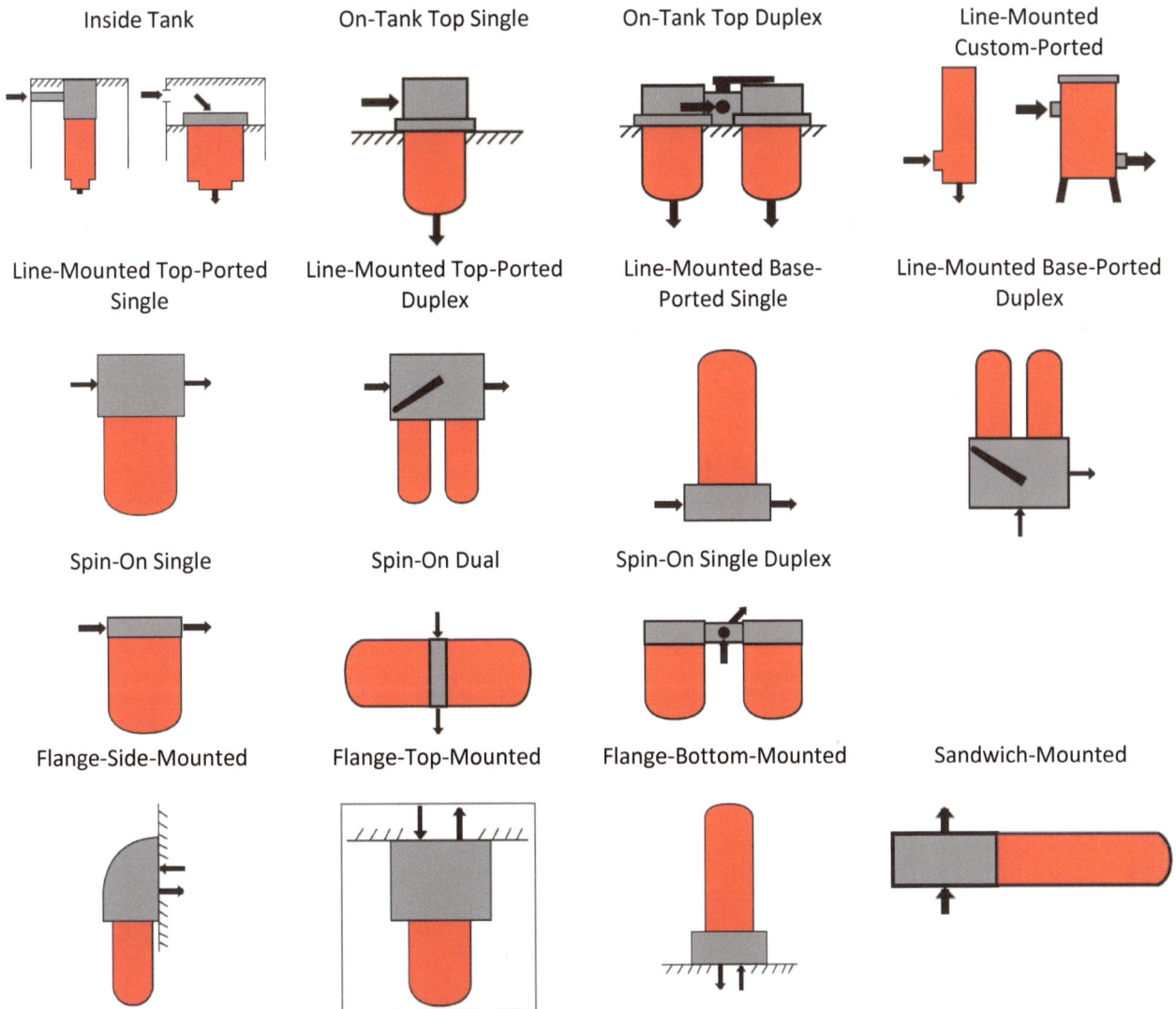

Fig. 6.10- Types of Filters Based on the Filter Body Style (Courtesy of Hydac)

6.7.1 – Inside Tank Filters

Figure 6.11 shows a unique design of *Inside Tank* filters that allow filter to be completely installed inside the reservoir. This saves space, protects the filter, and reduces leak points. It has the concern of disassembling for maintenance purposes when replacing filter media.

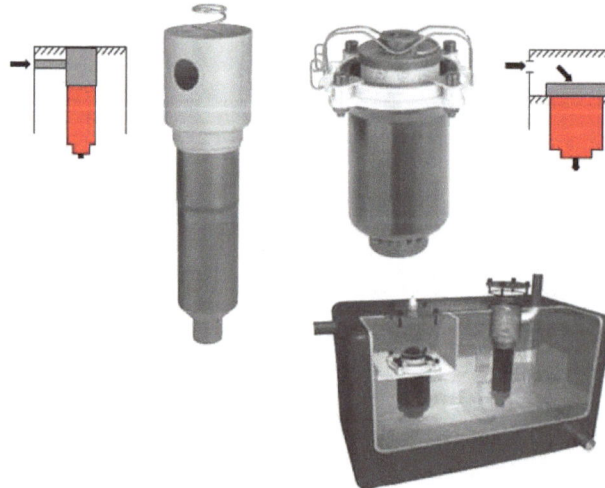

Fig. 6.11 Examples of Inside Tank Filters RFM-Series (Courtesy of Hydac)

6.7.2 – On-Tank Top Single Filters

Figure 6.12 shows various brands of *On-Tank Top Single* filters. Such filters installed where the head and the inlet/outlet ports are accessible from above the tank, while the housing remains inside the tank, offering system design flexibility.

SF Series
(Courtesy of Hydac)

ST Series
(Courtesy of Schroeder)

UT610 Series
(Courtesy of Pall)

Fig. 6.12- Examples of On-Tank Top Single Filters

Figure 6.13 shows a model of an *On-Tank Top Single* filter. This model offers a variety of options including aluminum or plastic access covers. Optional air breather featuring T.R.A.P.™ technology is available eliminating the cost associated with an additional penetration to the hydraulic tank for breather installation.

Multifunctional Ports (custom)
Contact your Donaldson sales representative for details
- Can be converted into auxiliary inlet ports
- The two secondary inlet ports can be used in conjunction with the main inlet port for higher flow rates

Flat Gasket Design
- For leak-tight operation

Service Indicator Ports
- Electrical, visual or pressure gauge options

T.R.A.P.™ Breather Technology
Breather ordered separately
Plug ships standard. Pressurized & atmospheric breathers available.
- Quick fit connection
- Anti-splash design allows smooth operation under tilt conditions
- Keeps reservoir free from condensation

Flexible Mounting Configurations
2 or 4 hole mounting option
- Better sealing and stability
- Enhanced stability on plastic tanks
- Reverse compatible – retrofit existing tanks with the new hole configuration

Built-In By-Pass Valve
- New by-pass valve installed with every filter replacement

Filter Media Technology
Wide range of Donaldson media offerings – to meet various performance targets and cleanliness standards

Fig. 6.13- Example of On-Tank Top Single Filter - FIK Series (Courtesy of Donaldson)

6.7.3 – On-Tank Top Duplex Filters

Both pressure and return filters are available in a duplex version. Such filters are made of two filter chambers and includes the necessary valving. One filter works at a time to allow for continuous, uninterrupted filtration. When a filter element needs servicing, the duplex valve is shifted, diverting flow to the opposite filter chamber. The dirty element can then be changed, while filtered flow continues to pass through the filter assembly. The duplex valve typically is an open cross-over type, which prevents any flow blockage. Figure 6.14 shows various models of *On-Top Tank Duplex* filters.

www.behringersystems.com   www.hydraulicoilfilters.com

Fig. 6.14- Examples of On-Tank Top Duplex Filters

6.7.4 – Line-Mounted Top-Ported Single Filters

Figure 6.15 shows various models for *Line-Mounted Top-Ported Single* filters.

LF Series
(Courtesy of Hydac)

LMP900 - 901 Series
(Courtesy of MP Filtri)

CTF60 Series
(Courtesy of Schroeder)

Fig. 6.15- Examples of Line-Mounted Top-Ported Single Filters

Figure 6.16 and 6.17 show detailed construction of line-mounted top-ported single filters.

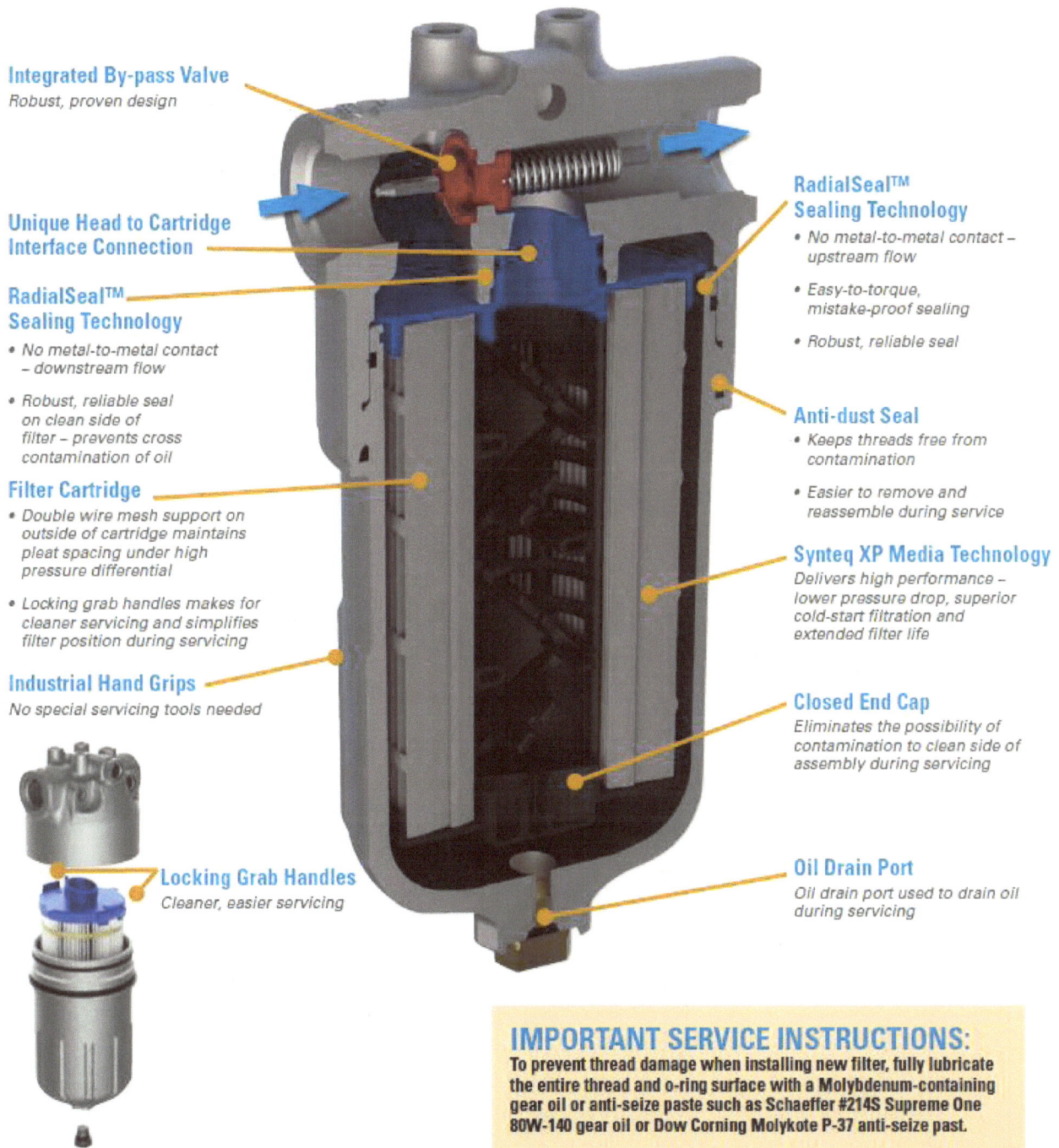

Integrated By-pass Valve
Robust, proven design

Unique Head to Cartridge Interface Connection

RadialSeal™ Sealing Technology
- *No metal-to-metal contact – downstream flow*
- *Robust, reliable seal on clean side of filter – prevents cross contamination of oil*

Filter Cartridge
- *Double wire mesh support on outside of cartridge maintains pleat spacing under high pressure differential*
- *Locking grab handles makes for cleaner servicing and simplifies filter position during servicing*

Industrial Hand Grips
No special servicing tools needed

Locking Grab Handles
Cleaner, easier servicing

RadialSeal™ Sealing Technology
- *No metal-to-metal contact – upstream flow*
- *Easy-to-torque, mistake-proof sealing*
- *Robust, reliable seal*

Anti-dust Seal
- *Keeps threads free from contamination*
- *Easier to remove and reassemble during service*

Synteq XP Media Technology
Delivers high performance – lower pressure drop, superior cold-start filtration and extended filter life

Closed End Cap
Eliminates the possibility of contamination to clean side of assembly during servicing

Oil Drain Port
Oil drain port used to drain oil during servicing

IMPORTANT SERVICE INSTRUCTIONS:
To prevent thread damage when installing new filter, fully lubricate the entire thread and o-ring surface with a Molybdenum-containing gear oil or anti-seize paste such as Schaeffer #214S Supreme One 80W-140 gear oil or Dow Corning Molykote P-37 anti-seize past.

Fig. 6.16- Example of Line-Mounted Top-Ported Single Filters - FLK Series (Courtesy of Donaldson)

Visual Indicator
P171945
5 bar, 72.5 psid

Plug
remove only when
installing indicator.

G 1/2"
threads

**AC/DC Electrical
Indicator**
P761056
5 bar, 72.5 psid

Head

O-Ring 2-140
P173382
Non-stock item

Filter

Back-up Ring
P173380
Non-stock item

Housing

- oil before assembling

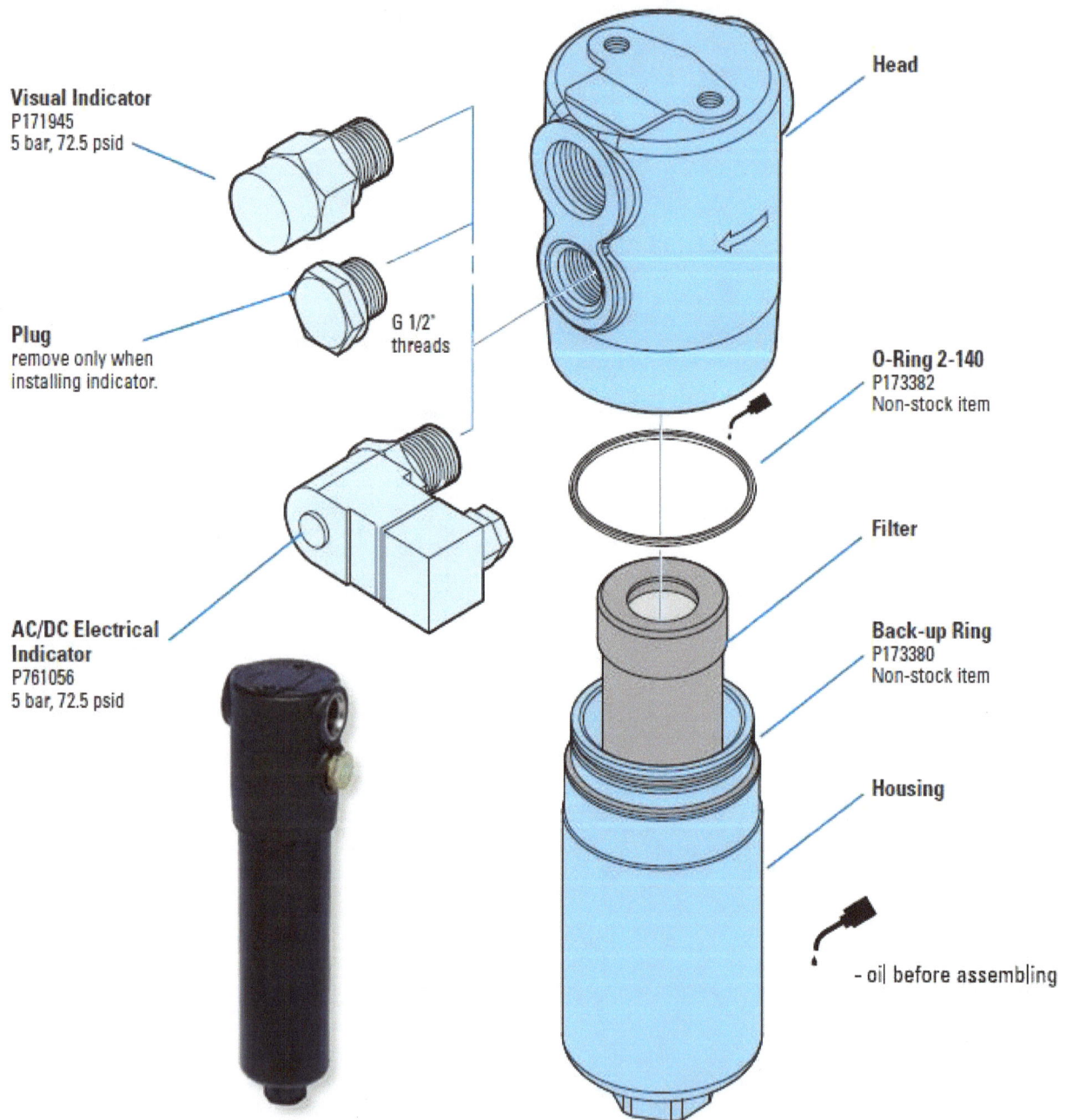

**Fig. 6.17- Example of Line-Mounted Top-Ported Single Filters - FPK02 Series
(Courtesy of Donaldson)**

6.7.5 – Line-Mounted Top-Ported Duplex Filters

Figure 6.18 shows various models of *Line-Mounted Top-Ported Duplex* filters.

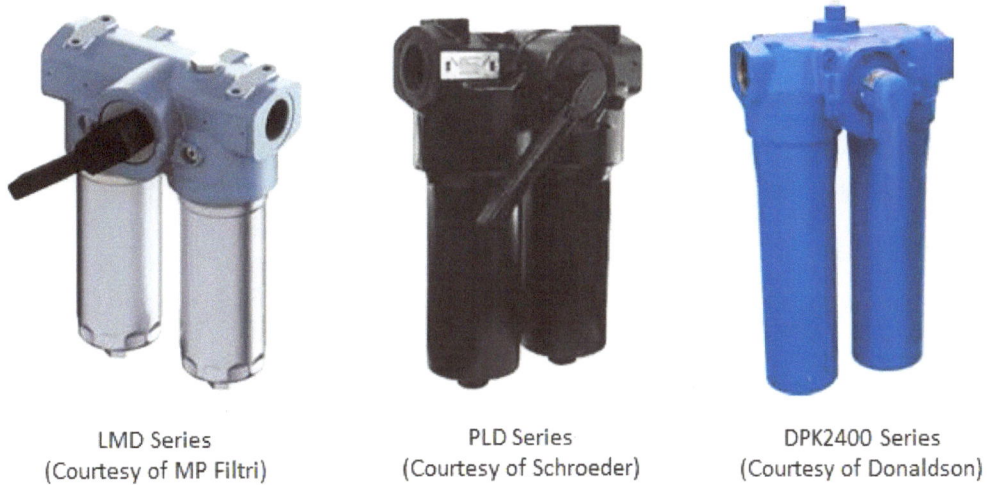

LMD Series
(Courtesy of MP Filtri)

PLD Series
(Courtesy of Schroeder)

DPK2400 Series
(Courtesy of Donaldson)

Fig. 6.18- Examples of Inline Top-Ported Duplex Filters

6.7.6 – Line-Mounted Base-Ported Single Filters

Figure 6.19 shows various models for *Line-Mounted Base-Ported Single* filters.

HF4P Series
(Courtesy of Hydac)

Series Athalon™ UH210
(Courtesy of Pall)

KF30 Series
(Courtesy of Schroeder)

Fig. 6.19- Examples of Line-Mounted Base-Ported Single Filters

6.7.7 – Line-Mounted Base-Ported Duplex Filters

Figure 6.20 shows an example of *Line-Mounted Base-Ported Duplex* filters.

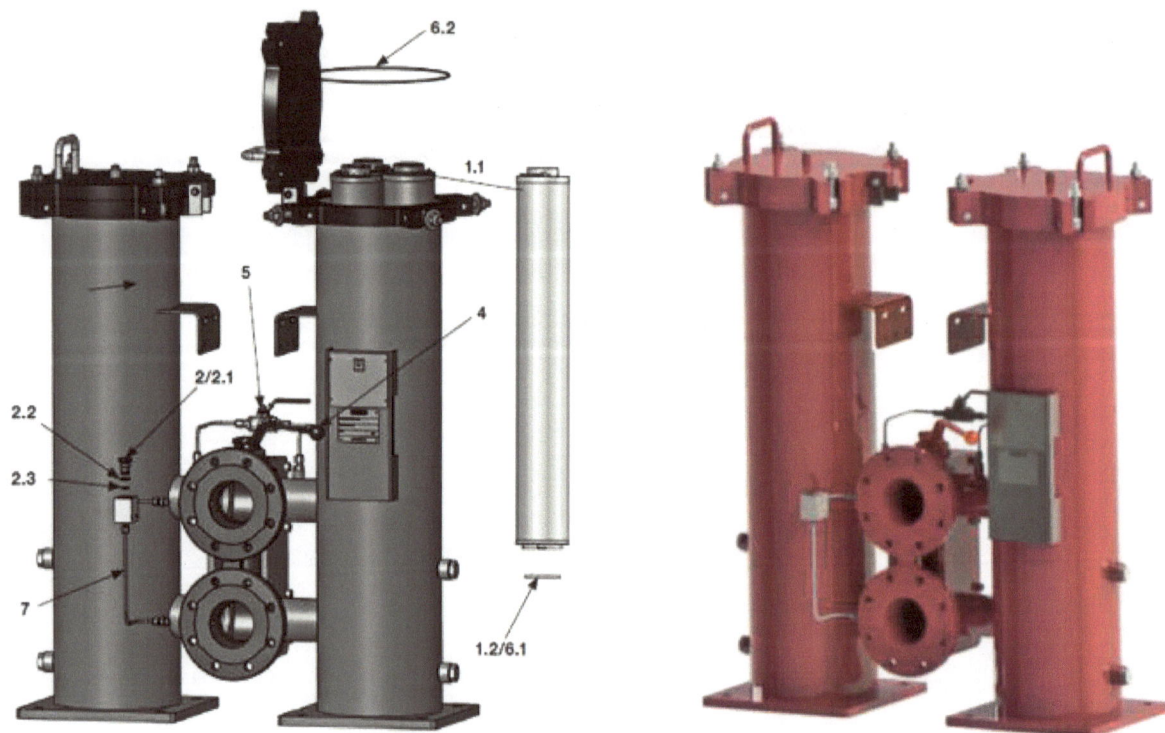

Item	Consists of	Designation
1.		Filter element
	1.1	Filter element
	1.2	O-ring
		No. of elements per filter side / size
2.		Indicator plug VD 0 A 1.0 /-V
	2.1	Clogging indicator or indicator plug
	2.2	Profile seal ring
	2.3	O-ring
3.		SEAL KIT VD/VM/VR/VR FKM
4.		Lever for change-over valve
5.		Equalization line ball valve
6.		SEAL KIT RFLD...FKM
	6.1	O-ring *(element)*
	6.2	Lid seal
7.		Indicator and equalization line pipe and plumbing

Fig. 6.20- Example of Line-Mounted Base-Ported Duplex Filters - RFLDH Series (Courtesy of Hydac)

Figure 6.21 shows another example of *Line-Mounted Base-Ported Duplex* filters.

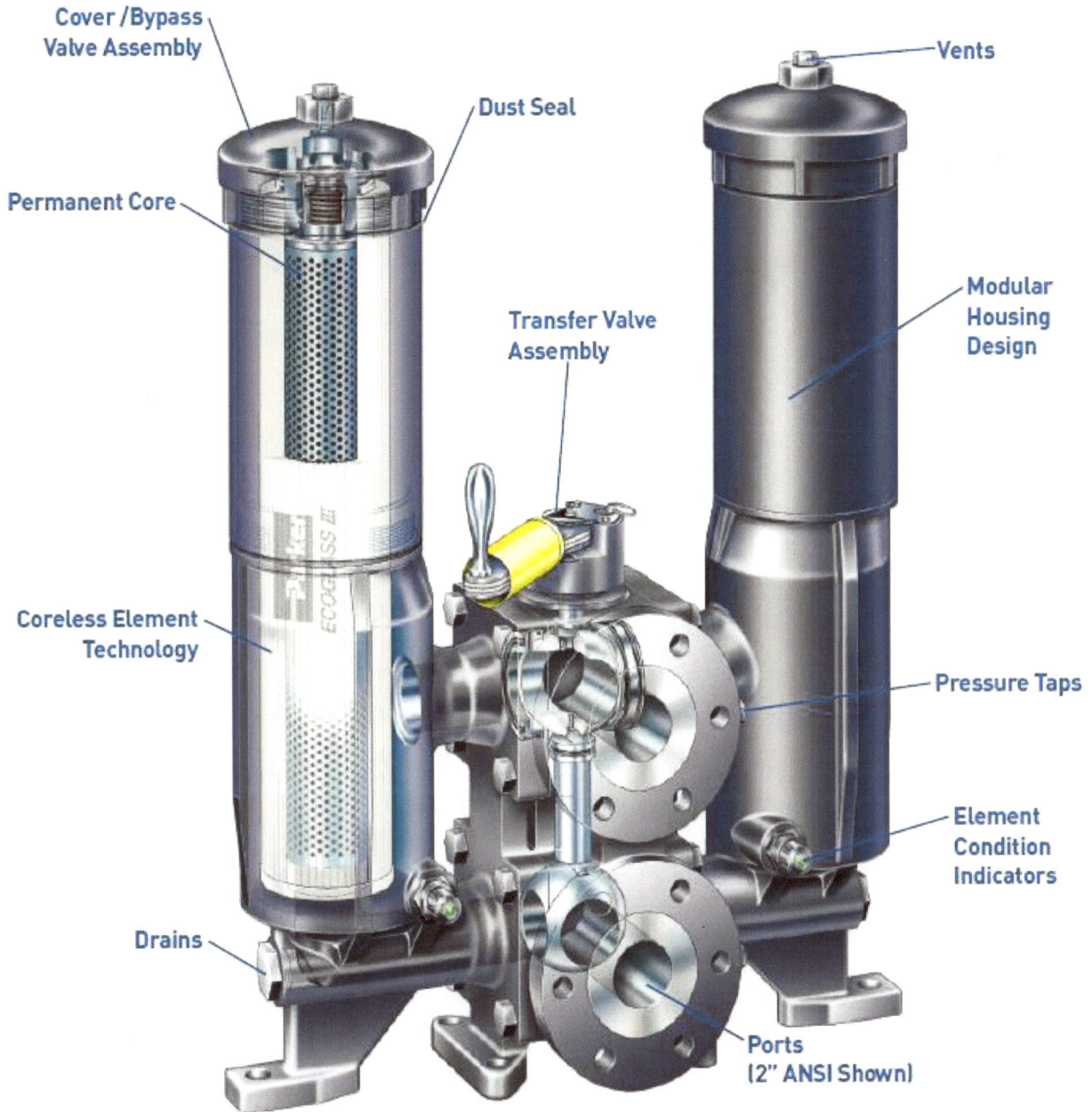

**Fig. 6.21- Example of Line-Mounted Base-Ported Duplex Filters - MPD Series
(Courtesy of Parker)**

6.7.8 – Line-Mounted Custom-Ported Filters

Figure 6.22 shows various models of *Line-Mounted Custom-Ported* filters. In such filters, the housing is made of rolled steel or stainless steel. ANSI flange connections for each filter size provide maximum connection flexibility eliminating additional adapters and intermediate flanges.

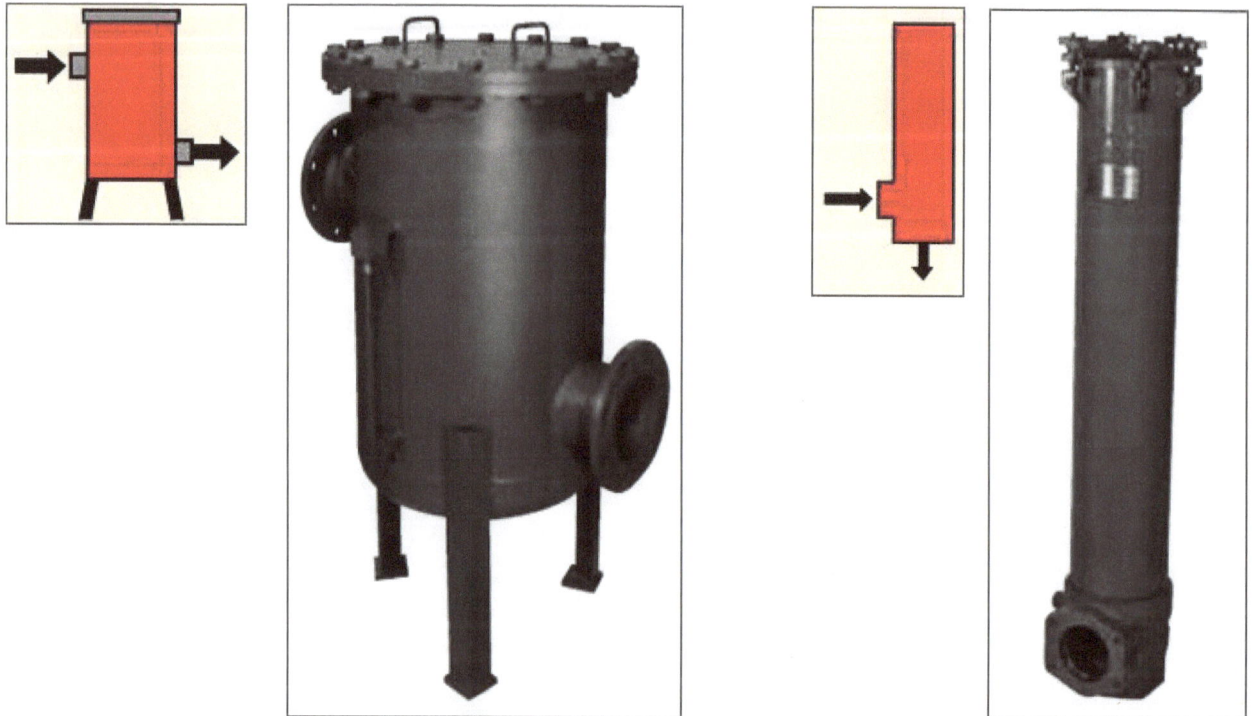

**Fig. 6.22- Examples of Line-Mounted Custom-Ported Filters – RFL Series
(Courtesy of Hydac)**

Figures 6.23 and 6.24, shows detailed construction of *Line-Mounted Custom-Ported* filters.

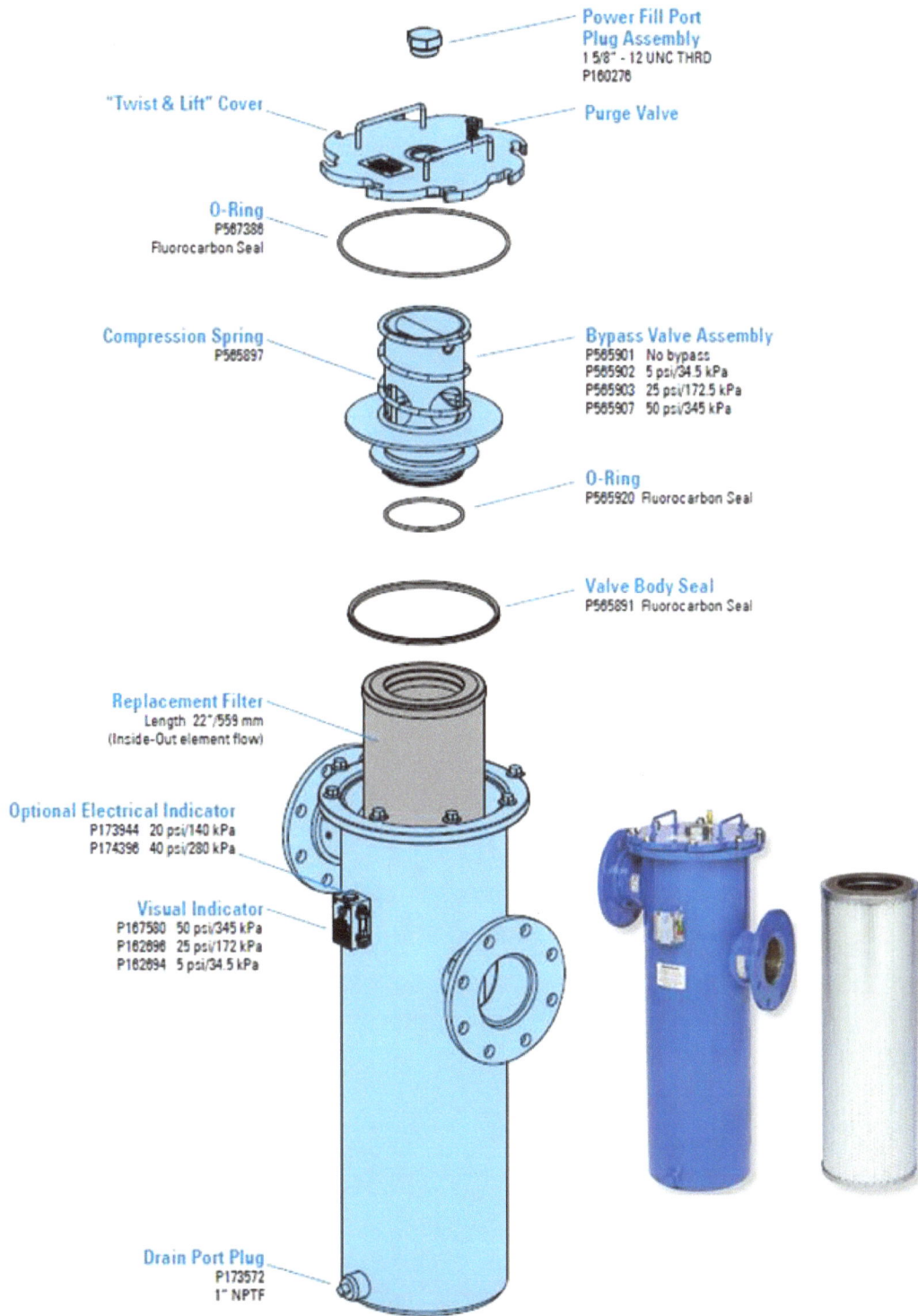

Power Fill Port Plug Assembly
1 5/8" - 12 UNC THRD
P160276

"Twist & Lift" Cover

Purge Valve

O-Ring
P567388
Fluorocarbon Seal

Compression Spring
P565897

Bypass Valve Assembly
P565901 No bypass
P565902 5 psi/34.5 kPa
P565903 25 psi/172.5 kPa
P565907 50 psi/345 kPa

O-Ring
P565920 Fluorocarbon Seal

Valve Body Seal
P565891 Fluorocarbon Seal

Replacement Filter
Length 22"/559 mm
(Inside-Out element flow)

Optional Electrical Indicator
P173944 20 psi/140 kPa
P174396 40 psi/280 kPa

Visual Indicator
P167580 50 psi/345 kPa
P162696 25 psi/172 kPa
P162694 5 psi/34.5 kPa

Drain Port Plug
P173572
1" NPTF

**Fig. 6.23- Example of Line-Mounted Custom-Ported Filters - HRK10 Series
(Courtesy of Donaldson)**

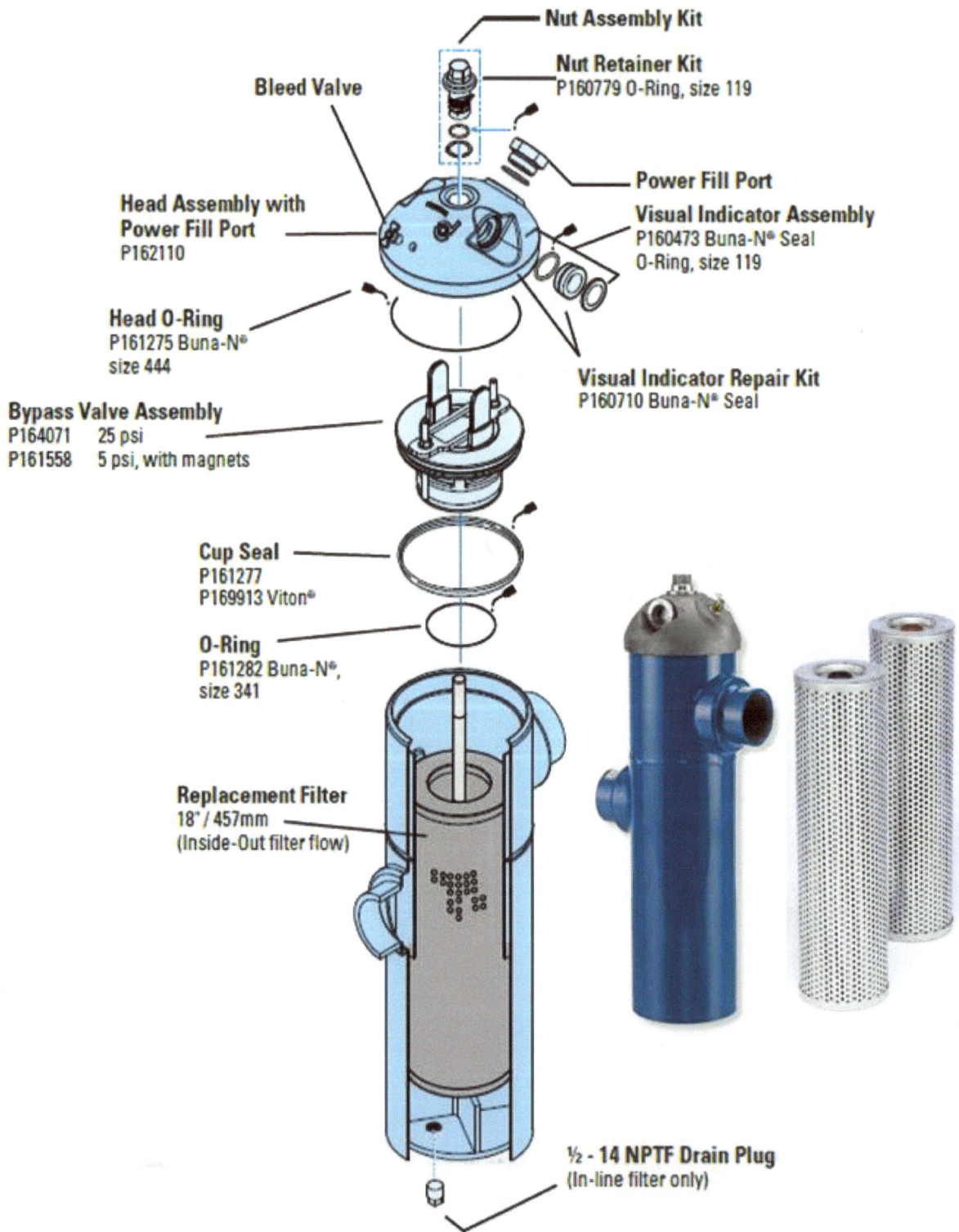

Nut Assembly Kit

Nut Retainer Kit
P160779 O-Ring, size 119

Bleed Valve

Power Fill Port

Head Assembly with Power Fill Port
P162110

Visual Indicator Assembly
P160473 Buna-N® Seal
O-Ring, size 119

Head O-Ring
P161275 Buna-N®
size 444

Visual Indicator Repair Kit
P160710 Buna-N® Seal

Bypass Valve Assembly
P164071   25 psi
P161558   5 psi, with magnets

Cup Seal
P161277
P169913 Viton®

O-Ring
P161282 Buna-N®,
size 341

Replacement Filter
18" / 457mm
(Inside-Out filter flow)

½ - 14 NPTF Drain Plug
(In-line filter only)

Fig. 6.24- Example of Line-Mounted Custom-Ported Filters - HFK08 Series
(Courtesy of Donaldson)

6.7.9 – Spin-On Single Filters

Figure 6.25 shows the exterior shape of *Spin-On* filters. Such filters feature non-welded screw-in housing design Figure 6.26 shows detailed construction of a single *Spin-On* filter consists.

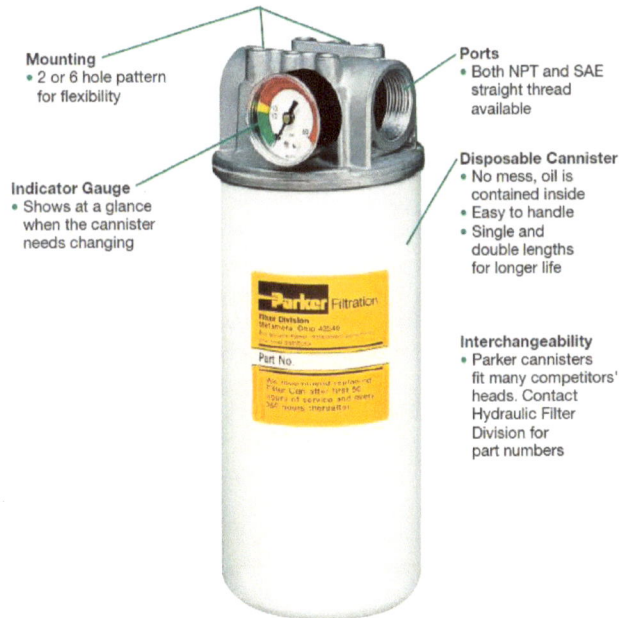

Fig. 6.25- Example of Spin-On Single Filters - 12AT/50AT Series (Courtesy of Parker)

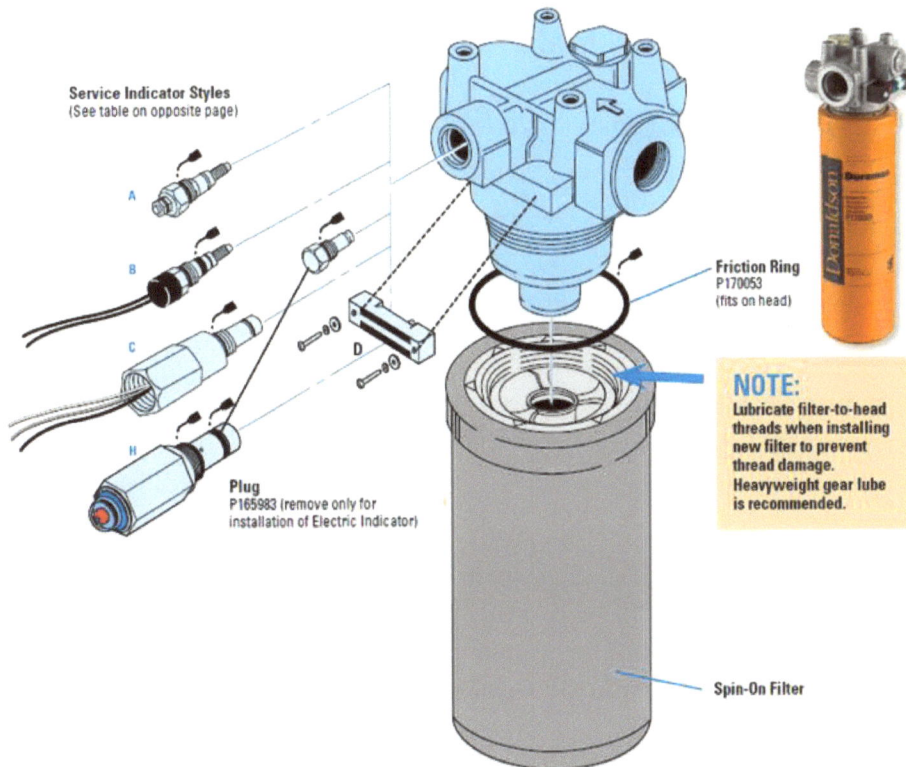

Fig. 6.26- Example of Spin-On Single Filters - HMK03 Series (Courtesy of Parker)

6.7.10 – Spin-On Dual Vertical Filters

Figure 6.27 shows an example of *Spin-On Dual Vertical* filters.

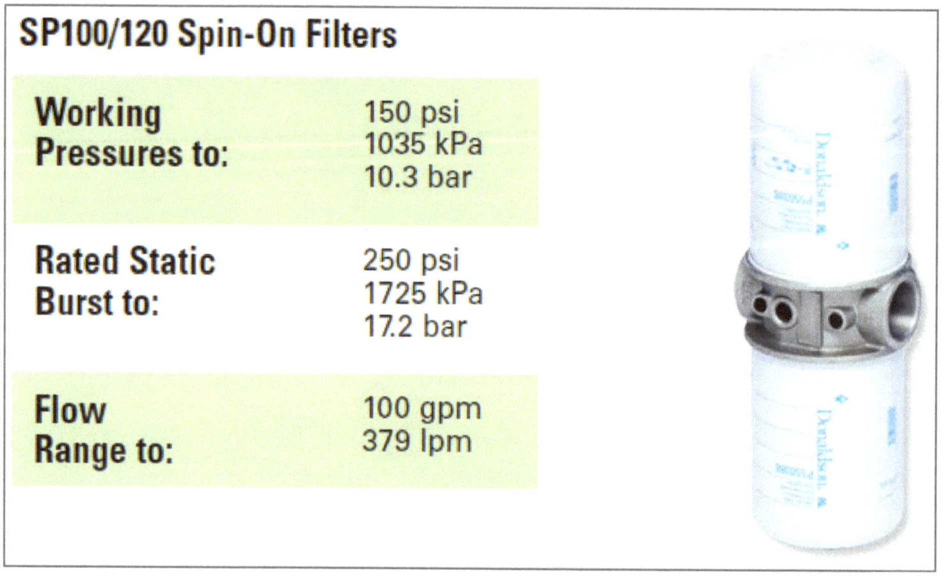

Fig. 6.27- Example of Spin-On Dual Vertical Filters (Courtesy of Donaldson)

6.7.11 – Spin-On Dual Horizontal Filters

Figure 6.28 shows an example of *Spin-On Horizontal Dual* filters.

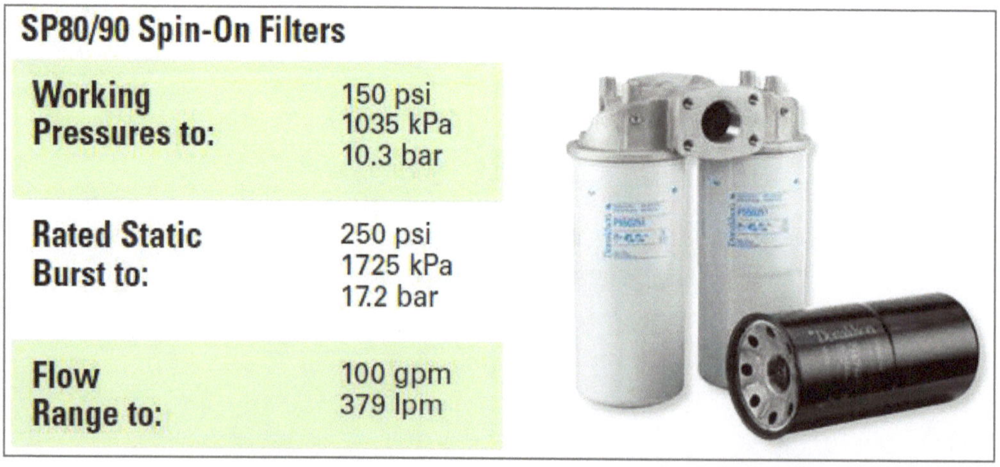

Fig. 6.28- Example of Spin-On Dual Horizontal Filters (Courtesy of Donaldson)

6.7.12 – Flange-Side-Mounted Filters

Figure 6.29 shows various models of of *Flange-Side-Mouned* filters.

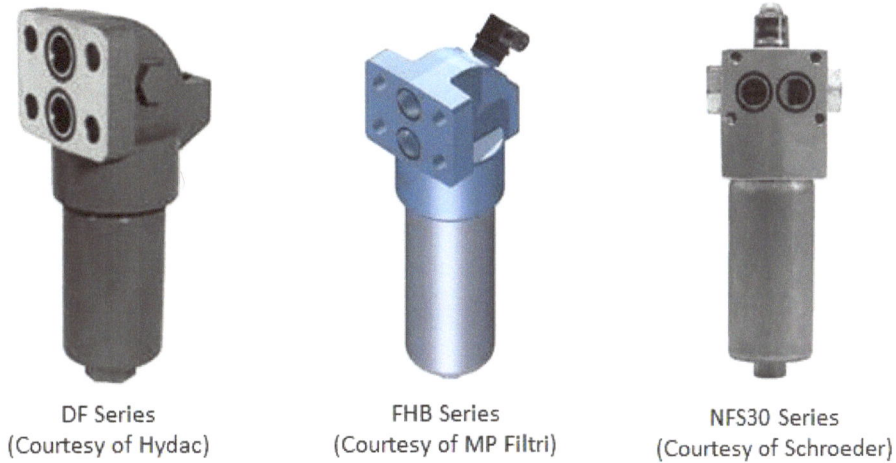

DF Series
(Courtesy of Hydac)

FHB Series
(Courtesy of MP Filtri)

NFS30 Series
(Courtesy of Schroeder)

Fig. 6.29- Examples of Flange-Side-Mounted Filters

6.7.13 – Flange-Top-Mounted Filters

Figure 6.30 shows various models of of *Flange-Top-Mouned* filters.

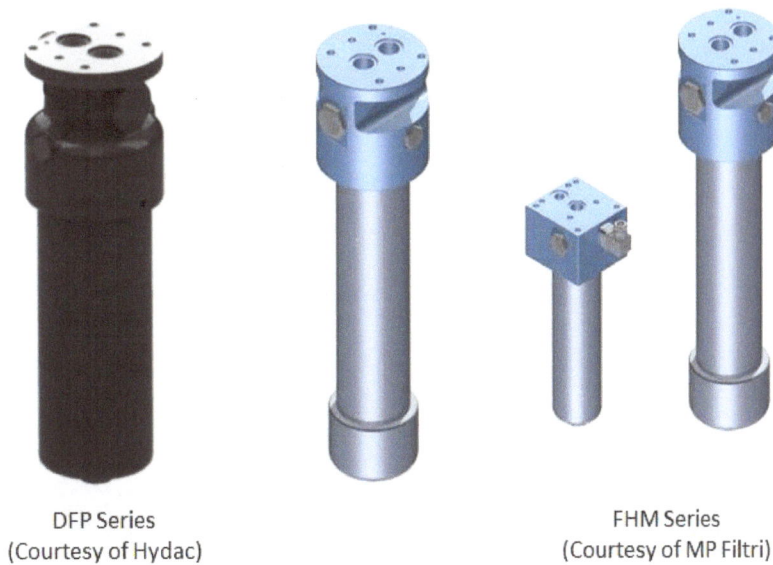

DFP Series
(Courtesy of Hydac)

FHM Series
(Courtesy of MP Filtri)

Fig. 6.30- Examples of Flange-Top-Mounted Filters

6.7.14 – Flange-Bottom-Mounted Filters

Figure 6.31 shows an example of of *Flange-Bottom-Mouned* filters.

Cover
- Handle protects indicators from damage
- Easy on, easy off, for fast service

Air Bleed
- Helps protect bearings and other sensitive components from trapped air

Fill Port
- Prefilter the fluid, before it gets into the machine's system
- Purge air while filling

Indicators
- You can tell element condition at a glance
- Both visual and electrical available

Bowl
- Rugged cold drawn steel— excellent fatigue resistance
- Three sizes for any application: Single (8"), Double (16"), and Triple (39")

Ports
- SAE straight thread or flange face

Bypass Valve (not visible)
- Soft seat design for zero internal leakage
- Located in cover assembly

Drain Port (not visible)
- Clean and easy servicing
- Lets you drain bowl of fluid- before element changes

Fig. 6.31- Examples of Flange-Bottom-Mounted Filters (Courtesy of Parker)

6.7.15 – Sandwich-Mounted Filters

Figure 6.32 shows various models of of *Sandwich-Mouned* filters.

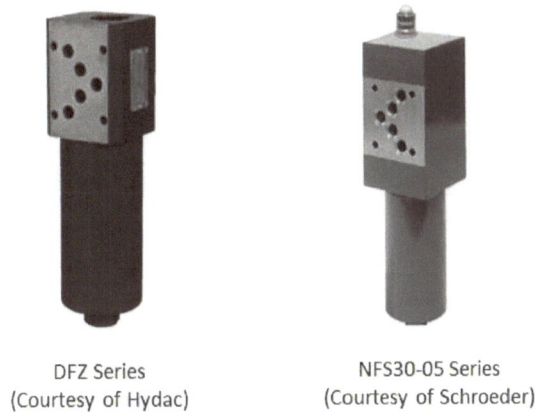

DFZ Series
(Courtesy of Hydac)

NFS30-05 Series
(Courtesy of Schroeder)

Fig. 6.32- Examples of Sandwich-Mounted Filters

6.7.16 – Screw-In Filters

Figure 6.33 shows an example of *Screw-In* filters. They also refrred to as *Manifold-Mounted* or *Cartridge-Style* filters. As shown in the figure, they are installed in special cavities to protect critical components from oil-born contaminants. They are not intended to replace main filters.

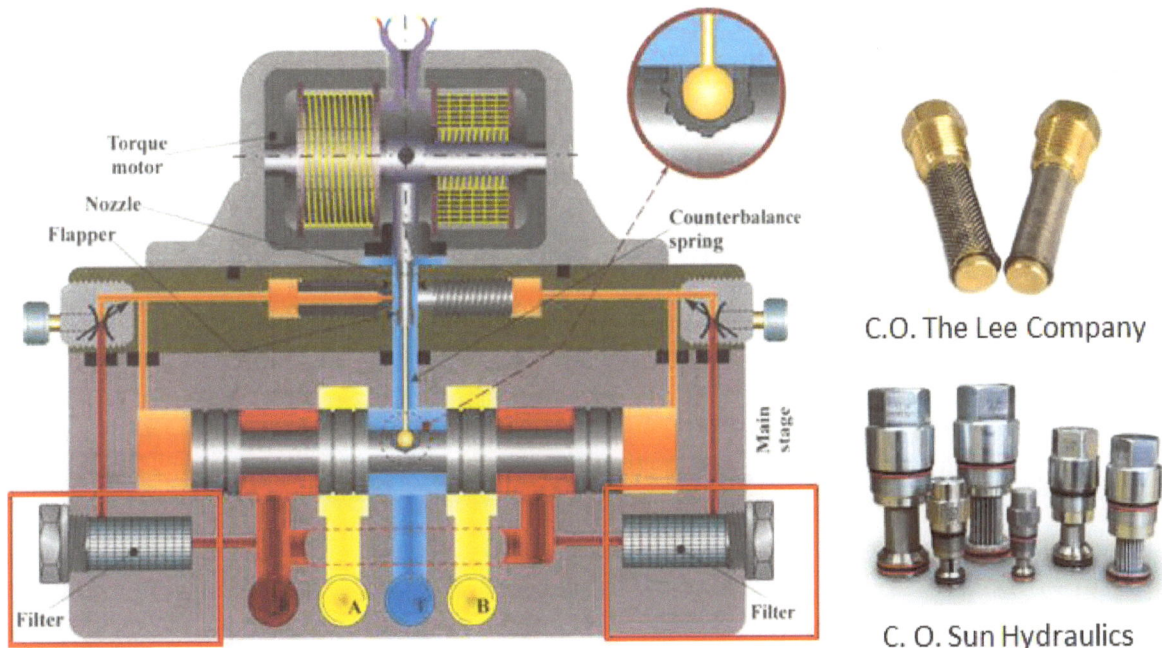

Fig. 6.33- Example of Screw-In Filters – CP-C16 Series (Courtesy of Hydeck)

6.8 - Filter Clogging Indicators

Purpose of Clogging Indicators: Some users install filters without indicators, preferring instead to change and/or clean elements according to a fixed time schedule or based on number of hours of operation. Considering this approach is either risky because a filter may be replaced while it is very dirty, or not cost effective because a filter may be replaced while it is still clean. It may be difficult to establish a reliable schedule for installing new elements because the rate of dirt ingression is not known and varies from time-to-time and from machine-to-machine. Clogging indicators are warning devices that signal visually and/or electrically that the filter element is filled with contaminants and should be changed or cleaned.

Advantages of using Clogging Indicators:
- Eliminates the need to guess when the element will clog.
- Avoids the unnecessary cost of replacing elements too soon.
- Shutdown a machine that has sensitive components that is intolerant to contamination.

Configurations of Clogging Indicators: As shown in Fig. 6.34, clogging indicators could be visual, electrical, or optoelectrical.

Fig. 6.34- Various Configurations of Clogging Indicators (Courtesy of MP filtri)

Differential vs. Static Pressure Visual Indicator:
- *Differential Pressure Indicators* react to the pressure drop across the filter that is caused by the flow of fluid through the filter housing and element. These devices measure the difference in pressure upstream and downstream of the filter element, regardless of the system pressure. They are utilized in most pressure and inline filters.

- *Static Pressure Indicators* measure only the build-up of pressure upstream of the filter element *(downstream pressure is ambient – tank vented to atmosphere)*. Consequently, if any components are located downstream of the filter, the indicator will generate a false reading of pressure at the filter entrance. As a result, static indicators are recommended only on filters that discharge directly to vented tanks (return and offline filters).

Clogging Indicator Settings: These devices activate *(trip)* when the flow of fluid causes a pressure drop across the filter element that exceeds the indicator setting. The indicator is set to trip well before the element becomes fully clogged and lower than bypass cracking pressure, thereby giving the operator sufficient time to take corrective action. The following are examples excerpted from Hydac literature:

- In a majority of applications, a HYDAC indicator is set to trip at 15 psi (1 bar) below the bypass valve cracking pressure; or, for a non-bypass filter, at 15 psid (1 bar) below the element design changeout pressure.

- Typically, a HYDAC pressure filter bypass valve begins to crack at 87 psid (6 bard), so the indicator is set to trip at 72 psid (5 bard).

- A HYDAC return filter ordinarily begins to bypass at 43 psi (3 bar), so the indicator is set to trip at 29 psi (2 bar).

Manual vs. Automatic Reset:
- *Electrical Clogging Indictors* reset automatically to their original position when the pressure across the filter drops below trip pressure.
- *Visual Clogging Indicator* are reset automatically or manually. The advantage of manual reset is that the indicator shows that the element is dirty even after the system is shut down.

Interchangeability: As shown in Fig. 6.35, a filter is manufactured to be equipped with various types of clogging indicators without the need for special mechanical arrangement.

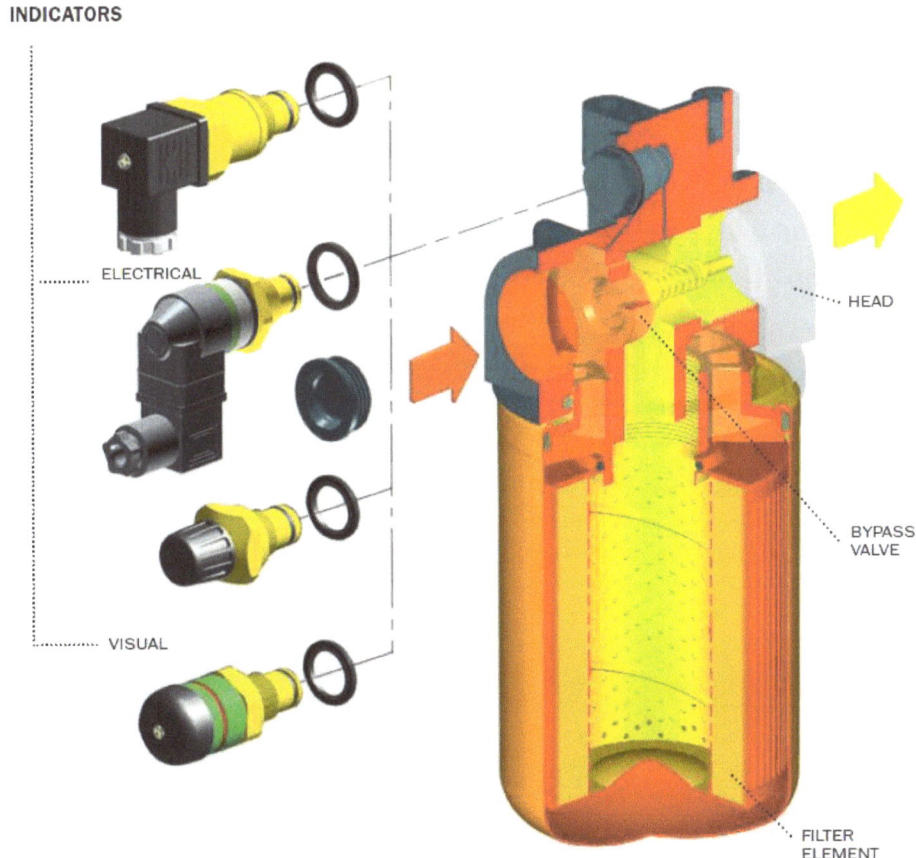

Fig. 6.35- Examples of Clogging Indicators (Courtesy of Assofluid)

6.8.1- Visual Clogging Indicators

Operation of Visual Differential Pressure Clogging Indicators: Figure 6.36 shows a typical *differential pressure* visual clogging indicator. The differential pressure across the filter increases, the piston/magnet assembly is driven down against a spring until the attractive force between the magnet and indicator pin is reduced sufficiently to allow the indicator to trip. Tripping results in the indicator pin rises giving visual indication that the filter must be serviced. Indicator is automatically reset when Δp < trip Δp.

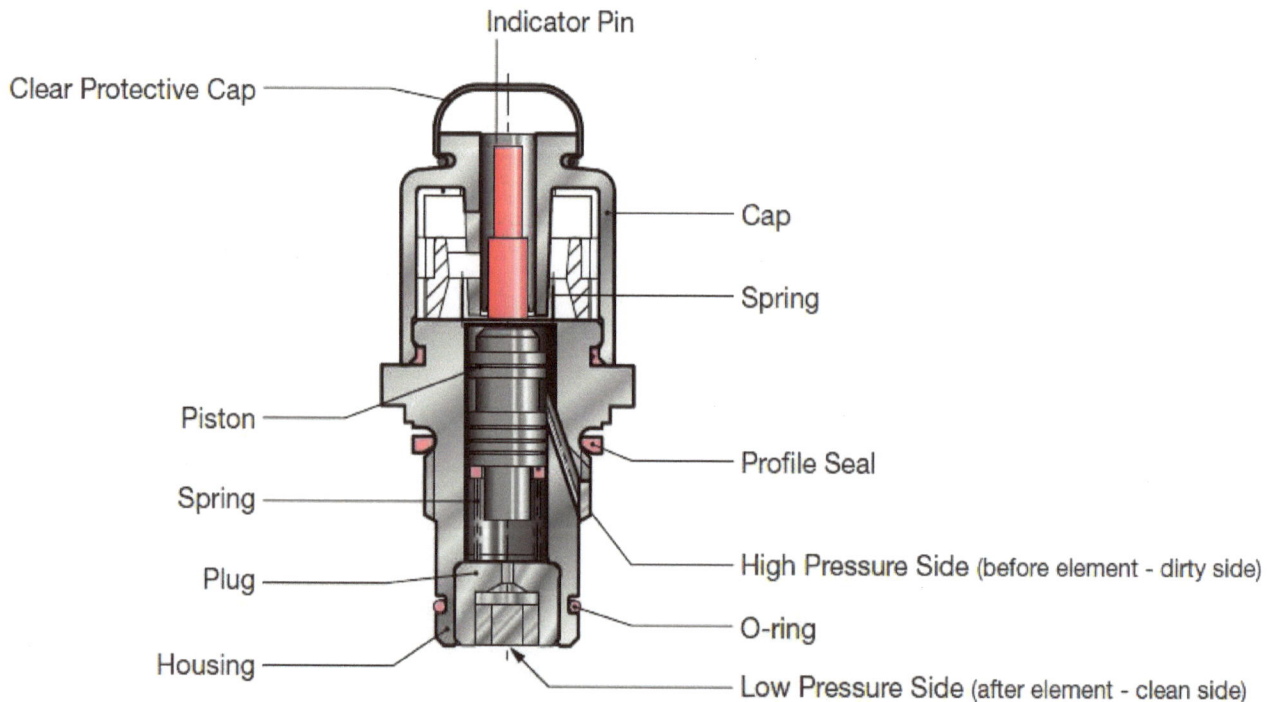

Fig. 6.36- Visual Differential Pressure Indicators (Courtesy of Hydac)

Operation of Static Pressure Visual Indicators: Figure 6.37 shows a typical *static pressure* visual clogging indicator. Increasing pressure upstream of the filter acts upon a diaphragm in the indicator and causes the indicator pin to rise opposing spring force until it trips at a pre-set pressure. The indicator pin automatically resets once pressure is reduced below the trip pressure.

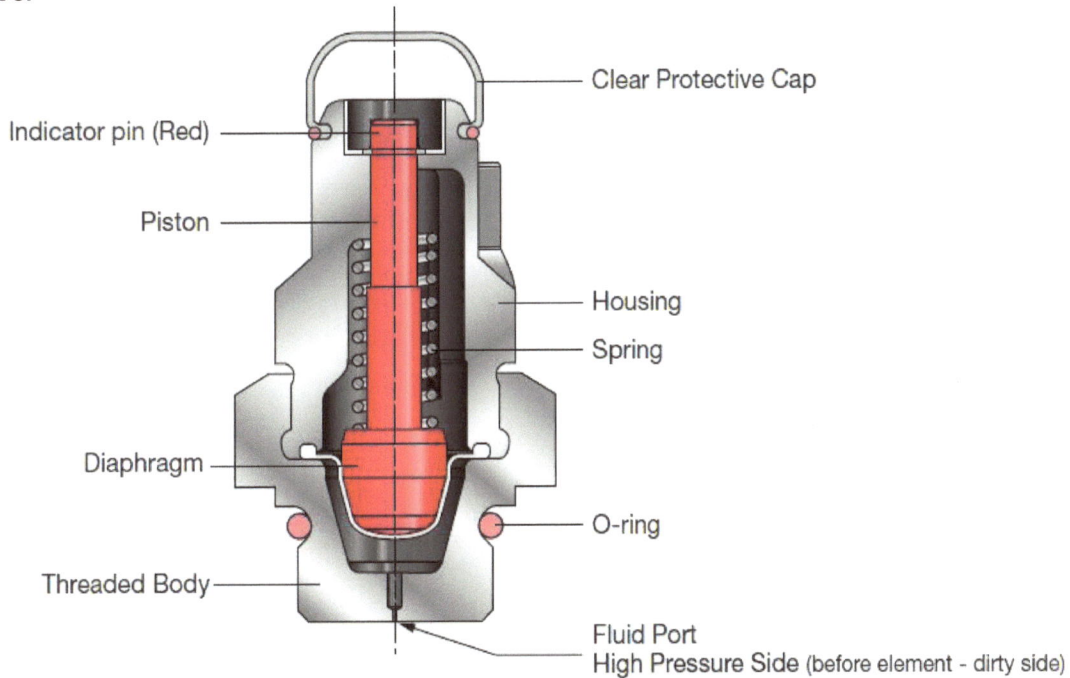

Fig. 6.37- Static Pressure Visual Indicators (Courtesy of Hydac)

6.8.2- Electrical Clogging Indicators

Concepts: In the *Electrical Clogging Indicator*, the pressure drop across the filter may also be used to actuate an electric switch. The resulting electric signal is used to actuate a remote light, bell or buzzer to indicate the need for service. In critical applications, such as when contamination sensitive expensive components are used, the resulting electric signal may be used to stop the machine to avoid damage that might occur due to a plugged or restricted filter.

Thermal Lockout: When mobile and other equipment is started in the cold, the hydraulic or lube fluid is likely to be highly viscous until it approaches normal operating temperature. The high pressure drop created by a highly viscous fluid can trip the indicator and falsely signify that the element is clogged. An optional thermal lockout device, available on many electric indicators, prevents the indicator from tripping until the fluid reaches a certain specified temperature. The device consists of a switch in series (AND function) in the indicator circuit, which is caused to make or break by a bi-metal strip that alters in shape according to temperature.

Single Pole, Double Throw Switches (SPDT): Differential pressure and most static pressure electrical indicators contain single-pole, double-throw switches. This provides the choice of *normally open (NO)* or *normally closed (NC)* contacts when the pressure differential is below trip-point.

Operation of Differential Pressure Electrical Clogging Indicator: Figure 6.38 shows a typical differential pressure electrical clogging indicator. The differential pressure across the filter increases, the piston is driven down against a spring. Tripping causes a switch to make or break, generating warning or control signal for servicing.

Electric static pressure indicators, which also operate mechanically, are available.

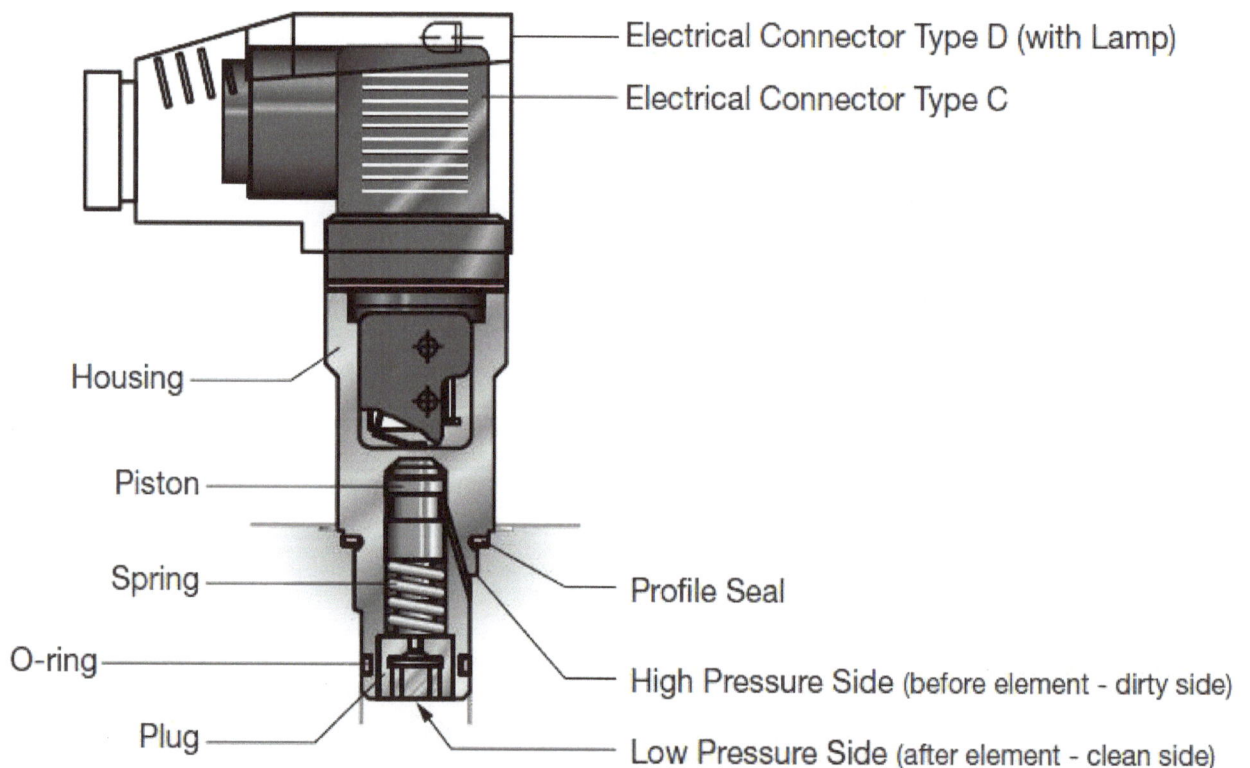

Electrical Connector Type D (with Lamp)
Electrical Connector Type C
Housing
Piston
Spring
Profile Seal
O-ring
Plug
High Pressure Side (before element - dirty side)
Low Pressure Side (after element - clean side)

Fig. 6.38- Differential Pressure Electrical Indicators (Courtesy of Hydac)

6.9 – Types of Filters Based on their Placement in the Circuit

Basically, filters shall be installed where they are readily accessible, the space around allows for replacing filter elements, and filter element is replaced without emptying the tank. Ideally, each component in a hydraulic system would be equipped with its own filter, but this is economically impractical in most cases. A more practical solution is to examine the system and identify the most contamination-sensitive components and provide filtration as close as possible to these components. As shown in Fig. 6.39 and Fig. 6.40, filters may be located in a circuit in one or more of the following locations:

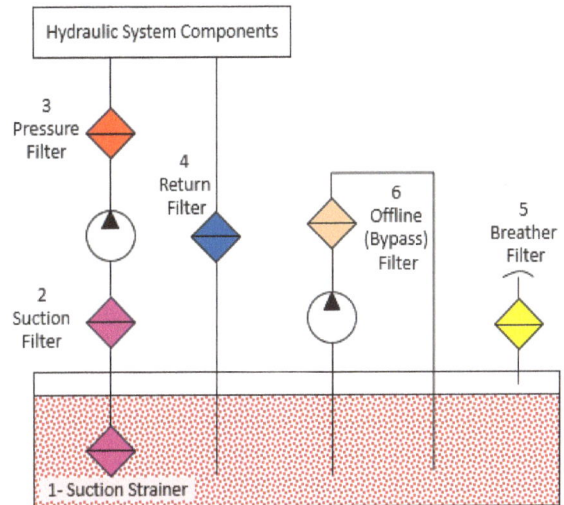

Fig. 6.39- Types of Filters Based on their Placement in the Circuits

- **Online Filters:** These filters are placed in series with the main pump as follows: Suction Strainer (1), Suction Filters (2), Pressure Filters (3), and Return Filters (4).

- **Offline Filters:** These filters are placed in parallel with the main pump as follows: Breather Filters (5), and Offline (Bypass) Filters (6).

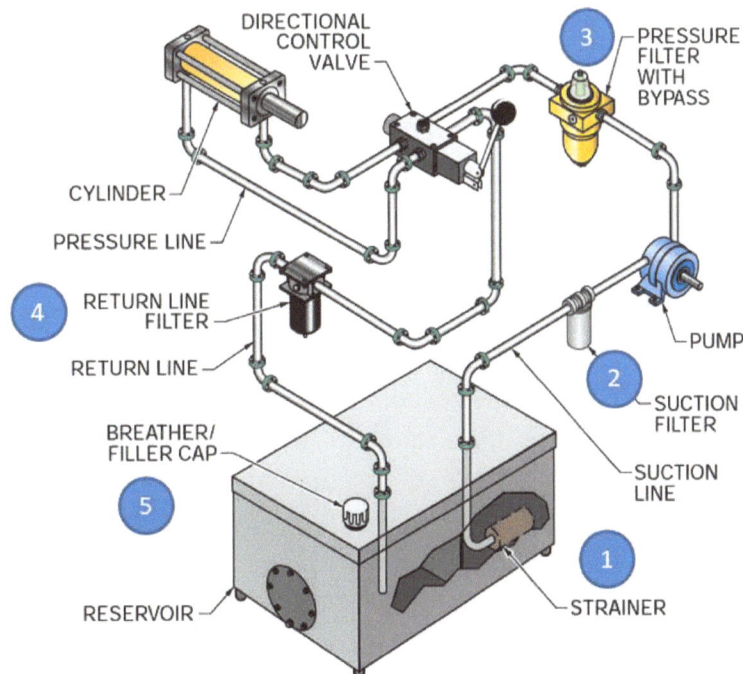

Fig. 6.40- Types of Filters Based on their Placement in the Circuits
(Courtesy of American Technical Publishers)

6.9.1- Suction Strainers

Placement: As shown in Fig. 6.41 and 6.42, *Suction Strainers* are connected to the beginning of the suction line before a suction filter (if found). As shown in the figure, a strainer could be aligned with the suction line or a 90° elbow at the bottom of the suction line.

Primary Duty: The primary duty of suction strainers is to capture the relatively large particles, chips or rags before getting into the pump. Generally speaking, because of the cavitation concerns, suction strainers are not recommended for large flow pumps. Installing a suction strainer on the suction line subject to approval from the pump manufacturer.

Cost: Suction strainers are relatively inexpensive as it does not have a complex housing.

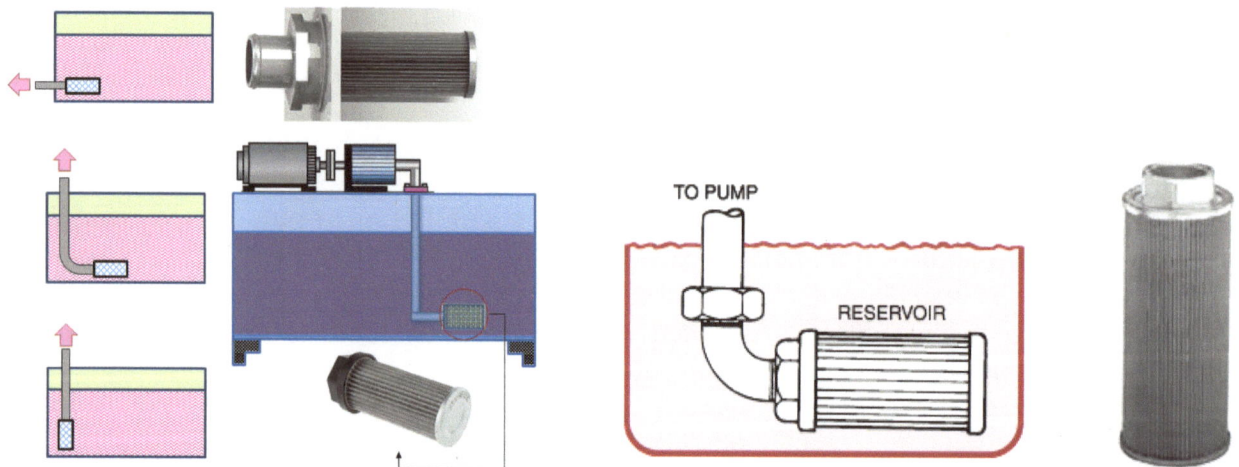

Fig. 6.41 – Suction Strainers   **Fig. 6.42 – Suction Strainers (Courtesy of Schroeder)**

Construction: As shown in Fig. 6.43, suction strainers are furnished with optimized pleat size and screen area for extended life and low pressure drop. Some suction strainers could be magnetic in order to capture metallic contaminants or wear products.

Bypass Valve: Suction strainers mounted at the beginning of the suction line inside the reservoir without a bypass valve or pressure drop indicator are not recommended. They are hidden from sight and there is no easy access to remove and clean them. Overtime, without proper cleaning, they will plug restricting flow and causing pump cavitation leading to pump failure.

One pieced high strength nylon hex cap for reduced cost.

Cap assembly epoxy bonded to body for superior strength.

Pleated stainless wire cloth provides excellent flow with minimum pressure drop. Choice of different mesh sizes.

Inner perforated steel support tube for added strength and rigidity.

Optional 3 or 5 PSI relief valve to prevent failure should the screen become clogged with debris.

Sides of end caps are reverse tapered to enhance epoxy bonding.

Fig. 6.43 – Suction Strainers (ohfab.com)

Micron Size vs. Mesh Size: *Micron size* is the size of the largest particles (in microns) that can pass through the screen. *Mesh Size* is the number of holes in one squared inch. Large mesh size means finer filter. Micron Size is also referred to as *Pore Size*.

Mesh Size: If no *mesh size* is reported by the pump manufacturer, 250-500 mesh size is recommended. Mesh size is also referred to as *Porosity*.

Flow: a strainer receives the full flow of the main pump flow.

Surface Area: A strainer is sized based on the pump flow. Size of the strainer should offer surface area to minimize the pressure drop across the strainer so that pump cavitation is avoided. Review the pump data sheet if found. Otherwise, consult the pump manufacturer in regard to recommended surface area. If no information is found, the following rule of thumb is applicable as a guideline. Surface area shouldn't be less than 2 square inches for every GPM of the pump flow ($\approx$ 3 cm^2 for every liters/min of the pump flow). Equations 1.1A and 1.1B are used to determine the surface area of the strainer in metric and English system of units, respectively.

Suction Strainer Surface Area (cm^2) = 3 x Qp (lit/min)   **1.1A**

Suction Strainer Surface Area (in^2) = 2 x Qp (gpm)   **1.1B**

Example: A 20 GPM pump. Eq. 1.1B → A minimum strainer surface area of 40 square inches.

Magnetic Suction Strainers: Figure 6.44 shows the specification for magnetic suction strainers. Magnetic suction strainers offer dual protection to the pump inlet without risk of cavitation. Powerful ceramic magnets located parallel to the pleated mesh attract and protect against damaging ferrous particles of all sizes. The pleated stainless-steel screen provides additional filtration protection for larger particles that would result in catastrophic failure. The generous open area of the stainless-steel pleated mesh screen eliminates the possibility of pump cavitation.

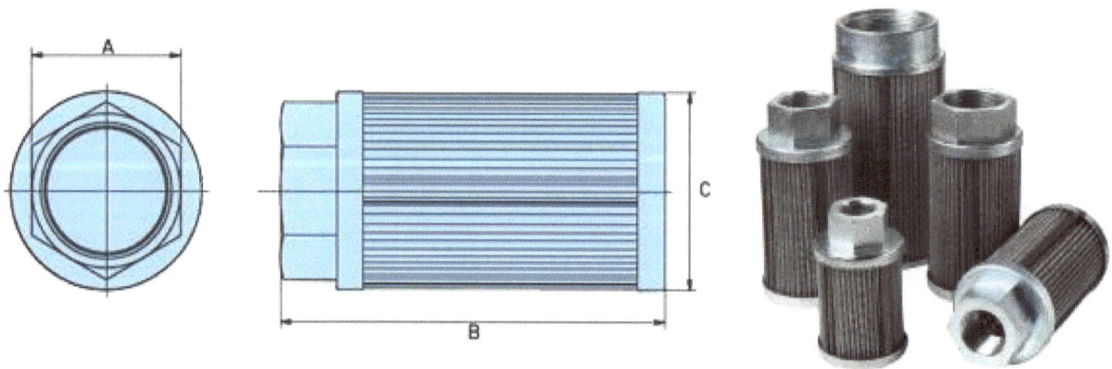

Flow Vs. Pressure Loss

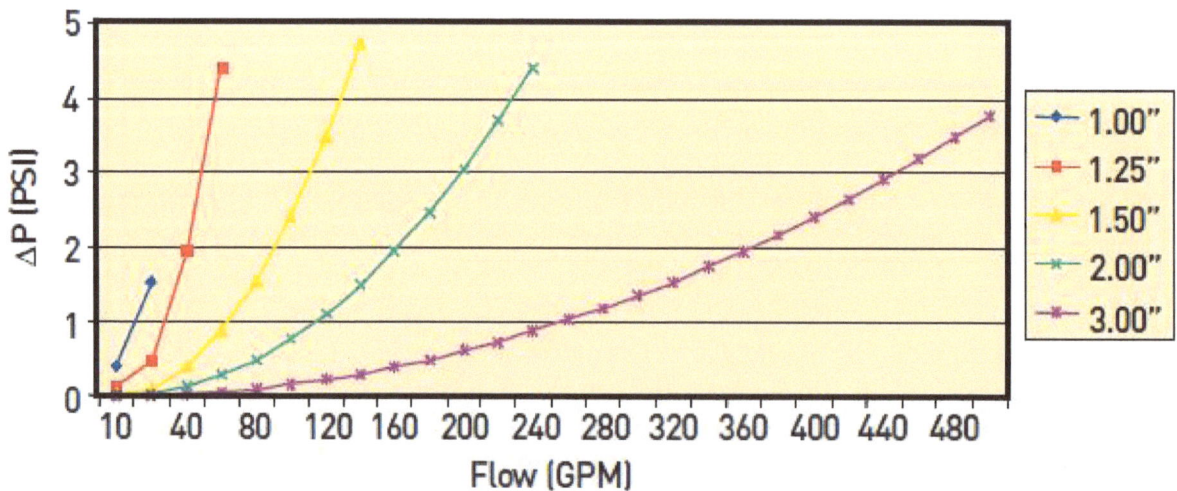

Fig. 6.44 – Magnetic Suction Strainers (Courtesy of Parker)

6.9.2- Suction Filters

Placement: As shown in Fig. 6.45, A *Suction Filter* is placed on the suction line before the inlet port of the main pump, between the pump and a suction strainer (if found). Unlike strainers, suction filters are mounted externally, i.e., outside the reservoir.

Primary Duty: The primary duty of suction filters is to protect the main pump. However, the system components downstream the pump aren't protected from any contamination that may be generated by the pump during operation. Like suction strainers, pressure drop across the filter and pump cavitation issue must be taken into consideration when sizing suction filters. Always consult the pump manufacturer for inlet restrictions. However, if no information is found, the maximum acceptable pressure drops must not be higher than 0.1 bar (1.45 psi).

Cost: Suction filters are more expensive than strainers, but less expensive than pressure filters since no special housing is needed for high pressure.

Construction: A typical suction filter consists of a housing, a filter element, and a filter head that contains filter ports.

Bypass Valve: Bypass valve is highly recommended in suction filters avoiding pump cavitation in case of filter clogging. When expensive pumps are used, adding a vacuum switch to stop the machine when the filter is clogged is an extra protection for the pump.

Mesh Size: Filter media in a suction filter is coarse filter, ranging from ranges from 60 to 250 micron, in order to limit the pressure, drop across the filter.

Flow: Suction filters receive the full flow of the main pump. So, they are sized based on the main pump flow.

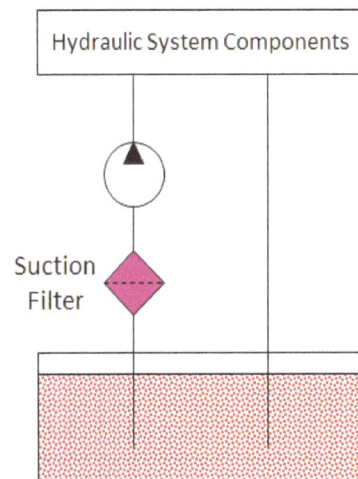

Fig. 6.45 – Placement of Suction Filters

Figure 6.46 shows a typical example of an On-Tank-Top Single suction filter. The filter is equipped with a differential pressure switch to trip the machine if the filter is clogged. The data sheet associated with the filter provides full information about it.

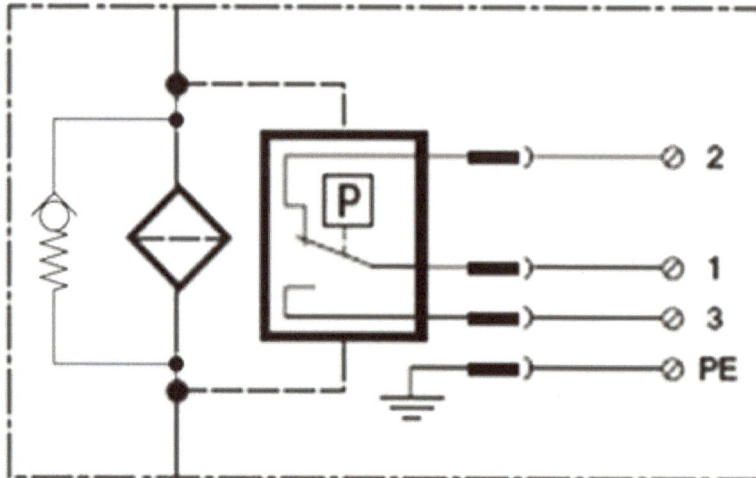

SF Series
In-tank Suction Filters
360 psi • up to 200 gpm

Technical Specifications

Mounting Method	4 mounting holes - filter head	
Flow Direction	Inlet: Bottom	Outlet: Side
Construc. Materials	Housing	Lid
SF 110-330	Aluminum	Aluminum
SF 950-1300	Ductile Iron	Ductile iron
Flow Capacity		
110	5 gpm (20 lpm)	
240	15 gpm (57 lpm)	
330	30 gpm (114 lpm)	
950	175 gpm (662 lpm)	
1300	200 gpm (757 lpm)	
Housing Pressure Rating		
Max. allowable working pressure	360 psi (25 bar)	
Fatigue Pressure	360 psi (25 bar) @ 700,000 cycles	
Burst Pressure	110	1080 psi (75 bar)
	240	1230 psi (85 bar)
	330	1440 psi (100 bar)
	950-1300	>1440 psi (100 bar)
Element Collapse Pressure Rating		
W/HC	290 psid (20 bar)	
Fluid Temp. Range	14°F to 212°F (-10°C to 100°C)	
Consult HYDAC for applications operating below 14°F (-10°C)		
Fluid Compatibility		
Compatible with all hydrocarbon based, synthetic, water glycol, oil/water emulsion, and high water based fluids when the appropriate seals are selected		
Indicator Trip Pressure		
ΔP = 3 psi (0.2 bar) -10% (standard)		
Bypass Valve Cracking Pressure		
ΔP = 3 psi (0.2 bar) +10% (standard - sizes 60, 950, 1300)		
ΔP = 4.4 psi (0.3 bar) +10% (standard - sizes 110,160,240,330)		

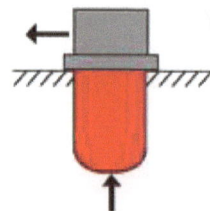

Fig. 6.46- Example of On-Tank Top Mounted Suction Filter (Courtesy of Hydac)

6.9.3- Pressure Filters

Placement: As shown in Fig. 6.47, In open hydraulic circuits, *Unidirectional Pressure Filters* are placed directly downstream of the main pump. In closed hydraulic circuits, *Bidirectional Pressure Filters* are used to handle reserve flow. This is accomplished using a check valve rectifier built into the filter head.

Primary Duty: The primary duty of pressure filters is to protect the sensitive components in the system such as proportional and servo valves. Since the pump produces wear debris, contamination is captured before it is spread to the rest of the system.

Cost: Pressure filters are the most expensive because it requires a special housing that is sealed and rated work under maximum system pressure.

Mesh Size: Filter media in a pressure filter is selected to maintain the cleanliness level required by the system manufacturer.

Flow: Pressure filters receive the full flow of the main pump. So, they are sized based on the main pump flow.

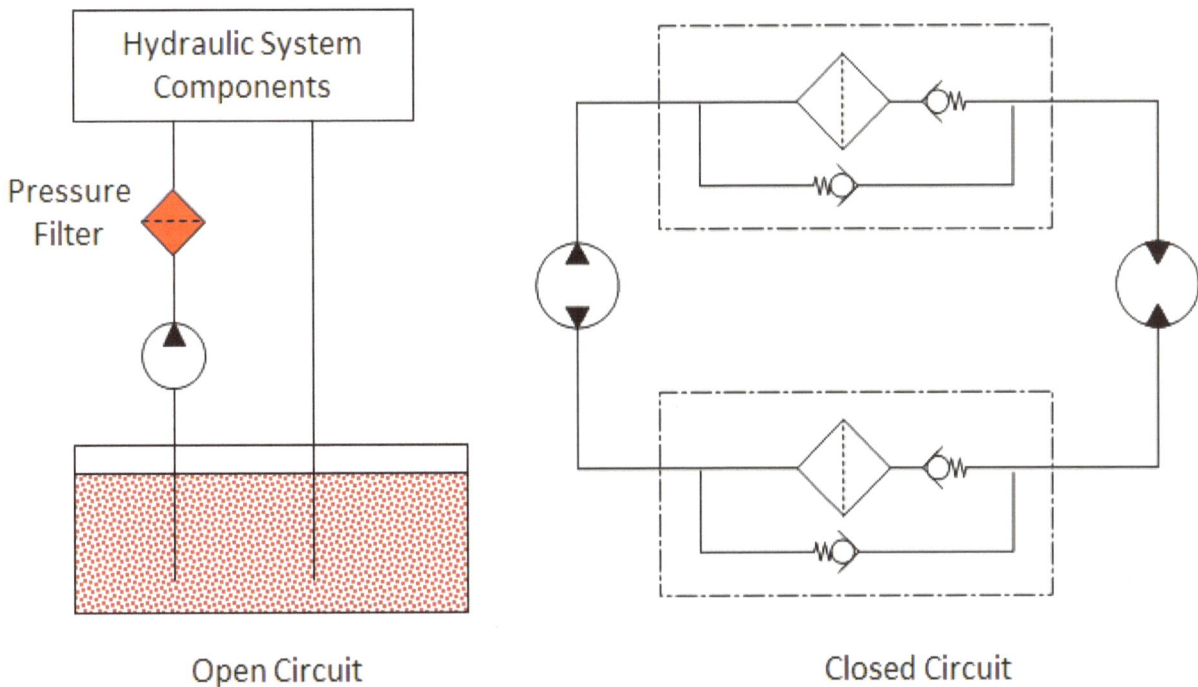

Open Circuit   Closed Circuit

Fig. 6.47 – Placement of Pressure Filters

Construction: As shown in Fig. 6.48, the construction of a pressure filter is typically consisted of same components as suction and return filters. However, housing of such filters must withstand the maximum pressure. Steady state pressure ratings are 100 - 400 bar (1500 psi to 6000 psi). Additionally, pressure distribution analysis must be made to make sure the pressure filter is capable to work under possible pressure spikes and pressure fluctuation without subjecting to collapse of fatigue failure.

Bypass Valve: in pressure filters, bypassing the filter media is an optional feature. Non-Bypass filters are used when sensitive components that are less tolerant to contamination are used. However, if such filters are used, filter media collapse pressure must be at least equal to maximum working pressure. Otherwise, Bypass-to-Tank filters could be used.

1. High strength ductile iron filter head with integral indicator port

2. Steel bowl with standard drain port

3. Proprietary element endcap assembly includes bypass and reverse flow valves

4. Patented deformable tangs secure element in bowl

5. Coreless element assembly

6. Re-usable element support core

Fig. 6.48- Sectional view of Pressure Line Filter WPF Series (Courtesy of Parker)

Figure 6.49 shows a typical example of a Line-Mounted Top Ported pressure filter. The filter is equipped with a differential pressure visual clogging indicator. The specification sheet shows that this filter can work in a pressure up to 420 bar (6000 psi) The data sheet associated with the pump provides full information about the filter.

Technical Specifications

Mounting Method	4 mounting holes
Port Connection	
30	SAE-8, 1/2" NPT, 1/2" BSPP
60/110	SAE-12, 3/4" NPT, 3/4" BSPP, 3/4" SAE, Code 62
160/240/280	SAE-20, 1 1/4" NPT, 1 1/4" BSPP, 1 1/4" SAE, Code 62
330/660/1320	SAE-24, 1 1/2" NPT, 1 1/2" BSPP, 2" SAE Flange Code 62
Flow Direction	Inlet: Side Outlet: Side
Flow Capacity	
30	8 gpm (30 lpm)
60	16 gpm (60 lpm)
110	29 gpm (110 lpm)
160	42 gpm (160 lpm)
240	63 gpm (240 lpm)
280	74 gpm (280 lpm)
330	87 gpm (330 lpm)
660	174 gpm (660 lpm)
1320	200 gpm (757 lpm)

Housing Pressure Rating

Max. Allowable Working Pressure: 6090 psi (420 bar)
Fatigue Pressure: 6090 psi (420 bar) @ 1 million cycles
Burst Pressure:
- 30: 15950 psi (1100 bar)
- 60/110: 17400 psi (1200 bar)
- 160/240/280: 17110 psi (1180 bar)
- 330/660/1320: 15080 psi (1040 bar)

Element Collapse Pressure Rating
- BH4HC, V: 3045 psid (210 bar)
- ON, W/HC: 290 psid (20 bar)

Fluid Temp. Range: 14°F to 212°F (-10°C to 100°C)
Consult HYDAC for applications operating below 14°F (-10°C)

Fluid Compatibility
Compatible with all hydrocarbon based, synthetic, water glycol, oil/water emulsion, and high water based fluids when the appropriate seals are selected.

Indicator Trip Pressure
ΔP = 29 psid (2 bar) -10% (optional)
ΔP = 72 psid (5 bar) -10% (standard)
ΔP = 116 psid (8 bar) -10% (optional non bypass)

Bypass Valve Cracking Pressure
ΔP = 43 psid (3 bar) +10% (optional)
ΔP = 87 psid (6 bar) +10% (standard)
Non Bypass Available

DF Series
Inline Filters
6090 psi • up to 200 gpm

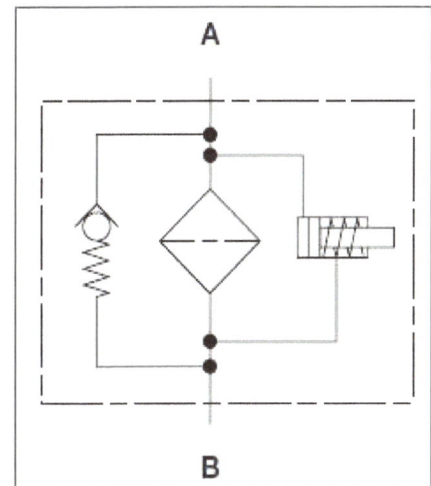

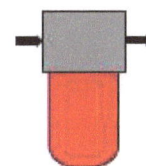

Fig. 6.49- Example of Line-Mounted Top-Ported Pressure Filter (Courtesy of Hydac)

Pressure Drop: As shown in Fig. 6.50, it is to be noted that the overall pressure drop across the filter is the sum of the pressure drop across the housing and the pressure drop across the filter element. As shown in the figure, for the same flow through the filter, the pressure drop across the filter element is inversely proportional to its micron size (i.e. directly proportional to mesh siz).

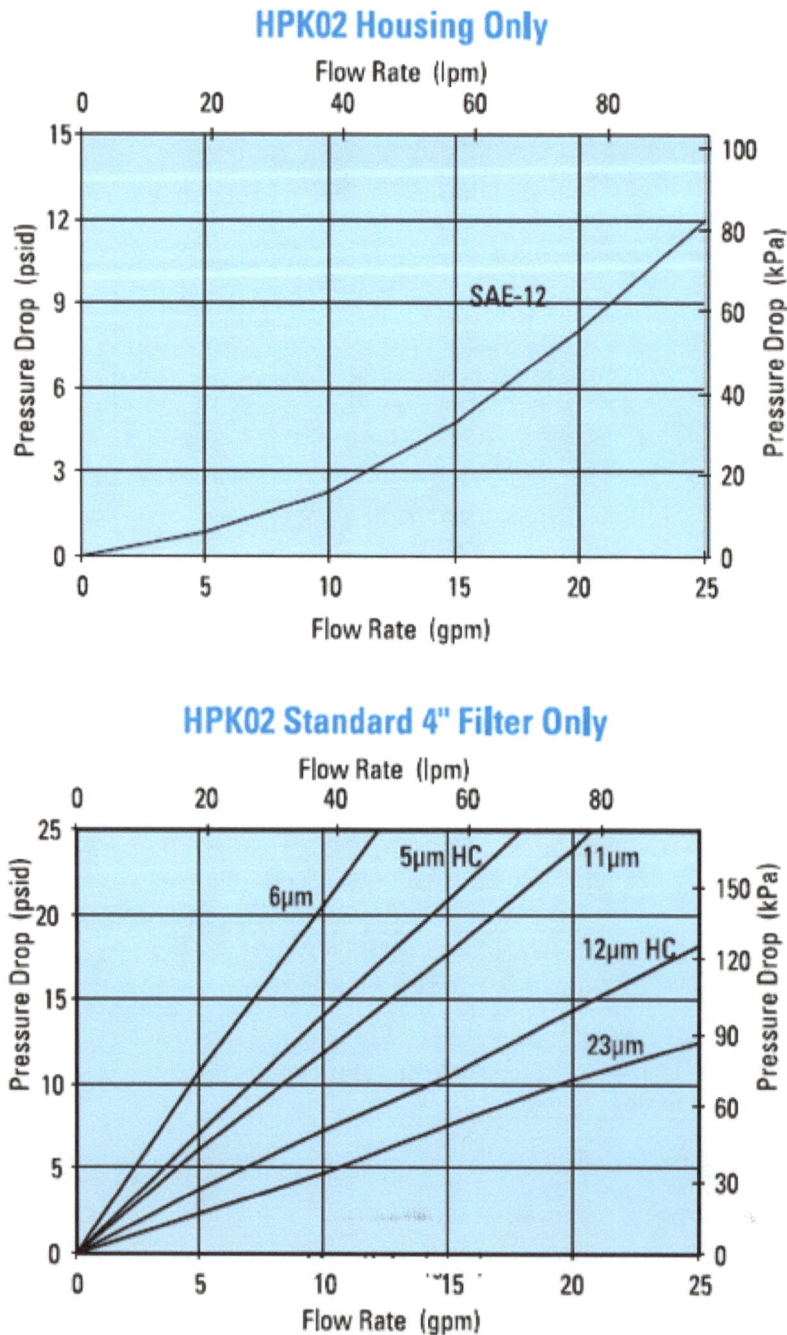

HPK02 Housing Only

HPK02 Standard 4" Filter Only

Fig. 6.50- Example of Line-Mounted Top-Ported Pressure Filter (Courtesy of Hydac)

6.9.4- Last Chance Filters

Primary Duty: If abrasive particles pass through main system filters and enter a hydraulic system, they may damage expensive components. These contaminants may prevent hydraulic components from operating properly by causing them to respond slowly or stick open. *Last Chance Filters* are used to protect critical components from catastrophic failure. However, they are not intended to replace the system filter.

Placement: Because they are placed in series and exposed to system pressure, they have a high collapse pressure.

Applications: They are recommended in the following market applications: servo circuits, precision machinery, variable-displacement pump and motor systems, hydrostatic drives, and machinery in dirty or dusty environments.

Figure 6.51 shows an example of last chance filters.

CP-C16 Series

Circuit Protector Manifold Cartridge Filters
3000 psi • up to 12 gpm

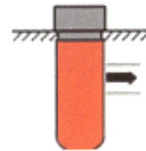

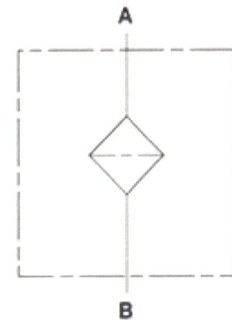

Technical Specifications

Mounting Method	C16-2 Cavity (SAE-16 Threaded Port)	
Flow Direction	Inlet: Bottom	Outlet: Side
Construction Materials	Steel	
Flow Capacity	12 gpm (45 lpm)	
Housing Pressure Rating		
Max. Allowable Working Pressure	3000 psi (210 bar)	
Fatigue Pressure	Contact HYDAC Office	
Burst Pressure	Contact HYDAC Office	
Element Collapse Pressure Rating		
W/HC	250 psid (17 bar)	
Fluid Temperature Range 14°F to 212°F (-10°C to 100°C) Consult HYDAC for applications operating below 14°F (-10°C)		
Fluid Compatibility		
Compatible with all petroleum oils rated for use with Nitrile rubber (NBR) seals.		

Fig. 6.51- Example of Last Chance Filters (Courtesy of Hydac)

6.9.5- Return Filters

Placement: As shown in Fig. 6.52, in open hydraulic circuits, *Return Filters* are placed on the main return line that collects return oil from actuators and other components back to the reservoir. As shown in the figure, in closed hydraulic circuits (hydrostatic Transmission), return filters are placed on the case drain line.

Primary Duty: The primary duty of return filters is to capture contaminates generated by all the components in the systems and leave only clean oil to get back to the reservoir. Return line filters are generally designed for lower pressures up to 34 bar (500 psi). Cooling can be integrated by installing an oil cooler downstream of the return filter.

Cost: Return filters are the most cost-effective filtration solution because return line pressure is low pressure, hence no special housing is required. Larger elements with a greater dirt-holding capacity can be used at a fraction of the cost.

Mesh Size: Filter media in return filters are selected to meet the cleanliness level required by system manufacturer. Common range is between 5 and 40 microns is acceptable.

Flow: When sizing a return filter, a through flow distribution analysis must be made to determine the maximum flow in the return line. The maximum flow in return line may be more than the main pump flow if a differential cylinder and/or accumulator are used.

Bypass Valve: Equipping return filters by bypass valve are highly recommended in order to limit the back pressure generated at the filter inlet port when the filter media is clogged. In closed circuits, the filter housing must incorporate a bypass valve with a cracking pressure lower than the maximum allowable case pressure for the pump or the motor, typically 0.5-1 bar (7-15 psi).

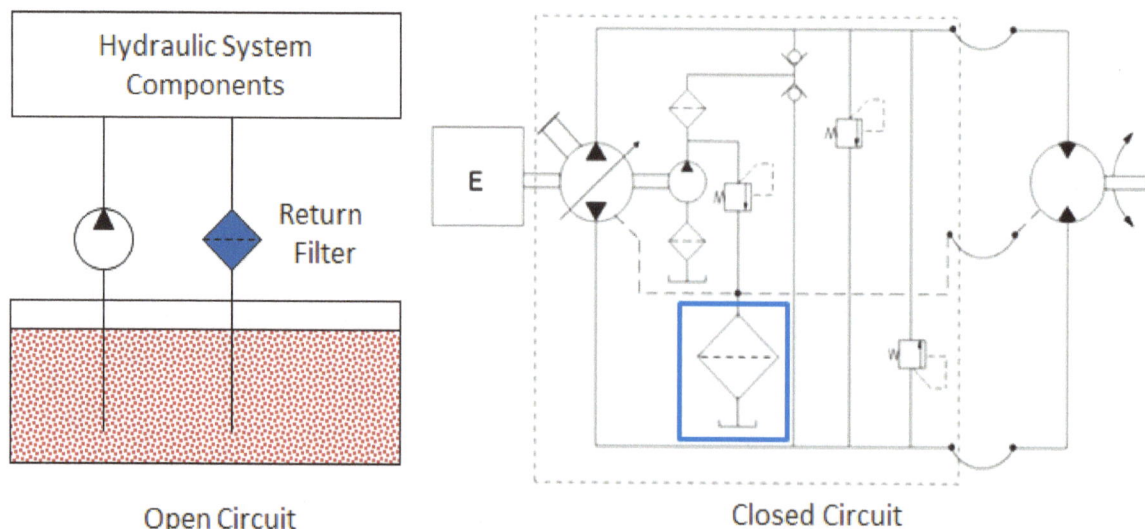

Open Circuit   Closed Circuit

Fig. 6.52 – Placement of Return Filters

Figure 6.53 shows an example of On-Tank Top single filters. As shown in the figure, the filter can be equipped with a clogging indicator and a built-in filter breather. The associated specification sheet shows full details of the filter specifications.

Technical Specifications

Mounting Method

75/90/150/165/185	2 mounting holes - filter housing
50/75/90/150/165/185/210/270/ 330/500/661/851/975/1100	4 mounting holes - filter housing

Flow Capacity

50 - 13 gpm (50 lpm)	270 - 71 gpm (270 lpm)
75 - 20 gpm (75 lpm)	330 - 87 gpm (330 lpm)
90 - 24 gpm (90 lpm)	500 - 132 gpm (500 lpm)
150 - 40 gpm (150 lpm)	661 - 174 gpm (660 lpm)
165 - 43 gpm (165 lpm)	851 - 225 gpm (850 lpm)
185 - 49 gpm (185 lpm)	975 - 258 gpm (950 lpm)
210 - 55 gpm (210 lpm)	1100 - 300 gpm (1100 lpm)

Housing Pressure Rating

Max. Allowable Working Pressure*	145 psi (10 bar), 101.5 psi (7 bar) *(Sizes 975 & 1100)*
Fatigue Pressure	145 psi (10 bar) @ 1 million cycles
Burst Pressure	75-500 >580 psi (40 bar)
	50, 661/851 536 psi (37 bar)
	975/1100 Consult Factory

Element Collapse Pressure Rating

BN4HC *(size 50, 975 & 1100 only)*	145 psid (10 bar)
ON *(size 50-851 only)*, W/HC	290 psid (20 bar)
ECON2, BN4AM, AM, P/HC, MM	145 psid (10 bar)
V	435 psid (30 bar)

Fluid Temperature Range   -22°F to 212°F (-30°C to 100°C)

Consult HYDAC for applications below -22°F (-30°C)

Fluid Compatibility

Compatible with all hydrocarbon based, synthetic, water glycol, oil/water emulsion, and high water based fluids when the appropriate seals are selected.

Indicator Trip Pressure

P = 20 psi (1.4 bar) - 10%
P = 29 psi (2 bar) -10% *(standard)*
P = 72 psi (5 bar) -10% *(optional)*

Bypass Valve Cracking Pressure

ΔP = 43 psid (3 bar) +10% *(Standard - All sizes except 50, 975, 1100)*
ΔP = 87 psid (6 bar) +10% *(Optional - Sizes 50, 975 & 1100 not available)*
ΔP = 25 psid (1.7 bar) +10% *(Standard for Sizes 50, 975 & 1100)*

RFM Series
In-Tank Return Line Filters
145 psi • up to 224 gpm

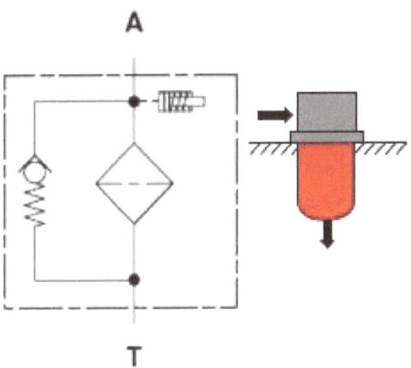

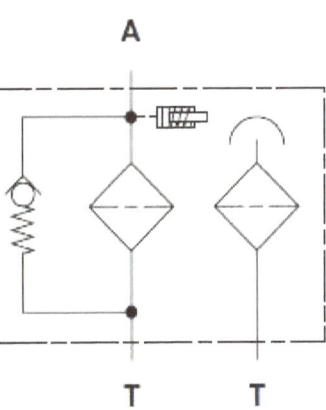

Fig. 6.53- Example of Return Filters (Courtesy of Hydac)

6.9.6- Combined Return and Suction Booster Filter

Description: Many hydraulic circuits may have more than one operating pump. As shown in Fig. 6.54, a return filter is used for the open circuit and a suction filter is used for the boosting pump in the closed circuit.

Advantage: A *Combined Return and Suction Booster Filter* is two filters in one that have the advantages:
- Cost and space saving.
- Easy maintenance.
- Meets automotive standard.
- Offered in pipe, SAE straight thread, flange and ISO 228 porting.
- Available with NPTF inlet and outlet female test ports.
- Available with magnet inserts.
- Various Dirt Alarm options.
- Available with housing drain plug.

Typical Applications: Such a unique design is used in machines with two or more circuits, such as in mobile working machines with hydrostatic traction drives (wheel loaders, forklifts) and automotive engineering.

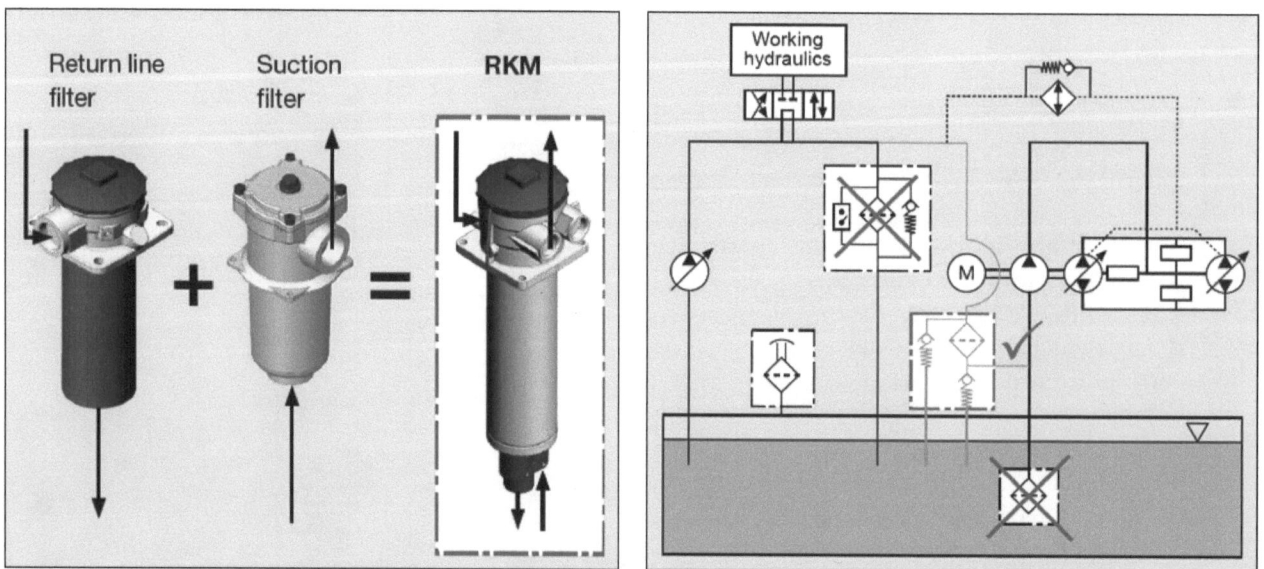

Fig. 6.54- Combined Return and Suction Booster Filter (Courtesy of Hydac)

Function (Refer to Fig. 6.55):

- <u>Inlet to Return Filter:</u>
  - Q_R is supplied via port **A**.
  - Q_R from the outside to the inside the element.
- <u>Inlet to Boosting Pump of a Closed Circuit:</u>
  - Q_s from inside the element to inlet of booting pump.
- <u>Inlet to Suction Filters:</u>
  - Back pressure valve **V1** builds 0.5 bar positive pressure.
  - Filtered oil is supplied to suction ports (**B1, B2**, etc.).
- <u>Bypass-to-Tank **V2**:</u>
  - Backpressure → surplus flow drains (bypassing the element) to port **T**.
- <u>Bypass Valve **V3** (optional):</u>
  - Oil can be drawn from the tank for short periods (e.g. for initial filling and for venting).

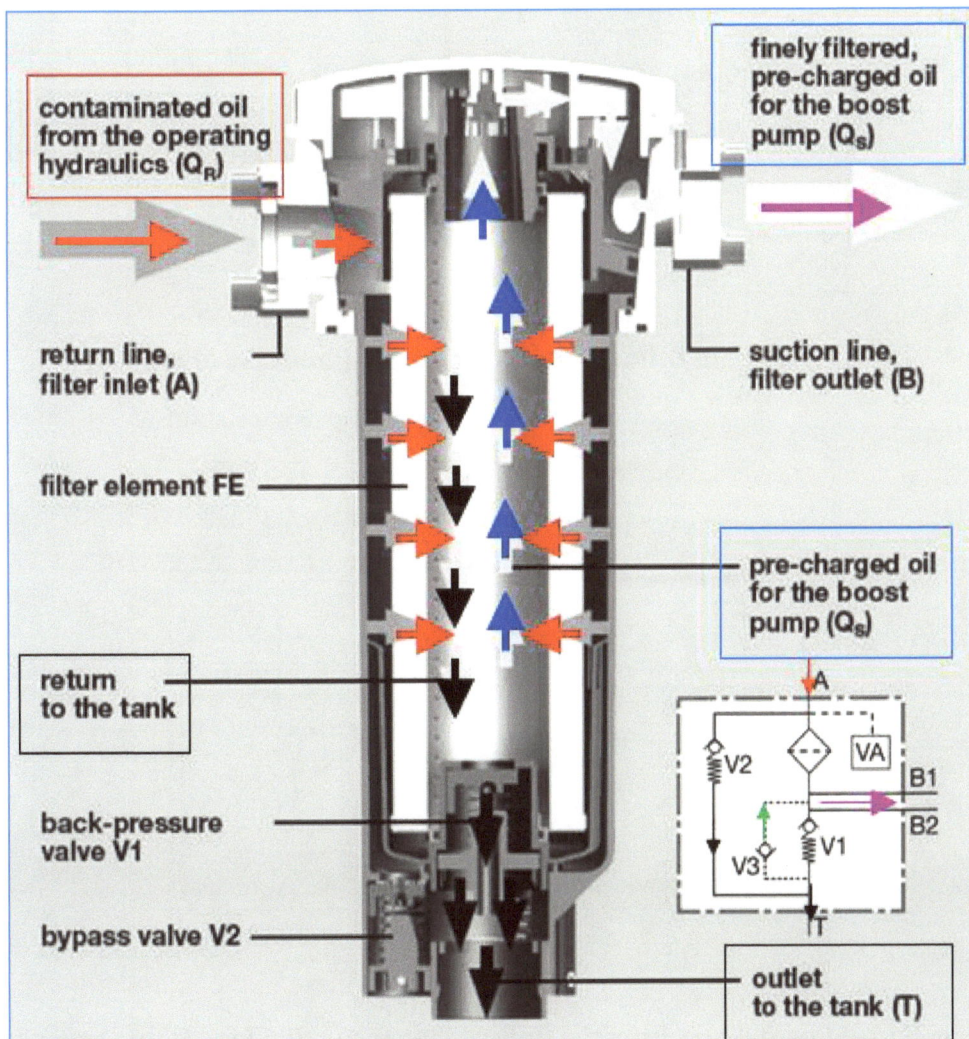

Fig. 6.55- Function of Combined Return and Suction Booster Filter (Courtesy of Hydac)

6.9.7- Diffusers

As shown in Fig. 6.56 installing a *diffuser* in a hydraulic reservoir is a simple addition that makes a big difference in system performance. With special concentric tubes designed with discharge holes 180° opposed, fluid aeration, foaming and reservoir noise are reduced. Using diffusers improves system performance and extend component life. Figure 6.57 shows the typical stream of flow around the baffle plate between the return line through a diffuser and the suction line through a suction strainer.

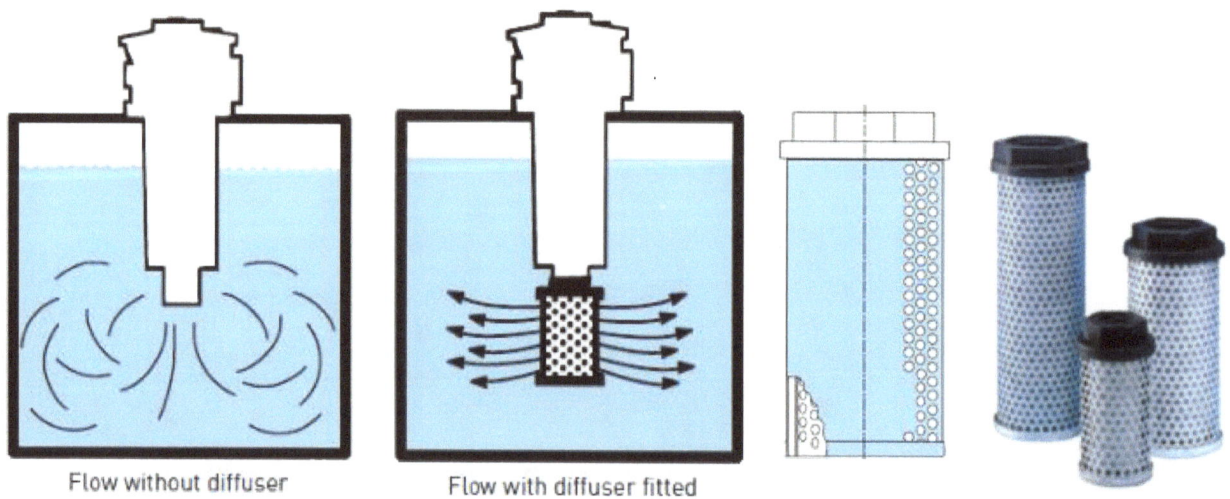

Flow without diffuser   Flow with diffuser fitted

Fig. 6.56- Diffusers on a return line (Courtesy of Parker)

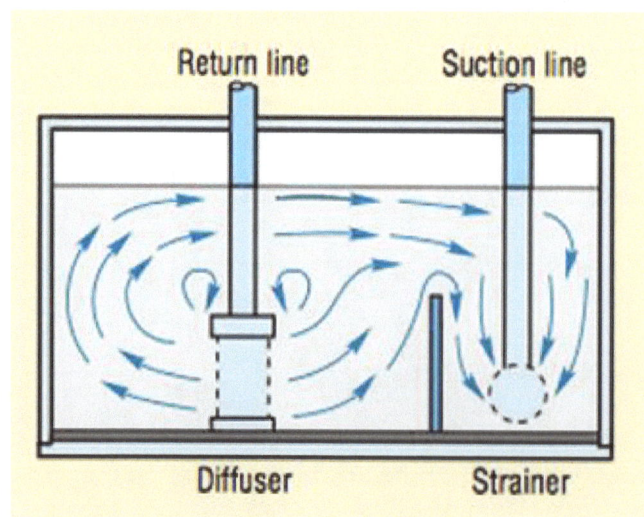

Fig. 6.57- Flow Streams from the Return Line to Suction Line (Courtesy of Parker)

6.9.8- Filler Caps

Figure 6.58 shows the least expensive traditional filler caps. A *filling cap* should be fitted with a sealed cover to prevent the ingress of contaminants when closed. It may contain a filling screen to catch relatively large contaminants during filling. Filler caps should be chained to the reservoir to keep them captive. For more protection, the cover should be lockable. Figure 6.59 shows a typical filler cap that allows reservoir breathing (but without filter breather) as follows:

1- Air intake to reservoir through vacuum breaker when pressure decreases (0.435 psi)
2- Venting to atmosphere through relief valve to maintain a 5 or 10 psi.

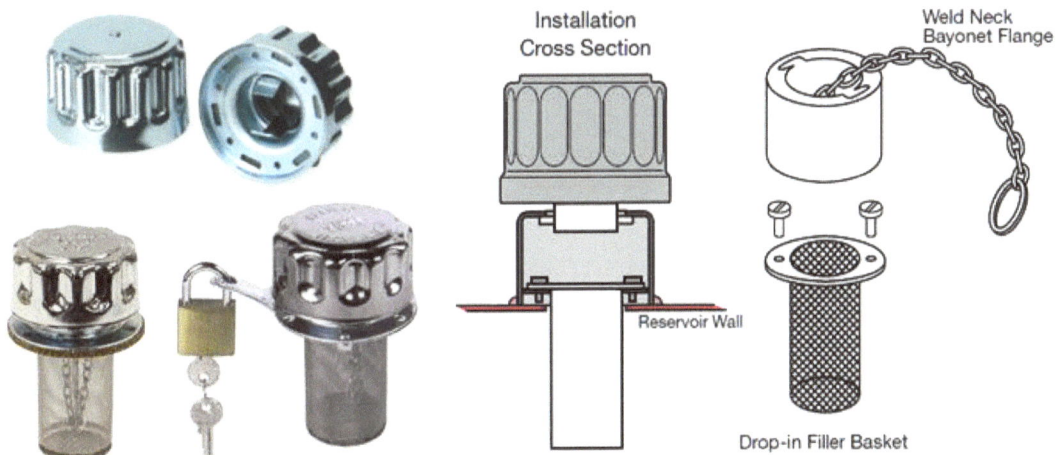

Fig. 6.58- Filler Caps

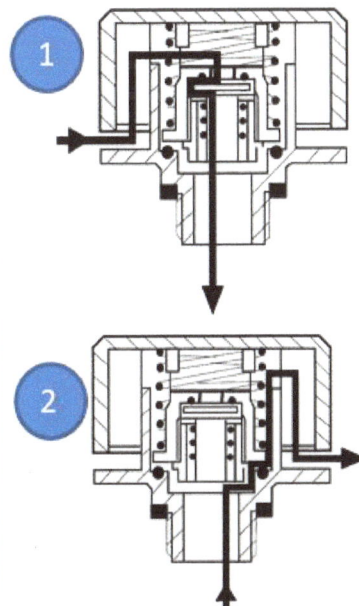

Specifications
- Chrome plated, epoxy coated or zinc plated steel cap
- Airflow to 30 cfm/850 lpm
- Compatible with petroleum based fluids
- Temperature to 212°F / 100°C
- 1/2", 3/4" and 1" NPT on ABS
- 1/4" and 3/8" NPT on MBS

Options
- 3, 10 and 40 micron (ABS), 10 and 40 micron (MBS)
- Zinc and epoxy coated weather-proof cap versions

Fig. 6.59- Example of Filler Caps (Courtesy of Donaldson)

6.9.9- Filter Breathers

Concept: In all systems using accumulators, single-acting cylinders or double-acting differential cylinders, the reservoir oil level falls as the cylinders extend and raises up as they retract. So, an open reservoir breathes during the machine operation like a human and hence a very large air volume passes in and out to the air space in the reservoir. If the air is allowed to freely move in and out of the reservoir, everything contained within that air is exchanged with the oil. This can include air-born contaminants and moisture, that both have severe effect on the oil lifetime and the hydraulic system reliability. Therefore, to prevent the dust from getting into the tank, *air breathers* are used. Breathers are available in various configurations, sizes, and working features. Filtration rating of an air breather must be equal to or better than main system filter.

Standard Filler Breathers: As shown in Fig. 6.60, *Standard Filter Breathers* are very similar in shape to the filler caps. They are usually screwed onto a threaded pipe that provides air exchange through the top of the reservoir. Other styles can look like a spin-on oil filter. They are available in various forms such as metallic or non-metallic, flange-mount or thread-mount, and with 10 microns size of a conventional or telescopic strainer. As shown in the figure, manufacturers report the differential pressure across the breather versus the air flow.

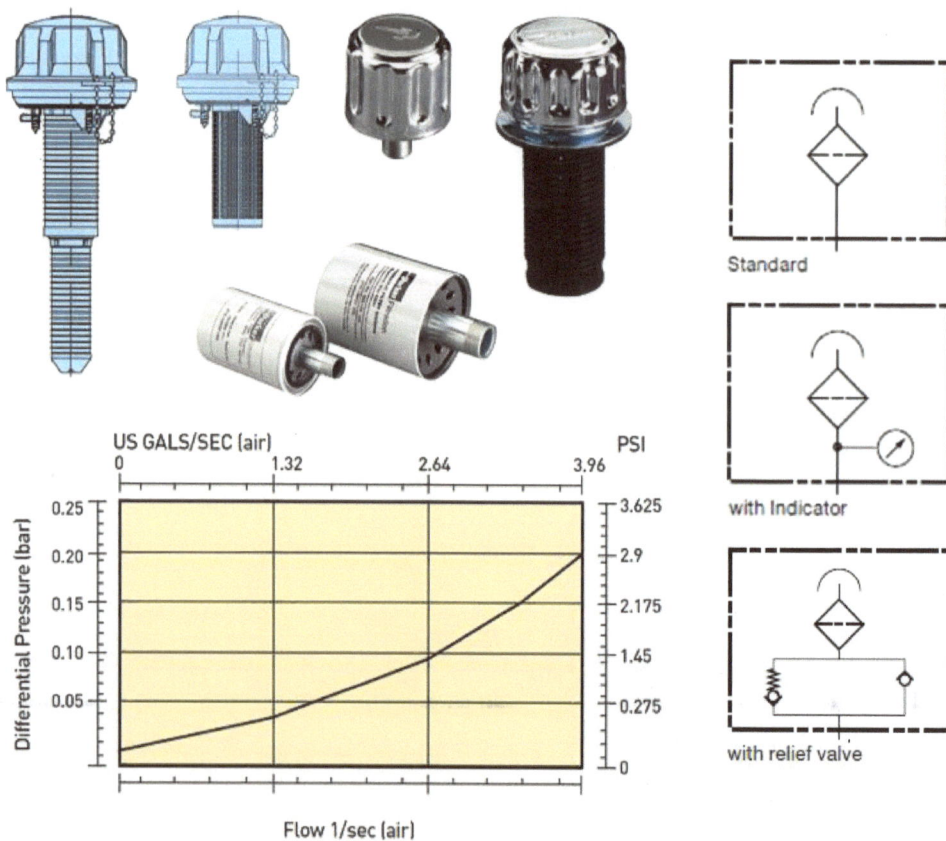

Fig. 6.60- Standard Filter Breathers (Courtesy of Parker)

Desiccant Breathers: In hydraulic systems that use petroleum-based hydraulic fluids, water ingress into the tank is a familiar problem. In such systems, if water content in oil increases above the allowable limit, system faces frequent breakdowns and high maintenance costs. For detailed information about contamination by water, review Volume 3 of this textbook series. Therefore, using *Desiccant Breathers* is a must for applications that work in very humid environments such as marine and offshore applications.

Figure 6.61 shows a typical example from industry for a desiccant breather and its hydraulic symbol. As shown in the figure, desiccant breathers are designed with a transparent body filled with silica-gel that are designed to absorb as much as 40% of its weight. The gel's color changes from a blue to light pink color when saturated. The unit also contains regular filter element to capture contaminants as small as 3 microns. As an option, an inlet check valve can be assembled on the air inlet to prevent the saturation of the desiccants during the system shutdown. An outlet check valve can be assembled on the air outlet to prevent exhaust air from the tank from flowing back through the desiccant so that the desiccant is protected from oil mist. Replace breather when desiccant color changes or when a built-in clogging indicator shows expiration of the drying material.

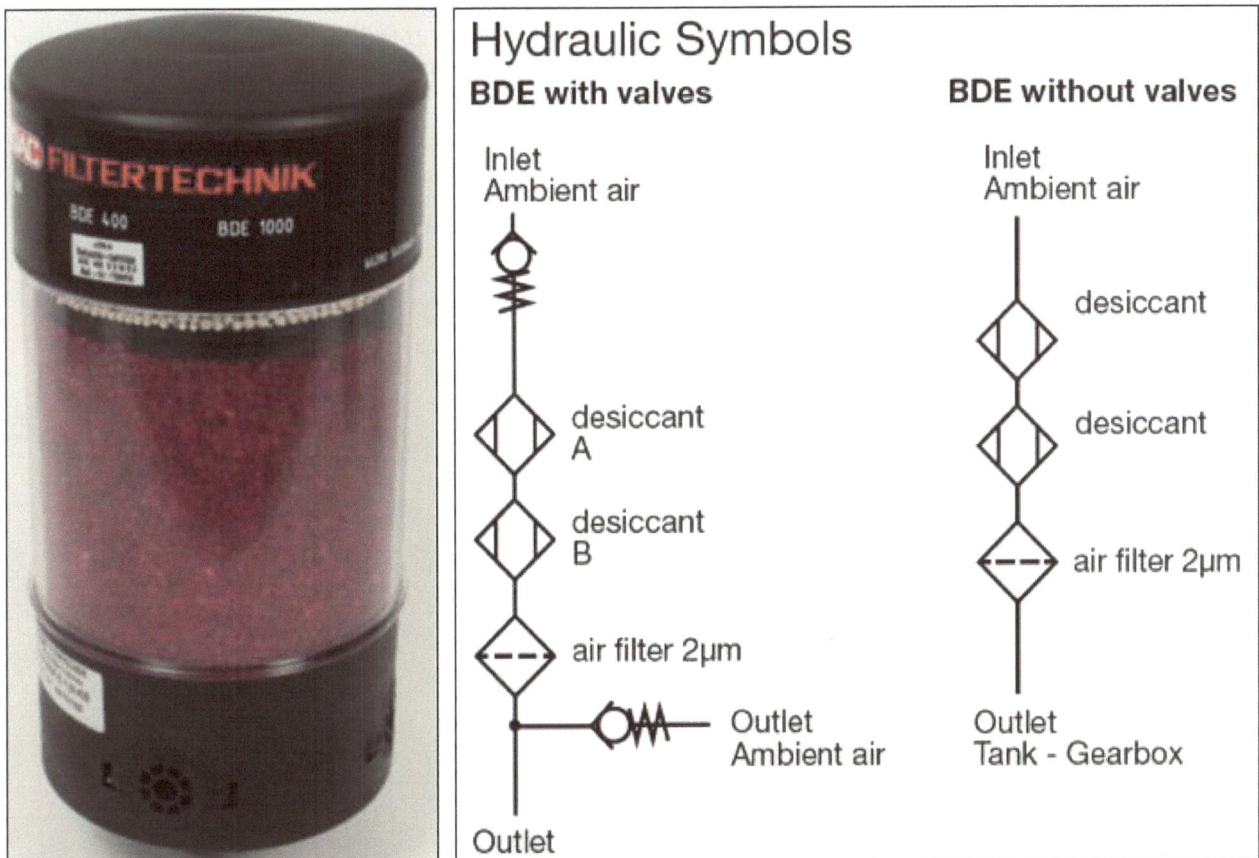

Fig. 6.61 – Desiccant Breather Dryer (Courtesy of HYDAC)

Construction and Operation of Desiccant Breather: As shown in Fig. 6.62, the desiccant breather consists of two separate chambers which can be filled with two desiccants, which in combination increase total water retention because of two-stage dewatering. The figure shows a built-in pleated air filter element (absolute filtration of particles > 2 μm) provides the filter with a very high contamination retention capacity (26 g). Such breather dryers can work in temperatures range -30 ^{o}C to 100 ^{o}C (-22 ^{o}F to 212 ^{o}F).

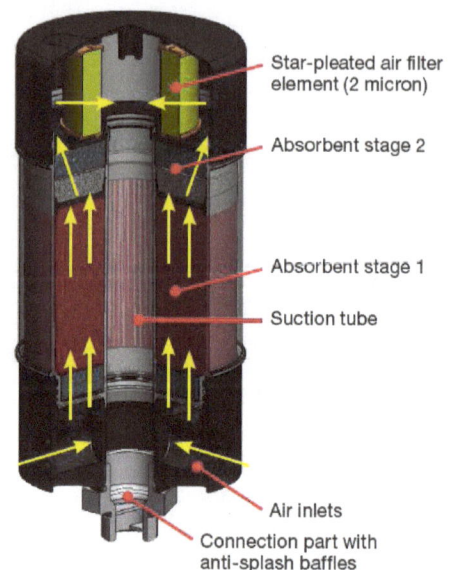

Star-pleated air filter element (2 micron)

Absorbent stage 2

Absorbent stage 1

Suction tube

Air inlets

Connection part with anti-splash baffles

Fig. 6.62 – Construction and Operation of the Desiccant Breather Dryer (Courtesy of HYDAC)

Breathers Dryers: Alternative to using desiccant breathers, *Breather Dryer* can be used. They are breathers with water absorption cartridges. Breather dryers collect and expel moisture out of reservoirs. This means that, unlike desiccant filters, breather dryers will not be changed due to water saturation. Figure 6.63 shows how breather dryers work. An indicator shows when maintenance is required and a new cartridge shall be installed.

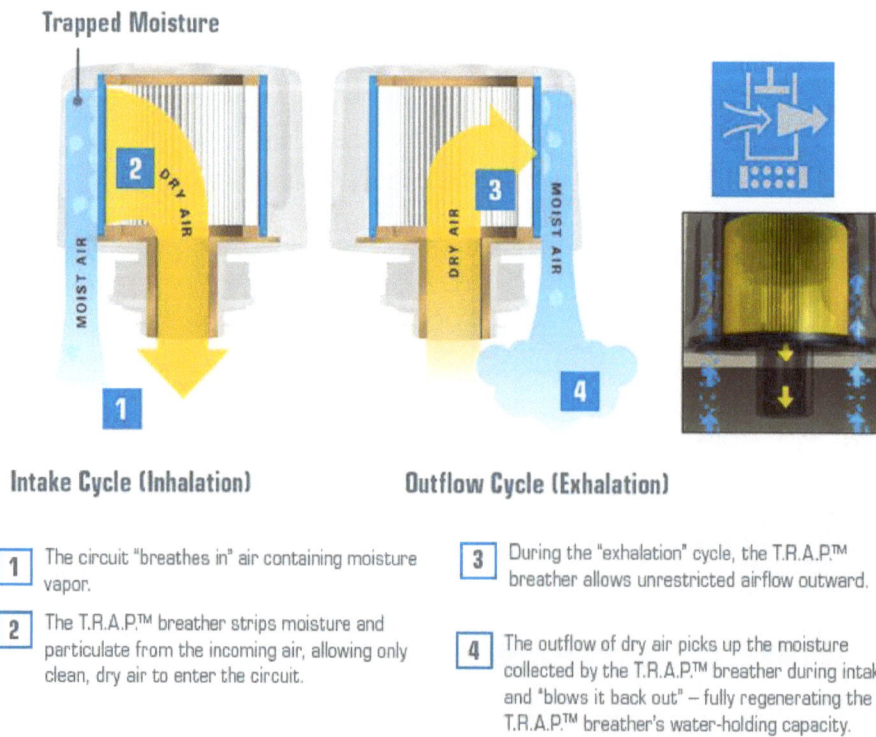

Trapped Moisture

Intake Cycle (Inhalation)   **Outflow Cycle (Exhalation)**

| 1 | The circuit "breathes in" air containing moisture vapor. |
| 2 | The T.R.A.P.™ breather strips moisture and particulate from the incoming air, allowing only clean, dry air to enter the circuit. |

| 3 | During the "exhalation" cycle, the T.R.A.P.™ breather allows unrestricted airflow outward. |
| 4 | The outflow of dry air picks up the moisture collected by the T.R.A.P.™ breather during intake, and "blows it back out" – fully regenerating the T.R.A.P.™ breather's water-holding capacity. |

Fig. 6.63 – Construction and Operation of Breather Dryers (Courtesy of Donaldson)

Sizing of Breather Dryer: Breather dryers are provided in different sizes. Undersized tank breather filters can place additional strain on the system and reduce the service life of breather. As shown in Fig. 6.64, larger size has a better water retention, but a relatively larger pressure drops.

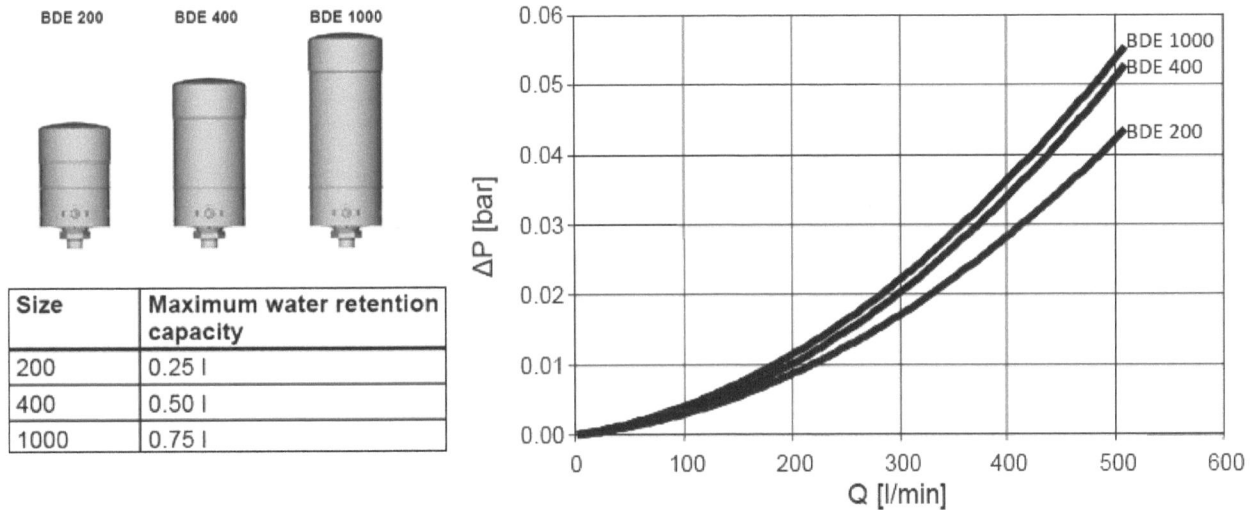

Size	Maximum water retention capacity
200	0.25 l
400	0.50 l
1000	0.75 l

Fig. 6.64 – Sizing of the Breather Dryer (Courtesy of HYDAC)

Air Breathers for Closed Tanks: In some highly contaminated applications, such as mills and foundries, closed reservoirs (pressure-sealed) are recommended. In such reservoirs, the inside pressure is increased above atmospheric by gas bladders and no conventional air breather can be used because the pressure inside the tank is always higher than the atmospheric. As shown in Fig. 6.65, in some cases a tank may be completely sealed from the atmosphere. Any air returns with the oil back to the tank will be accumulated on top of the oil surface. An air check valve is used to protect the tank against accidental over-pressure. Such a valve should be connected to any point on the tank above oil level. The valve allows free one air flow direction toward the atmosphere. Cracking pressure of 1 to 3 PSI is common.

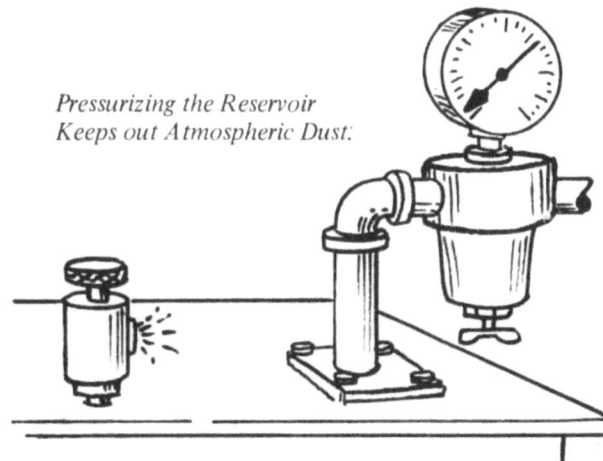

Pressurizing the Reservoir Keeps out Atmospheric Dust.

Fig. 6.65 – Sizing of the Breather Dryer (Courtesy of Womack)

6.9.10- Offline (Bypass) Filtration Units

Placement: *Offline Filters* are placed in a separate circuit. They are also referred to as *Bypass Filters.* Offline filtration can be done by a portable unit or by a permanently installed unit.

Primary Duty: The primary duty of offline filters is to filter the full volume of oil in the reservoir apart from the oil circulation in the main circuit. It has the advantage of the filter isn't subjected to surge flows, filter can be replaced without interrupting the system, and fine filtration is possible.

Cost: Offline filtration has high initial cost. However, the cost is justified through extended system life.

Sizing: Flow rate is controlled based on the oil volume in the reservoir and how many fluid circulations is required per hour. However, commonly the pump in such a loop is rated at 5%-10% of the oil volume in the reservoir. For example, if a reservoir has 100 liters of oil, pump is rated for 5-10 liter/min so that the whole reservoir will be filtered in 10-20 minutes. In other words, the whole reservoir is filtered 3-6 times per hour.

Filtration Rating: Filtration rates of 2 microns or less are possible, and polymeric (water-absorbent) filters and heat exchangers can be included in the circuit for total fluid conditioning.

Construction: Offline filtration units are available in various styles and sizes. Figure 6.66 shows an example of hand-portable offline filtration unit. The unit consists of a pump/motor unit, a filter, and the supporting frame.

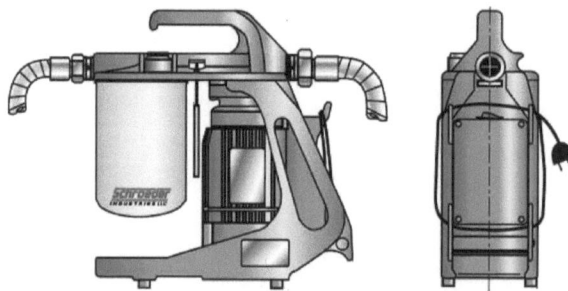

Fig. 6.66 – Example of Hand-Portable Offline Filtration Unit (Courtesy of Schroeder)

Figure 6.67 shows an example of portable-cart offline filtration unit. This provides a convenient portable mode of kidney loop filtration, flushing and fluid transfer. It can be used for in-plant machinery and hydraulic equipment to achieve and maintain proper ISO cleanliness levels. The *Filter Cart* includes a pump/motor unit, a filter, and the wheeled supporting frame. The cart is used in many cases such as *kidney loop filtration*, transferring new oil, cleaning stored oil, system filling/draining, line flushing, and flushing equipment after commissioning or rebuild.

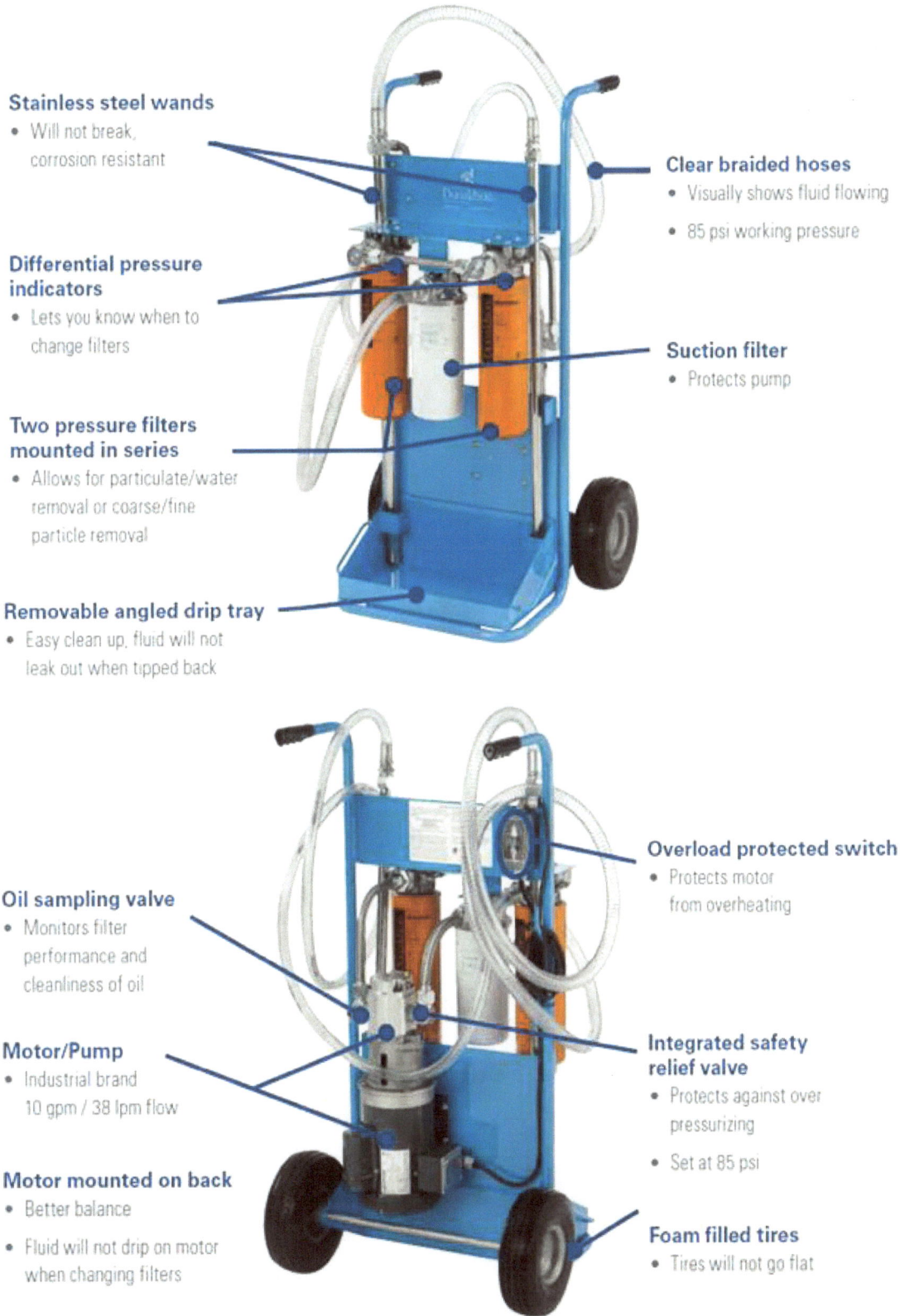

Stainless steel wands
- Will not break, corrosion resistant

Differential pressure indicators
- Lets you know when to change filters

Two pressure filters mounted in series
- Allows for particulate/water removal or coarse/fine particle removal

Removable angled drip tray
- Easy clean up, fluid will not leak out when tipped back

Clear braided hoses
- Visually shows fluid flowing
- 85 psi working pressure

Suction filter
- Protects pump

Oil sampling valve
- Monitors filter performance and cleanliness of oil

Motor/Pump
- Industrial brand 10 gpm / 38 lpm flow

Motor mounted on back
- Better balance
- Fluid will not drip on motor when changing filters

Overload protected switch
- Protects motor from overheating

Integrated safety relief valve
- Protects against over pressurizing
- Set at 85 psi

Foam filled tires
- Tires will not go flat

Fig. 6.67 – Example of Cart-Portable Offline Filtration Unit (Courtesy of Donaldson)

Figure 6.68 shows an example of <u>Fixed-Mounted</u> offline filtration unit. This unit is permanently mounted to offer supplemental filtration for in-plant machinery and hydraulic equipment helping to reduce costs and achieve and maintain proper ISO cleanliness levels. Figure 6.69 shows an example of a hydraulic system with offline filtration.

Fig. 6.68 – Example of Fixed-Mounted Offline Filtration Unit (Courtesy of Donaldson)

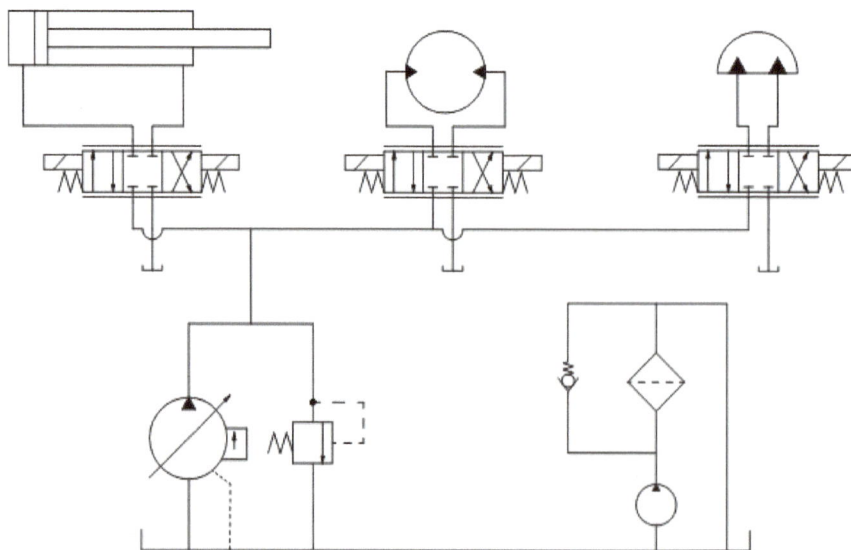

Fig. 6.69 – Example of Hydraulic Systems with Offline Filtration

Chapter 7

Filter Media and Filtration Mechanisms

Objectives

This chapter presents an overview of filter elements including the construction and material of the filter media. This chapter discusses surface filters versus depth filters. The chapter discusses also the principles of various filtration mechanisms that are applicable in hydraulic filters such as direct interception, absorption, adsorption, and magnetic separation.

Brief Contents

Chapter 7 – Filter Media and Filtration Mechanisms

7.1- Filtration Mechanisms

Capturing and retaining the particulate contaminants depends on one of the following mechanisms:

Retaining Large Size Particles Mechanically by Inertia: As shown in Fig. 7.1, when the fluid is accelerated between fibers and when the fluid changes direction to enter the fiber space, large and heavy particles suspended in the flow stream are slower than the fluid surrounding them because of particles *Inertia*. As a result, the particle continues in a straight line and is trapped by the media fibers where it is held and retained.

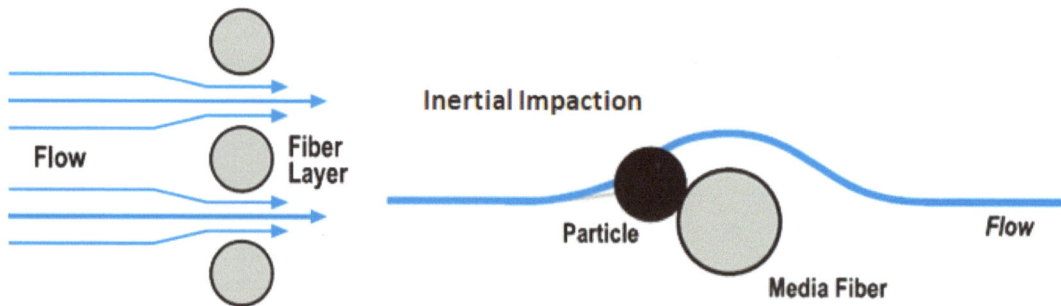

Fig. 7.1- Retaining Large Size Particles by Inertia (Courtesy of Donaldson)

Retaining Medium Size Particles Mechanically by Direct Interception: As shown in Fig. 7.2, particles are retained by *Direct Interception*. The mid-range size particles that are neither quite large enough to have inertia nor small enough to diffuse within the flow stream. These mid-sized particles are mechanically captured and retained just because they are larger than the micron size in the filter media. Direct interception s also referred as *"Sieving"*.

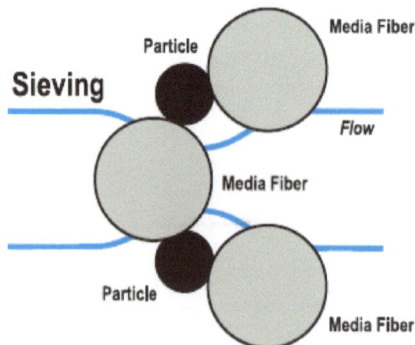

Fig. 7.2- Retaining Medium Size Particles by Direct Interception (Courtesy of Donaldson)

Retaining Small Size Particles by Absorption: As shown in Fig. 7.3, *Absorption* filtration mechanism works on the smallest particles. Absorption is also referred to as *Diffusion*. Small particles are not held in place by the viscous fluid and diffuse within the flow stream. As the particles traverse the flow stream, they are captured and collected by the fiber. Best interpretation of diffusion is that filter media attracts and retain particles by *electrostatic forces* or molecular attraction. Obviously special material is used in developing such filter media.

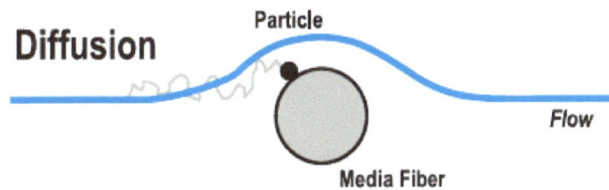

Fig. 7.3- Retaining Small Size Particles by Diffusion (Courtesy of Donaldson)

Retaining Particles by Adsorption: *Adsorbent filter media* use chemically treated filter media to remove contaminants. Charcoal, chemically treated paper, and other materials are used in this process. Adsorbent filter media are not typically used in hydraulic systems as they may remove desirable additives from the system fluid.

Oil Cleaning by Centrifugal Separators (explained in Volume 3): Water and solid particles with a density higher than that of oil can be removed by *Centrifugal Separators*. However, this method can't guarantee 100@ water removal and it can remove oil additives too.

Oil Cleaning by Vacuum Dehydration (explained in Volume 3): Air and water in oil can also be removes by *Vacuum Filters*.

Oil Cleaning by Magnetic Separation: In the control of contamination in hydraulic fluids, *Magnetic Separation* useful in separation of ferrous solids from fluid streams.

7.2- Materials for Filter Media

Cellulose Fibers Filter Media (Traditional): As shown in Fig. 7.4, *Cellulose Fibers* filter media has the following characteristics:
- **Material:** Wooden fibers held together by resin.
- **Shape:** Irregular shape.
- **Size:** Irregular small (microscopic) pores size.
- **Flow and ΔP:** Has more flow resistance, resulting higher pressure drop.
- **Filtration:** Good in catching contamination through the <u>depth</u> of the media. Poor filtration performance as compared to synthetic media.
- **Fluid:** Provides effective filtration for a wide variety of petroleum-based fluids.

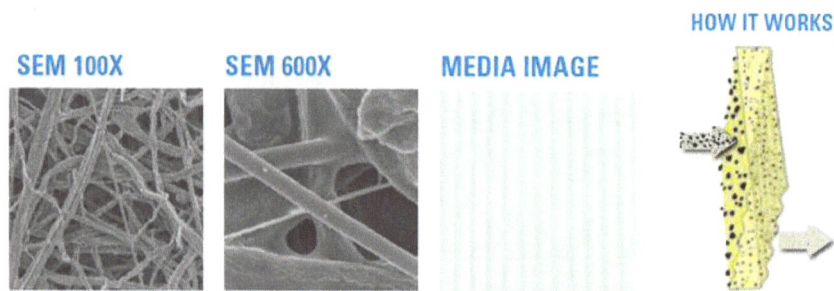

Fig. 7.4- Cellulose Fibers Filter Media (Courtesy of Donaldson)

Synthetic Fibers Filter Media (Fully Synthetic): As shown in Fig. 7.5, *Synthetic Fibers* filter media has the following characteristics:
- **Material:** Man-made, smooth, rounded fibers.
- **Shape:** Consistent shape.
- **Size:** Controlled size and distribution pattern through the media.
- **Flow and ΔP:** Provides low flow resistance, and consequently low pressure drop.
- **Filtration:** Consistency of fiber shape improves contaminant-catching ability on the <u>surface</u> and increases dirt holding capacity.
- **Fluid:** Ideal for use with synthetic fluids, water glycols, water/oil emulsions, HWCF and petroleum-based fluids.

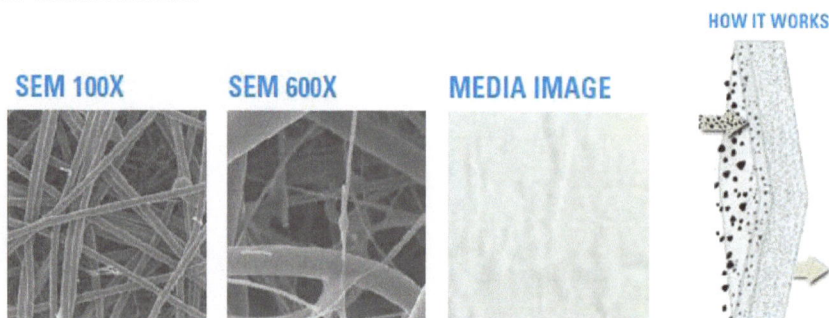

Fig. 7.5- Synthetic Fibers Filter Media (Courtesy of Donaldson)

Combined Fibers Filter Media (Cellulose & Synthetic): As shown in Fig. 7.6, *Combined Fibers filter media* is developed to provide effective fuel filtration performance for optimal protection.

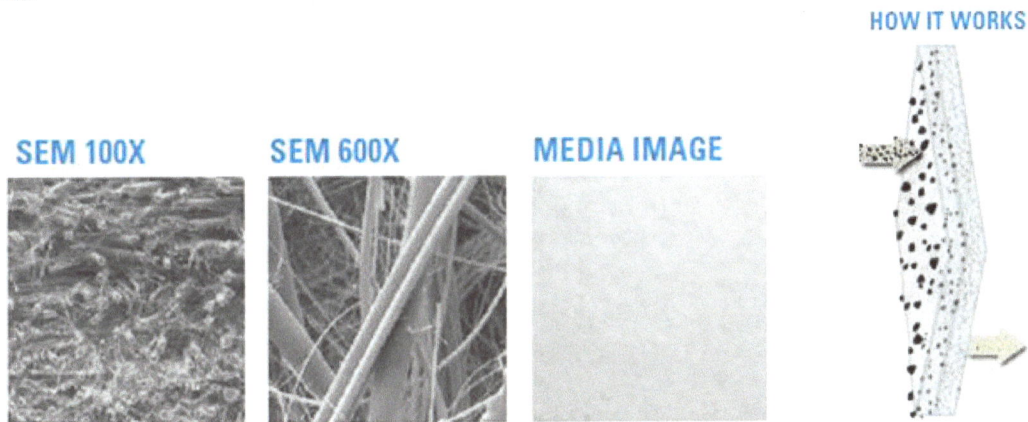

Fig. 7.6- Combined Fibers Filter Media (Courtesy of Donaldson)

High Performance Synthetic Fibers Filter Media: Today's fluid systems are often tailored towards the special needs of fire resistance, biodegradability, chemical and thermal resistances, and electrical insulating ability. As shown in Fig. 7.7, *High Performance Synthetic Fibers* filter media has the following characteristics:

- **Material:** A blend of borosilicate *Glass Fiber* whose matrix is bonded together with epoxy-based resin system.
- **Flow and ΔP:** Provides high flow resistance, and consequently high pressure drop.
- **Fluid:** They provide the best chemical resistance for the broadest array of hydraulic fluid. Ideal for use with phosphate ester and water glycol fluids.
- **Filtration:** Ideal for fine filtration and precision components.

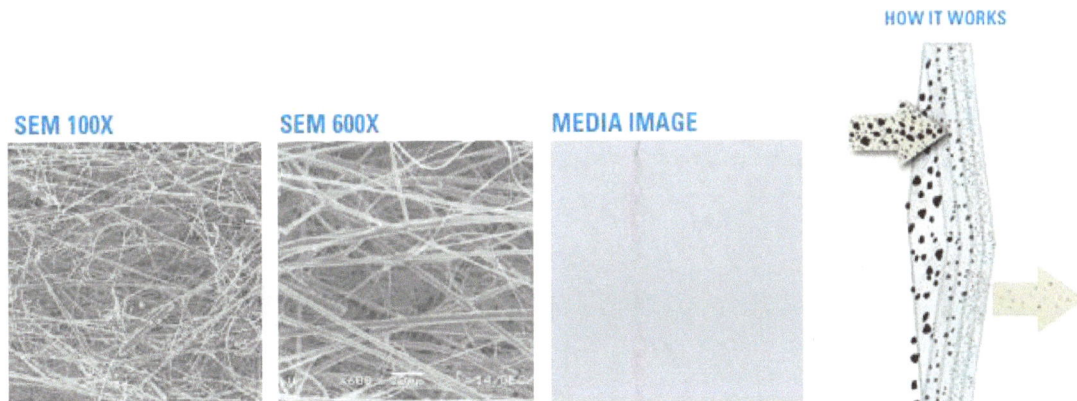

Fig. 7.7- High Performance Synthetic Fibers Filter Media (Courtesy of Donaldson)

Wire Mesh Filter Media: As shown in Fig. 7.8, *Wire Mesh* filter media has the following characteristics:

- **Material:** Stainless steel, epoxy-coated wire mesh.
- **Shape:** Consistent shape.
- **Size:** Available in different sizes.
- **Flow and ΔP:** Provide the least flow resistance, and consequently lowest pressure drop.
- **Filtration:** Available in various micron sizes and ranging from 100 to 500 microns. Typically wire-mesh filters will be applied to catch large and harsh particles that would plug up a normal filter. Generally used in suction strainers.

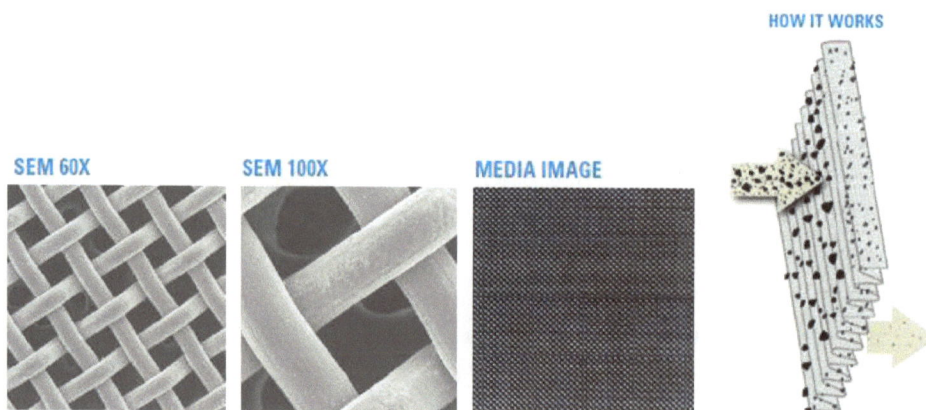

Fig. 7.8- Wire Mesh Filter Media (Courtesy of Donaldson)

Water Absorption Filter Media: As shown in Fig. 7.9, *Water Absorption* filter media quickly and effectively removes free water from hydraulic systems. Using super-absorbent polymer technology, with a high affinity for water absorption, prevents many of the problems associated with water contamination found in petroleum-based fluids.

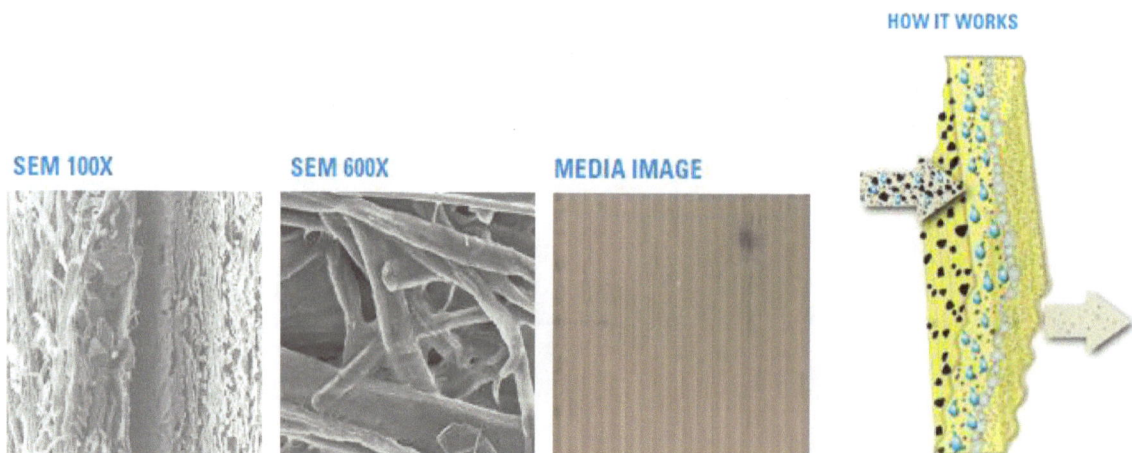

Fig. 7.9- Water Absorption Filter Media (Courtesy of Donaldson)

7.3- Filter Media Structure

Fiber Structure (Fig. 7.10 - 1): For fluid to pass through, the media must have pores or channels to direct the fluid flow and allow it to pass. That's why filter media is a porous material made of fibers that structured to twist, turn, and accelerate during passage.

Uniform vs. Graded Pore Size (Fig. 7.10 - 2): Based on the pore size along the depth of the filter media, it can be constructed to form uniform or graded pore size. Graded pore size with larger pore size on the surface. Graded pore size allows holding more dirt, but it causes higher pressure-drop across the filter media.

Fixed vs. Non-Fixed Pore Size (Fig. 7.10 - 3): Based on the method of bonding the fibers together on each layer, filter media can be constructed to form fixed or non-fixed pore size. In fixed pore media, fibers are bonded with specifically formulated resin to resist deterioration from pressure and flow fluctuations, temperature and aging conditions. Fibers in non-fixed pore media are inconsistently or poorly bonded. This facilitates movement of fibers under pressure and flow surges allowing media migration.

Surface vs. Depth Filters (Fig. 7.11): Hence, filter media are structured as *Surface Filter Media* and *Depth Filters Media*. As it works its way through the depths of the layers of fibers, the fluid becomes cleaner and cleaner. Generally, the thicker the media, the greater the dirt-holding capacity it has.

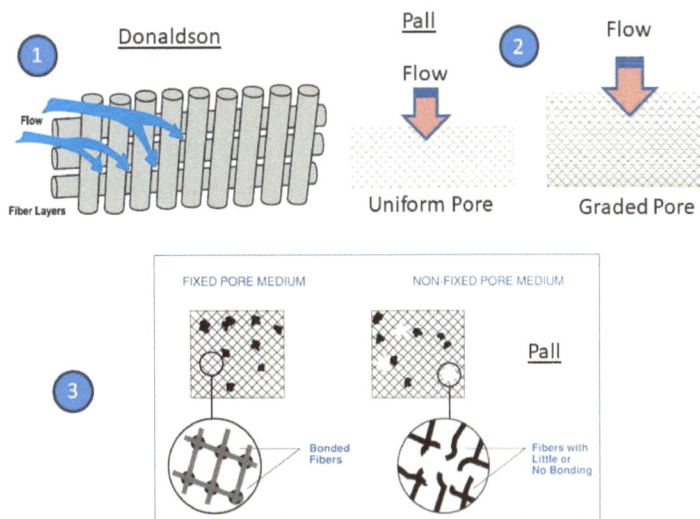

Fig. 7.10- Filter Media Fibers

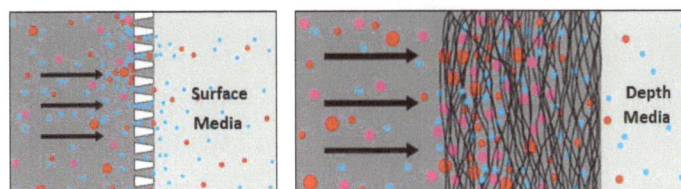

Fig. 7.11- Surface versus Depth Filter Media (Courtesy of Bosch Rexroth)

7.3.1- Surface Filter Media

Filtration Process: As shown in Fig. 7.12, primary filtration mechanism of a *Surface Filters* is direct interception. Most surface-type filters are exposed to the flow of contaminated fluid. Pore size of surface filter media is gradually reduced due to intrusion of soft and deformable particles. Over the time, the filter is completely clogged.

Applications: *Surface Filter* media is used commonly for strainers or suction filters.

Advantages of Surface Filter Media:
- Are washable and cleanable.
- No media migration with the oil.
- Low flow resistance and pressure drop.
- High fatigue and corrosion resistant.
- Work at high temperature.

Disadvantages of Surface Filter Media:
- Catch only relatively large contaminants.
- Can't be used to maintain high cleanliness Level.
- Needle-shaped contaminants that have less diameter than the pore size, even if its length is larger than the pore size, can pass through these filters.

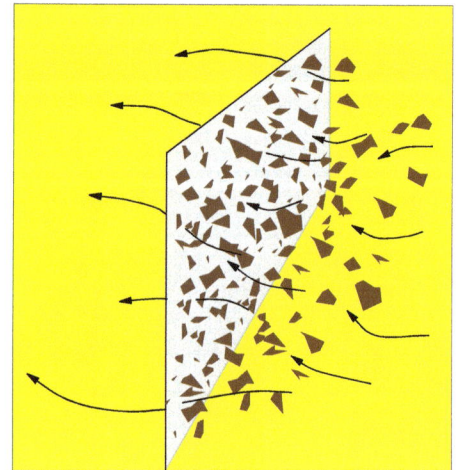

Fig. 7.12- Filtration using Surface Filters

Material: Various materials are used for surface filters such as stainless-steel wires, galvanized iron, or phosphor bronze, accordion-pleated paper, ribbon-shaped metal, and stacked metal disks.

Structure of Metallic Surface Filters: As shown in Fig. 7.13, *Square Wire Mesh* and *Braided Wire Mesh* are most commonly used materials for surface filters. Both are made from stainless steel, Both are washable, and provide low pressure drop. They are commonly used for coarse filters, lubricating systems, and suction filters. Braided mesh wire has better filter rating.

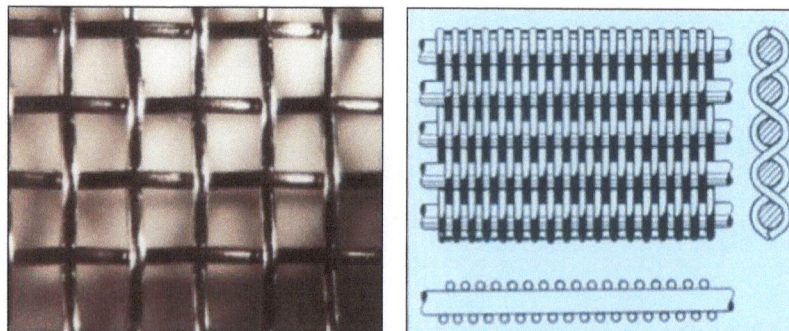

Fig. 7.13- Square versus Braided Wire Mesh Surface Filters Media (Courtesy of Bosch Rexroth)

7.3.2- Depth Filter Media

Filtration Process: As shown in Fig. 7.14, *Depth Filter* media particles are removed by *Direct Interception* and *Absorption.* Fine filtration is done by trapping solid dirt particles within the depth material. Also, water and water-soluble contaminants suspended in the hydraulic fluid in one of the many flow routes in the porous material. Such depth filters are classified as *Absorbent* filters.

Applications: *Depth Filter* media is used commonly for pressure, return, and offline filters.

Advantages of Surface Filter Media:

- Effective filtration for small contaminants.
- Used to maintain high cleanliness Level.
- Has large dirt holding capacity.

Disadvantages of Surface Filter Media:

- Are not washable or cleanable.
- Possible media migration with the oil.
- High flow resistance and pressure drop.

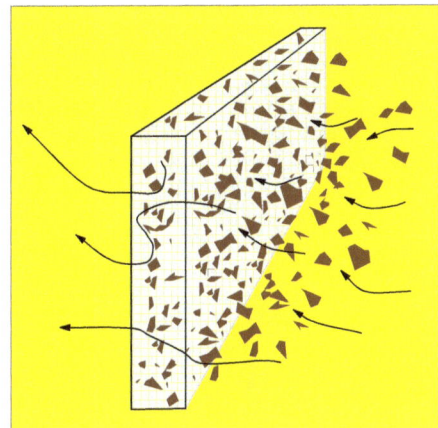

Fig. 7.14- Filtration using Depth Filters

Material: Various materials are used for depth filters such as organic (such as Cellulose, or Cotton) or synthetic fibers. Some depth filters are made from randomly oriented steel wires. However, the materials from which the depth media is constructed must be compatible with the hydraulic fluid and operating temperature of the system.

Structure: As shown in Fig. 7.15, depth filter media is composed of layers of porous material such fibers. It does not have 100% consistent pore size that is why it is rated based on average pore size. Fiber diameter is from 0.5 to 30 microns. Fibers are wounded in layers on top of each other. Each layer depth is 0.25 – 2 mm (0.01-0.08 in). The quality of the elements varies considerably between manufacturers depending on:

- Fiber bonding and the ability to prevent media migration.
- Central support of the filter element pleats.
- Sealing of the filter media to the end caps.
- Consistency of pore size through the media.

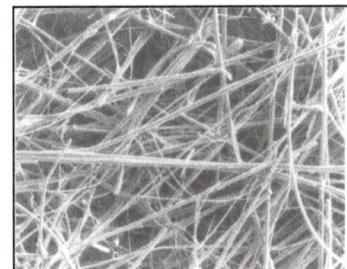

Fig. 7.15- Structure of Depth Filters Media

Example 1– Conventional Glass Fiber Pressure Filter: Conventional inline pressure filters are typically *glass fiber* based, because they need to operate under high pressure and high flow conditions, while creating as little restriction as possible.

As shown in Fig. 7.16, the filter element is pleated, conventionally using Fan-Pleating method, in order to increase the surface area and reduce the pressure drop. Since they are installed after the main system, they often live a tough life with cyclic flows and many stops and starts, which is very harmful for the efficiency of any filter. Capturing and retaining fine silt particles is therefore very difficult, which is why most of these inline filters have a rating of 5 – 50 microns.

However, many captured particles will be released again when the filter is exposed to pressure shocks at stop/start. The glass fiber-based pressure filter is capable of removing solid particles only, and due to the relatively small filter depth and volume, it has a limited dirt holding capacity (1 – 100 grams).

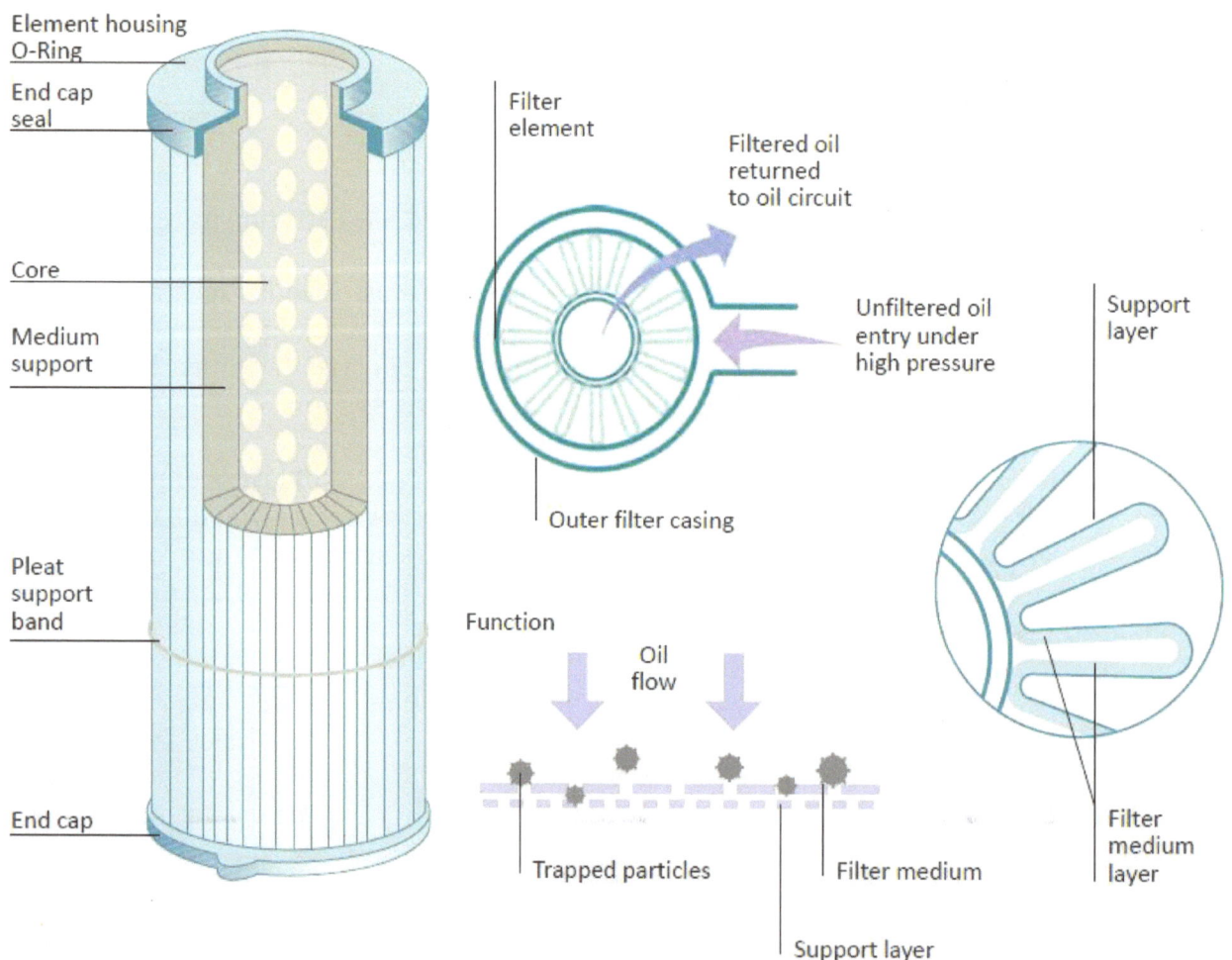

Fig. 7.16- Example of Glass Fiber Pressure Filter (Courtesy of C.C. Jensen Inc.)

Example 2– Conventional Micro Glass Fiber Pressure Filter: Figure 7.17 shows construction of a special class of micro-glass and other fibers depth filter Z-Media®. As shown in the figure, the filter media constructed from multiple layers, each successive layer performs a distinct and necessary function. Filter manufacturer reported the following features:

- Manufactured with utmost precision, to specific thicknesses and densities.
- Layers are bonded with select resins to create material with extra fine passages.
- Maximum dirt-holding capacity and superior particle capture.
- Excellent beta ratio (filter efficiency) stability.
- Minimum pressure drop.
- High flow rate and low operating cost.

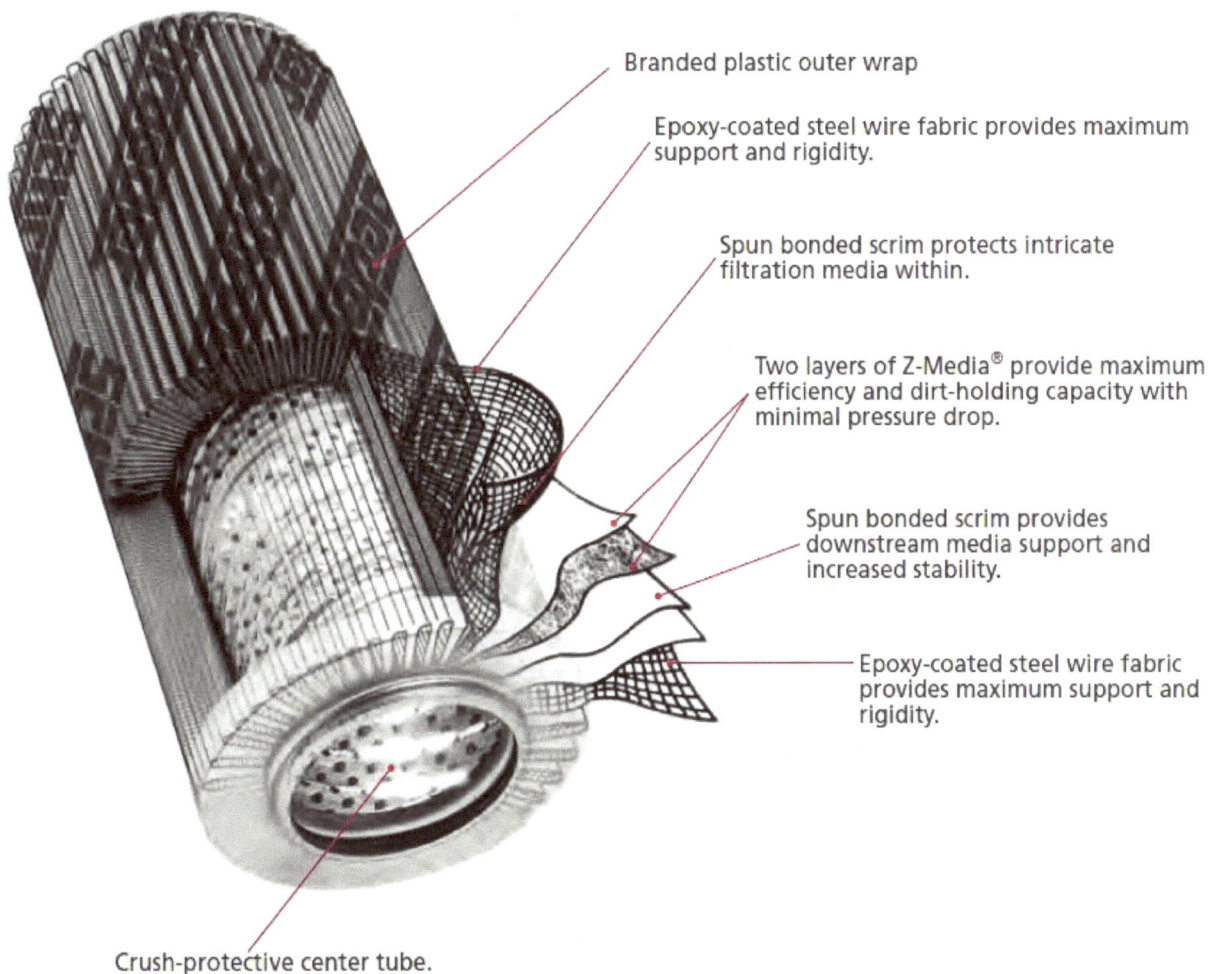

Branded plastic outer wrap

Epoxy-coated steel wire fabric provides maximum support and rigidity.

Spun bonded scrim protects intricate filtration media within.

Two layers of Z-Media® provide maximum efficiency and dirt-holding capacity with minimal pressure drop.

Spun bonded scrim provides downstream media support and increased stability.

Epoxy-coated steel wire fabric provides maximum support and rigidity.

Crush-protective center tube.

Fig. 7.17- Example of Glass Fiber Pressure Filter "Z-Media®" (Courtesy of Schroeder)

Examples 3– Laid-Over Pleating in Depth Filter Element Technology: Figure 7.18 shows description of new depth filter technology (Ultipleat). As shown in the figure, the conventional *fan-pleating* method results in nonuniform volumes between the pleats, and consequently nonuniform flow distribution. The laid-over pleating new pleating technology (Ultipleat) has the following features:

- Allows more filtration area to be packed into a given filter element envelope.
- Creates uniform flow distribution through the filter element.
- Protects pleat against collapse and bunching.
- Anti-static construction minimizes static discharges.
- Resistance to cyclic flow and pressure.

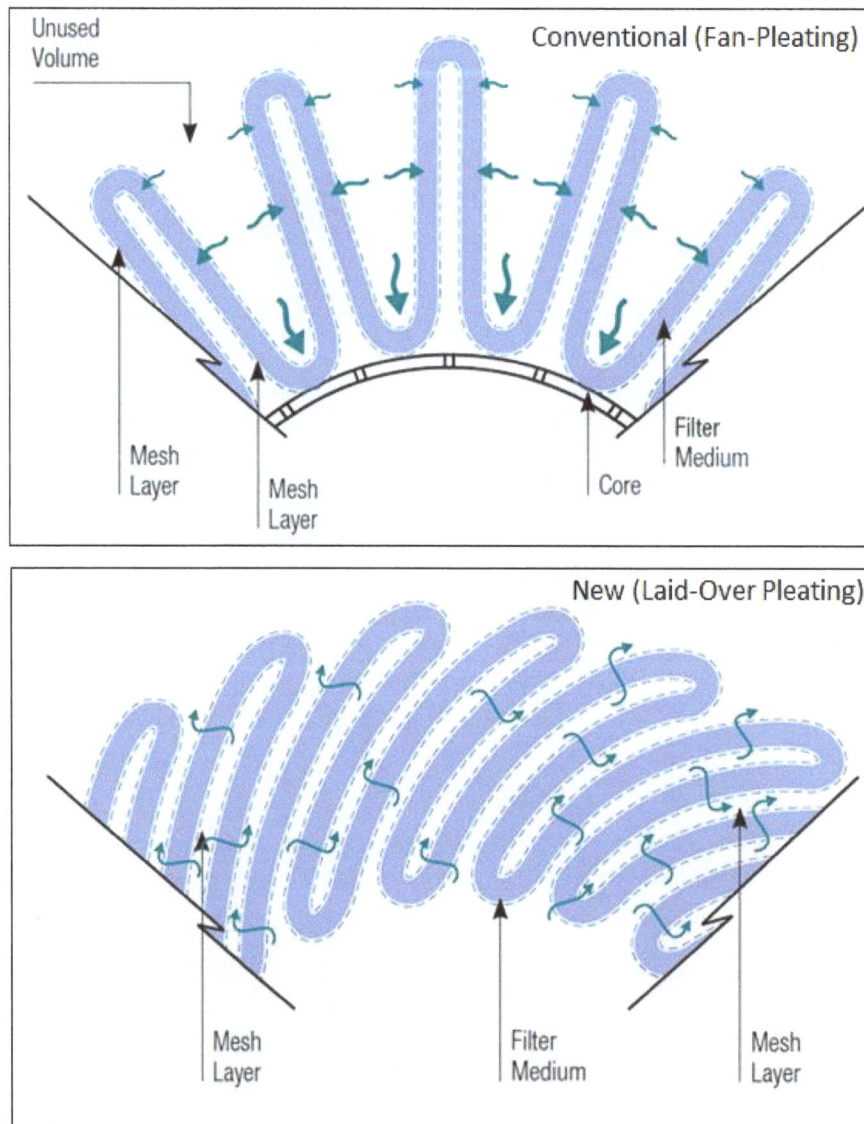

Fig. 7.18- Example of New Depth Filter Element Technology "Ultipleat" (Courtesy of Pall)

Figure 7.19 shows a typical depth filter using laid-over pleating technology. Athalon™ Filter is the next generation in *Anti-Static*, stress-resistant filters. This type of filter has enhanced performance that ensures equipment protection and extends component and fluid life.

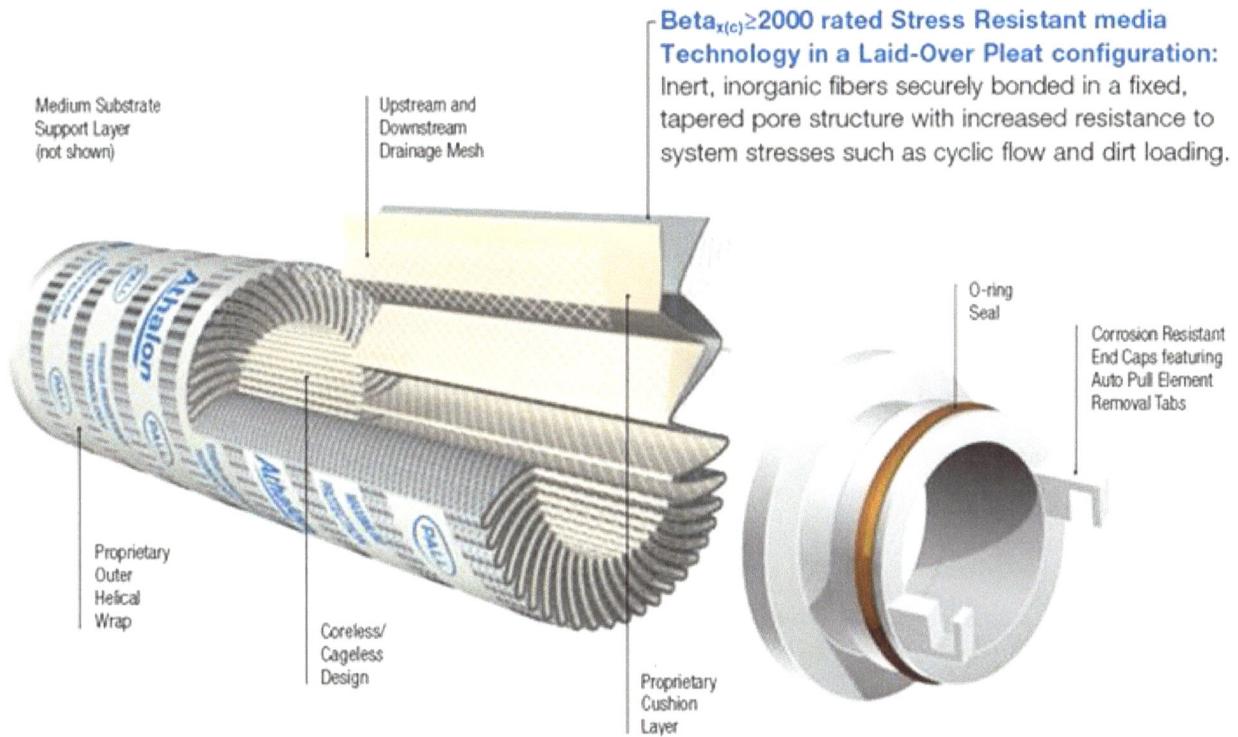

Medium Substrate
Support Layer
(not shown)

Upstream and
Downstream
Drainage Mesh

Beta$_{x(c)}$≥2000 rated Stress Resistant media Technology in a Laid-Over Pleat configuration: Inert, inorganic fibers securely bonded in a fixed, tapered pore structure with increased resistance to system stresses such as cyclic flow and dirt loading.

O-ring
Seal

Corrosion Resistant
End Caps featuring
Auto Pull Element
Removal Tabs

Proprietary
Outer
Helical
Wrap

Coreless/
Cageless
Design

Proprietary
Cushion
Layer

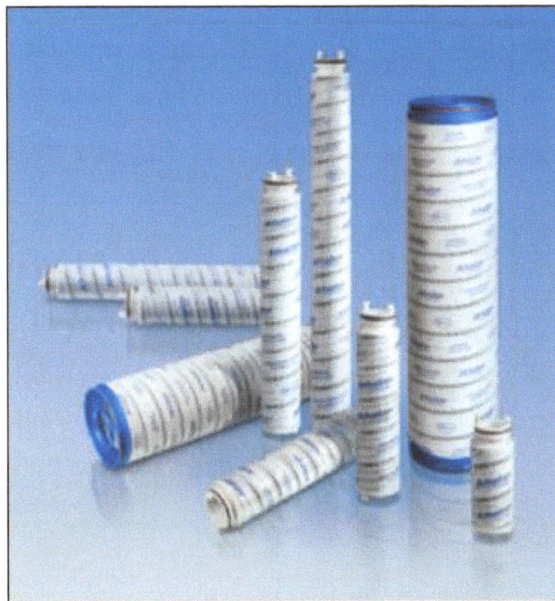

Fig. 7.19- Example of New Depth Filter Element Technology "Athalon" (Courtesy of Pall)

Example 4– New Depth Filter Element Technology: Figure 7.20 shows the of new depth filter technology (Optimicron). The filter has the following features:

- The filter element has an outer rap (1) around the outer surface to protect the sensitive filter media from fluid flow and increase the filter media robustness.
- Optimized crosssection (2) with new pleating shape. This new shape doubles the flow surface, lowers the flow velocity and ensures lower pressure drop across the media.
- Filter media consists of 7 consecutive layers (3) to increase the effectiveness of filtration.

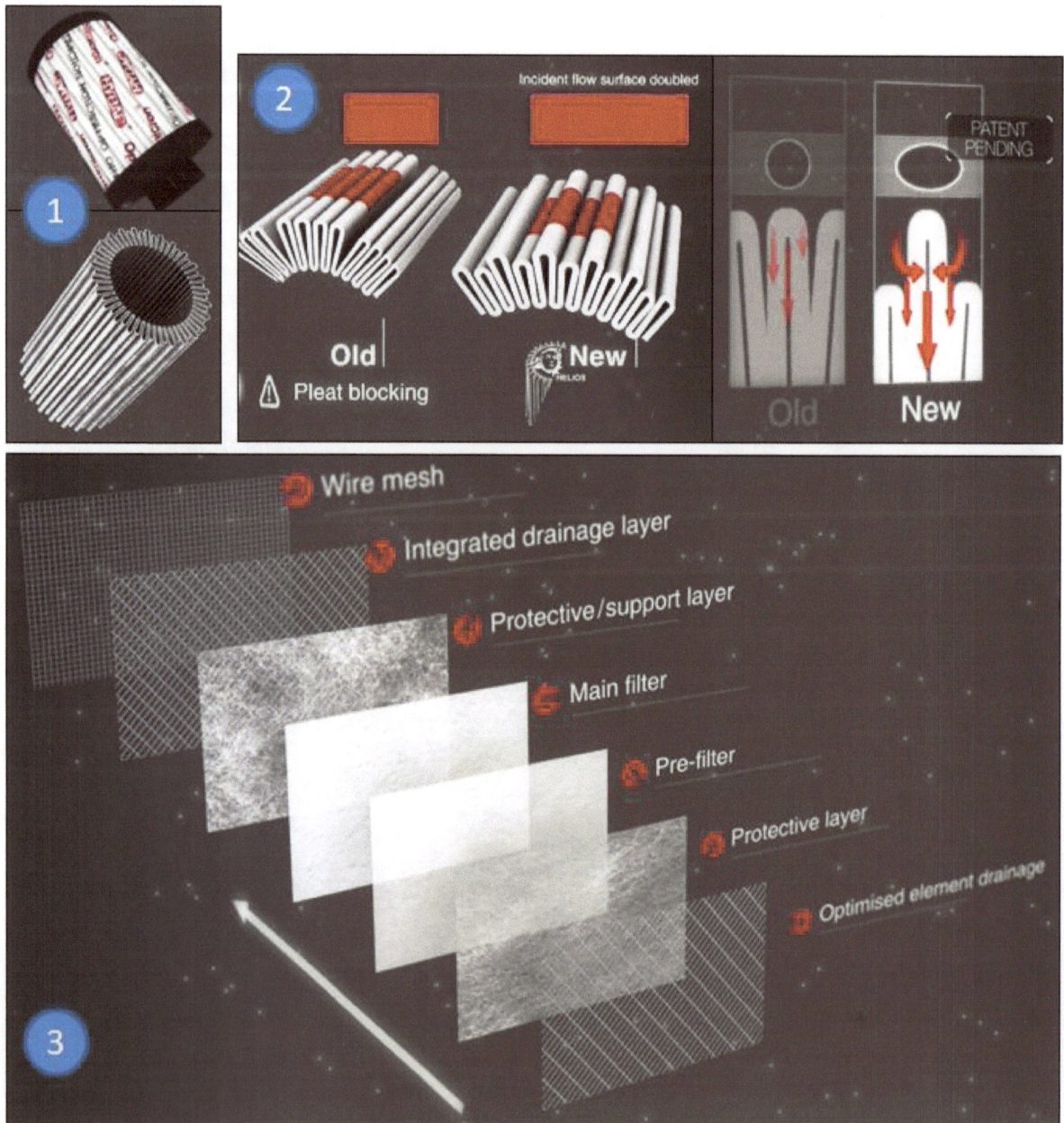

Fig. 7.20- Example of New Depth Filter Element Technology "Optimicron" (Courtesy of Hydac)

Example 5– Synthetic Depth Filter: Figure 7.21 shows a synthetic depth filter that has the following features:

- High-efficiency filtration rating.
- Exceptionally low flow resistance
- Consistent performance throughout filter life.
- Excellent fluid compatibility.
- Ideally suited for a variety of demanding applications, including heavy-duty mobile equipment, in-plant hydraulics, transmissions, and bearing lube oil systems.

The filter consists of the following elements:

Epoxy-Coated Steel Support Mesh (1): These two layers at the upstream and downstream sides provide excellent pleat support and spacing, which allows for maximum effective media area. They protect the media against damage during handling and installation.

Media Support Layers (2): These two layers at the upstream and downstream sides protect media during pressure surges.

Synteq™ Media Technology (3): Synthetic filter media has smooth, rounded fibers for low resistance to fluid flow. This media is ideal for filtering synthetic fluids, water glycols, water oil emulsions, HWCF (high water content fluids), and petroleum-based fluids.

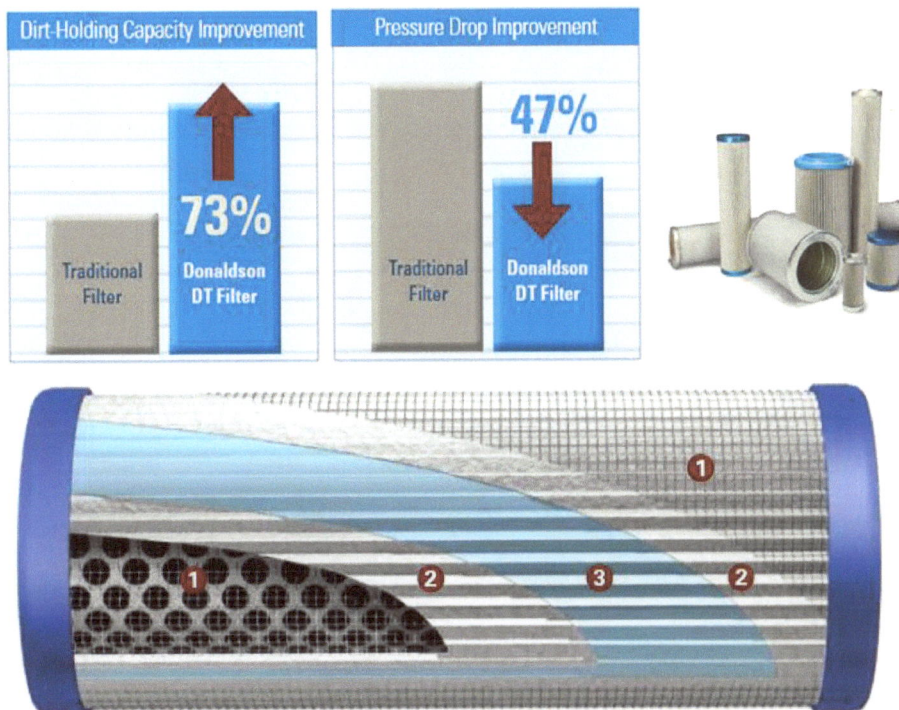

Fig. 7.21- Example of Synthetic Filter Element "DT Filters" (Courtesy of Donaldson)

Examples 6– Offline Filter for High Dirt Holding Capacity: Figure 7.22 shows a typical example of offline filters. Offline filters generally have a large dirt holding capacity of approximately 4 liters solid, 2 liters water, and 4 liters of oil degradation products (*Varnish, Sludge, and Oxidation*). They typically are replaced only on annual bases. Such filters are good to filter particles as small as 3 microns.

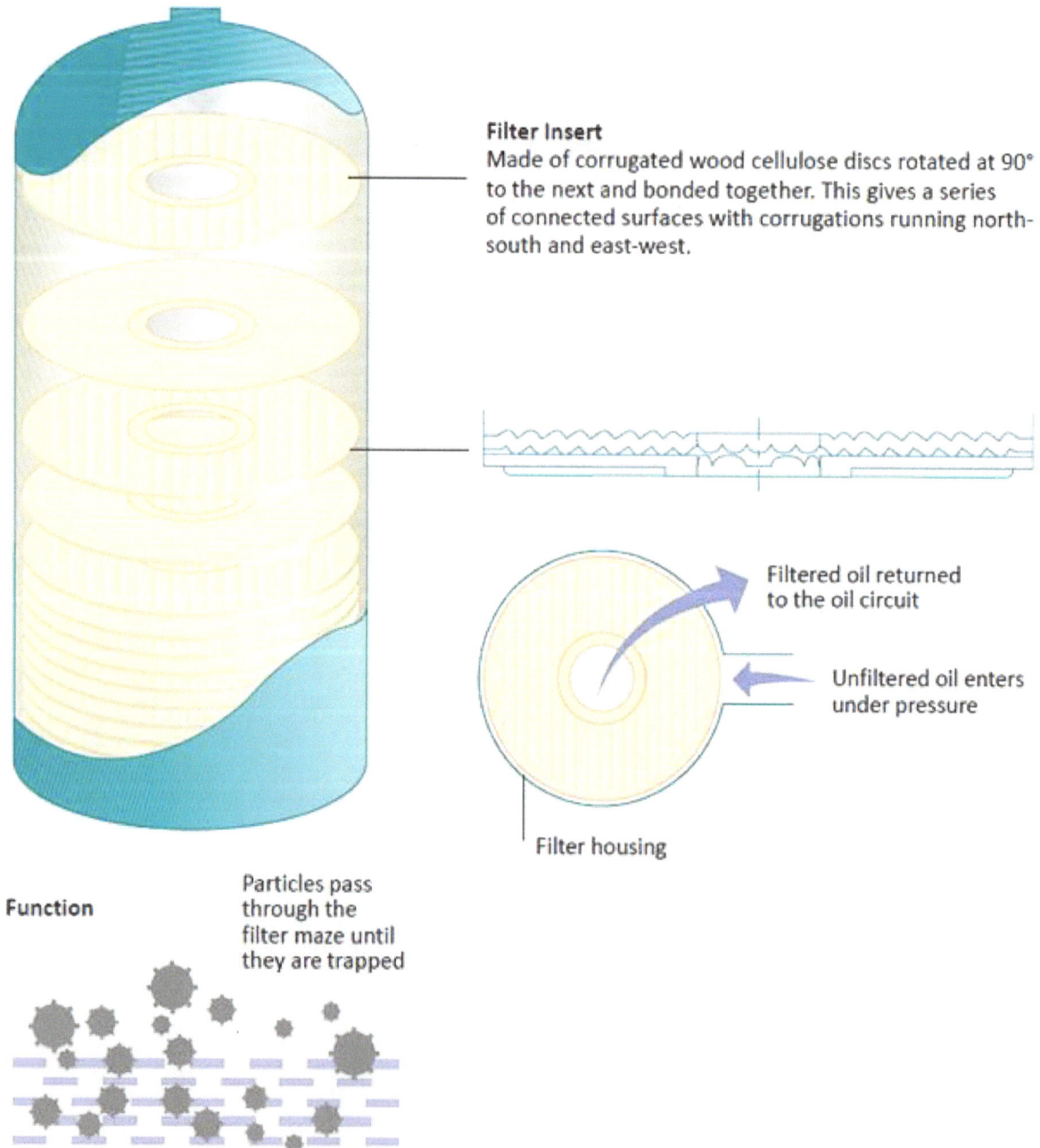

Filter Insert
Made of corrugated wood cellulose discs rotated at 90° to the next and bonded together. This gives a series of connected surfaces with corrugations running north-south and east-west.

Filtered oil returned to the oil circuit

Unfiltered oil enters under pressure

Filter housing

Function

Particles pass through the filter maze until they are trapped

Fig. 7.22- Example of Offline Filter for High Dirt Holding Capacity
(Courtesy of C.C. Jensen Inc.)

Example 7- Water Removal Filer Element: Figure 7.23 shows specification for a water removal filter element. Aquamicron® filter elements are specially designed to separate free water from mineral oils. Pressure drop is monitored by clogging indicator. A bypass valve is used to limit the pressure drop across the filter element. The figure shows the pressure drop versus the flow rate for mineral oils with a specific gravity of 0.86. Correction must be applied for fluid with different specific gravity.

Technical Specifications

Collapse Rating	145 psid (10 bar)
Temperature range	32°F to 212°F (0°C to 100°C)
Compatibility with hydraulic media	Mineral oils: Test criteria to ISO 2943 Lubricating oils: Test criteria to ISO 2943 Other media available on request
Opening pressure of by-pass valves	$\Delta P0 = 43$ psid ±7 psi (3 bar ±0.5 bar)
Bypass valve curves	The bypass valve curves apply to mineral oils with a specific gravity of 0.86. The differential pressure of the valve changes proportionally with the specific gravity.

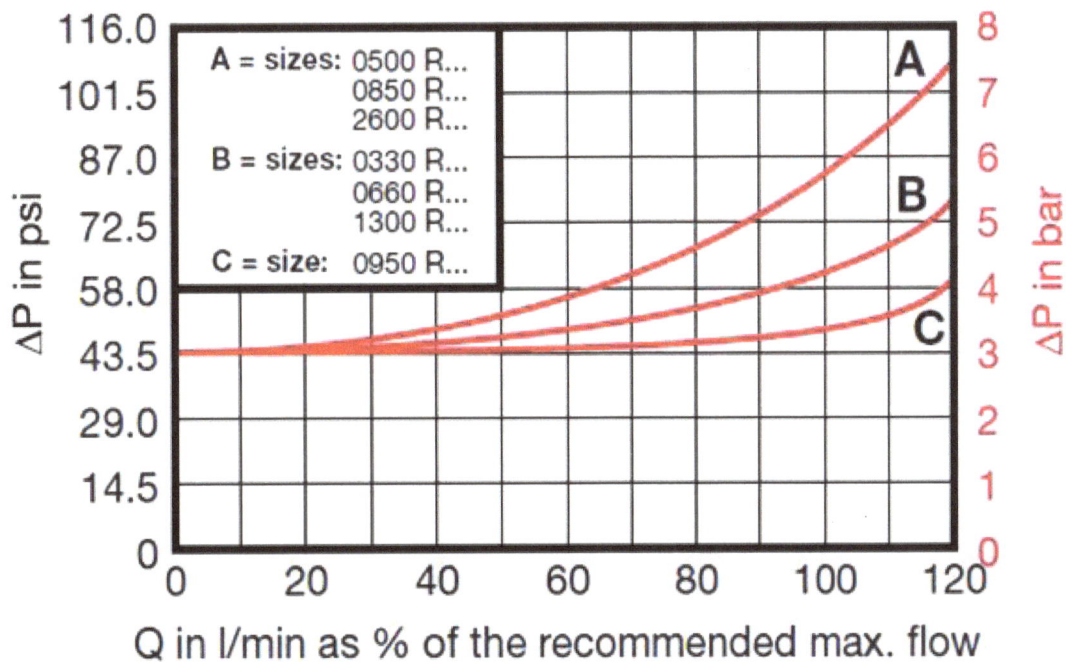

Fig. 7.23- Example of Water Removal Filter Element "Aquamicron®" (Courtesy of Hydac)

Example 8- Water Removal Filer Elements: Figure 7.24 shows dry water removal filter media. When it becomes wet (swollen) with absorbed water. The shown water removal filter media is an effective way of removing free water contamination from hydraulic systems. It is highly effective at removing free water from mineral-base and synthetic fluids. This filter media is a highly absorbent copolymer laminate with an affinity for water. The water is bonded to the filter media and forever removed from the system. It cannot even be squeezed out.

The figure also shows a conversion factor table to calculate the water content in a specific volume of oil. For example, assuming a reservoir stores 200 gallons of oil that is highly water contaminated (1000 ppm), then:
- %Water content 1000 x 0.0001 = 0.1% of the oil volume →
- Water Volume = (0.1 x 200)/100 gallon = 0.2 gallons of water.

If the acceptable water content in the oil is 300 ppm, then:
- %Water that should be removed = 700 ppm = 700 x 0.0001 = 0.07 % →
- Water volume that should be removed = (0.07x200)/100 = 0.14 gallons.

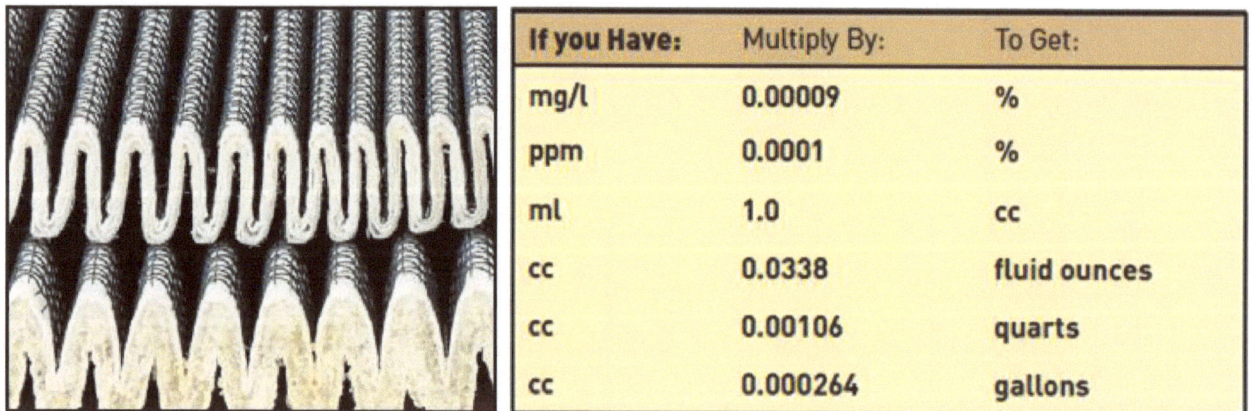

If you Have:	Multiply By:	To Get:
mg/l	0.00009	%
ppm	0.0001	%
ml	1.0	cc
cc	0.0338	fluid ounces
cc	0.00106	quarts
cc	0.000264	gallons

Fig. 7.24- Example of Water Removal Filter Element "Par-Gel" (Courtesy of Parker)

Example 9– Oil Cleaning by Magnetic Separation: Hydraulic fluid analysis shows that in some machines as much as 90% of all particles suspended in the oil can be ferromagnetic (iron or steel particles). Therefore, currently, there are a number of conventional and advanced products on the market that employ the use of magnets in various configurations and geometry.

While it is true that conventional mechanical filters can remove particles in the same size range as magnetic filters, these conventional filters have the following challenges:
- Cost of filter element disposal.
- Cost of Energy wasted due to pressure drop across the filter.
- Possibilities of filter burst under high pressure.
- Possibility of filter media collapse and migration when it becomes clogged.

Figure 7.25 shows the common magnetic products used in lubricating oil and hydraulic fluid applications.

Magnetic Drain Plug (1): The most basic type of magnetic filter is a *Magnetic Drain Plug.* It should be periodically removed and inspected for ferromagnetic particles, which are then wiped from the plug. such plugs are commonly used in engine oil pans, gearboxes and occasionally in hydraulic reservoirs.

Magnetic Rods (2): *Magnetic Rods* are immersed inside the reservoir. Magnetic rods can hold more particles than the drain plugs.

Flow-through Magnetic Filters (3): As fluid passes through the slots, ferromagnetic particles accumulate in the gap between the plates. The cleaning process typically involves removing the filter core and blowing the debris out from between the collection plates with an air hose.

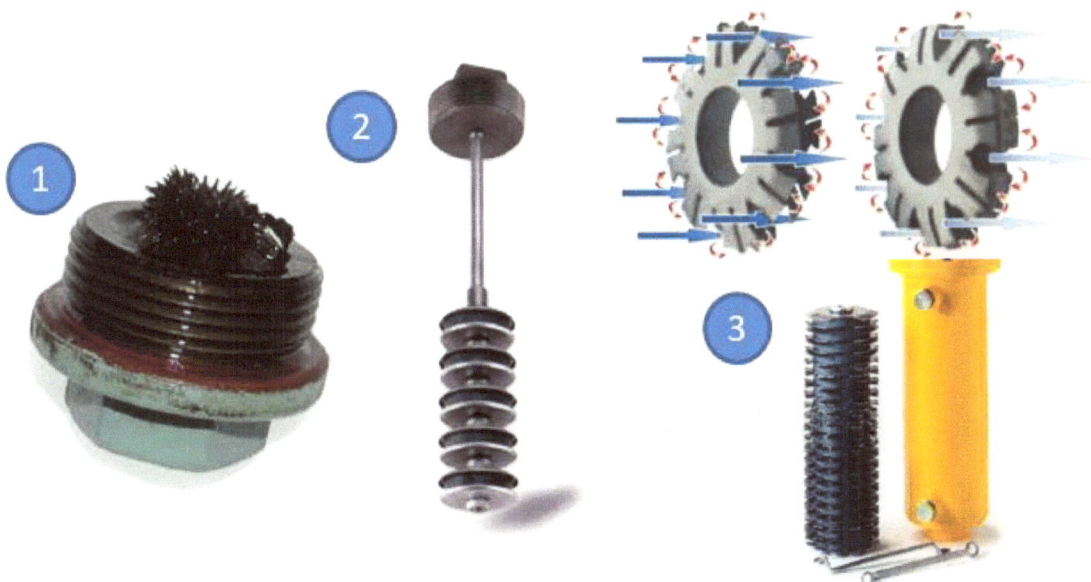

Fig. 7.25- Example of Magnetic Filters (Courtesy of Noria)

Example 10– Combo Mechanical and Magnetic Filter: Figure 7.26 shows a Spin-on mechanical filter with steel housing (Bowl). *Magnetic Wraps* are held on the exterior wall of the housing. These wraps transmit a magnetic field through the steel filter housing (bowl). A high-power magnet is installed at the bottom of the housing. The filter operates normally while the ferromagnetic debris are held tightly against the internal surface and at the bottom of the housing. The magnetic filter wraps can be used repeatedly.

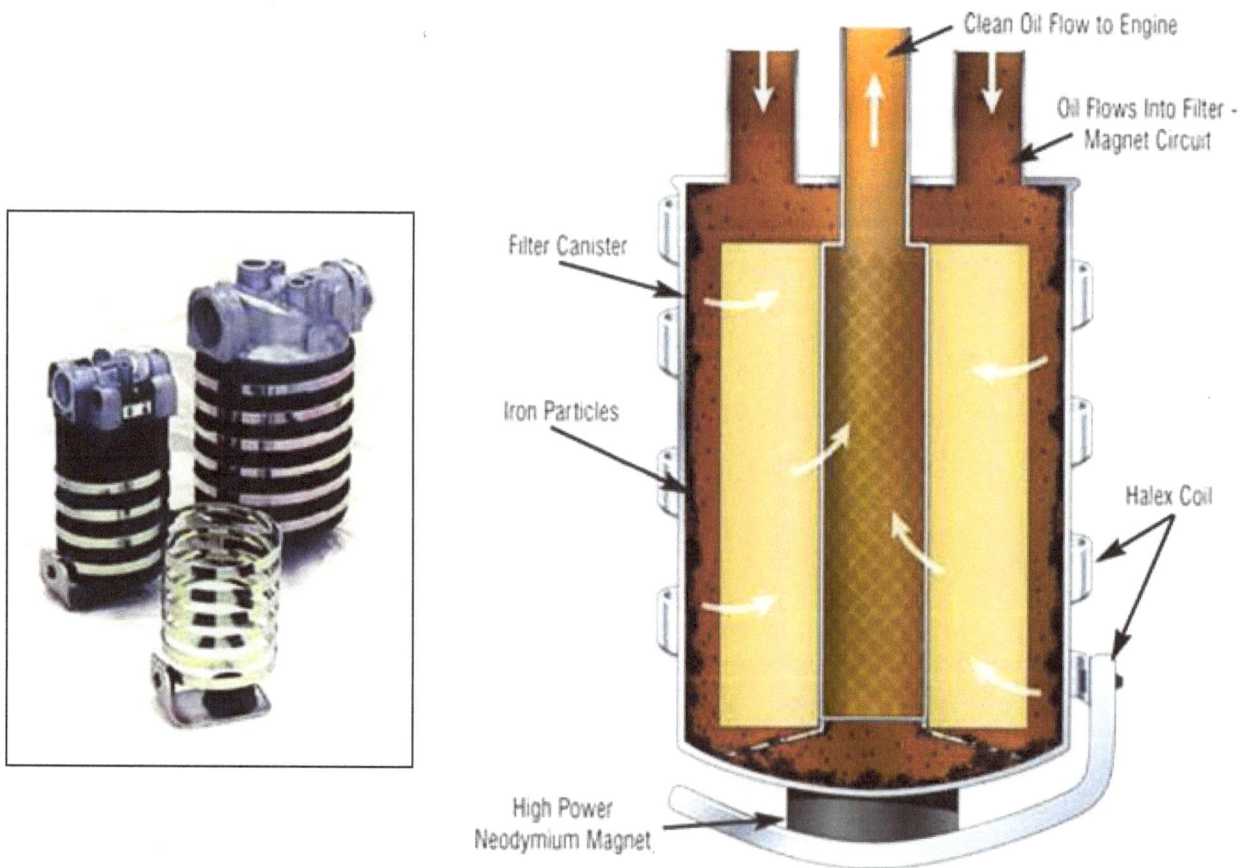

Fig. 7.26- Example of Mechanical and Magnetic Filters (Courtesy of Noria)

Chapter 8
Filter Selection Criteria

Objectives

This chapter presents a selection checklist as a guide for selecting proper filters. The chapter also discusses briefly the concepts for cost-effective filtration and selecting a filter cleanliness level based on system requirements. This chapter presents several examples of filtration solution for hydraulic systems.

Brief Contents

8.1- Filter Selection Checklist

8.2- Cost-Effective Filtration

8.3- Filter Selection Based on Cleanliness Requirements

8.4- Examples of Filtration Solutions

Chapter 8 – Filter Selection Criteria

8.1- Filter Selection Checklist

The proper choice of a filter is essential as early as the design stage of a system. When selecting a filter, the following questions in *Filter Selection Checklist* must be answered in order determine the proper filter

- Filter Purpose:
  - For regular operation, for flushing, for water or varnish removal?
  - For mobile or industrial application?

- Filter Location:
  - Inline filter (suction -pressure – return), or offline filter?

- System Operating Conditions:
  - What is the maximum system pressure, temperature, and flow?
  - Are there possible pressure fluctuation and/or pressure spikes?
  - Are there possible flow fluctuation or flow surges?
  - Are there sensitive components in the system that are intolerant to contamination?
  - What type of the hydraulic fluid is used and its viscosity?

- System Cleanliness Requirements:
  - What is the required absolute/nominal beta ratio and filter efficiency?
  - What is the mesh size (for screens)?
  - Requirements for bypass?
  - Requirements for clogging indicators/alarms?
  - What is the anticipated dirt holding capacity?

- Filter Media:
  - Collapse pressure and flow fatigue resistance?
  - Moisture absorbance characteristics and Anti-Static characteristics?

- Filter Housing:
  - Burst pressure and fatigue pressure?
  - Method of mounting and body style (inside tank, line mounted, on-tank top)
  - Port size?

8.2- Cost-Effective Filtration

The initial and running cost of efficient filtration in hydraulic systems are paid back by increasing the service life of important and expensive components and improving system reliability. The following bullets discuss solutions for *cost-effective* filtration.

Service Life Versus Filter Area: Dirt holding capacity and service life will vary greatly between filter types, media and manufacturers. The service life of a filter is the length of time that a filter element will last in actual system service before the allowable differential pressure is reached. Dirt holding capacity alone may not indicate which filter would give the best service life. In order to achieve the highest service life, element area should be increased. As shown in Fig. 8.1, a filter element with three times the area will yield 4 to 6 times the service life. Then, price of element with large area must be compromised with the cost paid back by extended filter lifetime.

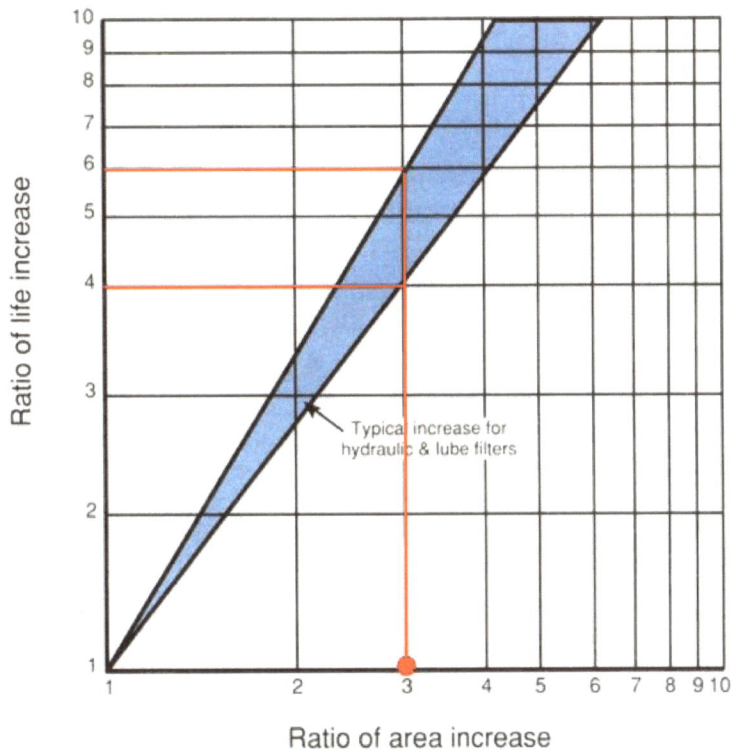

Fig. 8.1- Filter Element Service Life versus Filter Area

Staged Filtration: Some system designers prefer to use "*staged*" filtration rather than rely on a single filter element. Staged filtration utilizes two or more filters in series with the lower or finer filtration rating downstream of the first filter. The first filter acts as a pre-filter for the fine filter, removing the majority of coarse particulates. The initial cost of staged filtration is greater due to multiple filter housings. However, this initial cost is compensated with the extended filter element service life and clean operating system.

Filter Location: By selecting the proper location of filters in the system, filter costs can be minimized. Most cost-effective filtration solutions are either return filters or offline filters. However, a pressure filter might be mandatory in cases where sensitive components are used such as servo valves. If pressure filters are used, the filter housing must be selected for the maximum system operating pressure. Pressure filters are the most expensive.

Bypass OR Non-Bypass: Having a bypass valve in the filter construction limits the maximum differential pressure across the filter element. That helps reducing cost by selecting a filter element with relatively low collapse pressure. However, collapse pressure of the filter element must be higher than the bypass valve cracking pressure. However, non-bypass filters are required for cases where sensitive components are located downstream of the filter. An alternative solution is to use a bypass-to-tank filter.

8.3- Filter Selection Based on Cleanliness Requirements

It is the responsibility of the machine user to make sure that the oil continues to be clean as per the manufacturer' recommendation during machine operation. Filtration system in the machine should comply with the component that is the most sensitive (least tolerant) to contamination. In case of component failure, warranty may be voided if fluid cleanliness is greater than recommendations. If there are no recommendations given by the manufacturer, the following section provide some general guidelines.

Table 8.1 shows system-based and component-based cleanliness level recommendations, respectively. Table 8.2 shows pressure-based cleanliness requirements. Presented information shows that electro-hydraulic servo and proportional valves and variable pumps, particularly piston type, are components that are most sensitive to contamination. Additionally, high working pressure conditions requires cleaner oil.

Application	Oil cleanliness required in accordance with ISO 4406
Systems with extremely high dirt sensitivity and very high availability requirements	≤ 16/12/9
Systems with high dirt sensitivity and high availability requirements, such as servo valve technology	≤ 18/13/10
Systems with proportional valves and pressures > 160 bar	≤ 18/14/11
Vane pumps, piston pumps, piston engines	≤ 19/16/13
Modern industrial hydraulic systems, directional valves, pressure valves	≤ 20/16/13
Industrial hydraulic systems with large tolerances and low dirt sensitivity	≤ 21/17/14

Pumps	ISO Ratings
Fixed Gear Pump	19/17/15
Fixed Vane Pump	19/17/14
Fixed Piston Pump	18/16/14
Variable Vane Pump	18/16/14
Variable Piston Pump	17/15/13
Valves	
Directional (solenoid)	20/18/15
Pressure (modulating)	19/17/14
Flow Controls (standard)	19/17/14
Check Valves	20/18/15
Cartridge Valves	20/18/15
Load-sensing Directional Valves	18/16/14
Proportional Pressure Controls	18/16/13
Proportional Cartridge Valves	18/16/13
Servo Valves	16/14/11*
Actuators	
Cylinders	20/18/15
Vane Motors	19/17/14
Axial Piston Motors	18/16/13
Gear Motors	20/18/15
Radial Piston Motors	19/17/15

Table 8.1- System-Based and Component-Based Cleanliness Requirements

	ISO Target Levels		
	Low/Medium Pressure Under 2000 psi (moderate conditions)	High Pressure 2000 to 2999 psi (low/medium with severe conditions†)	Very High Pressure 3000 psi and over (high pressure with severe conditions†)
Pumps			
Fixed Gear or Fixed Vane	20/18/15	19/17/14	18/16/13
Fixed Piston	19/17/14	18/16/13	17/15/12
Variable Vane	18/16/13	17/15/12	not applicable
Variable Piston	18/16/13	17/15/12	16/14/11
Valves			
Check Valve	20/18/15	20/18/15	19/17/14
Directional (solenoid)	20/18/15	19/17/14	18/16/13
Standard Flow Control	20/18/15	19/17/14	18/16/13
Cartridge Valve	19/17/14	18/16/13	17/15/12
Proportional Valve	18/16/13	17/15/12	16/14/11
Servo Valve	16/14/11	16/14/11	15/13/10
Actuators			
Cylinders, Vane Motors, Gear Motors	20/18/15	19/17/14	18/16/13
Piston Motors, Swash Plate Motors	19/17/14	18/16/13	17/15/12
Hydrostatic Drives	16/15/12	16/14/11	15/13/10
Test Stands	15/13/10	15/13/10	15/13/10
Bearings			
Journal Bearings	17/15/12	not applicable	not applicable
Industrial Gearboxes	17/15/12	not applicable	not applicable
Ball Bearings	15/13/10	not applicable	not applicable
Roller Bearings	16/14/11	not applicable	not applicable

Table 8.2- Pressure-Based Cleanliness Requirements (Courtesy of Hydac)

8.4- Examples of Filtration Solutions

Example 1 (Fig. 8.2): Filtration Solutions for Injection Molding Machines.

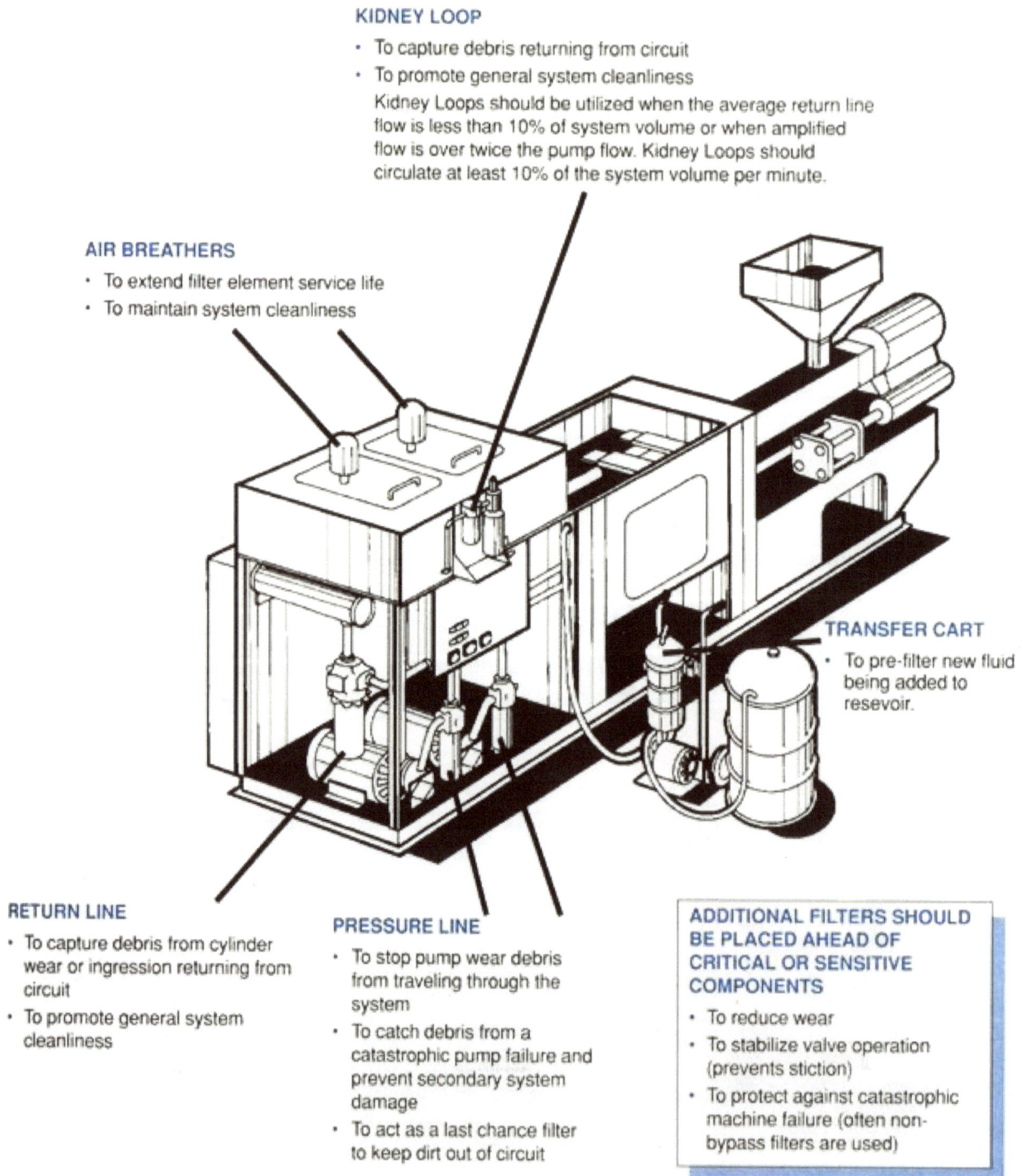

KIDNEY LOOP
- To capture debris returning from circuit
- To promote general system cleanliness
  Kidney Loops should be utilized when the average return line flow is less than 10% of system volume or when amplified flow is over twice the pump flow. Kidney Loops should circulate at least 10% of the system volume per minute.

AIR BREATHERS
- To extend filter element service life
- To maintain system cleanliness

TRANSFER CART
- To pre-filter new fluid being added to resevoir.

RETURN LINE
- To capture debris from cylinder wear or ingression returning from circuit
- To promote general system cleanliness

PRESSURE LINE
- To stop pump wear debris from traveling through the system
- To catch debris from a catastrophic pump failure and prevent secondary system damage
- To act as a last chance filter to keep dirt out of circuit

ADDITIONAL FILTERS SHOULD BE PLACED AHEAD OF CRITICAL OR SENSITIVE COMPONENTS
- To reduce wear
- To stabilize valve operation (prevents stiction)
- To protect against catastrophic machine failure (often non-bypass filters are used)

Fig. 8.2- Filtration Solutions for Injection Molding Machines (Courtesy of Pall)

Example 2 (Fig. 8.3) - Open Circuit Systems with Solenoid Valves:

System Parameters:
- Pump flow = 30 gpm.
- Maximum pressure = 1800 psi.
- Fluid type is petroleum-based fluid.

Selected Filters:
1- Pressure Line Filter (HH9660A20DNTBPT).
2- Return Line Filter (HH8800A2DNTBPL).
3- Air Breather Filter (HC7500S038H-B).

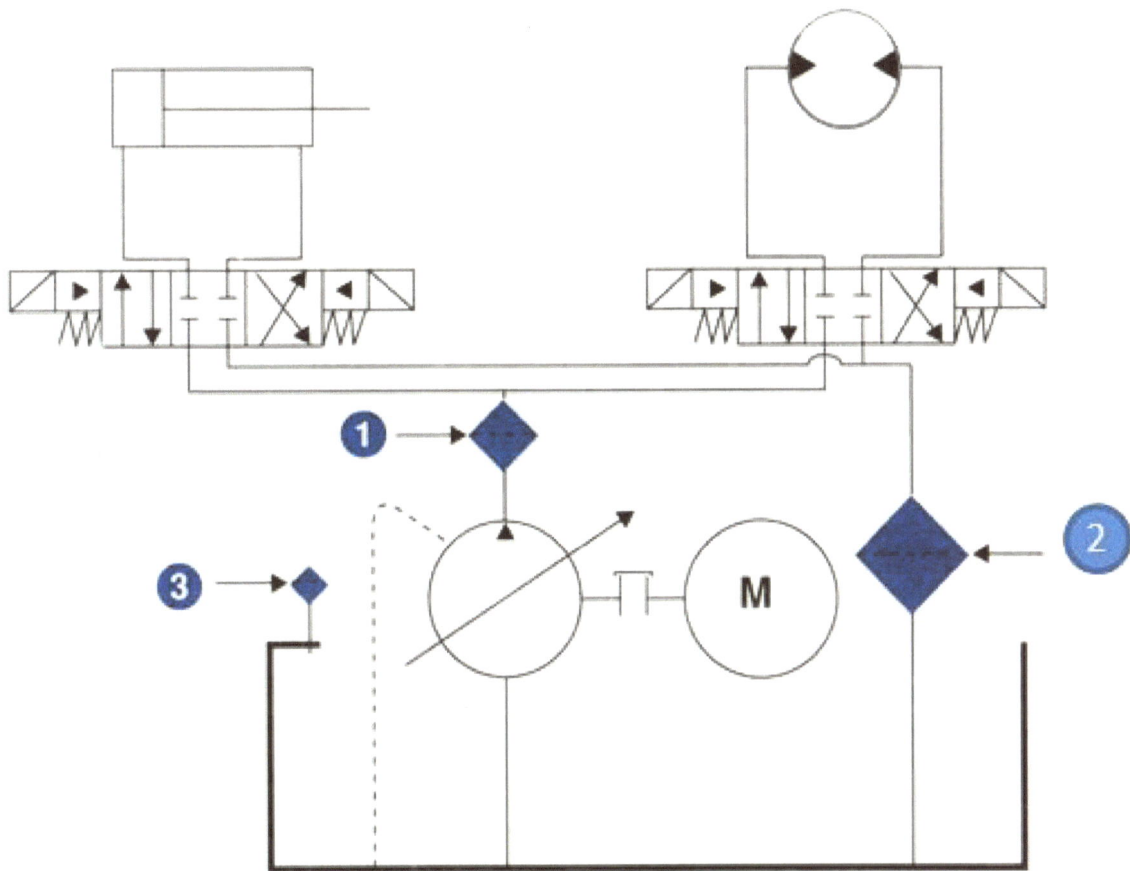

Fig. 8.3- Filtration Solutions for Open Circuit with Solenoid Valves (Courtesy of Pall)

Example 3 (Fig. 8.4) - Open Circuit Systems with Servo Valve:

System Parameters:
- Pump flow = 15 gpm.
- Maximum pressure = 1900 psi.
- Fluid type is petroleum-based fluid.

Selected Filters:
1- Pressure Line Filter (HH9850A 16DPSBPT).
2- Remote Mounted Non-Bypass Filter (HH9021 A 12DPRWPT).
3- Return Line Filter (HH8200A20DPSBPL).
4- Aire Breather Filter (HC7500S038H-B).

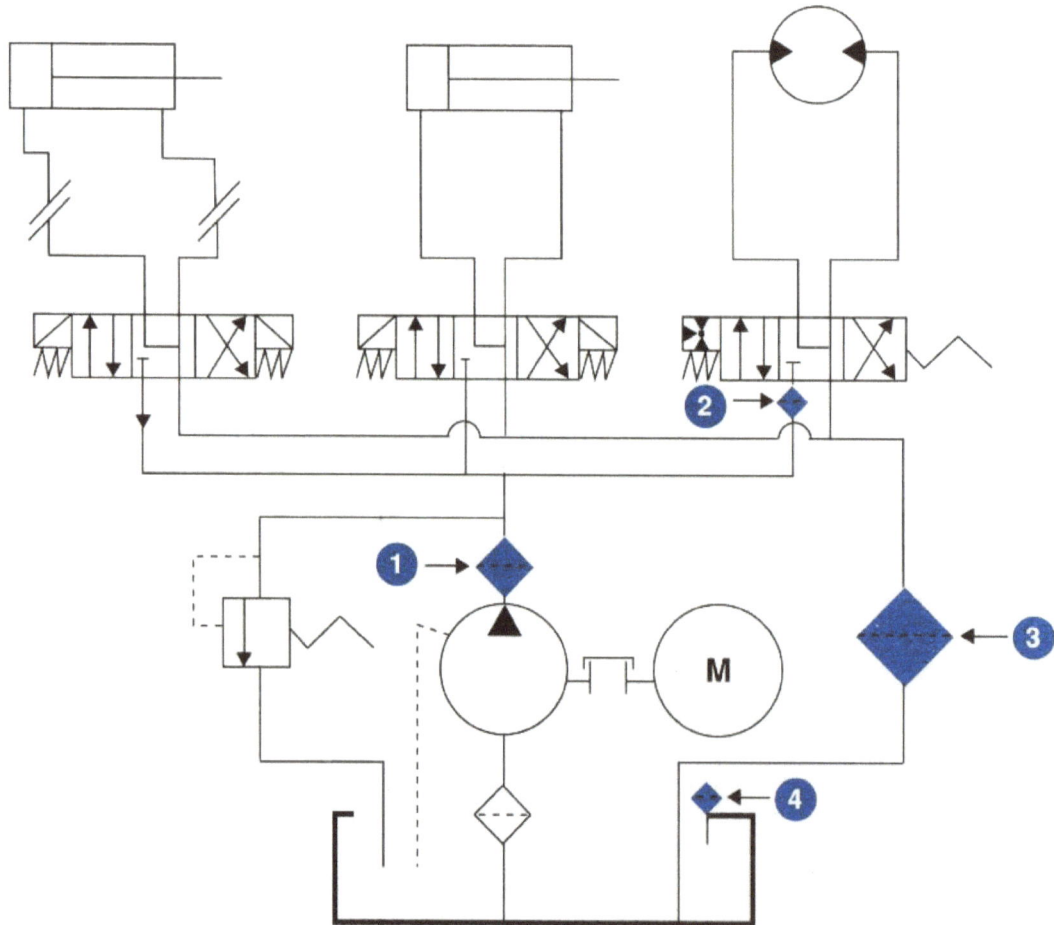

Fig. 8.4- Filtration Solutions for Open Circuit with Servo Valve (Courtesy of Pall)

413 | Hydraulic Systems Volume 4: Hydraulic Fluids Conditioning
Chapter 8- Filter Selection Criteria

Example 4 (Fig. 8.5) – Clamp and Hold Hydraulic Circuit:

<u>System Parameters:</u>
- Pump Type: Piston, variable displacement pressure compensated.
- Pump flow = 50 gpm.
- Maximum pressure = 2800 psi.
- Fluid type is Water Glycol.
- Total System Fluid Volume = 300 Gallons.

<u>Selected Filters:</u>
1- Pressure Line Filter (HH971 0A24DP2BDT).
2- Kidney Loop Filter (HH8900D32DPUBDL).
3- Aire Breather Filter (HC7500S038H-B).

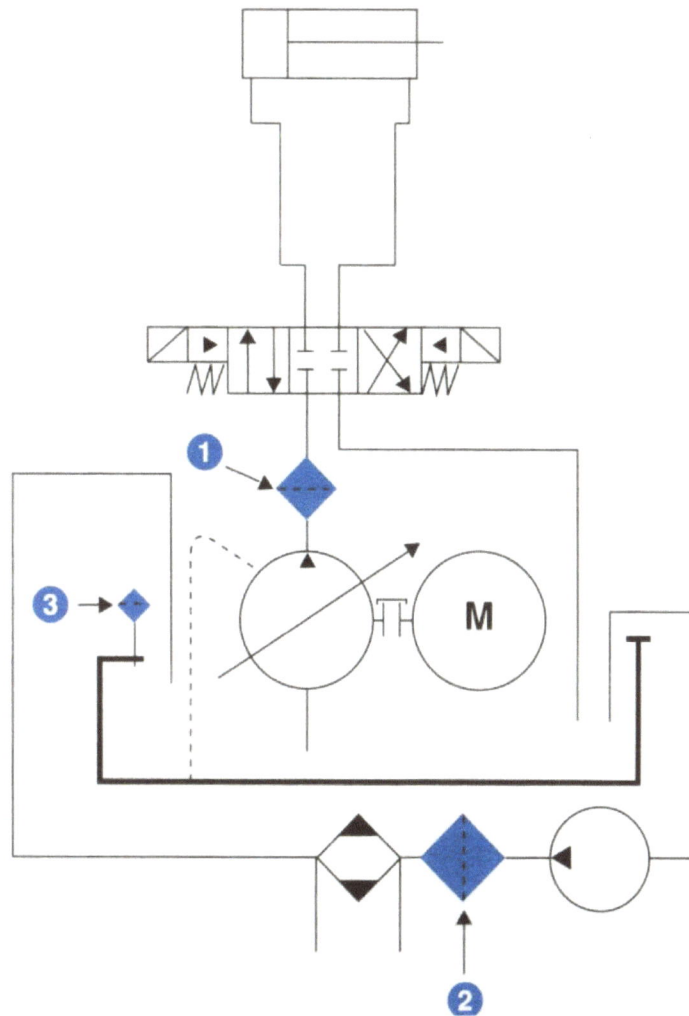

Fig. 8.5- Filtration Solutions for Clamp and Hold Circuit (Courtesy of Pall)

Example 5 (Fig. 8.6) – Closed Circuit Hydrostatic Transmission:

System Parameters:
- Charge pressure = 125 psi.
- Charge flow = 4 gpm.
- System pressure = 3200 psi.
- System flow = 32 gpm.
- System flow direction is bidirectional.
- Fluid type is petroleum-based.

Selected Filters:
1 and 2 in-loop filters with reverse flow valve (HH9660E20DPTCPT).

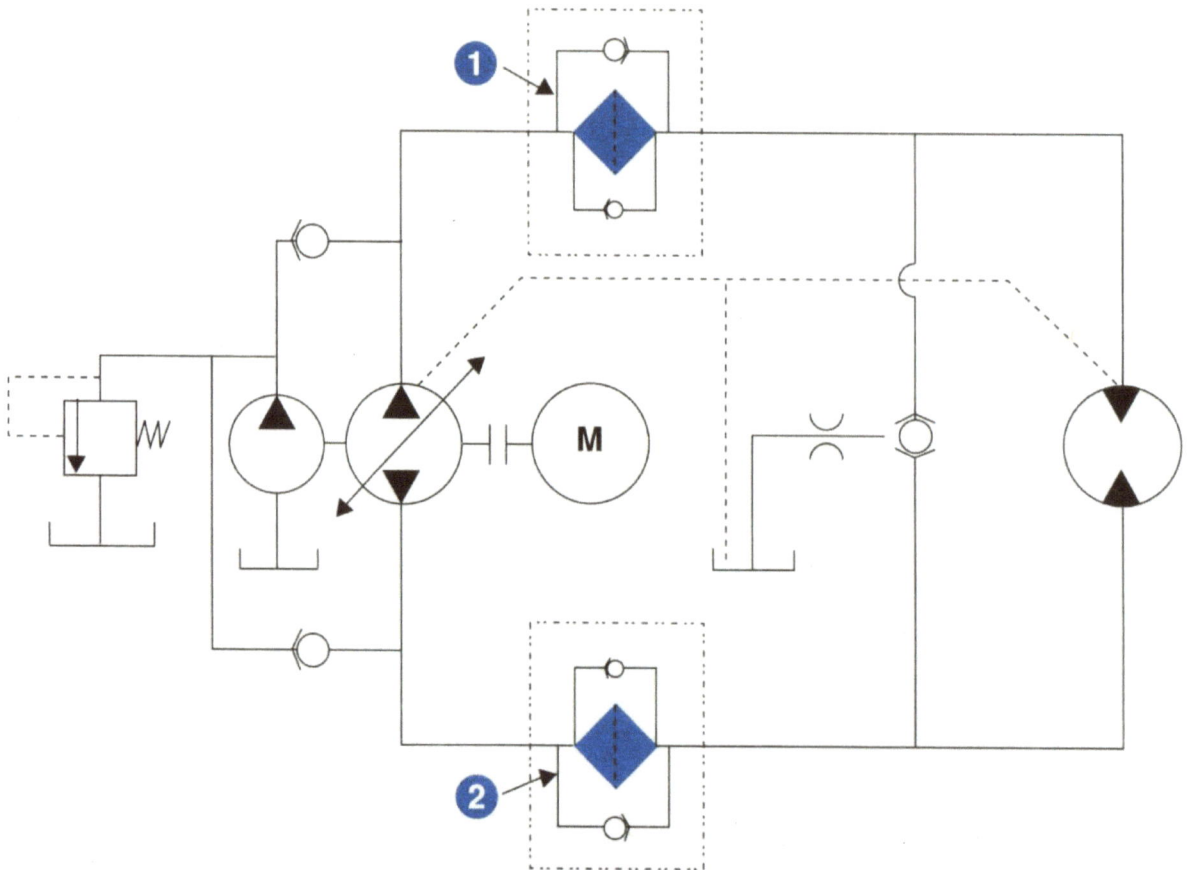

Fig. 8.6- Filtration Solutions for Hydrostatic Transmission (Courtesy of Pall)

Example 6 (Fig. 8.7) – Bearing Lubrication Circuit:

System Parameters:
- Pump flow = 120 gpm.
- System pressure = 45 psi.
- Maximum pressure at the pump = 120 psi
- Fluid type is petroleum-based.

Selected Filters:
1- Duplex filter in Pressure Line (HH8342D64DNXAPT)
2- Air Breathers - (2) HC7500S038H-B

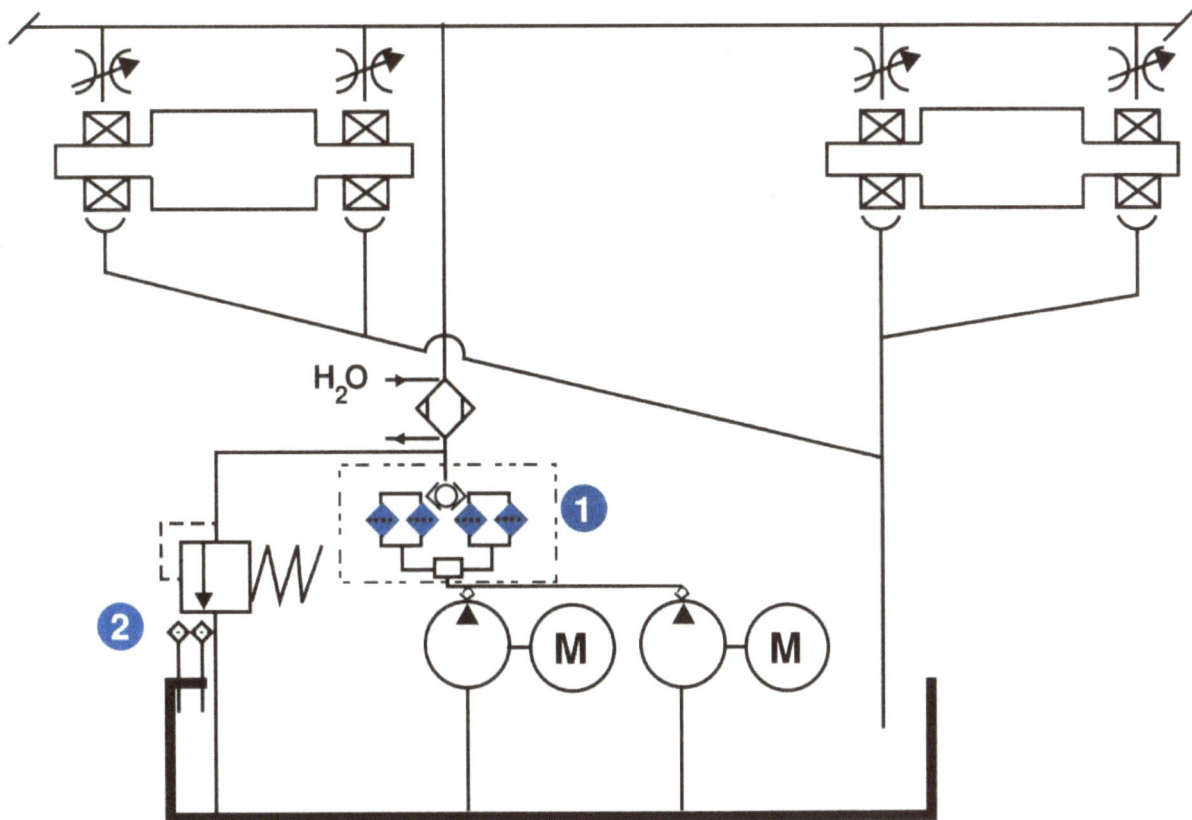

Fig. 8.7- Filtration Solutions for Bearing Lubrication Circuit (Courtesy of Pall)

Example 7 (Fig. 8.8) – Turbine System Lubrication:

System Parameters:
- Main reservoir volume = 6000 Gallons.
- Operating pressure = 15 psi.
- Kidney loop flow (added) = 100 Gallons.
- Fluid type is petroleum-based.

Selected Filters:
1 and 2 kidney loop filters (HH8300D40DPXBPT).

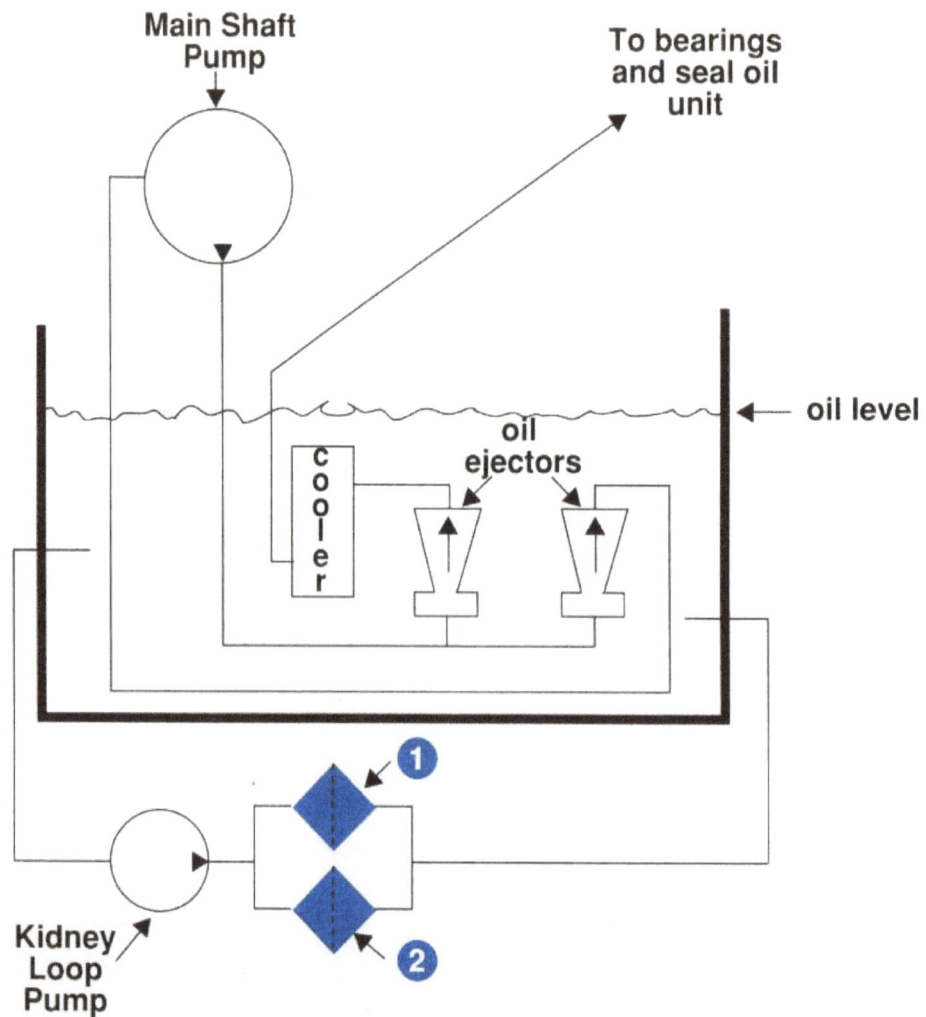

Fig. 8.8- Filtration Solutions for Turbine System Lubrication (Courtesy of Pall)

APPENDIXES

APPENDIX A: LIST OF FIGURES

Chapter 3: Hydraulic Transmission Lines

Fig. 3.7 - Pipe Connection by National Thread

Fig. 3.8 - Standard Spiral Thread versus Dryseal Thread

Fig. 3.9- Hydraulic Pipe Fittings

Fig. 3.10 - Use of Hydraulic Tubes in Industrial (Left) and Mobile (right) Applications

Fig. 3.11 - Material of Hydraulic Tubes

Fig. 3.12 - Manual Mandrills of Different Sizes for Tube Bending

Fig. 3.13 - 370 Single and Double Flaring According to SAE J533B

Fig. 3.14 - 450 Single and Double Flaring According to SAE J533B

Fig. 3.15 - Fitting Body Configurations (Courtesy of Bosch Rexroth)

Fig. 3.16 - Fitting Seal Configurations (Courtesy of Bosch)

Fig. 3.17 – Spiral Clearance on NPT Fittings (Courtesy of Brennan Industries)

Fig. 3.18 - National Pipe Thread (Courtesy of Gates)

Fig. 3.19 – Crossectional Views of NPTF and NPSM Fittings (Courtesy of Brennan Industries)

Fig. 3.20 - SAE Inverted Flare Thread (Courtesy of Gates)

Fig. 3.21 - British Standard Pipe Parallel Thread (Courtesy of Gates)

Fig. 3.22 - British Standard Pipe Tapered Thread (Courtesy of Gates)

Fig. 3.23 - Sectional Views for BSPT and PSPP Fittings (Courtesy of Brennan Industries)

Fig. 3.24 - GAZ 24° Metric French Thread (Courtesy of Gates)

Fig. 3.25 - DIN 24° Cone Metric German Thread (Courtesy of Gates)

Fig. 3.26 - EO-2 Fitting (Courtesy of Parker)

Fig. 3.27 - DIN 60° Cone Metric German Thread (Courtesy of Gates)

Fig. 3.28 - DIN 3852 Thread (Courtesy of Gates)

Fig. 3.29 - Japanese 300 Flare Thread (Courtesy of Gates)

Fig. 3.30 - Japanese Tapered Pipe Thread (Courtesy of Gates)

Fig. 3.31 - SAE JI926 Straight Thread O-Ring Boss (ORB)

Fig. 3.32 - Typical O-Ring Boss Fittings of Different Configurations (www.redl.com)

Fig. 3.33 – Sectional View of an ORB Fitting (Curtesy of Brennan Industries)

Fig. 3.34 - O-Ring Face Seal SAE J1453 (Courtesy of Gates)

Fig. 3.35 - Typical O-Ring Face Seal Fittings of Different Configurations (Courtesy of Parker)

Fig. 3.36 - O-Lok Fitting (Courtesy of Parker)

Fig. 3.37 - North American or Metric Flareless Assembly (Courtesy of Gates)

Fig. 3.38 - Typical North American or Metric Flareless Assembly (Courtesy of Parker)

Fig. 3.39 - How Cutting Rings Work (www.stauffusa.com)

Fig. 3.40 – Characteristics of 370 Flare Fitting (Courtesy of Gates)

Fig. 3.41 – Typical 370 Flared Fitting Configurations (Courtesy of Parker)

Fig. 3.42 – Sectional View of a 370 Flare Fitting Configurations
(Courtesy of Brennan Industries)

Fig. 3.43 - Triple-Lok Fitting (Courtesy of Parker)

Fig. 3.44 - SAE 450 Flare (Courtesy of Gates)

Fig. 3.45 - Typical JIC 450 Flare (Courtesy of Parker)

Fig. 3.46 - Japanese 300 Flare Thread (Courtesy of Gates)

Fig. 3.47 - Adaptors to Connect Fittings from Foreign Standards to SAE Standards
(Courtesy of Gates)

Fig. 3.48 - Typical Adapters (Courtesy of Parker)

Chapter 4: Hydraulic Sealing Elements

Chapter 5: Hydraulic Heat Exchangers

Chapter 6-Introduction to Hydraulic Filters

APPENDIX B: LIST OF TABLES

APPENDIX C: LIST OF REFERENCES

Hydraulic Systems Volume 1- Introduction to Hydraulics for Industry Professionals
Author: Dr. Medhat Kamel Bahr Khalil, 2016.
Publisher: Compudraulic, USA.
ISBN 978-0-692-62236-0

Hydraulic Systems Volume 2- Electro-Hydraulic Components and Systems
Author: Dr. Medhat Kamel Bahr Khalil, 2016.
Publisher: Compudraulic, USA.
ISBN: 978-0-9977634-2-3

Hydraulic Systems Volume 3- Hydraulic Fluids and Contamination Control
Author: Dr. Medhat Kamel Bahr Khalil, 2016.
Publisher: Compudraulic, USA.
ISBN: 978-0-9977816-3-2

Hydraulic Systems Volume 5- Safety and Maintenance
Author: Dr. Medhat Kamel Bahr Khalil, 2022.
Publisher: Compudraulic, USA.
ISBN: 978-0-9977816-5-6

Hydraulic Systems Volume 6- Troubleshooting and Failure Analysis
Author: Dr. Medhat Kamel Bahr Khalil, 2022.
Publisher: Compudraulic, USA.
ISBN: 978-0-9977634-6-1

Hydraulic Systems Volume 7- Modeling and Simulation for Application Engineers
Author: Dr. Medhat Kamel Bahr Khalil, 2016.
Publisher: Compudraulic, USA.
ISBN: 978-0-9977816-3-2

R01- Basic Electronics for Hydraulic Motion Control
Author: Jack L. Johnson, PE 1992.
Publisher: Penton Publishing Inc. 1100 Superior Avenue. Cleveland, OH 44114.
ISBN No. 0-932905-07-2.

R02- Closed Loop Electro-hydraulics Systems Manual
Author: Vickers/Eaton.
Publisher: Vickers Inc. 1992.
Training Center, 2730 Research Drive, Rochester Hills, MI 48309-3570.
ISBN 0-9634162-1-9

R03- Bosch Automation Technology
Author: Werner Gotz, Steffen Haack, Ralph Mertlick.
Publisher: Bosch.
ISBN 3-933698-05-7.

R04- Electrohydraulic Proportional and Control Systems
Publisher: Bosch Automation 1999.
ISBN 0-7680-0538-8.

R05- Proportional and Servo Valve Technology – The Hydraulic Trainer Volume 2
Author: R. Edwards, J. Hunter, D. Kretz, F. Liedhegener, W. Schenkel, A. Schmitt.
Publisher: Mannesman Rexroth AG 1988. D-8770 Lohr a. Main.
ISBN 3-8023-0266-4.

R06- Proportional Hydraulics
Author: D. Scholz.
Publisher: Festo Didactic KG, Esslingen, Germany.

R07- Electricity, Fluid Power, and Mechanical Systems for Industrial Maintenance
Author: Thomas Kissell.
Publisher: Prentice Hall, Inc. 1999, Upper Saddle River, NJ 07458.
ISBN 0-13-896473-4.

R08- Fluid Power in Plant and Field – First Edition
Author: Charles S. Hedges, R.C. Womack.
Publisher: Womack Machine Supply Co. 1968.
Womack Educational Publication, 2010 Shea Road, Dallas, TX 75235.
ISBN 68-22573 (Library of Congress Card Catalog No.).

R09- Hydraulics, Fundamentals of Service
Author: Deere and Company.
Publisher: John Deere Publishing 1999.
Almon TIAC Bldg. Suite 104, 1300-19th Street, East Moline, IL 61244.
ISBN 0-86691-265-7.

R10- Industrial Hydraulics Troubleshooting
Author: James E. Anders, Sr.
Publisher: McGraw-Hill, Inc.
ISBN 0-07-001592-9.

R11- Power Hydraulics
Author: John Ashby.
Publisher: Prentice Hall 1989. Prentice Hall International, (UK) Ltd.
66 Wood Lane End, Hemel Hempstead, Hertfordshire, HP2 4RG.
ISBN 0-13-687443-6.

R12- Fluid Power with Application
Author: Anthony Esposito.
Publisher: Prentice Hall.
ISBN 0-13-060899-8.

R13- Hydraulic Component Design and Selection
Author: E.C. Fitch.
Publisher: BarDyne Inc. 5111 North Perkins Rd. Stillwater, OK 74075.
ISBN 0-9705922-3-X.

R14- Planning and Design of Hydraulic Power Systems – The Hydraulic Trainer, Vol. 3
Author: Mannesmann Rexroth GmbH.
Publisher: Mannesman Rexroth AG 1988.
D-97813 Lhr a. Main, Jahnsrtrabe 3-5 D-97816 Lohr a. Main.
ISBN   3-8023-0266-4.

R15- Logic Element Technology: Hydraulic Trainer, Volume 4
Author: Mannesmann Rexroth GmbH.
Publisher: Mannesmann Rexroth GmbH 1989.
.Postfach 340, D 8770 Lohr am Main, Telefon (09352) 180.
ISBN   3-8023-0291-5.

R16- Hydrostatic Drives with Control of the Secondary Unit. The Hydraulic Trainer, Volume 6
Author: Dr. Alfred Feuser, Rolf Kordak, Gerold Liebler.
Publisher: Mannesmann Rexroth GmbH 1989.
Postfach 340, D 8770 Lohr am Main.

R17- Control Strategies for Dynamic Systems: Design and Implementation
Author: John H. Lumkes, Jr.
Publisher: Marcel Dekker, Inc. 2002.
Marcel Dekker, Inc. 270 Madison Avenue, New York, NY 10016.
ISBN 0-8247-0661-7.

R18- Feedback Control Of Dynamic Systems
Author: Gene F. Franklin, J. David Powell, Abbas Emami-Naeini.
Publisher: Prentice-Hall, Inc.
Upper Saddle River, New Jersey.
ISBN 0-13-032393-4.

R19- Modeling and Analysis of Dynamic Systems
Author: Charles M. Close, Dean. Frederick
Rensselaer Polytechnic Institute
Publisher: John Wiley & Sons, Inc.
ISBN 0-471-12517-2.

R20- Design of Electrohydraulic Systems For Industrial Motion Control
Author: Jack L. Johnson, PE.
Milwaukee School of Engineering.
Publisher: Parker.
Copyright © Jack L. Johnson, PE 1991.

R21- Basic Pneumatics
Author: Kjell Evensen & Jul Ruud.
Publisher: AB Mecmann Stockholm 1991.
S-125 81 Stockholm, Sweden.
ISBN 91-85800*21-X.

R22- Basic Pneumatics: The Pneumatic Trainer, Volume 1
Author: Ing. –Buro J.P. Hasebrink.
D7761 Moos.
Editor: Mannesmann Rexroth Pneumatik GmbH.
Bartweg 13, W 3000 Hannover 91.

R23- Electro-Pneumatics: The Pneumatic Trainer, Volume 2
Author: Rolf Balla.
Publisher: Mannesmann Rexroth 1990, Pneumatik GmbH.
Publication No: RE 00 262/01.92.

R24- Pneumatics Theory and Applications
Author: Bosch Automation.
Publisher: Robert Bosch GmbH 1998.
Automation Technology Division, Training (AT/VSZ)
ISBN 1-85226-135-8.

R25- Fluid Power Engineering
Author: M. Galal Rabie.
Publisher: McGraw-Hill.
ISBN 978-0-07-162246-2.

R26- Air Motors Ideas with Air
Author: GAST Mfg. Co.
Publisher: GAST Mfg. Co. 1978.
P.O. Box 97, Benton Harbor, MI 49022.
Book No: Booklet #100.

R27- Air Motor Handbook
Author: GAST Mfg. Co.
Publisher: GAST Mfg. Co. 1978.
P.O. Box 117, Benton Harbor, MI 49022.

R28- Troubleshooting Hydraulic Components: Using Leakage Path Analysis Methods
Author: Rory S. McLaren.
Publisher: Rory McLaren Fluid Power Training 1993.
562 East 7200 South, Salt Lake City, UT 84171.
ISBN No. 0-9639619-1-8.

R29- Hydraulics Theory and Application From Bosch
Author: Werner Gotz.
Publisher: Robert Bosch GmbH.
Hydraulics Division K6, Postfach 30 02 40, D-7000 Stuttgart 30.
Federal Republic of Germany, Technical Publications Department, K6/VKD2.

R30- A Complete Guide to ISO and ANSI Fluid Power Symbols
Author: Fluid Power Training Institute.
Publisher: Fluid Power Training Institute 200.
562 East Fort Union Boulevard, Midvale, Utah 84047.
R31- How to Work Safely with Hydraulics
Author: Fluid Power Training Institute.
Publisher: Fluid Power Training Institute 2004.
562 East7200 South, Midvale, Utah 84047.

R32- How to Interpret Fluid Power Symbols
Author: Rory S. McLaren.
Publisher: Fluid Power Training Institute.
Rory S. McLaren 1995.
ISBN 0-9639619-2-6.

R33- Safe Hydraulics
Editor: Gates Rubber Company.
Copyright 1995.
Denver, CO 80217.

R34- Electronically Controlled Proportional Valves. Selection and Application
Author: Michael J. Tonyan.
Publisher: Marcel Dekker, Inc. 1985.
Marcel Dekker, Inc., 270 Madison Avenue, New York, NY 10016.
ISBN 0-8247-7431-0.

R35- Introduction to Closed-Loop Oil Systems
Author: Rory S. McLaren.
Publisher: Rory McLaren Fluid Power Training Institute.
7050 Cherry Tree Lane, P.O. Box 711201, Salt Lake City, UT 84171.

R36- Industrial Hydraulic Technology, Second Edition
Author: Parker Hannifin Corporation.
Publisher: Parker Hannifin Corporation 1997.
6035 Parkland Blvd, Cleveland, OH 44124-4141.
Publication No: Bulletin 0231-B1.

R37- Basic Principle and Components of Fluid Technology – The Hydraulic Trainer, Volume 1
Author: Mannesman Rexroth.
Publisher: Mannesman Rexroth AG 1988.
D-97813 Lhr a. Main, Jahnsrtrabe 3-5 D-97816 Lohr a. Main.
ISBN 3-8023-0266-4.

R38- Safe-T-Bleed Corporation Catalog
Publisher: Safe-T-Bleed Corporation 2001.
Catalog No. STB-PC-1201-1
R39- Industrial Hydraulics Manual – EATON
Publisher: Eaton Fluid Power Training.
ISBN: 0-9788022-0-9.

R40- Vickers-Mobile Hydraulic Manual – Fourth Edition 1998
Author: Vickers.
Publisher: Vickers Inc. 1999.
Training Center, 2730 Research Drive, Rochester Hills, MI 48309-3570.
ISBN No. 0-9634162-5-1.

R41- Industrial Fluid Power Text, Volume 2
Author: Charles S. Hedges, R.C. Womack.
Publisher: Womack Machine Supply Company 1972.
Womack Educational Publications, 2010 Shea Road, Dallas, TX 75235.
ISBN 66-28254 (Library of Congress Card Catalog No.).
R42- Fluid Power Hydraulics and Pneumatics
Author: R. Daines.
Publisher: The Good-heart Willcox Company, Inc.

R43- Hydraulics in Industrial and Mobile Applications
Publisher: ASSOFLUID, Italian Association of Manufacturing and Trading Companies in Fluid Power Equipment and Components

R44- Fluid Power in Plant and Field – Second Edition
Author: Charles S. Hedges, R.C. Womack.
Publisher: Womack Machine Supply Co. 1968.
Womack Educational Publication, 2010 Shea Road, Dallas, TX 75235.
ISBN 68-22573.

R45- Mobile Hydraulics Manual
Author: Eaton.
Publisher: Eaton Corporation Training.
Eden Prairie, Minnesota.
ISBN 0-9634162-5-1.

R46- EH Control Systems
Author: F.D. Norvelle.

R47- Fluid Power Journal
Publisher: International Fluid Power Society.

R48- Fundamentals of Industrial Controls and Automation
Author: Lonnie L. Smith and Mike J. Rowlett.
Publisher: Womack Educational Publications.
Dallas, Texas.
ISBN: 0-943719-04-6.

R49- Lightning Reference Handbook
Publisher: Berendsen Fluid Power.

R50- Pneumatics Basic Level
Author: P. Croser, F. Ebel.
Publisher: Festo Didactic GmbH & Co.
R51- Electro-pneumatics Basic Level
Author: F. Ebel, G. Prede, D. Scholz.
Publisher: Festo Didactic GmbH & Co.

R52- Mechanical System Components
Author: James F. Thorpe.
Publisher: Allyn and Bacon.
Needham Heights, Massachusetts.
ISBN: 0-205-11713-9.

R53- Electrical Motor Controls for Integrated Systems, Third Edition
Author: Gary J. Rockis, Glen A. Mazur.
Publisher: American Technical Publishers, Inc.
ISBN: 0-8269-1207-9.

R54- Instrumentation, Fourth Edition
Author Franklyn W. Kirk, Thomas A. Weedon, Philip Kirk.
Publisher American Technical Publishers, Inc.
ISBN: 0-8269-3423-4.

R55- Introduction to Mechatronics and Measurement Systems, Second Edition
Author David G. Alciatore, Michael B. Histand.
Publisher McGraw-Hill, Inc.
ISBN: 0-07-240241-5.

R56- Study Guides for IFPS Certification

R57- Work Books from Coastal Training Technologies

R58- Industrial Hydraulic Manual – Fourth Edition 1999
Author: Vickers.
Publisher: Vickers Inc. 1999.
 Training center, 2730 Research Drive, Rochester hills, Michigan 48309-3570.
ISBN 0-9634162-0-0.

R59- Industrial Automation and Process Control
Author: John Stenerson.
Publisher: Prentice Hall.
ISBN 0-13-033030-2.

R60- Industrial Automated Systems
Author: Terry Bartelt.
Publisher: Delmar Cengage Learning.
ISBN: 10-1-4354-888-1.

R61- Introduction to Fluid Power
Author: James L. Johnson.
Publisher: Delmar Cengage Learning.
ISBN: 10-0-7668-2365-2.

R62- Summary for Engineers
Author: Dr. Abdel Nasser Zayed.
Publisher: Dr. Abdel Nasser Zayed .
ISBN: 977-03-0647-9.

R63- Mechanics of Materials
Author: Ferdinand P.Beer, E. Russell Johnston Jr., John T DeWolf.
Publisher: McGraw Hill Publishing .
ISBN: 0-07-365935-5.

R64- Oil Hydraulic System, Principles and Maintenance
Author: S. R. Majumdar.
Publisher: McGraw Hill.
ISBN 10: -0-07-140669-7.

R65- Contamination Control in Hydraulic and Lubricating Systems
Publisher: Pall

R66- Diagnosing Hydraulic Pump Failure
Publisher: Caterpillar.

R67- Oil Service Products Catalog
Publisher: Schroder Industries.

R68- Industrial Fluid Power Volume 1
Author: Charles S. Hedges.
Publisher: Womack Educational Publication.
ISBN: 0-9605644-5-4.

R69- Industrial Fluid Power Volume 2
Author: Charles S. Hedges.
Publisher: Womack Educational Publication.
ISBN: 0-943719-01-1.

R70- Industrial Fluid Power Volume 3
Author: Charles S. Hedges.
Publisher: Womack Educational Publication.
ISBN: 0-943719-00-3.

R71- Electrical Control of Fluid Power
Author: Charles S. Hedges.
Publisher: Womack Educational Publication.
ISBN 0-9605644-9-7.

R72- Hydraulic Cartridge Valve Technology
Author: John J. Pippenger, P.E.
Publisher: Amalgam Publishing Company.
Post Office Box 617, Jenks, OK 74037 USA.
ISBN: 0-929276-01-9.

R73- Noise Control of Hydraulic Machinery
Author: Stan Skaistis.
Publisher: Marcel Dekker, 270 Madison Avenue, New York, NY 10016.
ISBN: 0-8247-7934-7.

R74-Solenoid Valves
Author: Hydraforce

R75-HF Proportional Valve Manual
Author: Hydraforce

R76-Automatic Control for Mechanical Engineers
Author: M. Galal Rabie, Professor of Mechanical Engineering
ISBN: 977-17-9869-3,2010.

R77-Fluid Power System Dynamics
Author: W. Durfee, Z. Sun

Index

■Index ───

www.ingramcontent.com/pod-product-compliance
Lightning Source LLC
Chambersburg PA
CBHW051600190326

41458CB00029B/6486